Cell Survival and Cell Death

A subject collection from *Cold Spring Harbor Perspectives in Biology*

OTHER SUBJECT COLLECTIONS FROM *COLD SPRING HARBOR PERSPECTIVES IN BIOLOGY*

Immune Tolerance
DNA Replication
Endoplasmic Reticulum
Wnt Signaling
Protein Synthesis and Translational Control
The Synapse
Extracellular Matrix Biology
Protein Homeostasis
Calcium Signaling
The Golgi
Germ Cells
The Mammary Gland as an Experimental Model
The Biology of Lipids: Trafficking, Regulation, and Function
Auxin Signaling: From Synthesis to Systems Biology
The Nucleus
Neuronal Guidance: The Biology of Brain Wiring
Cell Biology of Bacteria
Cell–Cell Junctions
Generation and Interpretation of Morphogen Gradients
Immunoreceptor Signaling
NF-κB: A Network Hub Controlling Immunity, Inflammation, and Cancer
Symmetry Breaking in Biology
The Origins of Life
The p53 Family

SUBJECT COLLECTIONS FROM *COLD SPRING HARBOR PERSPECTIVES IN MEDICINE*

Cystic Fibrosis: A Trilogy of Biochemistry, Physiology, and Therapy
Hemoglobin and its Diseases
Addiction
Parkinson's Disease
Type 1 Diabetes
Angiogenesis: Biology and Pathology
HIV: From Biology to Prevention and Treatment
The Biology of Alzheimer Disease

Cell Survival and Cell Death

A subject collection from *Cold Spring Harbor Perspectives in Biology*

EDITED BY

Eric H. Baehrecke
University of Massachusetts

Douglas R. Green
St. Jude Children's Research Hospital

Sally Kornbluth
Duke University

Guy S. Salvesen
Sanford-Burnham Medical Research Institute

COLD SPRING HARBOR LABORATORY PRESS
Cold Spring Harbor, New York • www.cshlpress.org

Cell Survival and Cell Death

A Subject Collection from *Cold Spring Harbor Perspectives in Biology*
Articles online at www.cshperspectives.org

All rights reserved
© 2013 by Cold Spring Harbor Laboratory Press, Cold Spring Harbor, New York
Printed in the United States of America

Executive Editor	Richard Sever
Managing Editor	Maria Smit
Project Manager	Barbara Acosta
Permissions Administrator	Carol Brown
Production Editor	Diane Schubach
Production Manager/Cover Designer	Denise Weiss
Publisher	John Inglis

Front cover artwork: Programmed cell death ensures that our bodies contain just the right number of cells. This tightly regulated process removes damaged cells, shapes our organs and digits, and refines our immune systems. Here, multiphoton fluorescence imaging reveals an apoptotic HeLa cell (*middle*) among nondying neighbors. HeLa cells expressing green fluorescent protein targeted to the Golgi apparatus (yellow) are stained to reveal the distribution of microtubules (red) and cell nuclei (blue). Image kindly provided by Thomas Deerinck and Mark Ellisman, NCMIR, UCSD.

Library of Congress Cataloging-in-Publication Data

Cell survival and cell death : a subject collection from Cold Spring Harbor perspectives in biology / edited by Eric H. Baehrecke, University of Massachusetts, Douglas R. Green, St. Jude Children's Research Hospital, Sally Kornbluth, Duke University, Guy S. Salvesen, Sanford-Burnham Medical Research Institute.
      pages cm. -- (Cold Spring Harbor perspectives in biology)
  Includes bibliographical references and index.
  ISBN 978-1-936113-31-6 (hardcover : alk. paper)
 1. Cell death. 2. Cell physiology. I. Baehrecke, Eric H., 1962-

  QH671.C453 2012
  571.9'36--dc23
                                                                                    2012029706

10 9 8 7 6 5 4 3 2 1

All World Wide Web addresses are accurate to the best of our knowledge at the time of printing.

Authorization to photocopy items for internal or personal use, or the internal or personal use of specific clients, is granted by Cold Spring Harbor Laboratory Press, provided that the appropriate fee is paid directly to the Copyright Clearance Center (CCC). Write or call CCC at 222 Rosewood Drive, Danvers, MA 01923 (978-750-8400) for information about fees and regulations. Prior to photocopying items for educational classroom use, contact CCC at the above address. Additional information on CCC can be obtained at CCC Online at www.copyright.com.

For a complete catalog of all Cold Spring Harbor Laboratory Press publications, visit our website at www.cshlpress.org.

Contents

Preface, vii

Evolution of the Animal Apoptosis Network, 1
Christian M. Zmasek and Adam Godzik

Caspase Functions in Cell Death and Disease, 13
David R. McIlwain, Thorsten Berger, and Tak W. Mak

Apoptotic and Nonapoptotic Caspase Functions in Animal Development, 41
Masayuki Miura

Cellular Mechanisms Controlling Caspase Activation and Function, 57
Amanda B. Parrish, Christopher D. Freel, and Sally Kornbluth

Caspase Substrates and Inhibitors, 81
Marcin Poręba, Aleksandra Stróżyk, Guy S. Salvesen, and Marcin Drąg

Death Receptor–Ligand Systems in Cancer, Cell Death, and Inflammation, 101
Henning Walczak

Mitochondrial Regulation of Cell Death, 119
Stephen W.G. Tait and Douglas R. Green

Mechanisms of Action of Bcl-2 Family Proteins, 135
Aisha Shamas-Din, Justin Kale, Brian Leber, and David W. Andrews

Multiple Functions of BCL-2 Family Proteins, 157
J. Marie Hardwick and Lucian Soane

Inhibitor of Apoptosis (IAP) Proteins—Modulators of Cell Death and Inflammation, 179
John Silke and Pascal Meier

Clearing the Dead: Apoptotic Cell Sensing, Recognition, Engulfment, and Digestion, 199
Amelia Hochreiter-Hufford and Kodi S. Ravichandran

The Endolysosomal System in Cell Death and Survival, 219
Urška Repnik, Maruša Hafner Česen, and Boris Turk

Metabolic Stress in Autophagy and Cell Death Pathways, 233
Brian J. Altman and Jeffrey C. Rathmell

Contents

mTOR-Dependent Cell Survival Mechanisms, 249
Chien-Min Hung, Luisa Garcia-Haro, Cynthia A. Sparks, and David A. Guertin

Oncogenes in Cell Survival and Cell Death, 267
Jake Shortt and Ricky W. Johnstone

The Role of the Apoptotic Machinery in Tumor Suppression, 277
Alex R.D. Delbridge, Liz J. Valente, and Andreas Strasser

The Role of Apoptosis-Induced Proliferation for Regeneration and Cancer, 291
Hyung Don Ryoo and Andreas Bergmann

Fueling the Flames: Mammalian Programmed Necrosis in Inflammatory Diseases, 309
Francis Ka-Ming Chan

Regulation and Function of Autophagy during Cell Survival and Cell Death, 321
Gautam Das, Bhupendra V. Shravage, and Eric H. Baehrecke

Autophagy and Cancer, 335
Li Yen Mah and Kevin M. Ryan

Autophagy and Neuronal Cell Death in Neurological Disorders, 349
Ralph A. Nixon and Dun-Sheng Yang

Index, 373

Preface

THE REGULATED DEATH OF CELLS IS PART OF EVERYDAY LIFE, and inappropriate regulation of programmed cell death is associated with multiple human diseases. Like all science, our understanding of cell death is constantly evolving. In planning this book, we hoped to portray the current knowledge of this process and to provide some perspective on the future of cell death research.

At least three morphological forms of cell death occur, including apoptosis, autophagic cell death, and necrosis. Of these, our understanding of apoptosis is best; many of the molecules that control apoptosis are known, their mechanisms of action have been a subject of interest for many years, and drugs targeting apoptotic factors are of clinical interest. By contrast, far less is known about programmed cell death that involves either autophagy or necrosis, and understanding of the mechanisms of these areas of cell death research is evolving rapidly.

We thank all of our chapter authors for their attempts to be comprehensive, conform to style, and meet deadlines and for their patience for any delays that we may have caused. We also thank Barbara Acosta, Richard Sever, and others at Cold Spring Harbor Laboratory Press for their efforts in putting this book together. Each of the editors thank the past and current members of their laboratories. In addition, such a book would be impossible without the vibrant field of cell death research and the many scientists that have influenced our thinking. Those readers who have not participated in meetings and symposia involving this area may not be aware of the excitement and commitment that invigorated the earliest days and continue unabated to this day. Most of all, it is great *fun*, and this is very much a part of what makes this important research so enjoyable. We hope that at least some of this excitement is evident in this volume.

Finally, we thank and dedicate this book to our families for their support and patience. We hope that the collective efforts of the authors and editors help to illuminate an understanding of programmed cell death, as well as of diseases that are influenced by the elimination of cells.

ERIC H. BAEHRECKE
University of Massachusetts

DOUGLAS R. GREEN
St. Jude Children's Research Hospital

SALLY KORNBLUTH
Duke University

GUY SALVESEN
Sanford-Burnham Medical Research Institute

Evolution of the Animal Apoptosis Network

Christian M. Zmasek and Adam Godzik

Program on Bioinformatics and Systems Biology, Sanford-Burnham Medical Research Institute, La Jolla, California 92037

Correspondence: adam@sanfordburnham.org

The number of available eukaryotic genomes has expanded to the point where we can evaluate the complete evolutionary history of many cellular processes. Such analyses for the apoptosis regulatory networks suggest that this network already existed in the ancestor of the entire animal kingdom (Metazoa) in a form more complex than in some popular animal model organisms. This supports the growing realization that regulatory networks do not necessarily evolve from simple to complex and that the relative simplicity of these networks in nematodes and insects does not represent an ancestral state, but is the result of secondary simplifications. Network evolution is not a process of monotonous increase in complexity, but a dynamic process that includes lineage-specific gene losses and expansions, protein domain reshuffling, and emergence/reemergence of similar protein architectures by parallel evolution. Studying the evolution of such networks is a challenging yet interesting subject for research and investigation, and such studies on the apoptosis networks provide us with interesting hints of how these networks, critical in so many human diseases, have developed.

Apoptosis, a prominent form of programmed cell death (PCD), is one of the major evolutionary innovations of multicellular eukaryotes and in some form is found in representatives of all multicellular eukaryotes (Koonin and Aravind 2002). In the animal kingdom, besides the well-studied apoptosis of mammals, insects, and nematodes, apoptosis has also been described in some of the most basal forms, such as sponges and cnidarians (Wiens et al. 2000a,b, 2001, 2003; Seipp et al. 2001; David 2005; Chipuk and Green 2006; Pankow and Bamberger 2007; Oberst et al. 2008; Lasi et al. 2010; Pernice et al. 2011). Although processes bearing similarities to apoptosis have also been observed in plants (Reape and McCabe 2010), fungi (Madeo et al. 2002; Hamann et al. 2008; Sharon et al. 2009), and even in some single-cellular eukaryotes (Bidle and Falkowski 2004; Deponte 2008; Pollitt et al. 2010; Kaczanowski et al. 2011), in this review we focus specifically on apoptosis in animals. One of the most interesting questions concerning apoptosis is: How did the complex molecular network responsible for regulating and executing apoptotic cell death, which, for instance, in mammals involves around 300 genes (Doctor et al. 2003), evolve? Did this network monotonously increase in complexity over time? Did it evolve in bursts? Was the evolution linear, from a simple to a complex form, or did it involve dead ends, secondary simplifications, or other unexpected evolutionary events? These questions are of much more than academic interest—many diseases that are caused by errors in the apoptotic network are studied primarily in model systems, and the validity of observations in such systems for our understanding of apoptosis in human cells depends on the extent of evolutionary conservation of the specific elements of the apoptotic network (Zmasek et al. 2007).

The molecular mechanism of apoptosis was discovered and studied for the first time in the nematode *Caenorhabditis elegans*, which subsequently became one of the most popular model systems for the investigation of apoptosis. In this simple roundworm, three proteins (CED-3, CED-4, CED-9) and later a fourth protein (EGL-1) were shown to be directly involved in apoptosis (Yuan and Horvitz 2004). Homologs of the first three *C. elegans* genes were subsequently identified in all vertebrates and in the fruit fly *Drosophila melanogaster* (Richardson and Kumar 2002), which quickly became a second favorite model system to study apoptosis. At the same time, a much larger apoptotic network was identified in human and mouse cells (Meier et al. 2000; Koonin and Aravind 2002; Manoharan et al. 2006). Therefore, the natural assumption was that a simple, nematode-like apoptotic network was likely to have been present in the last common ancestor of nematodes, insects, and vertebrates. However, recent sequencing of genomes of several animal species representing deeper branches of the animal phylogenetic tree forced us to reevaluate this assumption.

A PROTEIN DOMAIN–CENTRIC VIEW OF FUNCTIONAL NETWORKS

Groups of proteins involved in a specific pathway, such as apoptosis, are usually described as a "network," in which links between proteins signify their direct or indirect interactions. An overview of the core human apoptotic network is shown in Figure 2A.

Because most eukaryotic proteins are composed of multiple domains, a simplified overview and comparison of protein networks and genomes can be provided by listing domains present in the proteins forming the network. Protein domains are minimal evolutionary and structural units in proteins, retaining their structure and usually their function even when being a part of proteins with different domain architectures (Ponting and Russell 2002). Using functional assignment of protein domains, we can compare and study the evolution of the functional profiles of entire genomes (Zmasek and Godzik 2011). Here, we adopt such a domain-centric view to study and compare apoptotic networks in different species, including ancestral ones, a task made easier by the fact that many of the domains present in proteins forming this network are very characteristic and are often referred to as "apoptotic domains" (Doctor et al. 2003). For example, mammalian Apaf-1, a central gene regulating the mitochondria-dependent apoptotic pathways, is composed of one CARD domain, one NB-ARC domain, and one domain with WD40 repeats, with the first two being specific to the apoptosis network. In the Amphioxus (*Branchiostoma floridae*) and purple sea urchin (*Strongylocentrotus purpuratus*) genomes, the CARD domain in the Apaf-1 homolog(s) can be replaced by a death domain (DD) and, in the starlet sea anemone (*Nematostella vectensis*) (Darling et al. 2005) genome, by a death effector domain (DED) (Zmasek et al. 2007), both being specific to the apoptosis network.

When investigating the evolution of proteins or protein domains, one key question is how "old" a certain protein or protein domain is, in other words, in which ancestral species it is likely to have originated. To make such inferences regarding the presence of proteins or protein domains in ancestral species, one can use the principle of Dollo parsimony, which, when applied to protein domains, states that each domain has evolved only once and seeks to minimize domain losses (Farris 1977; Zmasek and Godzik 2011). On the other hand, Dollo parsimony analysis does not allow one to estimate how many paralogs (homologous sequences that are related by a gene duplication event) of a given gene or domain existed in ancestral species. To make such inferences, explicit phylogenetic tree inference is required.

ANCESTRAL COMPLEXITY OF THE APOPTOSIS NETWORK

The last common ancestor of all extant metazoans (animals) (dark blue in Fig. 1), in all likelihood a multicellular organism, lived around 600 million years ago (Conway Morris 2000; Hedges et al. 2004; Schierwater et al. 2009). It gave rise to the wide variety of animals alive today, ranging from "basal" organisms such as sponges and Placozoa—represented here by *Trichoplax adhaerens* (Schierwater 2005), one of the simplest of all animals, with a body made up of a few thousand cells of just four types, with no internal organs—to vertebrates with multiple organs and hundreds of cell types.

Recently, an increasing number of studies show that this metazoan ancestor, and to some degree even the ancestor of all eukaryotes (the so-called LECA, the last eukaryotic common

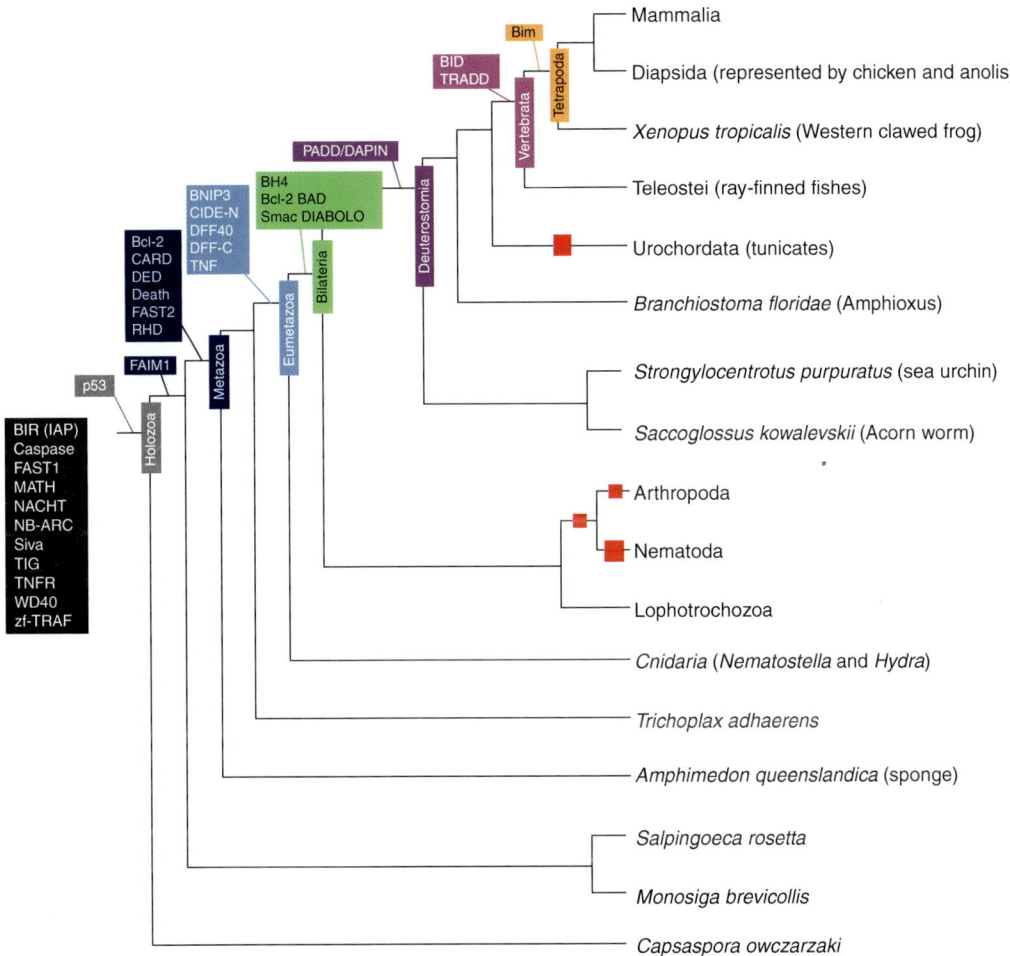

Figure 1. An overview of animal evolution overlaid with gains in apoptotic domains. Inferred gains of apoptotic domains are shown. Although domain losses are not shown in detail, the locations of significant domain losses in nematodes, insects, and tunicates are marked by red squares. Domains in the black box are estimated to have been present in the last eukaryotic common ancestor. Data are based on domain definitions from the Pfam 25.0 database and 174 completely sequenced eukaryotic genomes. The topology of the underlying eukaryotic evolutionary tree is based on data from Cameron et al. (2000), Halanych (2004), Adl et al. (2005), and Hampl et al. (2009).

ancestor), was complex in its genome content, if not in its morphology (Zmasek and Godzik 2011, and references therein). In particular, availability of genomes of "basal" marine invertebrates, such as the demosponge *Amphimedon queenslandica* (the largest class in the phylum Porifera [sponges] with spicules [structural elements] consisting of fibers of the protein spongin, the mineral silica, or both) (Srivastava et al. 2010), the placozoan *T. adhaerens* (Srivastava et al. 2008), the hydra *Hydra magnipapillata* (Chapman et al. 2010), and the starlet sea anemone *N. vectensis* (Putnam et al. 2007), as well as representatives of nonmetazoan holozoa such as the filose amoeboid *Capsaspora owczarzaki* (Ruiz-Trillo et al. 2004), forced us to reconsider certain long-held assumptions regarding eukaryotic evolution. (Nonmetazoan holozoa are a group of organisms that includes animals and their closest single-celled relatives (such as Choanoflagellata) but excludes fungi.) For instance, many genes with regulatory functions not only appeared, but were also present in large families of paralogs earlier than previously thought. In part, this erroneous view of a simple LECA and a simple animal ancestor was caused by the predominant usage in molecular and cell biology research of ecdysozoan (molting animals—a group of protostome animals that includes insects and nematodes) (Aguinaldo et al. 1997) model organisms, in particular the fruit fly *D. melanogaster* and the nematode *C. elegans*. With the availability of the new genomes, it has become increasingly apparent that the genomic simplicity (such as lack of or minimal size of many pathways and smaller gene families compared with vertebrates) in these organisms does not represent an ancestral simplicity but is the result of extensive gene loss in ecdysozoans (Kortschak et al. 2003). An impressive example of this is the analysis of *Nematostella* Wnt genes, which revealed an unforeseen ancestral diversity: *Nematostella* and bilaterians share at least 11 of the 12 known Wnt subfamilies, whereas five subfamilies appear to be lost in nematodes/insects (Kusserow et al. 2005). Similarly, it has been recently shown that the filose amoeboid *C. owczarzaki* (Ruiz-Trillo et al. 2004), a unicellular organism closely related to choanoflagellates (a group of unicellular eukaryotes, the closest living relatives of animals) and metazoans (see Fig. 1), possesses a strikingly complex repertoire of transcription factors (Sebé-Pedrós et al. 2011). Another example is the genome content of the demosponge *A. queenslandica*, which has been shown to be similar to more complex animals and possesses the necessary genes for most signaling pathways and transcription factors commonly associated with complex multicellular animals (Srivastava et al. 2010). Ecdysozoans (and nematodes, in particular) are not the only group of animals showing massive gene (and protein domain) loss; another example of a group of animals that show a secondarily reduced apoptotic network, as well as other reductions and loss of features, is tunicates (also known as urochordates) (Holland and Gibson-Brown 2003). This observation is based on analyzing the genomic contents of three tunicates for apoptotic domains—*Ciona intestinalis* (Dehal et al. 2002), *Ciona savignyi*, and *Oikopleura dioica*. In the following, we show that the same phenomenon of ancestral complexity applies also to the apoptotic regulatory network.

APOPTOSIS IN THE METAZOAN ANCESTOR

For every domain, we can reason about the ancestral species in which this domain first appeared by analyzing a given domain distribution on the organismal "tree of life." More precisely, we can provide an upper estimate, as in each case some newly sequenced genome can push the time of emergence of this domain earlier. For example, the fact that sponge and mammals, but not any species outside the metazoan clade, possess death domains means that this domain probably evolved in one of the first metazoans. Applying this principle (in phylogenetic analysis called Dollo parsimony) (Farris 1977) to the set of apoptotic domains allows inference of a domain repertoire of a putative apoptosis network of the last common metazoan ancestor and of the ancestral species for major evolutionary branches. The results for such an analysis using 174 completely sequenced eukaryotic genomes, including 49 from metazoans, is shown in Figures 1 and 2.

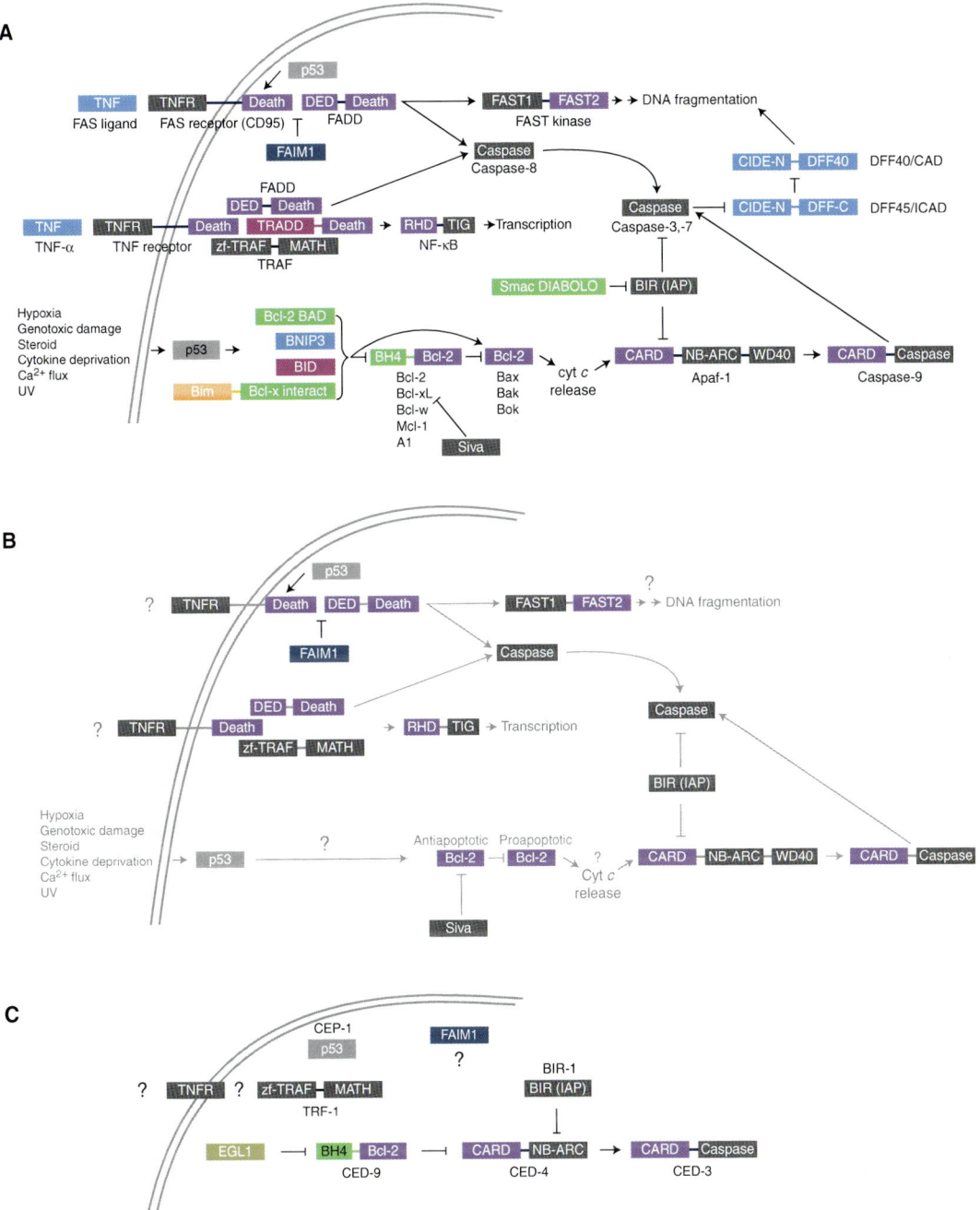

Figure 2. Evolution of the apoptosis network in animals. (A) Mammalian intrinsic and extrinsic apoptosis networks. (B) Inferred apoptosis networks of the last common metazoan ancestor. (C) The apoptosis network of C. elegans as an example of a secondarily reduced network. CEP-1 is the C. elegans p53 homolog necessary for DNA damage-induced apoptosis (Schumacher et al. 2001, 2005). TRF-1 is a C. elegans TRAF homolog (Wajant et al. 1998). Data are based on domain definitions from the Pfam 25.0 database and 174 completely sequenced eukaryotic genomes. Color coding corresponds to that in Figure 1. cyt c, cytochrome c.

It is apparent that the last common metazoan ancestor already had most components that make up extant intrinsic and extrinsic vertebrate-like apoptotic pathways: at least one TNF receptor; p53; DD (death) and DED (death effector) domains; FAIM (Fas apoptotic inhibitory molecule); Fas-activated serine/threonine (FAST) kinase; TRAIL (TNF receptor-associated factor), composed of zf-TRAF and MATH domains; NF-κB with Rel homology domain (RHD) and TIG domains; Bcl-2 (with BH1, BH2, and BH3 motifs, but lacking BH4); Siva; BIR (baculoviral inhibition of apoptosis protein repeat) of IAP (inhibitor of apoptosis) proteins; CARD (caspase recruitment) domains; NB-ARC; and caspases (Tian et al. 1995; Rothstein 2000; Locksley et al. 2001; Gilmore 2006; Py et al. 2007; Cao et al. 2008; Belyi et al. 2009). These domains are nearly sufficient to build a simple functional apoptotic network similar to the ones found in present-day animals, with only two core components seemingly having appeared later in eumetazoans: the DNase CAD (caspase-activated DNase) responsible for apoptotic DNA fragmentation and its inhibitor ICAD (inhibitor of CAD), and the TNF ligand (Enari et al. 1998; Locksley et al. 2001). (Eumetazoans are a clade of animals encompassing Ctenophora [comb jellies], Cnidaria [sea anemones, jellyfish, coral], and Bilateria [animals with bilateral symmetry], but excluding sponges.) It is not unlikely that the availability of more metazoan genomes outside the eumetazoan clade (such as additional sponge genomes or a first ctenophoran ["comb jelly"] genome) would push the evolution of TNF ligands and CAD/ICAD further back in time. Many of the apoptotic domains present in the last eumetazoan ancestor did not evolve in this organism and are significantly older. Examples of such proteins/domains are caspases and related proteases, as well as the NB-ARC domain, which are found in a wide variety of eukaryotes and, as some argue, also in bacteria and archaea (Aravind et al. 1999). On the other hand, the transcription factors p53 and NF-κB, which were initially thought to be animal specific, appeared on the path leading toward metazoans, for it seems they have already arisen in holozoa.

EXPANSIONS OF PARALOGOUS FAMILIES OF APOPTOTIC DOMAINS SUPPORT THE PICTURE OF THE ANCESTRAL COMPLEXITY OF THE APOPTOSIS NETWORK

Most apoptotic domains are phylogenetically old, but the complexity of the apoptotic network is also created by the interplay between multiple paralogs (such as the various types of Bcl-2 or CARD family members). This expansion was also argued to be specific to "higher" animals (Aravind et al. 2001). Using the data from recently sequenced genomes, we can show that this hypothesis is also incorrect. For instance, analysis of the *S. purpuratus* (purple sea urchin) genome showed that many groups of proteins related to apoptosis underwent major expansions in this organism compared not only with ecdysozoans, but also with vertebrates (Sodergren et al. 2006). In fact, some groups of apoptosis-related proteins have 10 times more members in sea urchin than in corresponding families in vertebrates (Robertson et al. 2006; Zmasek et al. 2007). The *B. floridae* (Amphioxus), as well as some other genomes, show similar expansion (Zmasek et al. 2007; Zhang et al. 2008). In the following, we discuss some specific cases of apoptotic domains, focusing on paralogous expansions and secondary reduction.

p53

p53 homologs are not only present in all animals sequenced so far, but also in the choanoflagellate *Monosiga brevicollis* (King et al. 2008; Belyi et al. 2009) and, according to our analysis, even in the filose amoeboid *C. owczarzaki* (Ruiz-Trillo et al. 2004). Therefore, the origin of p53 is likely to predate the rise of animals. Humans as well as most other mammals have three p53 family members (p53, p63, and p73). The choanoflagellates *M. brevicollis* and *Salpingoeca rosetta* (data not shown), as well as the sea anemone *N. vectensis*, have between two and three paralogs that cannot be assigned to any of the mammalian subtypes but that have been argued to resemble a hybrid of the mammalian types, whereas most insects and *C. elegans* contain one homolog that is a combination of p63 and p73 (Belyi et al. 2009). Despite this, functions of the early p53 homologs are close to those of the mammalian ones, because the *C. elegans* homolog is necessary for DNA damage-induced apoptosis (Schumacher et al. 2001, 2005) and the *Nematostella* ones for UV-induced cell death (Pankow and Bamberger 2007).

BIR

BIR domains are present in the inhibitors of apoptotic proteins (Salvesen and Duckett 2002). Homologs of these domains are found in animals, fungi, and various single-cell eukaryotes, but so far they have not been found in plant genomes. The evolutionary history of BIR-containing proteins is complex, with numerous domain duplications and insertions, resulting in many proteins with different domain combinations (Verhagen et al. 2001; Cao et al. 2008). In particular, proteins with BIR and NACHT domains appear to be specific to amniotes (mammals, birds, reptiles), whereas the BIR and CARD combinations are likely to be specific for vertebrates. On the other hand, most BIR-containing proteins outside the bilaterian clade are single-domain proteins (based on the analysis of 174 eukaryotic genomes for domain content) (CM Zmasek and A Godzik, unpubl.). Although this family shows no significant secondary reduction in ecdysozoans, basal metazoans also possess multiple BIR domain-containing proteins. For example, *A. queenslandica* is estimated to contain about six proteins with BIR domains, and the cnidarians *N. vectensis* and *H. magnipapillata* both possess three proteins with BIR domains (data not shown).

Apaf-1

The Apaf-1 central nucleotide binding (NB-ARC, or nucleotide-binding adaptor shared by Apaf-1, R proteins, and CED-4) domain is a member of a specific branch of the very large family of AAA$^+$ ATPases and is distantly homologous to, but distinctively different from, other nucleotide-binding domains present in other apoptosis-related proteins, such as the NACHT domain present in families of proteins involved in immunity (van der Biezen and Jones 1998; Neuwald et al. 1999; Inohara et al. 2005). Detailed phylogenetic analysis of the NB-ARC domain shows an unexpected picture of a paralogous expansion in early animals, followed by contraction in both ecdysozoans and vertebrates. The full set of paralogs is present in *Nematostella*, with different branches being lost in the nematode/insect branch and in vertebrates, thus leaving nematodes/insects without orthologs of human Apaf-1. In addition, several *Nematostella* and amphioxus homologs form additional subfamilies, which were lost in both nematodes/insects and vertebrates, indicating an evolutionary history for Apaf-1 predecessors rich in gene duplications and gene losses (Zmasek et al. 2007).

The functional variations among different branches of the Apaf-1 family are illustrated by their different domain organizations. Human Apaf-1 and its *Nematostella*, Amphioxus, and sea urchin orthologs show the same or similar domain organization (CARD–NB-ARC–WD40 repeats). Nematode and most, but not all, insect sequences seem to lack WD40 repeats, suggesting that the loss of the receptor domain of CED-4 is a (relatively) recent event, specific to the nematode/insect branch. The full repertoire of CED-4/Apaf-1 homologs in sea urchin, Amphioxus, and *Nematostella* contains proteins with additional domain combinations, including replacement of the single CARD domain at the amino terminus with pairs of CARD domains (*Nematostella* and Amphioxus), death domains (Amphioxus, sea urchin, and *H. magnipapillata*), death effector domains (*Nematostella*, Amphioxus, and the hemichordate [Cameron et al. 2000] *Saccoglossus kowalevski* [data not shown]), and TIR domains (Amphioxus), all of which function as protein–protein interaction facilitators. At the carboxyl terminus, the WD40 repeats are occasionally missing, replaced by TPR repeats or supplemented by double death domain repeats (Robertson et al. 2006; Zmasek et al. 2007).

The presence of numerous CED-4/Apaf-1 homologs in the common ancestor of Bilateria and Cnidaria suggests that initially there might have been several mechanisms to activate the intrinsic apoptosis pathways and/or several downstream pathways activated by similar signals and that the mechanism of human Apaf-1 and its vertebrate orthologs present only one of several possibilities. ("Orthologs" are homologous sequences that are related by a speciation event; in contrast to a common misuse of this term, orthology does not imply functional equivalence, and functional equivalence does not imply orthology.) The diverse domain architecture of Apaf-1 homologs also suggests that functional differences in this family could include both the sensing mechanism (carboxy-terminal receptor domains) and the downstream recruitment function (amino-terminal protein–protein interaction domains). These findings also explain why the biochemical/structural mechanism of *C. elegans* CED-4 and *Drosophila* Dark can be significantly different from human Apaf-1.

Bcl-2

In the case of Bcl-2, members of major subfamilies were most likely already present in the early metazoan ancestors but were subsequently lost in nematodes and insects. Detailed phylogenetic analysis of the roughly 11 multimotif Bcl-2 family members of the sea anemone *N. vectensis* has shown that the Bax, Bak, and Bok groups of proapoptotic Bcl-2 homologs appear to be ancient and that each has at least one well-supported ortholog in *Nematostella*. The other *Nematostella* Bcl-2 family members are difficult to assign to a specific subtype, although one of them has been shown to contain a region with a weak similarity to the BH4 motif, thus making it similar to the Bcl-2/Bcl-x branch of the Bcl-2 family. Recently, a very similar finding has been reported for another cnidarian species—the freshwater polyp *Hydra*. This animal has been shown to possess seven Bcl-2–like and two Bak-like proteins (Lasi et al. 2010). This is in sharp contrast to the model organisms *D. melanogaster*, which contains only two Bcl-2 family genes belonging to the Bok group (Debcl and Buffy), and *C. elegans*, which has one (CED-9) (Zmasek et al. 2007).

Caspases

The final step in apoptosis is proteolysis of a variety of target proteins in the cell by "effector" caspases, which are activated in a proteolytic cascade by several "apical" ("initiator") caspases (Riedl and Shi 2004). Both types are clearly present in all animals. Yet, again, *Nematostella*, *Hydra*, Amphioxus, and sea urchin have representatives of more subtypes than nematodes and insects (Zmasek et al. 2007; Lasi et al. 2010). Although the caspase family appears to be specific to animals, related cysteine proteases (para- and metacaspases) can be found in a broad range of eukaryotes (Lamkanfi 2002).

REFINEMENT OF APOPTOSIS IN MAMMALS

Although most key apoptotic domains were already present in the last metazoan ancestor, the path toward the vertebrate network is accompanied by the further acquisition of domains with regulatory functions such as various proapoptotic regulators from the Bcl-2 extended family, characterized by the presence of the BH3 motif (and lack of BH1, BH2, and BH4 motifs—e.g., Bad, Bid, and Bim). Another mammal-specific addition to the apoptosis network includes Smac DIABLO (see Figs. 1 and 2A) (Du et al. 2000; Verhagen et al. 2000), a mitochondrial protein that positively regulates apoptosis by eliminating the inhibitory effect of IAPs (inhibitor of apoptosis proteins). In *Drosophila*, the antiapoptotic activity of IAPs is removed by the three proteins Reaper, Grim, and Hid, which therefore have been termed functional homologs of the mammalian protein Smac DIABLO (Chai et al. 2000).

CELL DEATH IN FUNGI, PLANTS, AND UNICELLULAR EUKARYOTES

Fungi, plants, and even some unicellular eukaryotes have been shown to be able to execute forms of programmed cell death upon coming under various types of environmental stresses (Bidle and Falkowski 2004; Deponte 2008; Affenzeller et al. 2009; Pollitt et al. 2010). As mentioned above, some domains with key functions in the metazoan apoptosis program, such as caspases, BIR, and NB-ARC, have been shown to be present in various nonanimal eukaryotes, including fungi, plant, and unicellular eukaryote species—and thus they must have been present in the LECA (the last eukaryotic common ancestor). On the other hand, many other key apoptotic domains or proteins are animal specific (e.g., Bcl-2, DED, and DD). The set of domains absent in the LECA, fungi, plants, and unicellular eukaryotes is so large that at this point we cannot reconstruct a functional apoptotic network in the LECA. Of course, it is also entirely possible that cell death in these species relies on an entirely different molecular mechanism.

APOPTOSIS AND INNATE IMMUNITY

In mammals, apoptosis and innate immunity form distinct, but closely coupled networks. For example, numerous caspases act both as effectors and/or initiators of apoptosis but also have nonapoptotic functions (such as being components of the inflammasome, regulation of B cells) (Siegel 2006). Analysis of genomes of invertebrates suggests that the integration between these two networks could have been much closer in ancestral species and may be pointing to some as-yet-unrecognized connections between these networks in higher organisms. Such unexpected connections are continuously being discovered (Bertrand et al. 2009).

CONCLUSIONS

With the availability of new invertebrate genomes, we can now state unequivocally that even the last metazoan ancestor was likely to have had a complex apoptotic machinery, which could have been more similar to the mammalian than to the nematode form. Furthermore, an increasing number of studies show apoptosis-like cell death in unicellular organisms, and many domains involved in apoptosis have already been present in the last eukaryotic common ancestor and even in bacteria, even though some key domains or proteins appear to be animal specific (Bcl-2, death effector domain, death domain) (Koonin and Aravind 2002; Zmasek and Godzik 2011). It is clear that apoptosis may be as old as eukaryotes and that it existed in the form resembling that seen in extant animals already hundreds of millions of years ago. At the same time, we see dynamic evolution of this network, including massive losses in some lineages (e.g., ecdysozoans and tunicates), recruitment of novel domains (mammals), and dramatic expansions and contractions of individual families. When combined with frequent domain reshuffling and emergence/reemergence of similar protein architectures by parallel evolution, we can argue that evolution of pathways such as apoptosis does not proceed by a gradual increase from simpler to more complex forms, but instead involves frequent rewiring and reshuffling, expansions and contractions, making it more interesting, but also more challenging, to understand.

REFERENCES

Adl SM, Simpson AGB, Farmer Ma, Andersen Ra, Anderson OR, Barta JR, Bowser SS, Brugerolle G, Fensome Ra, Fredericq S, et al. 2005. The new higher level classification of eukaryotes with emphasis on the taxonomy of protists. *J Eukaryot Microbiol* **52:** 399–451.

Affenzeller MJ, Darehshouri A, Andosch A, Lütz C, Lütz-Meindl U. 2009. Salt stress-induced cell death in the unicellular green alga *Micrasterias denticulata*. *J Exp Bot* **60:** 939–954.

Aguinaldo AMA, Turbeville JM, Linford LS, Rivera MC, Garey JR, Raff RA, Lake JA. 1997. Evidence for a clade of nematodes, arthropods and other moulting animals. *Nature* **387:** 489–493.

Aravind L, Dixit VM, Koonin EV. 1999. The domains of death: Evolution of the apoptosis machinery. *Trends Biochem Sci* **24:** 47–53.

Aravind L, Dixit VM, Koonin EV. 2001. Apoptotic molecular machinery: Vastly increased complexity in vertebrates revealed by genome comparisons. *Science* **291:** 1279–1284.

Belyi VA, Ak P, Markert E, Wang H, Hu W, Puzio-Kuter A, Levine AJ. 2009. The origins and evolution of the p53 family of genes. *Cold Spring Harb Perspect Biol* **2:** a001198.

Bertrand MJM, Doiron K, Labbé K, Korneluk RG, Barker PA, Saleh M. 2009. Cellular inhibitors of apoptosis cIAP1 and cIAP2 are required for innate immunity signaling by the pattern recognition receptors NOD1 and NOD2. *Immunity* **30:** 789–801.

Bidle KD, Falkowski PG. 2004. Cell death in planktonic, photosynthetic microorganisms. *Nat Rev Microbiol* **2:** 643–655.

Cameron CB, Garey JR, Swalla BJ. 2000. Evolution of the chordate body plan: New insights from phylogenetic analyses of deuterostome phyla. *Proc Natl Acad Sci* **97:** 4469–4474.

Cao L, Wang Z, Yang X, Xie L, Yu L. 2008. The evolution of BIR domain and its containing proteins. *FEBS Lett* **582:** 3817–3822.

Chai J, Du C, Wu JW, Kyin S, Wang X, Shi Y. 2000. Structural and biochemical basis of apoptotic activation by Smac/DIABLO. *Nature* **406:** 855–862.

Chapman JA, Kirkness EF, Simakov O, Hampson SE, Mitros T, Weinmaier T, Rattei T, Balasubramanian PG, Borman J, Busam D, et al. 2010. The dynamic genome of *Hydra*. *Nature* **464:** 592–596.

Chipuk JE, Green DR. 2006. Dissecting p53-dependent apoptosis. *Cell Death Diff* **13:** 994–1002.

Conway Morris S. 2000. The Cambrian "explosion": Slow-fuse or megatonnage? *Proc Natl Acad Sci* **97:** 4426–4429.

Darling JA, Reitzel AR, Burton PM, Mazza ME, Ryan JF, Sullivan JC, Finnerty JR. 2005. Rising starlet: The starlet sea anemone, *Nematostella vectensis*. *BioEssays* **27:** 211–221.

David CN. 2005. *Hydra* and the evolution of apoptosis. *Integr Comp Biol* **45:** 631–638.

Dehal P, Satou Y, Campbell RK, Chapman J, Degnan B, De Tomaso A, Davidson B, Di Gregorio A, Gelpke M, Goodstein DM, et al. 2002. The draft genome of *Ciona intestinalis*: Insights into chordate and vertebrate origins. *Science* **298:** 2157–2167.

Deponte M. 2008. Programmed cell death in protists. *Biochim Biophys Acta* **1783:** 1396–1405.

Doctor KS, Reed JC, Godzik A, Bourne PE. 2003. The apoptosis database. *Cell Death Diff* **10:** 621–633.

Du C, Fang M, Li Y, Li L, Wang X. 2000. Smac, a mitochondrial protein that promotes cytochrome c–dependent caspase activation by eliminating IAP inhibition. *Cell* **102:** 33–42.

Enari M, Sakahira H, Yokoyama H, Okawa K, Iwamatsu A, Nagata S. 1998. A caspase-activated DNase that degrades DNA during apoptosis, and its inhibitor ICAD. *Nature* **391:** 43–50.

Farris JS. 1977. Phylogenetic analysis under Dollo's law. *Syst Zool* **26:** 77–88.

Gilmore TD. 2006. Introduction to NF-κB: Players, pathways, perspectives. *Oncogene* **25:** 6680–6684.

Halanych KM. 2004. The new view of animal phylogeny. *Annu Rev Ecol Evol Syst* **35:** 229–256.

Hamann A, Brust D, Osiewacz HD. 2008. Apoptosis pathways in fungal growth, development and ageing. *Trends Microbiol* **16:** 276–283.

Hampl V, Hug L, Leigh JW, Dacks JB, Lang BF, Simpson AGB, Roger AJ. 2009. Phylogenomic analyses support the monophyly of Excavata and resolve relationships among eukaryotic "supergroups." *Proc Natl Acad Sci* **106:** 3859–3864.

Hedges S, Blair J, Venturi M, Shoe J. 2004. A molecular timescale of eukaryote evolution and the rise of complex multicellular life. *BMC Evol Biol* **4:** 2.

Holland LZ, Gibson-Brown JJ. 2003. The *Ciona intestinalis* genome: When the constraints are off. *BioEssays* **25:** 529–532.

Inohara, Chamaillard, McDonald C, Nuñez G. 2005. NOD-LRR proteins: Role in host–microbial interactions and inflammatory disease. *Annu Rev Biochem* **74:** 355–383.

Kaczanowski S, Sajid M, Reece SE. 2011. Evolution of apoptosis-like programmed cell death in unicellular protozoan parasites. *Parasit Vectors* **4:** 44.

King N, Westbrook MJ, Young SL, Kuo A, Abedin M, Chapman J, Fairclough S, Hellsten U, Isogai Y, Letunic I, et al. 2008. The genome of the choanoflagellate *Monosiga brevicollis* and the origin of metazoans. *Nature* **451:** 783–788.

Koonin EV, Aravind L. 2002. Origin and evolution of eukaryotic apoptosis: The bacterial connection. *Cell Death Differ* **9:** 394–404.

Kortschak RD, Samuel G, Saint R, Miller DJ. 2003. EST analysis of the cnidarian *Acropora millepora* reveals extensive gene loss and rapid sequence divergence in the model invertebrates. *Curr Biol* **13:** 2190–2195.

Kusserow A, Pang K, Sturm C, Hrouda M, Lentfer J, Schmidt HA, Technau U, Haeseler Av, Hobmayer B, Martindale MQ, et al. 2005. Unexpected complexity of the Wnt gene family in a sea anemone. *Nature* **433:** 156–160.

Lamkanfi M. 2002. Alice in caspase land. A phylogenetic analysis of caspases from worm to man. *Cell Death Differ* **9:** 358–361.

Lasi M, Pauly B, Schmidt N, Cikala M, Stiening B, Kasbauer T, Zenner G, Popp T, Wagner A, Knapp RT, et al. 2010. The molecular cell death machinery in the simple cnidarian *Hydra* includes an expanded caspase family and pro- and anti-apoptotic Bcl-2 proteins. *Cell Res* **20:** 812–825.

Locksley RM, Killeen N, Lenardo MJ. 2001. The TNF and TNF receptor superfamilies: Integrating mammalian biology. *Cell* **104:** 487–501.

Madeo F, Herker E, Maldener C, Wissing S, Lächelt S, Herlan M, Fehr M, Lauber K, Sigrist SJ, Wesselborg S. 2002. A caspase-related protease regulates apoptosis in yeast. *Mol Cell* **9:** 911–917.

Manoharan A, Kiefer T, Leist S, Schrader K, Urban C, Walter D, Maurer U, Borner C. 2006. Identification of a "genuine" mammalian homolog of nematodal CED-4: Is the hunt over or do we need better guns? *Cell Death Differ* **13:** 1310–1317.

Meier P, Finch A, Evan G. 2000. Apoptosis in development. *Nature* **407:** 796–801.

Neuwald AF, Aravind L, Spouge JL, Koonin EV. 1999. AAA$^+$: A class of chaperone-like ATPases associated with the assembly, operation, and disassembly of protein complexes. *Genome Res* **9:** 27–43.

Oberst A, Bender C, Green DR. 2008. Living with death: The evolution of the mitochondrial pathway of apoptosis in animals. *Cell Death Differ* **15:** 1139–1146.

Pankow S, Bamberger C. 2007. The p53 tumor suppressor-like protein nvp63 mediates selective germ cell death in the sea anemone *Nematostella vectensis*. *PloS ONE* **2**: e782.

Pernice M, Dunn SR, Miard T, Dufour S, Dove S, Hoegh-Guldberg O. 2011. Regulation of apoptotic mediators reveals dynamic responses to thermal stress in the reef building coral *Acropora millepora*. *PLoS ONE* **6**: e16095.

Pollitt LC, Colegrave N, Khan SM, Sajid M, Reece SE. 2010. Investigating the evolution of apoptosis in malaria parasites: The importance of ecology. *Parasit Vectors* **3**: 105.

Ponting CP, Russell RR. 2002. The natural history of protein domains. *Annu Rev Biophys Biomol Struct* **31**: 45–71.

Putnam NH, Srivastava M, Hellsten U, Dirks B, Chapman J, Salamov A, Terry A, Shapiro H, Lindquist E, Kapitonov VV, et al. 2007. Sea anemone genome reveals ancestral eumetazoan gene repertoire and genomic organization. *Science* **317**: 86–94.

Py B, Bouchet J, Jacquot G, Sol-Foulon N, Basmaciogullari S, Schwartz O, Biard-Piechaczyk M, Benichou S. 2007. The Siva protein is a novel intracellular ligand of the CD4 receptor that promotes HIV-1 envelope-induced apoptosis in T-lymphoid cells. *Apoptosis* **12**: 1879–1892.

Reape TJ, McCabe PF. 2010. Apoptotic-like regulation of programmed cell death in plants. *Apoptosis* **15**: 249–256.

Richardson H, Kumar S. 2002. Death to flies: *Drosophila* as a model system to study programmed cell death. *J Immunol Methods* **265**: 21–38.

Riedl SJ, Shi Y. 2004. Molecular mechanisms of caspase regulation during apoptosis. *Nat Rev Mol Cell Biol* **5**: 897–907.

Robertson AJ, Croce J, Carbonneau S, Voronina E, Miranda E, McClay DR, Coffman JA. 2006. The genomic underpinnings of apoptosis in Strongylocentrotus purpuratus. *Dev Biol* **300**: 321–334.

Rothstein TL. 2000. Inducible resistance to Fas-mediated apoptosis in B cells. *Cell Res* **10**: 245–266.

Ruiz-Trillo I, Inagaki Y, Davis LA, Landfald B, Roger AJ. 2004. *Capsaspora owczarzaki* is an independent opisthokont lineage. *Curr Biol* **14**: R946–R947.

Salvesen GS, Duckett CS. 2002. IAP proteins: Blocking the road to death's door. *Nat Rev Mol Cell Biol* **3**: 401–410.

Schierwater B. 2005. My favorite animal, *Trichoplax adhaerens*. *BioEssays* **27**: 1294–1302.

Schierwater B, Eitel M, Jakob W, Osigus H-J, Hadrys H, Dellaporta SL, Kolokotronis S-O, Desalle R. 2009. Concatenated analysis sheds light on early metazoan evolution and fuels a modern "urmetazoon" hypothesis. *PLoS Biol* **7**: e20.

Schumacher B, Hofmann K, Boulton S, Gartner A. 2001. The *C. elegans* homolog of the p53 tumor suppressor is required for DNA damage-induced apoptosis. *Curr Biol* **11**: 1722–1727.

Schumacher B, Hanazawa M, Lee M-H, Nayak S, Volkmann K, Hofmann ER, Hofmann R, Hengartner M, Schedl T, Gartner A. 2005. Translational repression of *C. elegans* p53 by GLD-1 regulates DNA damage-induced apoptosis. *Cell* **120**: 357–368.

Sebé-Pedrós A, de Mendoza A, Lang BF, Degnan BM, Ruiz-Trillo I. 2011. Unexpected repertoire of metazoan transcription factors in the unicellular holozoan *Capsaspora owczarzaki*. *Mol Biol Evol* **28**: 1241–1254.

Seipp S, Schmich J, Leitz T. 2001. Apoptosis—A death-inducing mechanism tightly linked with morphogenesis in *Hydractina echinata* (Cnidaria, Hydrozoa). *Development* **128**: 4891–4898.

Sharon A, Finkelstein A, Shlezinger N, Hatam I. 2009. Fungal apoptosis: Function, genes and gene function. *FEMS Microbiol Rev* **33**: 833–854.

Siegel RM. 2006. Caspases at the crossroads of immune-cell life and death. *Nat Rev Immunol* **6**: 308–317.

Sodergren E, Weinstock GM, Davidson EH, Cameron RA, Gibbs RA, Angerer RC, Angerer LM, Arnone MI, Burgess DR, Burke RD, et al. 2006. The genome of the sea urchin *Strongylocentrotus purpuratus*. *Science* **314**: 941–952.

Srivastava M, Begovic E, Chapman J, Putnam NH, Hellsten U, Kawashima T, Kuo A, Mitros T, Salamov A, Carpenter ML, et al. 2008. The *Trichoplax* genome and the nature of placozoans. *Nature* **454**: 955–960.

Srivastava M, Simakov O, Chapman J, Fahey B, Gauthier MEA, Mitros T, Richards GS, Conaco C, Dacre M, Hellsten U, et al. 2010. The *Amphimedon queenslandica* genome and the evolution of animal complexity. *Nature* **466**: 720–726.

Tian Q, Taupin J, Elledge S, Robertson M, Anderson P. 1995. Fas-activated serine/threonine kinase (FAST) phosphorylates TIA-1 during Fas-mediated apoptosis. *J Exp Med* **182**: 865–874.

van der Biezen EA, Jones JD. 1998. The NB-ARC domain: A novel signalling motif shared by plant resistance gene products and regulators of cell death in animals. *Curr Biol* **8**: R226–R227.

Verhagen AM, Ekert PG, Pakusch M, Silke J, Connolly LM, Reid GE, Moritz RL, Simpson RJ, Vaux DL. 2000. Identification of DIABLO, a mammalian protein that promotes apoptosis by binding to and antagonizing IAP proteins. *Cell* **102**: 43–53.

Verhagen AM, Coulson EJ, Vaux DL. 2001. Inhibitor of apoptosis proteins and their relatives: IAPs and other BIRPs. *Genome Biol* **2**: REVIEWS3009.

Wajant H, Mühlenbeck F, Scheurich P. 1998. Identification of a TRAF (TNF receptor-associated factor) gene in *Caenorhabditis elegans*. *J Mol Evol* **47**: 656–662.

Wiens M, Krasko A, Blumbach B, Müller IM, Müller WE. 2000a. Increased expression of the potential proapoptotic molecule DD2 and increased synthesis of leukotriene B4 during allograft rejection in a marine sponge. *Cell Death Diff* **7**: 461–469.

Wiens M, Krasko A, Mu CI, Mu WEG. 2000b. Molecular Evolution of apoptotic pathways: Cloning of key domains from sponges (Bcl-2 homology domains and death domains) and their phylogenetic relationships. *J Mol Evol* **50**: 520–531.

Wiens M, Diehl-Seifert B, Müller WE. 2001. Sponge Bcl-2 homologous protein (BHP2-GC) confers distinct stress resistance to human HEK-293 cells. *Cell Death Differ* **8**: 887–898.

Wiens M, Krasko A, Perovic S, Müller WE. 2003. Caspase-mediated apoptosis in sponges: Cloning and function of the phylogenetic oldest apoptotic proteases from Metazoa. *Biochim Biophys Acta* **1593:** 179–189.

Yuan J, Horvitz HR. 2004. A first insight into the molecular mechanisms of apoptosis. *Cell* **116:** S53–S56.

Zhang Q, Zmasek CM, Dishaw LJ, Mueller MG, Ye Y, Litman GW, Godzik A. 2008. Novel genes dramatically alter regulatory network topology in amphioxus. *Genome Biol* **9:** R123.

Zmasek CM, Godzik A. 2011. Strong functional patterns in the evolution of eukaryotic genomes revealed by the reconstruction of ancestral protein domain repertoires. *Genome Biol* **12:** R4.

Zmasek CM, Zhang Q, Ye Y, Godzik A. 2007. Surprising complexity of the ancestral apoptosis network. *Genome Biol* **8:** R226.

Caspase Functions in Cell Death and Disease

David R. McIlwain[1,2], Thorsten Berger[1], and Tak W. Mak

The Campbell Family Institute for Breast Cancer Research and Ontario Cancer Institute, University Health Network, Toronto, Ontario M5G 2C1, Canada

Correspondence: tmak@uhnres.utoronto.ca

Caspases are a family of endoproteases that provide critical links in cell regulatory networks controlling inflammation and cell death. The activation of these enzymes is tightly controlled by their production as inactive zymogens that gain catalytic activity following signaling events promoting their aggregation into dimers or macromolecular complexes. Activation of apoptotic caspases results in inactivation or activation of substrates and the generation of a cascade of signaling events permitting the controlled demolition of cellular components. Activation of inflammatory caspases results in the production of active proinflammatory cytokines and the promotion of innate immune responses to various internal and external insults. Dysregulation of caspases underlies human diseases including cancer and inflammatory disorders, and major efforts to design better therapies for these diseases seek to understand how these enzymes work and how they can be controlled.

Caspases are a family of genes important for maintaining homeostasis through regulating cell death and inflammation. Here we will attempt to summarize what we currently know about how caspases normally work, and what happens when members of this diverse gene family fail to work correctly.

Caspases are endoproteases that hydrolyze peptide bonds in a reaction that depends on catalytic cysteine residues in the caspase active site and occurs only after certain aspartic acid residues in the substrate. Although caspase-mediated processing can result in substrate inactivation, it may also generate active signaling molecules that participate in ordered processes such as apoptosis and inflammation. Accordingly, caspases have been broadly classified by their known roles in apoptosis (caspase-3, -6, -7, -8, and -9 in mammals), and in inflammation (caspase-1, -4, -5, -12 in humans and caspase-1, -11, and -12 in mice) (Fig. 1). The functions of caspase-2, -10, and -14 are less easily categorized. Caspases involved in apoptosis have been subclassified by their mechanism of action and are either initiator caspases (caspase-8 and -9) or executioner caspases (caspase-3, -6, and -7).

Caspases are initially produced as inactive monomeric procaspases that require dimerization and often cleavage for activation. Assembly into dimers is facilitated by various adapter proteins that bind to specific regions in

---

[1]These authors contributed equally to this work.
[2]Present address: Department of Gastroenterology, Hepatology and Infectious Diseases, University of Düsseldorf, 40225 Düsseldorf, Germany.

Copyright © 2013 Cold Spring Harbor Laboratory Press; all rights reserved
Cite this article as *Cold Spring Harb Perspect Biol* doi: 10.1101/cshperspect.a008656

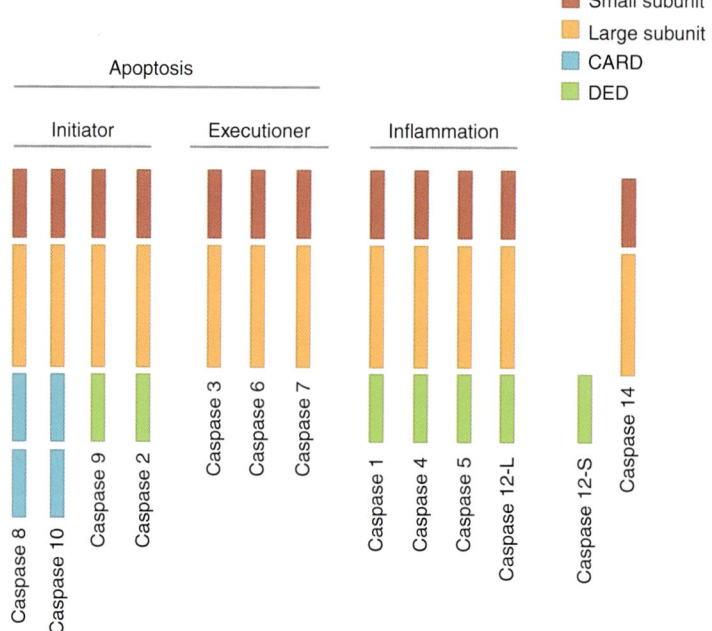

Figure 1. Domain structure of human caspases.

the prodomain of the procaspase. The exact mechanism of assembly depends on the specific adapter involved. Different caspases have different protein–protein interaction domains in their prodomains, allowing them to complex with different adapters. For example, caspase-1, -2, -4, -5, and -9 contain a caspase recruitment domain (CARD), whereas caspase-8 and -10 have a death effector domain (DED) (Taylor et al. 2008).

PHYSIOLOGICAL CASPASE FUNCTIONS

Caspases in Apoptosis

Apoptosis is programmed cell death that involves the controlled dismantling of intracellular components while avoiding inflammation and damage to surrounding cells. Initiator caspases activate executioner caspases that subsequently coordinate their activities to demolish key structural proteins and activate other enzymes. The morphological hallmarks of apoptosis result, including DNA fragmentation and membrane blebbing.

Initiator Caspases

The initiator caspases-8 and -9 normally exist as inactive procaspase monomers that are activated by dimerization and not by cleavage. This process is generalized as an "induced proximity model" (Muzio et al. 1998; Boatright et al. 2003; Chang et al. 2003), in which upstream signaling events induce caspase dimerization and activation. Dimerization also facilitates autocatalytic cleavage of caspase monomers into one large and one small subunit, which results in stabilization of the dimer.

Executioner Caspases

Inappropriate activation of the executioner caspases-3, -6, and -7 is prevented by their production as inactive procaspase dimers that must be cleaved by initiator caspases. This cleavage between the large and small subunits allows a conformational change that brings the two active sites of the executioner caspase dimer together and creates a functional mature protease (Riedl and Shi 2004). Once activated, a single executioner caspase can cleave and activate other executioner caspases, leading to an accelerated feedback loop of caspase activation.

Pathways of Apoptosis

Various apoptotic pathways exist that can be distinguished by the adapters and initiator caspases involved. Most apoptotic programs fall into either the extrinsic or intrinsic category (Fig. 2).

The Extrinsic Pathway of Apoptosis. Extrinsic apoptosis is triggered by extracellular cues delivered in the form of ligands binding to death receptors (DRs). Death receptors are members of the tumor necrosis factor (TNF) superfamily and include TNF receptor-1 (TNF-R1), CD95 (also called Fas and APO-1), death receptor 3 (DR3), TNF-related apoptosis-inducing ligand receptor-1 (TRAIL-R1; also called DR4), and TRAIL-R2 (also called DR5 in humans). Rodents have only one TRAIL-R protein and it resembles DR5. Death receptor ligands include TNF, CD95-ligand (CD95-L; also called Fas-L), TRAIL (also called Apo2-L), and TNF-like ligand 1A (TL1A). The binding of a DR ligand to a DR causes the monomeric procaspase-8 protein to be recruited via its DED to the death-inducing signaling complex (DISC) formed at the cytoplasmic tail of the engaged DR that also includes the adapter protein FAS-associated death domain (FADD) or TNFR-associated death domain (TRADD). Recruitment of caspase-8 monomers results in dimerization

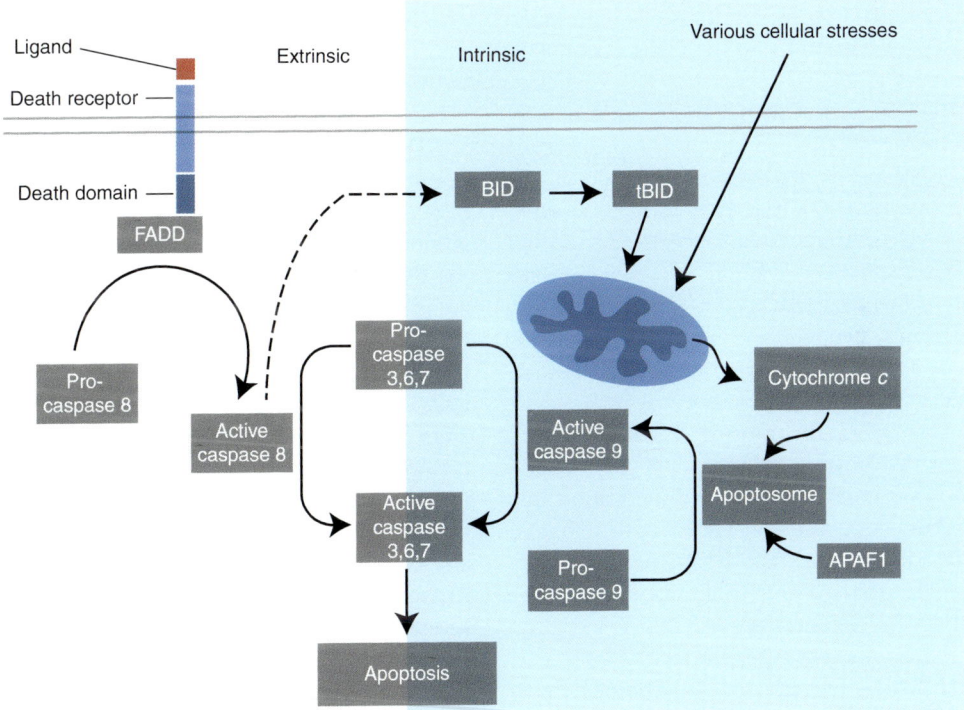

Figure 2. Extrinsic and intrinsic pathways of apoptosis. The extrinsic apoptosis pathway is activated through the binding of a ligand to a death receptor, which in turn leads, with the help of the adapter proteins (FADD/TRADD), to recruitment, dimerization, and activation of caspase-8. Active caspase-8 then either initiates apoptosis directly by cleaving and thereby activating executioner caspase (-3, -6, -7), or activates the intrinsic apoptotic pathway through cleavage of BID to induce efficient cell death. The intrinsic or mitochondrial apoptosis pathway can be activated through various cellular stresses that lead to cytochrome *c* release from the mitochondria and the formation of the apoptosome, comprised of APAF1, cytochrome *c*, ATP, and caspase-9, resulting in the activation of caspase-9. Active caspase-9 then initiates apoptosis by cleaving and thereby activating executioner caspases.

and activation. Cells from gene-targeted mice deficient for caspase-8 (casp8$^{-/-}$ mice) are thus resistant to DR-induced apoptosis (Juo et al. 1998; Varfolomeev et al. 1998; Kang et al. 2004), as are cells from mutant mice lacking either FADD (Yeh et al. 1998) or TRADD, which are specifically defective for TNF-α-mediated apoptosis (Chen et al. 2008b).

The outcome of DR-mediated activation of caspase-8 depends on the cell type. In so-called type I cells, caspase-8 initiates apoptosis directly by cleaving and thereby activating executioner caspases. In type II cells, caspase-8 must first activate the intrinsic apoptotic pathway (discussed below) to induce efficient cell death (Samraj et al. 2006). Type I and II cells differ in their content of intracellular inhibitor of apoptosis proteins (IAPs), which block executioner caspase function unless suppressed by proteins released from the mitochondria (Jost et al. 2009; Spencer et al. 2009).

The Intrinsic Pathway of Apoptosis. Intrinsic apoptosis is also known as mitochondrial apoptosis because it depends on factors released from the mitochondria. This pathway is activated by a vast array of cellular stresses, including growth factor deprivation, cytoskeletal disruption, DNA damage, accumulation of unfolded proteins, hypoxia, and many others. It can also be activated by developmental signals that instruct cells to die, such as hormones (Brenner and Mak 2009). The initiator caspase responsible for the intrinsic apoptosis pathway is caspase-9, which is activated by dimerization induced when the caspase-9 CARD domain binds to the adapter protein apoptotic protease-activating factor-1 (APAF1) (Shiozaki et al. 2002).

Both APAF1 and caspase-9 exist in a resting cell as cytosolic, inactive monomers. A cell experiencing stress first releases cytochrome c from the mitochondria. The binding of cytochrome c to the WD domain of the APAF1 monomer leads to a conformational change that exposes a nucleotide-binding site in the nucleotide-binding and oligomerization (NACHT) domain of APAF1. The nucleotide deoxy-ATP (dATP) binds to this site and induces a second conformational change in APAF1 that exposes both its oligomerization and CARD domains. Seven such activated APAF1 monomers then assemble into an oligomeric complex, the center of which contains the CARDs that recruit and activate caspase-9 (Acehan et al. 2002). The complex containing cytochrome c, APAF1, and caspase-9 has been termed the apoptosome (Cain et al. 2002).

Cytochrome c has a long established role in electron transport, and it was shown in 2000 that mammalian cells lacking cytochrome c could not activate caspases in response to mitochondrial pathway stimulation (Yeh et al. 2000). However, it was not until 2005 that Hao et al. (2005) formally established that the electron transport function of cytochrome c is independent of its ability to engage APAF1 and induce apoptosome formation and caspase activation. Cells from a knockin mouse mutant in which cytochrome c was mutated at lysine 72, a key residue for APAF1 interaction, were able to carry out electron transport but not apoptosis (Hao et al. 2005).

The critical role of intrinsic apoptosis in mammalian development is illustrated by the phenotypes of gene-targeted mice deficient for components of this pathway (Table 1). During the development of the normal brain, apoptosis is critical for culling massive amounts of brain cells to allow selection of those making the best neural connections (Madden and Cotter 2008). Caspase-9-deficient mice suffer from large brain outgrowths characterized by decreased apoptosis and excessive neurons (Hakem et al. 1998; Kuida et al. 1998), as do casp3$^{-/-}$ mice (Kuida et al. 1996; Woo et al. 1998). In vitro, embryonic stem cells and embryonic fibroblasts derived from casp9$^{-/-}$ mice are resistant to several intrinsic apoptotic stimuli, including UV and γ irradiation. Apaf1$^{-/-}$ mice show a similar phenotype including reduced brain cell apoptosis, as well as striking craniofacial abnormalities associated with neuronal cell hyperproliferation. Apaf1$^{-/-}$ cells cannot activate caspases in response to mitochondrial pathway stimulation, are resistant to many apoptotic stimuli, and display reduced processing of caspases-2, -3, and -8 (Yoshida et al. 1998).

Dual Role of Caspase-8 in Apoptosis and Necrosis. As mentioned earlier, caspase-8 plays an

Table 1. Summary of caspase-deficient mouse phenotypes

Caspase	Mouse mutant phenotype	Function derived from deficient phenotype	References
Caspase-1	Develop normally; have no defects in apoptosis	Are more susceptible to virus infection (Thomas et al. 2009); show enhanced tumor formation (Hu et al. 2010); have reduced apoptosis in several models such as neuronal cell death, myocardiac infarct, and heart failure (Frantz et al. 2003; Arai et al. 2006; Merkle et al. 2007); caspase-1-deficient mice are protected against cisplatin-induced apoptosis and acute tubular necrosis (Faubel et al. 2004)	Kuida et al. 1995; Li et al. 1995; Thomas et al. 2009; Hu et al. 2010
Caspase-2	Develop normally and are fertile; have only minor apoptotic defects in some cell types; MEFs show resistance to killing by HS and specific drugs	Caspase-2 has been proposed to be involved in different proapoptotic pathways, but the data from the gene-deficient mice do not support the majority of the in vitro results	Bergeron et al. 1998; O'Reilly et al. 2002; Tu et al. 2006
Caspase-3	Mice die perinatally in mixed background; some can survive to adulthood; show brain hyperplasia	Essential for neuronal cell death; caspase-3 is an essential component in some apoptosis pathways, dependent on the stimulus and cell type; essential for the regulation of B-cell homeostasis	Kuida et al. 1996; Woo et al. 1998, 2003
Caspase-6	Develop normally	No apoptotic defects	Unpublished (see Zheng et al. 1999)
Caspase-7	Develop normally	No apoptotic defects	Lakhani et al. 2006
Caspase-8	Embryonic lethal; defects in heart muscle development	Cells from caspase-8-deficient mice are resistant to death-receptor-induced apoptosis; inactivating mutation in humans shows immunodeficiency; tissue-specific deletion of caspase-8 revealed functions in T-cell homeostasis, in the generation of myeloid and lymphoid cells and the differentiation into macrophages, and in skin inflammation and wound healing; suppresses RIPK3-dependent necrosis	Juo et al. 1998; Varfolomeev et al. 1998; Chun et al. 2002; Salmena et al. 2003; Kang et al. 2004; Beisner et al. 2005; Kovalenko et al. 2009; Lee et al. 2009a; Li et al. 2010; Kaiser et al. 2011; Oberst et al. 2011; Zhang et al. 2011
Caspase-9	Perinatal lethal, but not 100% penetrant	Brain hyperplasia caused by decreased apoptosis and excess neurons; cells from caspase-9-deficient mice show resistance to apoptosis induced by a variety of cytotoxic drugs and irradiation	Hakem et al. 1998; Kuida et al. 1998
Caspase-10	No mouse homolog	Human inactivating mutations are associated with ALPS II	Wang et al. 1999
Caspase-11	Develop normally and are fertile	Mutant mice are resistant to endotoxic shock induced by LPS; IL-1 production after LPS stimulation is blocked; is necessary for caspase-1 activation; regulates cell migration in lymphocytes	Wang et al. 1998; Li et al. 2007

Continued

Table 1. Continued

Caspase	Mouse mutant phenotype	Function derived from deficient phenotype	References
Caspase-12	Develop normally	Mice are resistant to ER stress-induced apoptosis, but their cells undergo apoptosis in response to other death stimuli; thus, caspase-12 mediates an ER-specific apoptosis pathway; show an enhanced bacterial clearance and are more resistant to sepsis	Nakagawa et al. 2000; Saleh et al. 2006
Caspase-14	Develop normally and are fertile; their long-term survival was indistinguishable from that of wild-type mice	Mice show increased sensitivity to UVB irradiation; caspase-14-deficient epidermal cells show no defect in apoptosis; caspase-14 is responsible for the correct processing of (pro)filaggrin during cornification	Denecker et al. 2007

MEF, mouse embryonic fibroblast; HS heat shock; ALPS, autoimmune lymphoproliferative syndrome; LPS, lipopolysaccharide; ER, endoplasmic reticulum; RIPK3, receptor-interacting serine/threonine-protein kinase 3.

important role in extrinsic apoptosis, combining with FADD to form the DISC. Interestingly, the deletion in mice of caspase-8, FADD, or the DISC regulatory protein FLICE-like inhibitory protein (FLIP) leads to embryonic death caused by a variety of defects. Some of these defects appear to be related to apoptosis, such as impaired heart muscle development in the absence of caspase-8 (Varfolomeev et al. 1998), cardiac failure in the absence of FADD (Yeh et al. 1998; Zhang et al. 1998), and disrupted heart development in the absence of FLIP (Yeh et al. 2000). However, tissue-specific deletions of caspase-8 have revealed new roles for this caspase, which appear to be unrelated to apoptosis. Caspase-8 function is also critical for T-cell homeostasis (Salmena et al. 2003), the generation of myeloid and lymphoid cells and macrophage differentiation (Kang et al. 2004; Beisner et al. 2005), and skin inflammation and wound healing (Kovalenko et al. 2009; Lee et al. 2009a; Li et al. 2010). Recently, three reports provided evidence that some of the defects associated with loss of caspase-8, and result in embryonic death, are not owing to impaired apoptosis but rather to defective suppression of receptor-interacting serine-threonine kinase 3 (RIPK3)-dependent necrosis (Kaiser et al. 2011; Oberst et al. 2011; Zhang et al. 2011). Thus, caspase-8 appears to have dual roles in activation of apoptosis and suppression of necrosis, and caspase-8-dependent suppression of necrosis, but not caspase-8 activation of apoptosis, is critical for mouse embryonic survival.

Caspases in Inflammation

Inflammatory Caspases

Several caspases function as critical mediators of innate immune responses rather than pro-apoptotic factors. Caspase-1, -4, -5, and -12 comprise the inflammatory subset in humans, whereas caspase-1, -11, and -12 serve the same function in mice. Interestingly, the genes encoding inflammatory caspases are located in close proximity on human chromosome 11 and murine chromosome 9, suggesting that they may have arisen from gene duplication events. At the protein level, inflammatory caspases, like their proapoptotic counterparts, are produced as inactive procaspases in resting cells. Only after cellular stimulation via engagement of pattern-recognition receptors (see below) are inflammatory caspases activated through the formation of a cytosolic complex termed the inflammasome (Martinon et al. 2002).

Inflammasome Formation

Inflammasome formation resembles apoptosome formation and has been best studied for the nucleotide-binding domain, leucine-rich repeat-containing (NLR) proteins, which are a family of pattern-recognition receptors (PRRs), and other proteins (Fig. 3). In a resting cell, NLR monomers are held in an inactive conformation until an external or internal stimulus promotes their assembly (similar to the assembly of APAF1 monomers in the apoptosome). NLR monomers interact through their NACHT domains and bind to the adapter protein apoptosis-associated speck-like protein containing a CARD (ASC/PYCARD) (Ting et al. 2008b). The presence of ASC permits the recruitment of an inactive inflammatory procaspase, typically procaspase-1, to the inflammasome, followed by cleavage and activation of caspase-1 through induced proximity autocatalysis (Davis et al. 2011). Activated caspase-1 in turn cleaves pro-IL-1β and pro-IL-18, which facilitates the secretion of these proinflammatory cytokines (Ting et al. 2008b).

Different NLR-driven inflammasomes contain different NLR members and respond to different stimuli, so that inflammasome formation and the resulting immune response are appropriately tailored to the specific context. However, precisely how inflammasomes are activated in various situations remains poorly understood. It is known that signals initiating inflammasome formation can be delivered by environmental irritants, pathogen-derived molecules, self-derived molecules associated with cell damage, or inappropriate metabolite accumulation (Davis et al. 2011). Whether these signals are received directly by NLR proteins or relayed through secondary pattern-recognition receptors remains unclear and is likely to be context dependent (Monie et al. 2009).

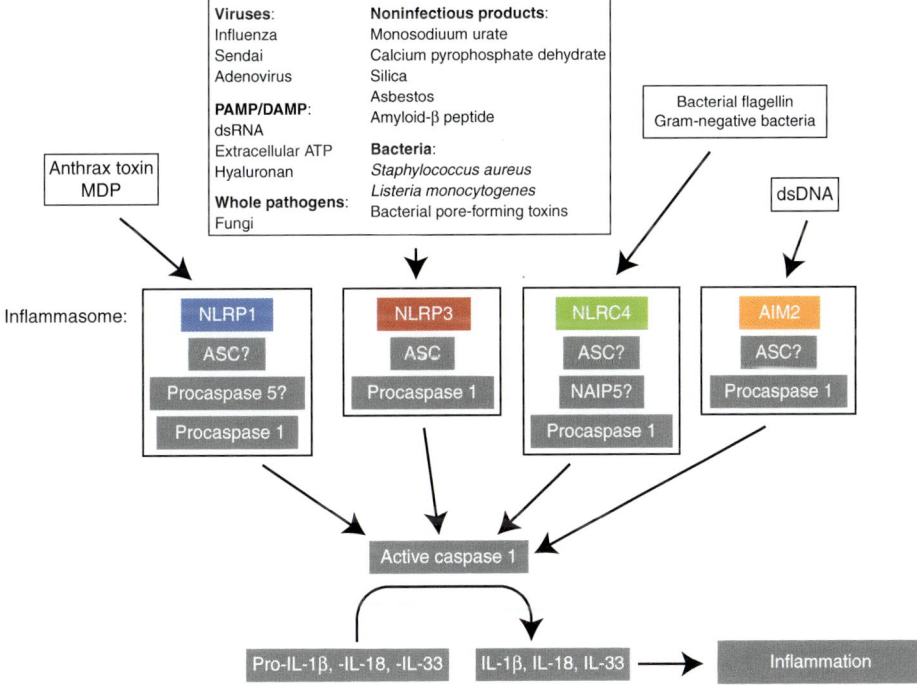

Figure 3. Signaling and composition of inflammasomes. Activation of inflammatory caspases such as caspase-1 is achieved through inflammasome formation. A multitude of cellular stimuli are recognized by a family of pattern-recognition receptors, engagement of which leads to the binding of the adapter protein ASC and the recruitment and activation of the inactive inflammatory procaspase, typically procaspase-1. Activated caspase-1 in turn cleaves pro-IL-1β, pro-IL-18, and pro-IL-33, which facilitates the secretion of these proinflammatory cytokines leading to inflammation.

Role of Pattern-Recognition Receptors. PRRs are molecules that can detect pathogen-associated molecular patterns (PAMPs) or damage-associated molecular patterns (DAMPs) and initiate inflammasome formation. PRRs are frequently expressed by cells that make contact with invading microbes, such as epithelial cells and cells of the innate and adaptive immune responses. Several different classes of PRRs exist, including Toll-like receptors (TLRs) that recognize a variety of PAMPs derived from bacteria, viruses, and fungi and work in synergy with the cytosolic, C-type lectin receptors (CLRs) (which sense fungi), retinoic acid-inducible gene (RIG)-I-like receptors (RLRs) (which sense viruses), and NLR proteins (which sense bacteria) (Davis et al. 2011) (for recent reviews on CLRs and RLRs see Loo and Gale 2011; Osorio and Reis 2011).

TLRs. TLRs were originally named for their similarity to the *Drosophila* protein Toll (Anderson et al. 1985a,b) and were first described in 1997 (Medzhitov et al. 1997). Different TLRs recognize different bacterial components, including lipopolysaccharides (LPS), flagellin, lipoproteins, double-stranded viral RNA, and the unmethylated CpG islands of bacterial and viral DNA. TLR engagement promotes inflammasome formation at least partially through signaling that activates the transcription factors NF-κB and AP-1. TLR-induced inflammasomes facilitate IL-1β and IL-18 secretion as well as the expression of interferon regulatory factor (IRF) transcription factors that mediate type I interferon (IFN)-dependent antiviral responses. Sometimes a second signal is required to complete TLR-mediated inflammasome formation and induce IL-1β secretion. For example, although monocytes circulating in the blood can secrete IL-1β in response to LPS stimulation alone (Netea et al. 2008), primary macrophages must undergo TLR engagement and receive an additional signal such as ATP before they can secrete IL-1β (Wewers and Herzyk 1989; Herzyk et al. 1992). It is unclear whether TLRs are inflammasome components, or whether they act as signaling molecules for inflammasome formation.

NLRs. Members of the highly conserved NLR family participate in innate immune defense against infection in all animals as well as in plants (Jones and Dangl 2006). There are 22 NLR proteins in humans and even more in mice (Schroder and Tschopp 2010). The largest NLR subclass (14 members) contains the NLR pyrin domain-containing proteins (NLRPs). Other NLR family members include the NODs (NOD1 and NOD2), class II transactivator (CIITA), NAIP, and NLRX (Ting et al. 2008a). The amino terminus of NRL proteins is comprised of either a caspase recruitment domain (CARD) or a pyrin domain (PyD) that permits the direct or indirect recruitment of inflammatory procaspases. At the carboxyl terminus, all NRL proteins (except NLRP10) have a central NACHT domain followed by leucine-rich repeats (LRRs). These LRRs are believed to confer specificity for a particular PAMP/DAMP.

The first NLR to be characterized was CIITA, which is essential for MHC class II gene expression and currently the only NRL member that functions as a transcriptional activator. "Bare lymphocyte syndrome" in humans is owing to a loss-of-function mutation in the *Ciita* gene (Steimle et al. 1993). All other NRLs are thought to perform cytoplasmic surveillance for PAMPs/DAMPs.

NOD1 and NOD2 were the first NLRs reported to be PAMP PRRs. Both recognize products of bacterial peptidoglycan degradation, with NOD1 binding to mesodiaminopimelic acid derived mainly from Gram-negative bacteria (Chamaillard et al. 2003; Girardin et al. 2003a), and NOD2 detecting the muramyl dipeptide common to both Gram-negative and Gram-positive bacteria (Girardin et al. 2003b; Inohara et al. 2003). After engagement, NOD1 and NOD2 oligomerize and transiently recruit receptor-interacting protein 2 (RIP2) through CARD–CARD interaction, leading to NF-κB activation and proinflammatory gene expression (Kufer et al. 2005).

Types of Inflammasomes

NLRP3 Inflammasome. NLRP3 is a scaffold protein that uses PyD–PyD interactions to bind to both ASC and procaspase-1, forming the NLRP3 inflammasome that promotes autocatalytic caspase-1 activation. The NLRP3 inflammasome is expressed by myeloid cells and is formed ("activated") in response to a broad range of PAMPs as well as whole pathogens, including fungi (Jin and Flavell 2010). In the latter case, inflammasome activation involves Syk tyrosine kinase activity, the production of reactive oxygen species (ROS), and potassium efflux, but is independent of its transcriptional regulation of *Il-1β* (Gross et al. 2009). NLRP3 inflammasome activation is also triggered by bacteria such as *Staphylococcus aureus* and *Listeria monocytogenes*, which produce pore-forming toxins (Mariathasan et al. 2006); influenza virus, Sendai virus, and adenovirus, double-stranded RNA (Kanneganti et al. 2006; Muruve et al. 2008); and host-derived molecules such as ATP (Mariathasan et al. 2006) and hyaluronan (Yamasaki et al. 2009) that are released by injured cells. The NLRP3 inflammasome is also a sensor for amyloid-β peptide, the accumulation of which is a hallmark of Alzheimer's disease (Halle et al. 2008). Noninfectious materials that activate the NLRP3 inflammasome include crystals of monosodium urate and calcium pyrophosphate dehydrate, which cause gout and pseudogout, respectively (Martinon et al. 2006); asbestos and crystalline silica (Cassel et al. 2008; Dostert et al. 2008); the skin irritants trinitrophenylcholride, trinitrochlorobenzene, and dinitrofluorobenzene (Sutterwala et al. 2006; Watanabe et al. 2007); and UVB radiation (Feldmeyer et al. 2007). It is surprising that a single molecule can "sense" all these different stimuli. A new hypothesis therefore states that this multitude of danger signals is integrated by mitochondria and that the NLRP3 monitors mitochondrial status, reacting to changes in mitochondrial activity that then trigger NLRP3 inflammasome formation (Tschopp 2011).

NLRC4 Inflammasome. The NLRC4 (NLR family, CARD domain-containing 4) inflammasome is formed in response to bacterial flagellin and conserved regions of the type III and type IV secretion systems of Gram-negative bacteria such as *Salmonella typhimurium*, *Burkholderia pseudomallei* (BsaK), *Escherichia coli* (EprJ and EscI), *Shigella flexneri* (MxiI), *Pseudomonas aeruginosa* (PscI), and *Legionella pneumophila* (Amer et al. 2006; Franchi et al. 2006, 2007; Miao et al. 2006, 2008, 2010; Molofsky et al. 2006; Sutterwala et al. 2007). The exact composition of the NLRC4 inflammasome is not fully understood and it is unclear if NLRC4 requires ASC for physiological caspase-1 activation (Poyet et al. 2001; Mariathasan et al. 2004; Franchi et al. 2007; Suzuki et al. 2007). The NLR member NAIP5 (NLR family, apoptosis inhibitory protein 5) is variably required for activation of this inflammasome (Lightfield et al. 2011).

NLRP1 Inflammasome. The NLRP1 inflammasome is activated in response to *Bacillus anthracis* lethal toxin (LeTx) (Boyden and Dietrich 2006) and muramyl dipeptide (MDP) (Faustin et al. 2007). The sequence of the *NLRP1* gene has diverged between humans and rodents, and the murine genome contains three orthologues that are highly polymorphic (Boyden and Dietrich 2006). This variation is presumably responsible for the differences in LeTx susceptibility observed among inbred mouse strains (Boyden and Dietrich 2006). The precise mechanism by which the NLRP1 inflammasome induces caspase-1 activation is still controversial. Human NLRP1 contains a carboxy-terminal CARD domain and so can interact directly with procaspase-1; however, the addition of ASC to this complex in vitro increases inflammasome activity (Faustin et al. 2007). Notably, human NLRP1 can also bind to caspase-5 and thereby contribute to the processing of pro-IL-1β and pro-IL-18 (Tschopp et al. 2003). In contrast, the mouse NLRP1 orthologues do not contain a functional PyD domain, and consequently NLRP1-associated caspase-1 activation in mouse macrophages is not dependent on ASC (Hsu et al. 2008).

Aim2 Inflammasome. Absent in melanoma 2 (AIM2) is a member of the pyrin and HIN domain-containing protein (PYHIN) family (Ludlow et al. 2005). AIM2 interacts with ASC through PyD–PyD interactions (Fernandes-Alnemri et al. 2009) to form an inflammasome that recruits and activates procaspase-1 in response to cytosolic double-stranded DNA (dsDNA) (Fernandes-Alnemri et al. 2009; Hornung et al. 2009). AIM2 senses cytosolic DNA through its carboxy-terminal HIN-200 domain, which contains two oligonucleotide/oligosaccharide-binding folds (Fernandes-Alnemri et al. 2010). Because AIM2 does not contain a central oligomerization domain equivalent to the NACHT domain in NLRs, it is believed that the dsDNA ligand itself, which can bind to multiple AIM2 molecules, mediates AIM2 oligomerization in the inflammasome (Fernandes-Alnemri et al. 2009). Studies of gene-targeted Aim2-deficient mice have shown that, in addition to dsDNA, AIM2 detects the cytosolic bacterial pathogen *Francisella tularensis* (live vaccine strain) as well as DNA viruses such as vaccinia and mouse cytomegalovirus (mCMV) (Fernandes-Alnemri et al. 2010; Rathinam et al. 2010). Because it recognizes dsDNA, the AIM2 inflammasome may also play a role in the autoimmune responses to dsDNA characteristic of systemic lupus erythematosis (SLE) and related diseases (Fernandes-Alnemri et al. 2010; Rathinam et al. 2010).

Caspase-1 and Cell Death

Although caspase-1 activation most often contributes to inflammation, excessive caspase-1 activity can cause pyroptosis, a nonapoptotic type of programmed cell death that is characterized by plasma membrane rupture and the release of proinflammatory intracellular contents (Cookson and Brennan 2001; Fink and Cookson 2006). Pyroptosis does not involve classical apoptotic caspases like caspase-3 and -8. Instead, activated caspase-1 activates caspase-7 and an unidentified nuclease that induces DNA cleavage and nuclear condensation without compromising nuclear integrity (Molofsky et al. 2006; Bergsbaken and Cookson 2007). Because caspase-1 activation is required for cell death in a variety of experimental settings, including in the immune system (Shi et al. 1996), the cardiovascular system (Kolodgie et al. 2000; Frantz et al. 2003), and the central nervous system (Liu et al. 1999; Yang et al. 1999; Zhang et al. 2003), pyroptosis has been thought to have an important physiological role. However, $casp1^{-/-}$ mice develop normally, implying that this protease is redundant in vivo during development (Kuida et al. 1995; Li et al. 1995). Additional reports suggest that caspase-1 is also capable of cleaving and activating BID and thereby engaging the mitochondrial pathway of apoptosis (Guegan et al. 2002; Zhang et al. 2003).

Recently it has been shown that caspase-1-deficient mice show enhanced tumor formation in an azoxymethane and dextran sodium sulfate colitis-associated colorectal cancer model. Interestingly the mechanism of caspase-1 tumor formation in this model was not through regulation of inflammation, but rather owing to increased colonic epithelial cell proliferation in the early stages of tumor formation and reduced apoptosis in advanced tumors in the caspase-1-deficient mice (Hu et al. 2010).

It should be noted that the interpretation of past data generated using caspase-1-deficient animals may need to be revisited in light of several new pieces of evidence. It has recently been shown that caspase-1-deficient mice generated from strain 129 embryonic stem cells also harbor a mutation in the caspase-11 locus, and so are de facto caspase-1/caspase-11 double-knockout mice (Kayagaki et al. 2011). Kayagaki et al. addressed this issue in their study by rescuing caspase-11 activity in their caspase-1-deficient mice via transgenic expression of a caspase-11 bacterial artificial chromosome. The in vivo data then generated indicate that caspase-11 rather than caspase-1 may be the critical effector caspase responsible for the inflammatory response, making human caspases-4 and -5 potential interesting targets for intervention in patients with sepsis (Kayagaki et al. 2011).

Caspase-12 and Anti-Inflammation

In mice, caspase-12 appears to abrogate the inflammatory response largely owing to an inhibitory effect on caspase-1 (Scott and Saleh 2007). Consequently, $casp12^{-/-}$ mice show enhanced bacterial clearance and resistance to sepsis (Saleh et al. 2006). Interestingly, the enzymatic function of caspase-12 is not required for caspase-1 inhibition (Saleh et al. 2006), suggesting that caspase-12 is more likely a protease regulator rather than a protease itself.

In most humans from Eurasia and a significant proportion of individuals from African populations, there exists a frameshift mutation in the caspase-12 gene (CASP12) that generates a premature stop codon and prevents expression of full-length caspase-12 leading to a shortened caspase-12 protein (caspase-12S) (Fischer et al. 2002). However, in about 20% of individuals of sub-Saharan African descent, a single nucleotide polymorphism (SNP) changes this stop codon to an arginine residue, resulting in successful readthrough and the synthesis of the full-length caspase-12 protein (caspase-12L) (Saleh et al. 2004). Individuals expressing the readthrough polymorphism show reduced inflammatory and innate responses to endotoxins and thus an increased risk of developing severe sepsis (Saleh et al. 2004). It has been suggested that the rise in infectious disease that accompanied the increased population density developing in Europe over time favored the survival of individuals expressing the truncated caspase-12 variant (Xue et al. 2006).

Caspases in Proliferation

Although caspases are most often associated with apoptosis, there has been persistent evidence that some of these enzymes can also influence proliferation. One of the earliest observations was that treatment of T cells with caspase inhibitors led to a surprising suppression of CD3-induced T-cell expansion (Alam et al. 1999; Kennedy et al. 1999). This growth-promoting caspase function was later attributed to caspase-8, because c-FLIP, a caspase-8 inhibitor, was shown to modulate T-cell proliferation (Lens et al. 2002). Similarly, caspase-8 and -6 can positively regulate B-cell proliferation (Olson et al. 2003; Beisner et al. 2005).

However, caspase-3 may have the opposite effect, as B cells lacking caspase-3 showed increased proliferation in vivo and hyperproliferation after mitogenic stimulation in vitro (Woo et al. 2003). This hyperproliferative B-cell phenotype was rescued in double-knockout mice lacking both caspase-3 and the cyclin-dependent kinase inhibitor p21 (encoded by *Cdkn1a*), which is a caspase-3 substrate (Woo et al. 2003).

As mentioned earlier, recent work has now provided convincing evidence that the suppression of RIPK signaling by caspase-8 and FADD accounts for the nonapoptotic roles of these proteins (Kaiser et al. 2011; Oberst and Green 2011; Oberst et al. 2011; Zhang et al. 2011). Another important question, namely, how is it possible that the proteolytic function of caspase-8, which normally leads to apoptosis, can suppress RIPK signaling without causing apoptotic cell death, can also now be explained. C-FLIP has a greater affinity for procaspase-8 than this proenzyme has for itself, which permits C-FLIP to inhibit the activation of apoptosis by caspase-8 while allowing caspase-8 to retain its catalytic activity (Boatright et al. 2004; Oberst and Green 2011).

Less Well-Categorized Caspases

Caspase-2

Caspase-2 is evolutionarily ancient, the most highly conserved caspase among animals, and one of the earliest caspases discovered (Kumar et al. 1994; Wang et al. 1994); its function resembles a more rudimentary type caspase similar to *Caenorhabditis elegans* in which it needs to fulfill multiple, sometimes opposing roles, that later during evolution have been taken over by other members of the caspase family.

The mammalian caspase-2 protein has a long prodomain containing a CARD sequence. In response to apoptotic stimuli such as DNA damage, cytoskeletal disruption, metabolic perturbation, or heat shock (Harvey et al. 1997), inactive procaspase-2 monomers are induced to oligomerize and are activated by induced proximity. The ensuing autocatalytic cleavage stabilizes the mature caspase-2 enzyme and enhances its activity (Baliga et al. 2004; Krumschnabel et al. 2009).

Procaspase-2 oligomerization is mediated by the adapter protein Rip-associated protein with a death domain (RAIDD), which binds to procaspase-2 via CARD–CARD interaction (Harvey et al. 1997; Baliga et al. 2004; Krumschnabel et al. 2009). Procaspase-2-bound RAIDD molecules form a complex via additional adapter molecules such as p53-induced protein with a DD (PIDD), which binds to RAIDD via DD–DD interaction. This PIDD–RAIDD–procaspase-2 complex has been termed the PIDDosome (Tinel and Tschopp 2004). Resolution of the crystal structure of the PIDDosome has revealed the presence of multiple PIDD and RAIDD subunits (Park and Wu 2006; Park et al. 2007) in a structure resembling the CD95–FADD complex involved in procaspase-8 activation.

Caspase-2 can also be activated by a mechanism that involves p53-dependent CD95 up-regulation and the recruitment of caspase-8 to the DISC complex. BID is cleaved by this coordination of caspase-2 and -8 and mitochondrial apoptosis is activated (Sidi et al. 2008; Olsson et al. 2009). However, caspase-2 is also involved in seemingly opposing functions such as protection against DNA damage (Shi et al. 2009) or cancer development (Ho et al. 2009). Caspase-2 is also important for programmed oocyte death during mouse development (Bergeron et al. 1998).

Caspase-10

Human caspase-10 is highly homologous to caspase-8 and is recruited to the DISC on DR engagement (Sprick et al. 2002). However, the role of caspase-10 in the extrinsic apoptotic cascade is still not clear. Reports have conflicted on the requirement for caspase-10 in CD95-mediated apoptosis in the absence of caspase-8 (Kischkel et al. 2001; Sprick et al. 2002). Although some recent findings suggest that caspase-10 acts in an atypical CD95-induced cell death pathway (Lafont et al. 2010), other evidence points to a role for caspase-10 in intrinsic apoptosis that is triggered by cytotoxic drugs in a fashion that is FADD-dependent but DR-independent (Park et al. 2004; Filomenko et al. 2006; Lee et al. 2007). To date, no mouse caspase-10 homolog has been reported.

Caspase-14

Caspase-14 is unique because it is found only in terrestrial mammals and does not seem to have evolved from orthologues in insects or nematodes like the apoptotic caspases (Lamkanfi et al. 2002). Furthermore, caspase-14 expression is restricted to cornifying epithelial cells, such as occur in the skin, and plays a role in terminal keratinocyte differentiation (Denecker et al. 2008). Studies of $casp14^{-/-}$ mice have shown that caspase-14 is responsible for both the correct processing of profilaggrin during cornification and the protection of mice against UVB irradiation (Denecker et al. 2007). However, caspase-14 is dispensable for keratinocyte apoptosis (Denecker et al. 2008).

CASPASES IN HUMAN DISEASE

Caspase activity is a double-edged sword. Although defective caspase activation and the inadequate cell death that results can promote tumorigenesis, extreme caspase activation and the excessive cell death that ensues can promote neurodegenerative conditions. Furthermore, insufficient activation of caspases involved in inflammation can lead to an increased susceptibility to infection, whereas hyperactivation of these caspases can promote inflammatory conditions.

Caspases and Cancer

Our bodies use several sophisticated mechanisms to safeguard against cancer development. These mechanisms recognize DNA mutations, and induce either the repair of the faulty DNA, or the death of the affected cell before it can become oncogenic. Because caspases are crucial for apoptosis, it is not surprising that deregulation of these enzymes and the pathways in which they are involved can aid in the persistence of mutated cells and promote tumorigenesis. However, although caspases are key players in the best documented mechanism of cancer cell death, unlike mutation of p53 or elements of the PI3K pathway, mutation of CASP genes is not frequent in human tumor cells. Genetic and inhibitor studies have shown that the inactivation of individual caspases is not usually sufficient to either prevent continuation of the caspase cascade, or to derail alternative nonapoptotic cell death mechanisms. Instead, malignant cells appear to more frequently gain a survival advantage by inactivating signaling mediators upstream of caspase activation.

Despite the above, the reduced expression of proapoptotic caspases has been reported in a variety of cancers (Philchenkov et al. 2004), and specific inactivating mutations have been linked to various tumor types and stages of transformation (discussed below). Moreover, although inherited mutations in the CASP genes are relatively rare, certain caspase polymorphisms thought to affect caspase abundance or activity have been associated with variable effects on tumorigenesis.

Caspase-8

Inactivating CASP8 mutations have been reported in various cancers, including childhood neuroblastoma. Wild-type caspase-8 acts as a tumor suppressor in neuroblastomas with amplification of N-myc (Teitz et al. 2000), thus mutations leading to loss of caspase-8 function render neuroblastoma cell lines resistant to DR-induced apoptosis (Hopkins-Donaldson et al. 2000; Teitz et al. 2000; Eggert et al. 2001; Yang et al. 2003). Furthermore, in an experimental neuroblastoma cell line model, caspase-8 deletion enhanced the metastatic potential of neuroblastoma cells in chick embryos (Stupack et al. 2006).

In a study of 180 human colorectal tumors (98 invasive carcinomas and 82 adenomas), somatic CASP8 mutations were detected in 5% of invasive carcinomas but in no adenomas. At least three of the mutations were confirmed to decrease caspase-8-mediated apoptosis by acting in a dominant–negative fashion (Kim et al. 2003). In a similar study of 69 hepatocellular carcinomas (HCCs), a single somatic CASP8 mutation was detected in nine cases (13.0%). This frameshift mutation resulted in a two base-pair deletion (1225_1226delTG) that caused premature termination of translation and loss of caspase-8 function (Soung et al. 2005b). In another study of 162 gastric carcinomas (40 early and 122 advanced cancers), 185 non-small-cell lung cancers, 93 breast carcinomas, and 88 acute leukemias, CASP8 mutations were detected only in advanced gastric cancers (10.7%) (Soung et al. 2005a). Again, these mutations led to markedly decreased caspase-8-dependent cell death in vitro (Soung et al. 2005b).

An interesting linkage between CASP8 and cancer occurs for inheritance of the D302H polymorphism in CASP8 (rs1045485), which substitutes histidine for aspartic acid and is associated with reduced breast cancer risk (MacPherson et al. 2004; Frank et al. 2005). An analysis of 16,423 cases and 17,109 controls from 14 studies conducted by the Breast Cancer Association Consortium (BCAC) has confirmed the dose-dependent protective effect of this allele [P trend $= 1.1 \times 10^{-7}$, per allele odds ratio (OR) $= 0.88$, with a 95% confidence interval (CI) of 0.84–0.92] (Cox et al. 2007). Another well-studied inherited CASP8 polymorphism is a six-nucleotide deletion (-652 6N del; 6N del, rs3834129) in the promoter region. A recent meta-analysis of 23 publications covering 55,174 cancer cases and 59,336 controls from 55 individual studies concluded that the D302H variant and the -652 6N del polymorphism were associated with a significantly reduced overall risk of cancer (Yin et al. 2010). It is assumed that these alterations to the mutated caspase-8 protein enhance its proapoptotic effects and prevent tumor cell persistence, although such a relationship has yet to be shown in vivo.

Caspase-9

Germline variation in the CASP9 gene has been linked to non-Hodgkin's lymphoma (NHL) (Kelly et al. 2010). In a study of 36 apoptosis pathway genes, alterations of CASP9 at both the gene and SNP levels were associated with NHL risk (Kelly et al. 2010). In another study of the impact on lymphomagenesis of genetic variation in 12 caspases, examination of 1946 NHL cases and 1808 controls showed significant associations for alterations of CASP8, CASP9, or CASP1 with NHL (Lan et al. 2009). An earlier smaller study of 461 NHL cases and 535 controls also showed a significant association between certain variants of CASP3 and CASP9 and NHL risk (Lan et al. 2007). In both studies, the caspase-9 SNPs associated with NHL showed decreased risk of NHL, whereas the other caspases showed increased NHL risk.

In an analysis of polymorphisms in the CASP9 promoter in 432 lung cancer patients and 432 matched controls, the −1263 GG genotype was linked to a significantly decreased risk of lung cancer compared with the −1263 AA genotype or the −1263 AA + AG genotype (Park et al. 2006). It was proposed that this protective effect might be owing to increased promoter activity of the G-C haplotype compared with the −1263G/−712T and −1263A/−712C haplotypes that enhances caspase-9 expression (Park et al. 2006).

Caspase-3

Many studies have analyzed whether alterations to the CASP3 gene encoding the crucial executioner caspase-3 might promote human tumorigenesis. One study examined the caspase-3 coding region in 944 tumors of 14 different types compared with healthy adjacent tissue. However, only 14 tumors (1.48%) showed somatic CASP3 mutations (Soung et al. 2004). In another study analyzing 930 squamous cell carcinomas of the head and neck (SCCHN) and 993 controls, the CASP3 rs4647601:TT variant was associated with an increased risk of SCCHN compared with the GG genotype (Chen et al. 2008a). This finding was most evident in certain subgroups, including younger (≤56 yr) subjects, males, and never smokers. Conversely, in an analysis of 582 lung cancer patients and 582 controls, individuals bearing at least one allele with a −928A > G, 77G > A, or 17532A > C polymorphism had a significantly decreased risk for lung cancer compared with individuals who were homozygous for the wild-type CASP3 allele (Jang et al. 2008).

An important study of 128 multiple myeloma cases and 516 controls analyzed five SNPs in various CASP genes. Compared with individuals with the TT genotype of CASP3 Ex8 + 567 T > C, subjects with the CC genotype had a fivefold lower risk of multiple myeloma (Hosgood et al. 2008). Multiple myeloma risk was also reduced in individuals with the AG and AA genotypes of CASP9 Ex5 + 32 G > A (Hosgood et al. 2008). An earlier study by the same group found a similar association between decreased risk of NHL and certain CASP3 variants (Lan et al. 2007). Finally, a study of 1028 endometrial cancer patients and 1003 healthy controls examined potential links between caspase-3, -7, and -8 variant alleles and risk of endometrial cancer. Compared with the CC genotype, the GG genotype of rs2705901 in CASP3 was significantly associated with increased cancer risk (Xu et al. 2009). Taken together, these results suggest that CASP3 polymorphisms and their haplotypes help to define an individual's genetic susceptibility to cancer development.

Caspase-7

In one analysis of multiple cancer types, somatic mutations in CASP7 were detected in two of 98 colon carcinomas (2.0%), one of 50 esophageal carcinomas (2.0%), and one of 33 head/neck carcinomas (3.0%), but not in stomach, urinary bladder, or lung cancers (Soung et al. 2003). When these tumor-derived caspase-7 mutants were overexpressed in 293T human kidney cells, the cells showed reduced apoptosis (Soung et al. 2003). In a different study of 720 lung cancer patients and 720 controls, certain CASP7 polymorphisms were found to promote susceptibility to lung cancer (Lee et al. 2009b). As mentioned earlier, a study of 1028 endometrial cancer patients and 1003 healthy controls examined potential links between caspase-3, -7, and -8 variant alleles and risk of endometrial cancer. Of 35 selected SNPs, four in CASP7 were in high linkage disequilibrium and associated with increased risk of endometrial cancer; two CASP7 SNPs were associated with reduced risk; and two CASP7 SNPs were associated with increased risk compared with individuals homozygous for the major CASP7 alleles. These findings suggest that mutations altering the executioner function of caspase-7 affect the pathogenesis of some human solid cancers.

Caspase-1, -4, -5

An evaluation of mutations in the inflammatory caspases-1, -4, and -5 in 337 samples of various types of human cancers showed that CASP1 mutations were present in two malignancies (0.6%), CASP4 mutations in two (0.6%), and CASP5 mutations in 15 (4.4%) (Soung et al. 2008). The highest prevalence of CASP5 mutations was in microsatellite instability (MSI)-positive gastric carcinomas, suggesting that caspase-5 activity may be important in the etiology of these tumors.

In a mouse model of colorectal cancer based on colitis induced by azoxymethane and dextran sodium sulfate treatment, $casp1^{-/-}$ mutants showed enhanced tumor formation owing to alterations to two different caspase functions. In early-stage tumors, the proliferation of colonic epithelial cells was increased in the absence of caspase-1, whereas in advanced tumors, apoptosis was reduced (Hu et al. 2010). Interestingly, despite the association of caspase-1 with inflammation, in neither early nor late colorectal tumors was defective regulation of inflammation observed.

Caspase-6

CASP6 mutations have been found in 2% of 150 human cancers of colonic or gastric origin (Lee et al. 2006). Furthermore, expression of caspase-6 in gastric cancer samples is decreased, suggesting that loss of caspase-6 expression might be involved in the mechanism of gastric cancer development (Yoo et al. 2004).

Caspase-10

An analysis of 117 NHL samples revealed that 17 (14.5%) contained inactivating CASP10 mutations (Shin et al. 2002). When overexpressed in 293T cells, these mutations suppressed apoptosis. Rare CASP10 mutations have also been detected in cases of T-cell acute lymphoblastic leukemia and multiple myeloma (Kim et al. 2009), as well as in colon, breast, lung, and hepatocellular carcinomas (Oh et al. 2010) and gastric cancers (Park et al. 2002).

Caspase-1 in Inflammatory Diseases

The production of IL-1, and thus caspase-1 activation, has been implicated in a wide variety of inflammatory and autoimmune diseases (Gabay et al. 2010). Researchers have frequently sought to confirm the involvement of caspase-1 in these conditions through the use of agents that attempt to modulate IL-1 production by blocking caspase-1, IL-1 functions, or IL-1 receptors. However, human trials of agents targeting IL-1 in rheumatoid arthritis (RA) (Drevlow et al. 1996; Bresnihan et al. 1998; Jiang et al. 2000; Cohen et al. 2002; Genovese et al. 2004; Alten et al. 2008), as well as in other rheumatic diseases such as SLE, psoriatic arthritis, and osteoarthritis (Finckh and Gabay 2008), have shown only modest efficacy or no improvement. Nevertheless, these drugs have improved the health of patients with several other hereditary and acquired conditions linked to elevated IL-1β levels, as outlined below.

Gout

Gout is a common autoinflammatory disorder characterized by chronic elevated blood uric acid levels (hyperuricemia) and the deposition of monosodium urate (MSU) crystals in joints. Patients experience severe pain and joint inflammation. The pathogenesis of this disease, as well as that of pseudogout (deposition of calcium pyrophosphate dihydrate crystals) and pulmonary silicosis, have been linked to inflammatory responses activated by the deposited crystals and mediated by the NLRP3 inflammasome (Cronstein and Terkeltaub 2006; Martinon et al. 2006; Hornung et al. 2008).

Cryopyrin-Associated Periodic Syndromes

Mutations in NLRP3 cause three rare inherited autoinflammatory diseases known collectively as cryopyrin-associated periodic syndromes (CAPS) (Hoffman et al. 2001; Aksentijevich et al. 2002; Feldmann et al. 2002). These disorders are, in order of increasing severity, familial cold autoinflammatory syndrome (FCAS), Muckle-Wells syndrome (MWS), and neonatal-onset multisystem inflammatory disease (NOMID), which is also referred to as chronic infantile neurologic cutaneous articular (CINCA) syndrome. Gene-targeted mice harboring Nlrp3 mutations equivalent to those found in FCAS and MWS patients have hyperactive NLRP3 inflammasome activity and elevated IL-1β levels (Brydges et al. 2009; Meng et al. 2009).

Type 2 Diabetes

Type 2 diabetes (T2D) occurs when insulin production by pancreatic islet β cells fails to compensate for insulin resistance. Elevated IL-1β levels are a risk factor for T2D development (Spranger et al. 2003) and contribute to insulin resistance (Maedler et al. 2009). Excessive caspase-1 activity has thus been implicated in T2D etiology.

Familial Mediterranean Fever

Familial Mediterranean fever (FMF) is an autoinflammatory disease caused by mutations in the Mediterranean fever gene (MEFV) that encodes the pyrin protein (Chae et al. 2008). The most severe form of FMF arises from missense mutations affecting the carboxy-terminal B30.2/SPRY domain of pyrin, which is important for its interaction with procaspase-1 (Chae et al. 2006; Papin et al. 2007). However, there is conflicting evidence on whether pyrin mutations affect IL-1β production (Chae et al. 2003, 2006; Yu et al. 2006; Seshadri et al. 2007). Treatment with an IL-1 targeting agent (see below) has induced symptom regression in FMF patients, implying a causative role for IL-1β in this disease (Roldan et al. 2008).

Pyogenic Sterile Arthritis, Pyoderma Gangrenosum, and Acne Syndrome

Pyogenic sterile arthritis, pyoderma gangrenosum, and acne (PAPA) syndrome is a rare autosomal-dominant genetic disorder caused by mutations in the CD2-binding protein 1 (CD2BP1) gene. Patients suffer from severe, juvenile-onset arthritis, pyoderma gangrenosum, and acne. These mutations disrupt the binding of (CD2BP1) to protein tyrosine phosphatase, nonreceptor type 12 (PTPN 12) (Wise et al. 2002), thereby increasing CD2BP1 binding to pyrin. Association with CD2BP1 reduces pyrin's ability to inhibit inappropriate inflammasome activation (Shoham et al. 2003).

Hyperimmunoglobulinemia D with Periodic Fever Syndrome

Hyperimmunoglobulinemia D with periodic fever syndrome (HIDS) is a rare autosomal-recessive disorder (van der Meer et al. 1984) that is thought to be caused by mutations in the gene encoding mevalonate kinase (Drenth et al. 1999; Houten et al. 1999). HIDS patients show increased blood levels of IL-1β and IgD. Mevalonate kinase deficiency (MKD), an autosomal-recessive disorder characterized by recurring episodes of inflammation, leads to decreased production of nonsterol isoprenoid end products, in particular, geranylgeranyl groups (Mandey et al. 2006). Isoprenoid deficiency can induce PI3K pathway-dependent procaspase-1 activation, leading to increased IL-1β production (Kuijk et al. 2008).

Systemic-Onset Juvenile Idiopathic Arthritis

The pathophysiology of systemic-onset juvenile idiopathic arthritis (sJIA), which affects an estimated 250,000 children in the United States alone, and which presents itself with initial systemic symptoms such as fever, anemia, leukocytosis, and elevated erythrocyte sedimentation rate (ESR), has been linked to elevated levels of IL-1β (Pascual et al. 2005).

Caspases in Other Diseases

Alzheimer's Disease

Neuronal death in a variety of neurodegenerative diseases, including Alzheimer's disease (AD), has been associated with deregulated caspase activation (Rohn and Head 2009). However, several lines of evidence suggest that the role of caspases in AD may involve more than just action as cellular executioners driven by upstream disease processes. Caspase-mediated cleavage of β-amyloid precursor protein (APP) has been reported (Rohn et al. 2001), as has caspase activation by amyloid-β peptide (O'Brien and Wong 2011). In one murine AD model, caspase activation associated with disease onset occurred earlier than the induction of neuronal apoptosis (D'Amelio et al. 2011). Similarly, caspase activation has been noted before the development of neurofibrillary tangles of Tau in the brain of tau transgenic mice (de Calignon et al. 2010).

Kawasaki Disease

Kawasaki disease (KD) is an acute vasculitis syndrome that predominantly affects arteries in young children (Kawasaki 1967; Burns 2002). In one study, a G to A substitution in a particular SNP located in the 5′ untranslated region of CASP3 abolished the binding of the nuclear factor of activated T cells (NFAT) transcription factor to the DNA sequence surrounding the SNP, suggesting that altered CASP3 expression in immune effector cells can influence KD susceptibility (Onouchi et al. 2010). However, another study of 341 KD patients and 751 controls found an association of only borderline significance between this CASP3 polymorphism and KD ($P = 0.0535$ under the dominant model; $P = 0.0575$ under the allelic model) (Kuo et al. 2011).

Autoimmune Lymphoproliferative Syndrome

Autoimmune lymphoproliferative syndrome (ALPS) causes lymphoadenopathy, splenomegaly, autoimmune hemolytic anemia, thrombocytopenia, and hypergammaglobulinemia in children (Lenardo et al. 1999; Straus et al. 1999). The majority of ALPS patients have dominant mutations in CD95, CD95L, or CASP10 (Lenardo et al. 1999; Straus et al. 1999; Wang et al. 1999). It is thought that ALPS may be caused by insufficient apoptosis of autoreactive T cells during negative thymic selection (Fleisher 2008).

CASPASES IN DISEASE THERAPY

Activating Caspases to Promote Cell Death

Cancer

Several attempts have been made within the last decade to develop molecules capable of directly activating caspase-3 for use in cancer therapy. A particular target suggested for intervention has been the "safety catch" sequence present in inactive procaspase-3 (Roy et al. 2001). This sequence is a triplet of aspartic acid residues that maintains the intramolecular electrostatic interactions that keep procaspase-3 in an inactive state in resting cells (Roy et al. 2001). High-throughput screening (HTS) projects have identified a series of molecules, including α-(trichloromethyl)-4-pyridineethanol (PETCM), gambonic acid, and the gambonic acid derivative MX-2060, that efficiently activate caspase-3 in vitro (Jiang et al. 2003; Zhang et al. 2004; Fischer and Schulze-Osthoff 2005). This series has shown promise in inducing the apoptosis of cancer cell lines, but no clinical development of these agents has been reported. Another promising caspase-3 activator identified by HTS is first procaspase-activating compound (PAC-1), which contains a zinc-chelating motif (Putt et al. 2006). This motif is critical to PAC-1's ability to activate caspase-3 (Peterson et al. 2009). Recent in vivo canine studies using a "next-generation" compound (S-PAC-1) have been efficacious, and treatments induced partial tumor regression (Peterson et al. 2010). However, the mechanism of PAC-1-mediated caspase activation is controversial because another group was unable to confirm the results obtained by Putt et al. (Putt et al. 2006). Denault et al. (2007) have suggested instead that PAC-1 cannot directly activate executioner caspases but rather uses an indirect and therefore blunted and less effective activation mechanism.

Another area of active research concentrates on compounds that activate caspases indirectly. Some of these agents block endogenous caspase inhibitors such as the Bcl-2 and IAP proteins (Vogler et al. 2009), whereas others are analogs of the endogenous IAP inhibitor Smac (Chen and Huerta 2009). Still others are activators and antibodies that engage DRs (Ying Lu 2011). Several of these compounds are currently under examination in clinical trials.

It should be noted that the use of apoptosis-inducing compounds for cancer treatment is not new, and most of these agents are subject to the same limitations of delivery and specificity as traditional chemotherapeutics. However, certain caspase activators, such as those that inhibit antiapoptotic molecules like Bcl-2, appear to have an enhanced therapeutic index when used to treat cancer cells that rely mainly on antiapoptotic proteins to stave off cell death (Certo et al. 2006).

Graft versus Host Disease

Another situation in which induction of cell death might be advantageous is the elimination of autoreactive lymphocytes in graft versus host disease (GVHD). Patients are currently being recruited for a phase I/II clinical trial of a modified version of caspase-9. This inducible caspase-9 "safety switch" agent consists of a truncated caspase-9 protein that lacks the CARD domain and is fused to a forkhead protein binding sequence. In the presence of a particular small molecule, the caspase-9 safety switch protein dimerizes and activates the hydrolytic function of the enzyme, triggering apoptosis (Straathof et al. 2005). When used as a therapy, the caspase-9 safety switch is virally transduced into allodepleted T cells, which are then administered to patients who have received a T-cell-depleted stem cell transplant (Tey et al. 2007). Should GVHD occur following the transplant, the small molecule is administered to the patient to induce caspase-9 activation and quickly eliminate autoreactive T cells via apoptosis.

Inhibiting Caspases to Prevent Cell Death

In general, the inhibition of caspase activity has had less striking therapeutic effects than has caspase activation. Nevertheless, there are several instances in which, regardless of whether caspases have been definitively implicated in the etiology or pathological consequences of a disease, caspase inhibition has ameliorated the symptoms of several conditions caused by inappropriate apoptotic cell death. For example, because chronic hepatitis virus C infection is accompanied by detrimental hepatocyte apoptosis, a recent clinical trial examined the therapeutic potential of a caspase inhibitor (Manns 2010). Similarly, the severity of ischemia reperfusion injury resulting from cell death that often follows a stroke (Renolleau et al. 2007), traumatic brain injury (Knoblach et al. 2004), or organ transplant (Baskin-Bey et al. 2007) can be reduced by caspase inhibition. Last, because the neuronal death characteristic of AD and other neurodegenerative diseases, as well as possibly other aspects of disease progression, are associated with caspase activation (see above), caspase inhibitors are under investigation in mouse models of AD and have already shown promising results (O'Brien and Wong 2011).

IL-1β Antagonism

A key component of many inflammatory disorders appears to be the activation of caspase-1 leading to the generation of active IL-1β. Accordingly, agents that can antagonize either the generation or function of IL-1β or its receptor (IL-1R) have been developed for patient treatment. An early such agent was Anakinra, a small molecule antagonist of IL-1R. Newer agents include monoclonal antibodies (mAbs) directed against IL-1β (canakinumab) (Alten et al. 2008; Church and McDermott 2009; Lachmann et al. 2009a), and IL-1Trap, a decoy receptor with high affinity for IL-1 (Kalliolias and Liossis 2008).

Clinical trials are currently under way to assess the efficacy of the above inhibitors and related molecules as treatment for several of the inflammatory diseases discussed above. For example, patients with gout, pseudogout, or pulmonary silicosis have shown great improvement after treatment with an IL-1β antagonist (McGonagle et al. 2007, 2008; So et al. 2007; Terkeltaub et al. 2009). CAPS patients also respond well to IL-1β antagonists (Hawkins et al. 2003, 2004a,b; Hoffman et al. 2004, 2008; Goldbach-Mansky et al. 2006, 2008; Hoffman 2009; Lachmann et al. 2009b), although the disease is also ameliorated by caspase-1 inhibition (Stack et al. 2005). IL-1β antagonists have also shown efficacy in clinical trials for the treatment of T2D (Larsen et al. 2007, 2009), confirming the important role of the NLRP3 inflammasome containing caspase-1 as a sensor of metabolic stress (Schroder and Tschopp 2010). Last, IL-1β antagonists have induced symptom regression in

FMF patients (Roldan et al. 2008), HIDS patients (Cailliez et al. 2006), and sJIA patients (Pascual et al. 2005; Kelly and Ramanan 2008; Lequerre et al. 2008).

CONCLUSION

In conclusion, caspase family members are at the nexus of critical regulatory networks controlling cell death and inflammation. We know that although caspase activity is critical for homeostasis of organisms, cells must take steps to protect themselves against unintended caspase activation through complex systems required to turn inactive caspase zymogens into functional proteases. The long list of diseases associated with caspases tells us that the inappropriate activation of caspases and dysregulation of the cell death and inflammatory pathways they control has dire consequences for human health. A growing body of research is providing us with ever increasing clarity about how these exciting proteases operate and how we might fight disease by manipulating their functions.

REFERENCES

Acehan D, Jiang X, Morgan DG, Heuser JE, Wang X, Akey CW. 2002. Three-dimensional structure of the apoptosome: Implications for assembly, procaspase-9 binding, and activation. *Mol Cell* **9:** 423–432.

Aksentijevich I, Nowak M, Mallah M, Chae JJ, Watford WT, Hofmann SR, Stein L, Russo R, Goldsmith D, Dent P, et al. 2002. De novo CIAS1 mutations, cytokine activation, and evidence for genetic heterogeneity in patients with neonatal-onset multisystem inflammatory disease (NOMID): A new member of the expanding family of pyrin-associated autoinflammatory diseases. *Arthritis Rheum* **46:** 3340–3348.

Alam A, Cohen LY, Aouad S, Sekaly RP. 1999. Early activation of caspases during T lymphocyte stimulation results in selective substrate cleavage in nonapoptotic cells. *J Exp Med* **190:** 1879–1890.

Alten R, Gram H, Joosten LA, van den Berg WB, Sieper J, Wassenberg S, Burmester G, van Riel P, Diaz-Lorente M, Bruin GJ, et al. 2008. The human anti-IL-1 β monoclonal antibody ACZ885 is effective in joint inflammation models in mice and in a proof-of-concept study in patients with rheumatoid arthritis. *Arthritis Res Ther* **10:** R67.

Amer A, Franchi L, Kanneganti TD, Body-Malapel M, Ozoren N, Brady G, Meshinchi S, Jagirdar R, Gewirtz A, Akira S, et al. 2006. Regulation of Legionella phagosome maturation and infection through flagellin and host Ipaf. *J Biol Chem* **281:** 35217–35223.

Anderson KV, Bokla L, Nusslein-Volhard C. 1985a. Establishment of dorsal-ventral polarity in the *Drosophila* embryo: The induction of polarity by the Toll gene product. *Cell* **42:** 791–798.

Anderson KV, Jurgens G, Nusslein-Volhard C. 1985b. Establishment of dorsal-ventral polarity in the *Drosophila* embryo: Genetic studies on the role of the Toll gene product. *Cell* **42:** 779–789.

Arai J, Katai N, Kuida K, Kikuchi T, Yoshimura N. 2006. Decreased retinal neuronal cell death in caspase-1 knockout mice. *Jpn J Ophthalmol* **50:** 417–425.

Baliga BC, Read SH, Kumar S. 2004. The biochemical mechanism of caspase-2 activation. *Cell Death Differ* **11:** 1234–1241.

Baskin-Bey ES, Washburn K, Feng S, Oltersdorf T, Shapiro D, Huyghe M, Burgart L, Garrity-Park M, van Vilsteren FG, Oliver LK, et al. 2007. Clinical trial of the pan-caspase inhibitor, IDN-6556, in human liver preservation injury. *Am J Transplant* **7:** 218–225.

Beisner DR, Ch'en IL, Kolla RV, Hoffmann A, Hedrick SM. 2005. Cutting edge: Innate immunity conferred by B cells is regulated by caspase-8. *J Immunol* **175:** 3469–3473.

Bergeron L, Perez GI, Macdonald G, Shi L, Sun Y, Jurisicova A, Varmuza S, Latham KE, Flaws JA, Salter JC, et al. 1998. Defects in regulation of apoptosis in caspase-2-deficient mice. *Genes Dev* **12:** 1304–1314.

Bergsbaken T, Cookson BT. 2007. Macrophage activation redirects yersinia-infected host cell death from apoptosis to caspase-1-dependent pyroptosis. *PLoS Pathog* **3:** e161.

Boatright KM, Renatus M, Scott FL, Sperandio S, Shin H, Pedersen IM, Ricci JE, Edris WA, Sutherlin DP, Green DR, et al. 2003. A unified model for apical caspase activation. *Mol Cell* **11:** 529–541.

Boatright KM, Deis C, Denault JB, Sutherlin DP, Salvesen GS. 2004. Activation of caspases-8 and -10 by FLIP(L). *Biochem J* **382:** 651–657.

Boyden ED, Dietrich WF. 2006. Nalp1b controls mouse macrophage susceptibility to anthrax lethal toxin. *Nat Genet* **38:** 240–244.

Brenner D, Mak TW. 2009. Mitochondrial cell death effectors. *Curr Opin Cell Biol* **21:** 871–877.

Bresnihan B, Alvaro-Gracia JM, Cobby M, Doherty M, Domljan Z, Emery P, Nuki G, Pavelka K, Rau R, Rozman B, et al. 1998. Treatment of rheumatoid arthritis with recombinant human interleukin-1 receptor antagonist. *Arthritis Rheum* **41:** 2196–2204.

Brydges SD, Mueller JL, McGeough MD, Pena CA, Misaghi A, Gandhi C, Putnam CD, Boyle DL, Firestein GS, Horner AA, et al. 2009. Inflammasome-mediated disease animal models reveal roles for innate but not adaptive immunity. *Immunity* **30:** 875–887.

Burns JC. 2002. Commentary: Translation of Dr. Tomisaku Kawasaki's original report of fifty patients in 1967. *Pediatr Infect Dis J* **21:** 993–995.

Cailliez M, Garaix F, Rousset-Rouviere C, Bruno D, Kone-Paut I, Sarles J, Chabrol B, Tsimaratos M. 2006. Anakinra is safe and effective in controlling hyperimmunoglobulinaemia D syndrome-associated febrile crisis. *J Inherit Metab Dis* **29:** 763.

Cain K, Bratton SB, Cohen GM. 2002. The Apaf-1 apoptosome: A large caspase-activating complex. *Biochimie* **84:** 203–214.

Cassel SL, Eisenbarth SC, Iyer SS, Sadler JJ, Colegio OR, Tephly LA, Carter AB, Rothman PB, Flavell RA, Sutterwala FS. 2008. The Nalp3 inflammasome is essential for the development of silicosis. *Proc Natl Acad Sci* **105:** 9035–9040.

Certo M, Del Gaizo Moore V, Nishino M, Wei G, Korsmeyer S, Armstrong SA, Letai A. 2006. Mitochondria primed by death signals determine cellular addiction to antiapoptotic BCL-2 family members. *Cancer Cell* **9:** 351–365.

Chae JJ, Komarow HD, Cheng J, Wood G, Raben N, Liu PP, Kastner DL. 2003. Targeted disruption of pyrin, the FMF protein, causes heightened sensitivity to endotoxin and a defect in macrophage apoptosis. *Mol Cell* **11:** 591–604.

Chae JJ, Wood G, Masters SL, Richard K, Park G, Smith BJ, Kastner DL. 2006. The B30.2 domain of pyrin, the familial Mediterranean fever protein, interacts directly with caspase-1 to modulate IL-1β production. *Proc Natl Acad Sci* **103:** 9982–9987.

Chae JJ, Wood G, Richard K, Jaffe H, Colburn NT, Masters SL, Gumucio DL, Shoham NG, Kastner DL. 2008. The familial Mediterranean fever protein, pyrin, is cleaved by caspase-1 and activates NF-κB through its N-terminal fragment. *Blood* **112:** 1794–1803.

Chamaillard M, Hashimoto M, Horie Y, Masumoto J, Qiu S, Saab L, Ogura Y, Kawasaki A, Fukase K, Kusumoto S, et al. 2003. An essential role for NOD1 in host recognition of bacterial peptidoglycan containing diaminopimelic acid. *Nat Immunol* **4:** 702–707.

Chang DW, Xing Z, Capacio VL, Peter ME, Yang X. 2003. Interdimer processing mechanism of procaspase-8 activation. *EMBO J* **22:** 4132–4142.

Chen DJ, Huerta S. 2009. Smac mimetics as new cancer therapeutics. *Anticancer Drugs* **20:** 646–658.

Chen K, Zhao H, Hu Z, Wang LE, Zhang W, Sturgis EM, Wei Q. 2008a. CASP3 polymorphisms and risk of squamous cell carcinoma of the head and neck. *Clin Cancer Res* **14:** 6343–6349.

Chen NJ, Chio II, Lin WJ, Duncan G, Chau H, Katz D, Huang HL, Pike KA, Hao Z, Su YW, et al. 2008b. Beyond tumor necrosis factor receptor: TRADD signaling in toll-like receptors. *Proc Natl Acad Sci* **105:** 12429–12434.

Chun HJ, Zheng L, Ahmad M, Wang J, Speirs CK, Siegel RM, Dale JK, Puck J, Davis J, Hall CG, et al. 2002. Pleiotropic defects in lymphocyte activation caused by caspase-8 mutations lead to human immunodeficiency. *Nature* **419:** 395–399.

Church LD, McDermott MF. 2009. Canakinumab, a fully-human mAb against IL-1β for the potential treatment of inflammatory disorders. *Curr Opin Mol Ther* **11:** 81–89.

Cohen S, Hurd E, Cush J, Schiff M, Weinblatt ME, Moreland LW, Kremer J, Bear MB, Rich WJ, McCabe D. 2002. Treatment of rheumatoid arthritis with anakinra, a recombinant human interleukin-1 receptor antagonist, in combination with methotrexate: Results of a twenty-four-week, multicenter, randomized, double-blind, placebo-controlled trial. *Arthritis Rheum* **46:** 614–624.

Cookson BT, Brennan MA. 2001. Pro-inflammatory programmed cell death. *Trends Microbiol* **9:** 113–114.

Cox A, Dunning AM, Garcia-Closas M, Balasubramanian S, Reed MW, Pooley KA, Scollen S, Baynes C, Ponder BA, Chanock S, et al. 2007. A common coding variant in CASP8 is associated with breast cancer risk. *Nat Genet* **39:** 352–358.

Cronstein BN, Terkeltaub R. 2006. The inflammatory process of gout and its treatment. *Arthritis Res Ther* **8:** S3.

D'Amelio M, Cavallucci V, Middei S, Marchetti C, Pacioni S, Ferri A, Diamantini A, De Zio D, Carrara P, Battistini L, et al. 2011. Caspase-3 triggers early synaptic dysfunction in a mouse model of Alzheimer's disease. *Nat Neurosci* **14:** 69–76.

Davis BK, Wen H, Ting JP. 2011. The inflammasome NLRs in immunity, inflammation, and associated diseases. *Annu Rev Immunol* **29:** 707–735.

de Calignon A, Fox LM, Pitstick R, Carlson GA, Bacskai BJ, Spires-Jones TL, Hyman BT. 2010. Caspase activation precedes and leads to tangles. *Nature* **464:** 1201–1204.

Denault JB, Drag M, Salvesen GS, Alves J, Heidt AB, Deveraux Q, Harris JL. 2007. Small molecules not direct activators of caspases. *Nat Chem Biol* **3:** 519; author reply 520.

Denecker G, Hoste E, Gilbert B, Hochepied T, Ovaere P, Lippens S, Van den Broecke C, Van Damme P, D'Herde K, Hachem JP, et al. 2007. Caspase-14 protects against epidermal UVB photodamage and water loss. *Nat Cell Biol* **9:** 666–674.

Denecker G, Ovaere P, Vandenabeele P, Declercq W. 2008. Caspase-14 reveals its secrets. *J Cell Biol* **180:** 451–458.

Dostert C, Petrilli V, Van Bruggen R, Steele C, Mossman BT, Tschopp J. 2008. Innate immune activation through Nalp3 inflammasome sensing of asbestos and silica. *Science* **320:** 674–677.

Drenth JP, Cuisset L, Grateau G, Vasseur C, van de Velde-Visser SD, de Jong JG, Beckmann JS, van der Meer JW, Delpech M, International Hyper-IgD Study Group. 1999. Mutations in the gene encoding mevalonate kinase cause hyper-IgD and periodic fever syndrome. *Nat Genet* **22:** 178–181.

Drevlow BE, Lovis R, Haag MA, Sinacore JM, Jacobs C, Blosche C, Landay A, Moreland LW, Pope RM. 1996. Recombinant human interleukin-1 receptor type I in the treatment of patients with active rheumatoid arthritis. *Arthritis Rheum* **39:** 257–265.

Eggert A, Grotzer MA, Zuzak TJ, Wiewrodt BR, Ho R, Ikegaki N, Brodeur GM. 2001. Resistance to tumor necrosis factor-related apoptosis-inducing ligand (TRAIL)-induced apoptosis in neuroblastoma cells correlates with a loss of caspase-8 expression. *Cancer Res* **61:** 1314–1319.

Faubel S, Ljubanovic D, Reznikov L, Somerset H, Dinarello CA, Edelstein CL. 2004. Caspase-1-deficient mice are protected against cisplatin-induced apoptosis and acute tubular necrosis. *Kidney Int* **66:** 2202–2213.

Faustin B, Lartigue L, Bruey JM, Luciano F, Sergienko E, Bailly-Maitre B, Volkmann N, Hanein D, Rouiller I, Reed JC. 2007. Reconstituted NALP1 inflammasome reveals two-step mechanism of caspase-1 activation. *Mol Cell* **25:** 713–724.

Feldmann J, Prieur AM, Quartier P, Berquin P, Certain S, Cortis E, Teillac-Hamel D, Fischer A, de Saint Basile G. 2002. Chronic infantile neurological cutaneous and articular syndrome is caused by mutations in CIAS1, a gene highly expressed in polymorphonuclear cells and chondrocytes. *Am J Hum Genet* **71:** 198–203.

Feldmeyer L, Keller M, Niklaus G, Hohl D, Werner S, Beer HD. 2007. The inflammasome mediates UVB-induced activation and secretion of interleukin-1β by keratinocytes. *Curr Biol* **17:** 1140–1145.

Fernandes-Alnemri T, Yu JW, Datta P, Wu J, Alnemri ES. 2009. AIM2 activates the inflammasome and cell death in response to cytoplasmic DNA. *Nature* **458:** 509–513.

Fernandes-Alnemri T, Yu JW, Juliana C, Solorzano L, Kang S, Wu J, Datta P, McCormick M, Huang L, McDermott E, et al. 2010. The AIM2 inflammasome is critical for innate immunity to *Francisella tularensis*. *Nat Immunol* **11:** 385–393.

Filomenko R, Prevotat L, Rebe C, Cortier M, Jeannin JF, Solary E, Bettaieb A. 2006. Caspase-10 involvement in cytotoxic drug-induced apoptosis of tumor cells. *Oncogene* **25:** 7635–7645.

Finckh A, Gabay C. 2008. At the horizon of innovative therapy in rheumatology: New biologic agents. *Curr Opin Rheumatol* **20:** 269–275.

Fink SL, Cookson BT. 2006. Caspase-1-dependent pore formation during pyroptosis leads to osmotic lysis of infected host macrophages. *Cell Microbiol* **8:** 1812–1825.

Fischer U, Schulze-Osthoff K. 2005. New approaches and therapeutics targeting apoptosis in disease. *Pharmacol Rev* **57:** 187–215.

Fischer H, Koenig U, Eckhart L, Tschachler E. 2002. Human caspase 12 has acquired deleterious mutations. *Biochem Biophys Res Commun* **293:** 722–726.

Fleisher TA. 2008. The autoimmune lymphoproliferative syndrome: An experiment of nature involving lymphocyte apoptosis. *Immunol Res* **40:** 87–92.

Franchi L, Amer A, Body-Malapel M, Kanneganti TD, Ozoren N, Jagirdar R, Inohara N, Vandenabeele P, Bertin J, Coyle A, et al. 2006. Cytosolic flagellin requires Ipaf for activation of caspase-1 and interleukin 1β in salmonella-infected macrophages. *Nat Immunol* **7:** 576–582.

Franchi L, Stoolman J, Kanneganti TD, Verma A, Ramphal R, Nunez G. 2007. Critical role for Ipaf in *Pseudomonas aeruginosa*-induced caspase-1 activation. *Eur J Immunol* **37:** 3030–3039.

Frank B, Bermejo JL, Hemminki K, Klaes R, Bugert P, Wappenschmidt B, Schmutzler RK, Burwinkel B. 2005. Re: Association of a common variant of the CASP8 gene with reduced risk of breast cancer. *J Natl Cancer Inst* **97:** 1012; author reply 1012–1013.

Frantz S, Ducharme A, Sawyer D, Rohde LE, Kobzik L, Fukazawa R, Tracey D, Allen H, Lee RT, Kelly RA. 2003. Targeted deletion of caspase-1 reduces early mortality and left ventricular dilatation following myocardial infarction. *J Mol Cell Cardiol* **35:** 685–694.

Gabay C, Lamacchia C, Palmer G. 2010. IL-1 pathways in inflammation and human diseases. *Nat Rev Rheumatol* **6:** 232–241.

Genovese MC, Cohen S, Moreland L, Lium D, Robbins S, Newmark R, Bekker P. 2004. Combination therapy with etanercept and anakinra in the treatment of patients with rheumatoid arthritis who have been treated unsuccessfully with methotrexate. *Arthritis Rheum* **50:** 1412–1419.

Girardin SE, Boneca IG, Carneiro LA, Antignac A, Jehanno M, Viala J, Tedin K, Taha MK, Labigne A, Zahringer U, et al. 2003a. Nod1 detects a unique muropeptide from gram-negative bacterial peptidoglycan. *Science* **300:** 1584–1587.

Girardin SE, Boneca IG, Viala J, Chamaillard M, Labigne A, Thomas G, Philpott DJ, Sansonetti PJ. 2003b. Nod2 is a general sensor of peptidoglycan through muramyl dipeptide (MDP) detection. *J Biol Chem* **278:** 8869–8872.

Goldbach-Mansky R, Dailey NJ, Canna SW, Gelabert A, Jones J, Rubin BI, Kim HJ, Brewer C, Zalewski C, Wiggs E, et al. 2006. Neonatal-onset multisystem inflammatory disease responsive to interleukin-1β inhibition. *N Engl J Med* **355:** 581–592.

Goldbach-Mansky R, Shroff SD, Wilson M, Snyder C, Plehn S, Barham B, Pham TH, Pucino F, Wesley RA, Papadopoulos JH, et al. 2008. A pilot study to evaluate the safety and efficacy of the long-acting interleukin-1 inhibitor rilonacept (interleukin-1 Trap) in patients with familial cold autoinflammatory syndrome. *Arthritis Rheum* **58:** 2432–2442.

Gross O, Poeck H, Bscheider M, Dostert C, Hannesschlager N, Endres S, Hartmann G, Tardivel A, Schweighoffer E, Tybulewicz V, et al. 2009. Syk kinase signalling couples to the Nlrp3 inflammasome for anti-fungal host defence. *Nature* **459:** 433–436.

Guegan C, Vila M, Teismann P, Chen C, Onteniente B, Li M, Friedlander RM, Przedborski S. 2002. Instrumental activation of bid by caspase-1 in a transgenic mouse model of ALS. *Mol Cell Neurosci* **20:** 553–562.

Hakem R, Hakem A, Duncan GS, Henderson JT, Woo M, Soengas MS, Elia A, de la Pompa JL, Kagi D, Khoo W, et al. 1998. Differential requirement for caspase 9 in apoptotic pathways in vivo. *Cell* **94:** 339–352.

Halle A, Hornung V, Petzold GC, Stewart CR, Monks BG, Reinheckel T, Fitzgerald KA, Latz E, Moore KJ, Golenbock DT. 2008. The NALP3 inflammasome is involved in the innate immune response to amyloid-β. *Nat Immunol* **9:** 857–865.

Hao Z, Duncan GS, Chang CC, Elia A, Fang M, Wakeham A, Okada H, Calzascia T, Jang Y, You-Ten A, et al. 2005. Specific ablation of the apoptotic functions of cytochrome c reveals a differential requirement for cytochrome c and Apaf-1 in apoptosis. *Cell* **121:** 579–591.

Harvey NL, Butt AJ, Kumar S. 1997. Functional activation of Nedd2/ICH-1 (caspase-2) is an early process in apoptosis. *J Biol Chem* **272:** 13134–13139.

Hawkins PN, Lachmann HJ, McDermott MF. 2003. Interleukin-1-receptor antagonist in the Muckle-Wells syndrome. *N Engl J Med* **348:** 2583–2584.

Hawkins PN, Bybee A, Aganna E, McDermott MF. 2004a. Response to anakinra in a de novo case of neonatal-onset multisystem inflammatory disease. *Arthritis Rheum* **50:** 2708–2709.

Hawkins PN, Lachmann HJ, Aganna E, McDermott MF. 2004b. Spectrum of clinical features in Muckle-Wells syndrome and response to anakinra. *Arthritis Rheum* **50:** 607–612.

Herzyk DJ, Allen JN, Marsh CB, Wewers MD. 1992. Macrophage and monocyte IL-1β regulation differs at multiple sites. Messenger RNA expression, translation, and post-translational processing. *J Immunol* **149:** 3052–3058.

Ho LH, Taylor R, Dorstyn L, Cakouros D, Bouillet P, Kumar S. 2009. A tumor suppressor function for caspase-2. *Proc Natl Acad Sci* **106:** 5336–5341.

Hoffman HM. 2009. Rilonacept for the treatment of cryopyrin-associated periodic syndromes (CAPS). *Expert Opin Biol Ther* **9:** 519–531.

Hoffman HM, Mueller JL, Broide DH, Wanderer AA, Kolodner RD. 2001. Mutation of a new gene encoding a putative pyrin-like protein causes familial cold autoinflammatory syndrome and Muckle-Wells syndrome. *Nat Genet* **29:** 301–305.

Hoffman HM, Rosengren S, Boyle DL, Cho JY, Nayar J, Mueller JL, Anderson JP, Wanderer AA, Firestein GS. 2004. Prevention of cold-associated acute inflammation in familial cold autoinflammatory syndrome by interleukin-1 receptor antagonist. *Lancet* **364:** 1779–1785.

Hoffman HM, Throne ML, Amar NJ, Sebai M, Kivitz AJ, Kavanaugh A, Weinstein SP, Belomestnov P, Yancopoulos GD, Stahl N, et al. 2008. Efficacy and safety of rilonacept (interleukin-1 Trap) in patients with cryopyrin-associated periodic syndromes: Results from two sequential placebo-controlled studies. *Arthritis Rheum* **58:** 2443–2452.

Hopkins-Donaldson S, Bodmer JL, Bourloud KB, Brognara CB, Tschopp J, Gross N. 2000. Loss of caspase-8 expression in highly malignant human neuroblastoma cells correlates with resistance to tumor necrosis factor-related apoptosis-inducing ligand-induced apoptosis. *Cancer Res* **60:** 4315–4319.

Hornung V, Bauernfeind F, Halle A, Samstad EO, Kono H, Rock KL, Fitzgerald KA, Latz E. 2008. Silica crystals and aluminum salts activate the NALP3 inflammasome through phagosomal destabilization. *Nat Immunol* **9:** 847–856.

Hornung V, Ablasser A, Charrel-Dennis M, Bauernfeind F, Horvath G, Caffrey DR, Latz E, Fitzgerald KA. 2009. AIM2 recognizes cytosolic dsDNA and forms a caspase-1-activating inflammasome with ASC. *Nature* **458:** 514–518.

Hosgood HD 3rd, Baris D, Zhang Y, Zhu Y, Zheng T, Yeager M, Welch R, Zahm S, Chanock S, Rothman N, et al. 2008. Caspase polymorphisms and genetic susceptibility to multiple myeloma. *Hematol Oncol* **26:** 148–151.

Houten SM, Kuis W, Duran M, de Koning TJ, van Royen-Kerkhof A, Romeijn GJ, Frenkel J, Dorland L, de Barse MM, Huijbers WA, et al. 1999. Mutations in MVK, encoding mevalonate kinase, cause hyperimmunoglobulinaemia D and periodic fever syndrome. *Nat Genet* **22:** 175–177.

Hsu LC, Ali SR, McGillivray S, Tseng PH, Mariathasan S, Humke EW, Eckmann L, Powell JJ, Nizet V, Dixit VM, et al. 2008. A NOD2-NALP1 complex mediates caspase-1-dependent IL-1β secretion in response to *Bacillus anthracis* infection and muramyl dipeptide. *Proc Natl Acad Sci* **105:** 7803–7808.

Hu B, Elinav E, Huber S, Booth CJ, Strowig T, Jin C, Eisenbarth SC, Flavell RA. 2010. Inflammation-induced tumorigenesis in the colon is regulated by caspase-1 and NLRC4. *Proc Natl Acad Sci* **107:** 21635–21640.

Inohara N, Ogura Y, Fontalba A, Gutierrez O, Pons F, Crespo J, Fukase K, Inamura S, Kusumoto S, Hashimoto M, et al. 2003. Host recognition of bacterial muramyl dipeptide mediated through NOD2. Implications for Crohn's disease. *J Biol Chem* **278:** 5509–5512.

Jang JS, Kim KM, Choi JE, Cha SI, Kim CH, Lee WK, Kam S, Jung TH, Park JY. 2008. Identification of polymorphisms in the Caspase-3 gene and their association with lung cancer risk. *Mol Carcinog* **47:** 383–390.

Jiang Y, Genant HK, Watt I, Cobby M, Bresnihan B, Aitchison R, McCabe D. 2000. A multicenter, double-blind, dose-ranging, randomized, placebo-controlled study of recombinant human interleukin-1 receptor antagonist in patients with rheumatoid arthritis: Radiologic progression and correlation of Genant and Larsen scores. *Arthritis Rheum* **43:** 1001–1009.

Jiang X, Kim HE, Shu H, Zhao Y, Zhang H, Kofron J, Donnelly J, Burns D, Ng SC, Rosenberg S, et al. 2003. Distinctive roles of PHAP proteins and prothymosin-α in a death regulatory pathway. *Science* **299:** 223–226.

Jin C, Flavell RA. 2010. Molecular mechanism of NLRP3 inflammasome activation. *J Clin Immunol* **30:** 628–631.

Jones JD, Dangl JL. 2006. The plant immune system. *Nature* **444:** 323–329.

Jost PJ, Grabow S, Gray D, McKenzie MD, Nachbur U, Huang DC, Bouillet P, Thomas HE, Borner C, Silke J, et al. 2009. XIAP discriminates between type I and type II FAS-induced apoptosis. *Nature* **460:** 1035–1039.

Juo P, Kuo CJ, Yuan J, Blenis J. 1998. Essential requirement for caspase-8/FLICE in the initiation of the Fas-induced apoptotic cascade. *Curr Biol* **8:** 1001–1008.

Kaiser WJ, Upton JW, Long AB, Livingston-Rosanoff D, Daley-Bauer LP, Hakem R, Caspary T, Mocarski ES. 2011. RIP3 mediates the embryonic lethality of caspase-8-deficient mice. *Nature* **471:** 368–372.

Kalliolias GD, Liossis SN. 2008. The future of the IL-1 receptor antagonist anakinra: From rheumatoid arthritis to adult-onset Still's disease and systemic-onset juvenile idiopathic arthritis. *Expert Opin Investig Drugs* **17:** 349–359.

Kang TB, Ben-Moshe T, Varfolomeev EE, Pewzner-Jung Y, Yogev N, Jurewicz A, Waisman A, Brenner O, Haffner R, Gustafsson E, et al. 2004. Caspase-8 serves both apoptotic and nonapoptotic roles. *J Immunol* **173:** 2976–2984.

Kanneganti TD, Body-Malapel M, Amer A, Park JH, Whitfield J, Franchi L, Taraporewala ZF, Miller D, Patton JT, Inohara N, et al. 2006. Critical role for Cryopyrin/Nalp3 in activation of caspase-1 in response to viral infection and double-stranded RNA. *J Biol Chem* **281:** 36560–36568.

Kawasaki T. 1967. Acute febrile mucocutaneous syndrome with lymphoid involvement with specific desquamation of the fingers and toes in children. *Arerugi* **16:** 178–222.

Kayagaki N, Warming S, Lamkanfi M, Vande Walle L, Louie S, Dong J, Newton K, Qu Y, Liu J, Heldens S, et al. 2011. Non-canonical inflammasome activation targets caspase-11. *Nature* **479:** 117–121.

Kelly A, Ramanan AV. 2008. A case of macrophage activation syndrome successfully treated with anakinra. *Nat Clin Pract Rheumatol* **4:** 615–620.

Kelly JL, Novak AJ, Fredericksen ZS, Liebow M, Ansell SM, Dogan A, Wang AH, Witzig TE, Call TG, Kay NE, et al. 2010. Germline variation in apoptosis pathway genes and risk of non-Hodgkin's lymphoma. *Cancer Epidemiol Biomarkers Prev* **19:** 2847–2858.

Kennedy NJ, Kataoka T, Tschopp J, Budd RC. 1999. Caspase activation is required for T cell proliferation. *J Exp Med* **190:** 1891–1896.

Kim HS, Lee JW, Soung YH, Park WS, Kim SY, Lee JH, Park JY, Cho YG, Kim CJ, Jeong SW, et al. 2003. Inactivating mutations of caspase-8 gene in colorectal carcinomas. *Gastroenterology* **125:** 708–715.

Kim MS, Oh JE, Min CK, Lee S, Chung NG, Yoo NJ, Lee SH. 2009. Mutational analysis of CASP10 gene in acute leukaemias and multiple myelomas. *Pathology* **41:** 484–487.

Kischkel FC, Lawrence DA, Tinel A, LeBlanc H, Virmani A, Schow P, Gazdar A, Blenis J, Arnott D, Ashkenazi A. 2001. Death receptor recruitment of endogenous caspase-10 and apoptosis initiation in the absence of caspase-8. *J Biol Chem* **276:** 46639–46646.

Knoblach SM, Alroy DA, Nikolaeva M, Cernak I, Stoica BA, Faden AI. 2004. Caspase inhibitor z-DEVD-fmk attenuates calpain and necrotic cell death in vitro and after traumatic brain injury. *J Cereb Blood Flow Metab* **24:** 1119–1132.

Kolodgie FD, Narula J, Burke AP, Haider N, Farb A, Hui-Liang Y, Smialek J, Virmani R. 2000. Localization of apoptotic macrophages at the site of plaque rupture in sudden coronary death. *Am J Pathol* **157:** 1259–1268.

Kovalenko A, Kim JC, Kang TB, Rajput A, Bogdanov K, Dittrich-Breiholz O, Kracht M, Brenner O, Wallach D. 2009. Caspase-8 deficiency in epidermal keratinocytes triggers an inflammatory skin disease. *J Exp Med* **206:** 2161–2177.

Krumschnabel G, Manzl C, Villunger A. 2009. Caspase-2: Killer, savior and safeguard—Emerging versatile roles for an ill-defined caspase. *Oncogene* **28:** 3093–3096.

Kufer TA, Fritz JH, Philpott DJ. 2005. NACHT-LRR proteins (NLRs) in bacterial infection and immunity. *Trends Microbiol* **13:** 381–388.

Kuida K, Lippke JA, Ku G, Harding MW, Livingston DJ, Su MS, Flavell RA. 1995. Altered cytokine export and apoptosis in mice deficient in interleukin-1β converting enzyme. *Science* **267:** 2000–2003.

Kuida K, Zheng TS, Na S, Kuan C, Yang D, Karasuyama H, Rakic P, Flavell RA. 1996. Decreased apoptosis in the brain and premature lethality in CPP32-deficient mice. *Nature* **384:** 368–372.

Kuida K, Haydar TF, Kuan CY, Gu Y, Taya C, Karasuyama H, Su MS, Rakic P, Flavell RA. 1998. Reduced apoptosis and cytochrome c-mediated caspase activation in mice lacking caspase 9. *Cell* **94:** 325–337.

Kuijk LM, Beekman JM, Koster J, Waterham HR, Frenkel J, Coffer PJ. 2008. HMG-CoA reductase inhibition induces IL-1β release through Rac1/PI3K/PKB-dependent caspase-1 activation. *Blood* **112:** 3563–3573.

Kumar S, Kinoshita M, Noda M, Copeland NG, Jenkins NA. 1994. Induction of apoptosis by the mouse Nedd2 gene, which encodes a protein similar to the product of the *Caenorhabditis elegans* cell death gene ced-3 and the mammalian IL-1 β-converting enzyme. *Genes Dev* **8:** 1613–1626.

Kuo HC, Yu HR, Juo SH, Yang KD, Wang YS, Liang CD, Chen WC, Chang WP, Huang CF, Lee CP, et al. 2011. CASP3 gene single-nucleotide polymorphism (rs72689236) and Kawasaki disease in Taiwanese children. *J Hum Genet* **56:** 161–165.

Lachmann HJ, Kone-Paut I, Kuemmerle-Deschner JB, Leslie KS, Hachulla E, Quartier P, Gitton X, Widmer A, Patel N, Hawkins PN. 2009a. Use of canakinumab in the cryopyrin-associated periodic syndrome. *N Engl J Med* **360:** 2416–2425.

Lachmann HJ, Lowe P, Felix SD, Rordorf C, Leslie K, Madhoo S, Wittkowski H, Bek S, Hartmann N, Bosset S, et al. 2009b. In vivo regulation of interleukin 1β in patients with cryopyrin-associated periodic syndromes. *J Exp Med* **206:** 1029–1036.

Lafont E, Milhas D, Teissie J, Therville N, Andrieu-Abadie N, Levade T, Benoist H, Segui B. 2010. Caspase-10-dependent cell death in Fas/CD95 signalling is not abrogated by caspase inhibitor zVAD-fmk. *PLoS ONE* **5:** e13638.

Lakhani SA, Masud A, Kuida K, Porter GA Jr, Booth CJ, Mehal WZ, Inayat I, Flavell RA. 2006. Caspases 3 and 7: Key mediators of mitochondrial events of apoptosis. *Science* **311:** 847–851.

Lamkanfi M, Declercq W, Kalai M, Saelens X, Vandenabeele P. 2002. Alice in caspase land. A phylogenetic analysis of caspases from worm to man. *Cell Death Differ* **9:** 358–361.

Lan Q, Zheng T, Chanock S, Zhang Y, Shen M, Wang SS, Berndt SI, Zahm SH, Holford TR, Leaderer B, et al. 2007. Genetic variants in caspase genes and susceptibility to non-Hodgkin lymphoma. *Carcinogenesis* **28:** 823–827.

Lan Q, Morton LM, Armstrong B, Hartge P, Menashe I, Zheng T, Purdue MP, Cerhan JR, Zhang Y, Grulich A, et al. 2009. Genetic variation in caspase genes and risk of non-Hodgkin lymphoma: A pooled analysis of 3 population-based case-control studies. *Blood* **114:** 264–267.

Larsen CM, Faulenbach M, Vaag A, Volund A, Ehses JA, Seifert B, Mandrup-Poulsen T, Donath MY. 2007. Interleukin-1-receptor antagonist in type 2 diabetes mellitus. *N Engl J Med* **356:** 1517–1526.

Larsen CM, Faulenbach M, Vaag A, Ehses JA, Donath MY, Mandrup-Poulsen T. 2009. Sustained effects of interleukin-1 receptor antagonist treatment in type 2 diabetes. *Diabetes Care* **32:** 1663–1668.

Lee JW, Kim MR, Soung YH, Nam SW, Kim SH, Lee JY, Yoo NJ, Lee SH. 2006. Mutational analysis of the CASP6 gene in colorectal and gastric carcinomas. *APMIS* **114:** 646–650.

Lee HJ, Pyo JO, Oh Y, Kim HJ, Hong SH, Jeon YJ, Kim H, Cho DH, Woo HN, Song S, et al. 2007. AK2 activates a novel apoptotic pathway through formation of a complex with FADD and caspase-10. *Nat Cell Biol* **9:** 1303–1310.

Lee P, Lee DJ, Chan C, Chen SW, Ch'en I, Jamora C. 2009a. Dynamic expression of epidermal caspase 8 simulates a wound healing response. *Nature* **458:** 519–523.

Lee WK, Kim JS, Kang HG, Cha SI, Kim DS, Hyun DS, Kam S, Kim CH, Jung TH, Park JY. 2009b. Polymorphisms in the Caspase7 gene and the risk of lung cancer. *Lung Cancer* **65:** 19–24.

Lenardo M, Chan KM, Hornung F, McFarland H, Siegel R, Wang J, Zheng L. 1999. Mature T lymphocyte apoptosis—Immune regulation in a dynamic and unpredictable antigenic environment. *Annu Rev Immunol* **17:** 221–253.

Lens SM, Kataoka T, Fortner KA, Tinel A, Ferrero I, MacDonald RH, Hahne M, Beermann F, Attinger A, Orbea HA, et al. 2002. The caspase 8 inhibitor c-FLIP(L) modulates T-cell receptor-induced proliferation but not activation-induced cell death of lymphocytes. *Mol Cell Biol* **22:** 5419–5433.

Lequerre T, Quartier P, Rosellini D, Alaoui F, De Bandt M, Mejjad O, Kone-Paut I, Michel M, Dernis E, Khellaf M, et al. 2008. Interleukin-1 receptor antagonist (anakinra) treatment in patients with systemic-onset juvenile idiopathic arthritis or adult onset Still disease: Preliminary experience in France. *Ann Rheum Dis* **67:** 302–308.

Li P, Allen H, Banerjee S, Franklin S, Herzog L, Johnston C, McDowell J, Paskind M, Rodman L, Salfeld J, et al. 1995. Mice deficient in IL-1β-converting enzyme are defective in production of mature IL-1β and resistant to endotoxic shock. *Cell* **80:** 401–411.

Li J, Brieher WM, Scimone ML, Kang SJ, Zhu H, Yin H, von Andrian UH, Mitchison T, Yuan J. 2007. Caspase-11 regulates cell migration by promoting Aip1-Cofilin-mediated actin depolymerization. *Nat Cell Biol* **9:** 276–286.

Li C, Lasse S, Lee P, Nakasaki M, Chen SW, Yamasaki K, Gallo RL, Jamora C. 2010. Development of atopic dermatitis-like skin disease from the chronic loss of epidermal caspase-8. *Proc Natl Acad Sci* **107:** 22249–22254.

Lightfield KL, Persson J, Trinidad NJ, Brubaker SW, Kofoed EM, Sauer JD, Dunipace EA, Warren SE, Miao EA, Vance RE. 2011. Differential requirements for NAIP5 in activation of the NLRC4 (IPAF) inflammasome. *Infect Immun* **79:** 1606–1614.

Liu XH, Kwon D, Schielke GP, Yang GY, Silverstein FS, Barks JD. 1999. Mice deficient in interleukin-1 converting enzyme are resistant to neonatal hypoxic-ischemic brain damage. *J Cereb Blood Flow Metab* **19:** 1099–1108.

Loo YM, Gale M Jr. 2011. Immune signaling by RIG-I-like receptors. *Immunity* **34:** 680–692.

Ludlow LE, Johnstone RW, Clarke CJ. 2005. The HIN-200 family: More than interferon-inducible genes? *Exp Cell Res* **308:** 1–17.

MacPherson G, Healey CS, Teare MD, Balasubramanian SP, Reed MW, Pharoah PD, Ponder BA, Meuth M, Bhattacharyya NP, Cox A. 2004. Association of a common variant of the CASP8 gene with reduced risk of breast cancer. *J Natl Cancer Inst* **96:** 1866–1869.

Madden SD, Cotter TG. 2008. Cell death in brain development and degeneration: Control of caspase expression may be key! *Mol Neurobiol* **37:** 1–6.

Maedler K, Dharmadhikari G, Schumann DM, Storling J. 2009. Interleukin-1β targeted therapy for type 2 diabetes. *Expert Opin Biol Ther* **9:** 1177–1188.

Mandey SH, Kuijk LM, Frenkel J, Waterham HR. 2006. A role for geranylgeranylation in interleukin-1β secretion. *Arthritis Rheum* **54:** 3690–3695.

Manns EL, Hoepelman AIM, Choi HJ, Lee JY, Cornpropst M, Liang W, King B, Hirsch KR, Oldach D, Rousseau FS. 2010. Short term safety, tolerability, pharmacokinetics and preliminary activity of GS 9450, a selective caspase inhibitor, in patients with chronic HCV infection. In *45th Annual Meeting of the European Association for the Study of the Liver (EASL 2010)*, Vienna, Austria.

Mariathasan S, Newton K, Monack DM, Vucic D, French DM, Lee WP, Roose-Girma M, Erickson S, Dixit VM. 2004. Differential activation of the inflammasome by caspase-1 adaptors ASC and Ipaf. *Nature* **430:** 213–218.

Mariathasan S, Weiss DS, Newton K, McBride J, O'Rourke K, Roose-Girma M, Lee WP, Weinrauch Y, Monack DM, Dixit VM. 2006. Cryopyrin activates the inflammasome in response to toxins and ATP. *Nature* **440:** 228–232.

Martinon F, Burns K, Tschopp J. 2002. The inflammasome: A molecular platform triggering activation of inflammatory caspases and processing of proIL-β. *Mol Cell* **10:** 417–426.

Martinon F, Petrilli V, Mayor A, Tardivel A, Tschopp J. 2006. Gout-associated uric acid crystals activate the NALP3 inflammasome. *Nature* **440:** 237–241.

McGonagle D, Tan AL, Shankaranarayana S, Madden J, Emery P, McDermott MF. 2007. Management of treatment resistant inflammation of acute on chronic tophaceous gout with anakinra. *Ann Rheum Dis* **66:** 1683–1684.

McGonagle D, Tan AL, Madden J, Emery P, McDermott MF. 2008. Successful treatment of resistant pseudogout with anakinra. *Arthritis Rheum* **58:** 631–633.

Medzhitov R, Preston-Hurlburt P, Janeway CA Jr. 1997. A human homologue of the *Drosophila* Toll protein signals activation of adaptive immunity. *Nature* **388:** 394–397.

Meng G, Zhang F, Fuss I, Kitani A, Strober W. 2009. A mutation in the Nlrp3 gene causing inflammasome hyperactivation potentiates Th17 cell-dominant immune responses. *Immunity* **30:** 860–874.

Merkle S, Frantz S, Schon MP, Bauersachs J, Buitrago M, Frost RJ, Schmitteckert EM, Lohse MJ, Engelhardt S. 2007. A role for caspase-1 in heart failure. *Circ Res* **100:** 645–653.

Miao EA, Alpuche-Aranda CM, Dors M, Clark AE, Bader MW, Miller SI, Aderem A. 2006. Cytoplasmic flagellin activates caspase-1 and secretion of interleukin 1β via Ipaf. *Nat Immunol* **7:** 569–575.

Miao EA, Ernst RK, Dors M, Mao DP, Aderem A. 2008. *Pseudomonas aeruginosa* activates caspase 1 through Ipaf. *Proc Natl Acad Sci* **105:** 2562–2567.

Miao EA, Mao DP, Yudkovsky N, Bonneau R, Lorang CG, Warren SE, Leaf IA, Aderem A. 2010. Innate immune detection of the type III secretion apparatus through the NLRC4 inflammasome. *Proc Natl Acad Sci* **107:** 3076–3080.

Molofsky AB, Byrne BG, Whitfield NN, Madigan CA, Fuse ET, Tateda K, Swanson MS. 2006. Cytosolic recognition of flagellin by mouse macrophages restricts *Legionella pneumophila* infection. *J Exp Med* **203:** 1093–1104.

Monie TP, Bryant CE, Gay NJ. 2009. Activating immunity: Lessons from the TLRs and NLRs. *Trends Biochem Sci* **34:** 553–561.

Muruve DA, Petrilli V, Zaiss AK, White LR, Clark SA, Ross PJ, Parks RJ, Tschopp J. 2008. The inflammasome recognizes cytosolic microbial and host DNA and triggers an innate immune response. *Nature* **452:** 103–107.

Muzio M, Stockwell BR, Stennicke HR, Salvesen GS, Dixit VM. 1998. An induced proximity model for caspase-8 activation. *J Biol Chem* **273:** 2926–2930.

Nakagawa T, Zhu H, Morishima N, Li E, Xu J, Yankner BA, Yuan J. 2000. Caspase-12 mediates endoplasmic-reticulum-specific apoptosis and cytotoxicity by amyloid-β. *Nature* **403:** 98–103.

Netea MG, van de Veerdonk FL, Kullberg BJ, Van der Meer JW, Joosten LA. 2008. The role of NLRs and TLRs in the activation of the inflammasome. *Expert Opin Biol Ther* **8:** 1867–1872.

Oberst A, Green DR. 2011. It cuts both ways: Reconciling the dual roles of caspase 8 in cell death and survival. *Nat Rev Mol Cell Biol* **12:** 757–763.

Oberst A, Dillon CP, Weinlich R, McCormick LL, Fitzgerald P, Pop C, Hakem R, Salvesen GS, Green DR. 2011. Catalytic activity of the caspase-8-FLIP(L) complex inhibits RIPK3-dependent necrosis. *Nature* **471:** 363–367.

O'Brien RJ, Wong PC. 2011. Amyloid precursor protein processing and Alzheimer's disease. *Ann Rev Neurosci* **34:** 185–204.

Oh JE, Kim MS, Ahn CH, Kim SS, Han JY, Lee SH, Yoo NJ. 2010. Mutational analysis of CASP10 gene in colon, breast, lung and hepatocellular carcinomas. *Pathology* **42:** 73–76.

Olson NE, Graves JD, Shu GL, Ryan EJ, Clark EA. 2003. Caspase activity is required for stimulated B lymphocytes to enter the cell cycle. *J Immunol* **170:** 6065–6072.

Olsson M, Vakifahmetoglu H, Abruzzo PM, Hogstrand K, Grandien A, Zhivotovsky B. 2009. DISC-mediated activation of caspase-2 in DNA damage-induced apoptosis. *Oncogene* **28:** 1949–1959.

Onouchi Y, Ozaki K, Buns JC, Shimizu C, Hamada H, Honda T, Terai M, Honda A, Takeuchi T, Shibuta S, et al. 2010. Common variants in CASP3 confer susceptibility to Kawasaki disease. *Hum Mol Genet* **19:** 2898–2906.

O'Reilly LA, Ekert P, Harvey N, Marsden V, Cullen L, Vaux DL, Hacker G, Magnusson C, Pakusch M, Cecconi F, et al. 2002. Caspase-2 is not required for thymocyte or neuronal apoptosis even though cleavage of caspase-2 is dependent on both Apaf-1 and caspase-9. *Cell Death Differ* **9:** 832–841.

Osorio F, Reis ESC. 2011. Myeloid C-type lectin receptors in pathogen recognition and host defense. *Immunity* **34:** 651–664.

Papin S, Cuenin S, Agostini L, Martinon F, Werner S, Beer HD, Grutter C, Grutter M, Tschopp J. 2007. The SPRY domain of Pyrin, mutated in familial Mediterranean fever patients, interacts with inflammasome components and inhibits proIL-1β processing. *Cell Death Differ* **14:** 1457–1466.

Park HH, Wu H. 2006. Crystal structure of RAIDD death domain implicates potential mechanism of PIDDosome assembly. *J Mol Biol* **357:** 358–364.

Park WS, Lee JH, Shin MS, Park JY, Kim HS, Kim YS, Lee SN, Xiao W, Park CH, Lee SH, et al. 2002. Inactivating mutations of the caspase-10 gene in gastric cancer. *Oncogene* **21:** 2919–2925.

Park SJ, Wu CH, Gordon JD, Zhong X, Emami A, Safa AR. 2004. Taxol induces caspase-10-dependent apoptosis. *J Biol Chem* **279:** 51057–51067.

Park JY, Park JM, Jang JS, Choi JE, Kim KM, Cha SI, Kim CH, Kang YM, Lee WK, Kam S, et al. 2006. Caspase 9 promoter polymorphisms and risk of primary lung cancer. *Hum Mol Genet* **15:** 1963–1971.

Park HH, Logette E, Raunser S, Cuenin S, Walz T, Tschopp J, Wu H. 2007. Death domain assembly mechanism revealed by crystal structure of the oligomeric PIDDosome core complex. *Cell* **128:** 533–546.

Pascual V, Allantaz F, Arce E, Punaro M, Banchereau J. 2005. Role of interleukin-1 (IL-1) in the pathogenesis of systemic onset juvenile idiopathic arthritis and clinical response to IL-1 blockade. *J Exp Med* **201:** 1479–1486.

Peterson QP, Hsu DC, Goode DR, Novotny CJ, Totten RK, Hergenrother PJ. 2009. Procaspase-3 activation as an anti-cancer strategy: Structure-activity relationship of procaspase-activating compound 1 (PAC-1) and its cellular co-localization with caspase-3. *J Med Chem* **52:** 5721–5731.

Peterson QP, Hsu DC, Novotny CJ, West DC, Kim D, Schmit JM, Dirikolu L, Hergenrother PJ, Fan TM. 2010. Discovery and canine preclinical assessment of a nontoxic procaspase-3-activating compound. *Cancer Res* **70:** 7232–7241.

Philchenkov A, Zavelevich M, Kroczak TJ, Los M. 2004. Caspases and cancer: Mechanisms of inactivation and new treatment modalities. *Exp Oncol* **26:** 82–97.

Poyet JL, Srinivasula SM, Tnani M, Razmara M, Fernandes-Alnemri T, Alnemri ES. 2001. Identification of Ipaf, a human caspase-1-activating protein related to Apaf-1. *J Biol Chem* **276:** 28309–28313.

Putt KS, Chen GW, Pearson JM, Sandhorst JS, Hoagland MS, Kwon JT, Hwang SK, Jin H, Churchwell MI, Cho MH, et al. 2006. Small-molecule activation of procaspase-3 to caspase-3 as a personalized anticancer strategy. *Nat Chem Biol* **2:** 543–550.

Rathinam VA, Jiang Z, Waggoner SN, Sharma S, Cole LE, Waggoner L, Vanaja SK, Monks BG, Ganesan S, Latz E, et al. 2010. The AIM2 inflammasome is essential for host defense against cytosolic bacteria and DNA viruses. *Nat Immunol* **11:** 395–402.

Renolleau S, Fau S, Goyenvalle C, Joly LM, Chauvier D, Jacotot E, Mariani J, Charriaut-Marlangue C. 2007. Specific caspase inhibitor Q-VD-OPh prevents neonatal stroke in P7 rat: A role for gender. *J Neurochem* **100:** 1062–1071.

Riedl SJ, Shi Y. 2004. Molecular mechanisms of caspase regulation during apoptosis. *Nat Rev Mol Cell Biol* **5:** 897–907.

Rohn TT, Head E. 2009. Caspases as therapeutic targets in Alzheimer's disease: Is it time to "cut" to the chase? *Int J Clin Exp Pathol* **2:** 108–118.

Rohn TT, Head E, Su JH, Anderson AJ, Bahr BA, Cotman CW, Cribbs DH. 2001. Correlation between caspase activation and neurofibrillary tangle formation in Alzheimer's disease. *Am J Pathol* **158:** 189–198.

Roldan R, Ruiz AM, Miranda MD, Collantes E. 2008. Anakinra: New therapeutic approach in children with Familial Mediterranean Fever resistant to colchicine. *Joint Bone Spine* **75:** 504–505.

Roy S, Bayly CI, Gareau Y, Houtzager VM, Kargman S, Keen SL, Rowland K, Seiden IM, Thornberry NA, Nicholson DW. 2001. Maintenance of caspase-3 proenzyme dormancy by an intrinsic "safety catch" regulatory tripeptide. *Proc Natl Acad Sci* **98:** 6132–6137.

Saleh M, Vaillancourt JP, Graham RK, Huyck M, Srinivasula SM, Alnemri ES, Steinberg MH, Nolan V, Baldwin CT, Hotchkiss RS, et al. 2004. Differential modulation of endotoxin responsiveness by human caspase-12 polymorphisms. *Nature* **429:** 75–79.

Saleh M, Mathison JC, Wolinski MK, Bensinger SJ, Fitzgerald P, Droin N, Ulevitch RJ, Green DR, Nicholson DW. 2006. Enhanced bacterial clearance and sepsis resistance in caspase-12-deficient mice. *Nature* **440:** 1064–1068.

Salmena L, Lemmers B, Hakem A, Matysiak-Zablocki E, Murakami K, Au PY, Berry DM, Tamblyn L, Shehabeldin A, Migon E, et al. 2003. Essential role for caspase 8 in T-cell homeostasis and T-cell-mediated immunity. *Genes Dev* **17:** 883–895.

Samraj AK, Keil E, Ueffing N, Schulze-Osthoff K, Schmitz I. 2006. Loss of caspase-9 provides genetic evidence for the type I/II concept of CD95-mediated apoptosis. *J Biol Chem* **281:** 29652–29659.

Schroder K, Tschopp J. 2010. The inflammasomes. *Cell* **140:** 821–832.

Scott AM, Saleh M. 2007. The inflammatory caspases: Guardians against infections and sepsis. *Cell Death Differ* **14:** 23–31.

Seshadri S, Duncan MD, Hart JM, Gavrilin MA, Wewers MD. 2007. Pyrin levels in human monocytes and monocyte-derived macrophages regulate IL-1β processing and release. *J Immunol* **179:** 1274–1281.

Shi L, Chen G, MacDonald G, Bergeron L, Li H, Miura M, Rotello RJ, Miller DK, Li P, Seshadri T, et al. 1996. Activation of an interleukin 1 converting enzyme-dependent apoptosis pathway by granzyme B. *Proc Natl Acad Sci* **93:** 11002–11007.

Shi M, Vivian CJ, Lee KJ, Ge C, Morotomi-Yano K, Manzl C, Bock F, Sato S, Tomomori-Sato C, Zhu R, et al. 2009. DNA-PKcs-PIDDosome: A nuclear caspase-2-activating complex with role in G2/M checkpoint maintenance. *Cell* **136:** 508–520.

Shin MS, Kim HS, Kang CS, Park WS, Kim SY, Lee SN, Lee JH, Park JY, Jang JJ, Kim CW, et al. 2002. Inactivating mutations of CASP10 gene in non-Hodgkin lymphomas. *Blood* **99:** 4094–4099.

Shiozaki EN, Chai J, Shi Y. 2002. Oligomerization and activation of caspase-9, induced by Apaf-1 CARD. *Proc Natl Acad Sci* **99:** 4197–4202.

Shoham NG, Centola M, Mansfield E, Hull KM, Wood G, Wise CA, Kastner DL. 2003. Pyrin binds the PSTPIP1/CD2BP1 protein, defining familial Mediterranean fever and PAPA syndrome as disorders in the same pathway. *Proc Natl Acad Sci* **100:** 13501–13506.

Sidi S, Sanda T, Kennedy RD, Hagen AT, Jette CA, Hoffmans R, Pascual J, Imamura S, Kishi S, Amatruda JF, et al. 2008. Chk1 suppresses a caspase-2 apoptotic response to DNA damage that bypasses p53, Bcl-2, and caspase-3. *Cell* **133:** 864–877.

So A, De Smedt T, Revaz S, Tschopp J. 2007. A pilot study of IL-1 inhibition by anakinra in acute gout. *Arthritis Res Ther* **9:** R28.

Soung YH, Lee JW, Kim HS, Park WS, Kim SY, Lee JH, Park JY, Cho YG, Kim CJ, Park YG, et al. 2003. Inactivating mutations of CASPASE-7 gene in human cancers. *Oncogene* **22:** 8048–8052.

Soung YH, Lee JW, Kim SY, Park WS, Nam SW, Lee JY, Yoo NJ, Lee SH. 2004. Somatic mutations of CASP3 gene in human cancers. *Hum Genet* **115:** 112–115.

Soung YH, Lee JW, Kim SY, Jang J, Park YG, Park WS, Nam SW, Lee JY, Yoo NJ, Lee SH. 2005a. CASPASE-8 gene is inactivated by somatic mutations in gastric carcinomas. *Cancer Res* **65:** 815–821.

Soung YH, Lee JW, Kim SY, Sung YJ, Park WS, Nam SW, Kim SH, Lee JY, Yoo NJ, Lee SH. 2005b. Caspase-8 gene is frequently inactivated by the frameshift somatic mutation 1225_1226delTG in hepatocellular carcinomas. *Oncogene* **24:** 141–147.

Soung YH, Jeong EG, Ahn CH, Kim SS, Song SY, Yoo NJ, Lee SH. 2008. Mutational analysis of caspase 1, 4, and 5 genes in common human cancers. *Hum Pathol* **39:** 895–900.

Spencer SL, Gaudet S, Albeck JG, Burke JM, Sorger PK. 2009. Non-genetic origins of cell-to-cell variability in TRAIL-induced apoptosis. *Nature* **459:** 428–432.

Spranger J, Kroke A, Mohlig M, Hoffmann K, Bergmann MM, Ristow M, Boeing H, Pfeiffer AF. 2003. Inflammatory cytokines and the risk to develop type 2 diabetes: Results of the prospective population-based European Prospective Investigation into Cancer and Nutrition (EPIC)-Potsdam Study. *Diabetes* **52:** 812–817.

Sprick MR, Rieser E, Stahl H, Grosse-Wilde A, Weigand MA, Walczak H. 2002. Caspase-10 is recruited to and activated at the native TRAIL and CD95 death-inducing signalling complexes in a FADD-dependent manner but can not functionally substitute caspase-8. *EMBO J* **21:** 4520–4530.

Stack JH, Beaumont K, Larsen PD, Straley KS, Henkel GW, Randle JC, Hoffman HM. 2005. IL-converting enzyme/caspase-1 inhibitor VX-765 blocks the hypersensitive response to an inflammatory stimulus in monocytes from familial cold autoinflammatory syndrome patients. *J Immunol* **175:** 2630–2634.

Steimle V, Otten LA, Zufferey M, Mach B. 1993. Complementation cloning of an MHC class II transactivator mutated in hereditary MHC class II deficiency (or bare lymphocyte syndrome). *Cell* **75:** 135–146.

Straathof KC, Pule MA, Yotnda P, Dotti G, Vanin EF, Brenner MK, Heslop HE, Spencer DM, Rooney CM. 2005. An inducible caspase 9 safety switch for T-cell therapy. *Blood* **105:** 4247–4254.

Straus SE, Sneller M, Lenardo MJ, Puck JM, Strober W. 1999. An inherited disorder of lymphocyte apoptosis: The autoimmune lymphoproliferative syndrome. *Ann Intern Med* **130:** 591–601.

Stupack DG, Teitz T, Potter MD, Mikolon D, Houghton PJ, Kidd VJ, Lahti JM, Cheresh DA. 2006. Potentiation of neuroblastoma metastasis by loss of caspase-8. *Nature* **439:** 95–99.

Sutterwala FS, Ogura Y, Szczepanik M, Lara-Tejero M, Lichtenberger GS, Grant EP, Bertin J, Coyle AJ, Galan JE, Askenase PW, et al. 2006. Critical role for NALP3/CIAS1/Cryopyrin in innate and adaptive immunity

through its regulation of caspase-1. *Immunity* **24:** 317–327.
Sutterwala FS, Mijares LA, Li L, Ogura Y, Kazmierczak BI, Flavell RA. 2007. Immune recognition of *Pseudomonas aeruginosa* mediated by the IPAF/NLRC4 inflammasome. *J Exp Med* **204:** 3235–3245.
Suzuki T, Franchi L, Toma C, Ashida H, Ogawa M, Yoshikawa Y, Mimuro H, Inohara N, Sasakawa C, Nunez G. 2007. Differential regulation of caspase-1 activation, pyroptosis, and autophagy via Ipaf and ASC in Shigella-infected macrophages. *PLoS Pathog* **3:** e111.
Taylor RC, Cullen SP, Martin SJ. 2008. Apoptosis: Controlled demolition at the cellular level. *Nat Rev Mol Cell Biol* **9:** 231–241.
Teitz T, Wei T, Valentine MB, Vanin EF, Grenet J, Valentine VA, Behm FG, Look AT, Lahti JM, Kidd VJ. 2000. Caspase 8 is deleted or silenced preferentially in childhood neuroblastomas with amplification of MYCN. *Nat Med* **6:** 529–535.
Terkeltaub R, Sundy JS, Schumacher HR, Murphy F, Bookbinder S, Biedermann S, Wu R, Mellis S, Radin A. 2009. The interleukin 1 inhibitor rilonacept in treatment of chronic gouty arthritis: Results of a placebo-controlled, monosequence crossover, non-randomised, single-blind pilot study. *Ann Rheum Dis* **68:** 1613–1617.
Tey SK, Dotti G, Rooney CM, Heslop HE, Brenner MK. 2007. Inducible caspase 9 suicide gene to improve the safety of allodepleted T cells after haploidentical stem cell transplantation. *Biol Blood Marrow Transplant* **13:** 913–924.
Thomas PG, Dash P, Aldridge JR Jr, Ellebedy AH, Reynolds C, Funk AJ, Martin WJ, Lamkanfi M, Webby RJ, Boyd KL, et al. 2009. The intracellular sensor NLRP3 mediates key innate and healing responses to influenza A virus via the regulation of caspase-1. *Immunity* **30:** 566–575.
Tinel A, Tschopp J. 2004. The PIDDosome, a protein complex implicated in activation of caspase-2 in response to genotoxic stress. *Science* **304:** 843–846.
Ting JP, Lovering RC, Alnemri ES, Bertin J, Boss JM, Davis BK, Flavell RA, Girardin SE, Godzik A, Harton JA, et al. 2008a. The NLR gene family: A standard nomenclature. *Immunity* **28:** 285–287.
Ting JP, Willingham SB, Bergstralh DT. 2008b. NLRs at the intersection of cell death and immunity. *Nat Rev Immunol* **8:** 372–379.
Tschopp J. 2011. Mitochondria: Sovereign of inflammation? *Eur J Immunol* **41:** 1196–1202.
Tschopp J, Martinon F, Burns K. 2003. NALPs: A novel protein family involved in inflammation. *Nat Rev Mol Cell Biol* **4:** 95–104.
Tu S, McStay GP, Boucher LM, Mak T, Beere HM, Green DR. 2006. In situ trapping of activated initiator caspases reveals a role for caspase-2 in heat shock-induced apoptosis. *Nat Cell Biol* **8:** 72–77.
van der Meer JW, Vossen JM, Radl J, van Nieuwkoop JA, Meyer CJ, Lobatto S, van Furth R. 1984. Hyperimmunoglobulinaemia D and periodic fever: A new syndrome. *Lancet* **1:** 1087–1090.
Varfolomeev EE, Schuchmann M, Luria V, Chiannilkulchai N, Beckmann JS, Mett IL, Rebrikov D, Brodianski VM, Kemper OC, Kollet O, et al. 1998. Targeted disruption of the mouse Caspase 8 gene ablates cell death induction by the TNF receptors, Fas/Apo1, and DR3 and is lethal prenatally. *Immunity* **9:** 267–276.
Vogler M, Dinsdale D, Dyer MJ, Cohen GM. 2009. Bcl-2 inhibitors: Small molecules with a big impact on cancer therapy. *Cell Death Differ* **16:** 360–367.
Wang L, Miura M, Bergeron L, Zhu H, Yuan J. 1994. Ich-1, an Ice/ced-3-related gene, encodes both positive and negative regulators of programmed cell death. *Cell* **78:** 739–750.
Wang S, Miura M, Jung YK, Zhu H, Li E, Yuan J. 1998. Murine caspase-11, an ICE-interacting protease, is essential for the activation of ICE. *Cell* **92:** 501–509.
Wang J, Zheng L, Lobito A, Chan FK, Dale J, Sneller M, Yao X, Puck JM, Straus SE, Lenardo MJ. 1999. Inherited human Caspase 10 mutations underlie defective lymphocyte and dendritic cell apoptosis in autoimmune lymphoproliferative syndrome type II. *Cell* **98:** 47–58.
Watanabe H, Gaide O, Petrilli V, Martinon F, Contassot E, Roques S, Kummer JA, Tschopp J, French LE. 2007. Activation of the IL-1β-processing inflammasome is involved in contact hypersensitivity. *J Invest Dermatol* **127:** 1956–1963.
Wewers MD, Herzyk DJ. 1989. Alveolar macrophages differ from blood monocytes in human IL-1β release. Quantitation by enzyme-linked immunoassay. *J Immunol* **143:** 1635–1641.
Wise CA, Gillum JD, Seidman CE, Lindor NM, Veile R, Bashiardes S, Lovett M. 2002. Mutations in CD2BP1 disrupt binding to PTP PEST and are responsible for PAPA syndrome, an autoinflammatory disorder. *Hum Mol Genet* **11:** 961–969.
Woo M, Hakem R, Soengas MS, Duncan GS, Shahinian A, Kagi D, Hakem A, McCurrach M, Khoo W, Kaufman SA, et al. 1998. Essential contribution of caspase 3/CPP32 to apoptosis and its associated nuclear changes. *Genes Dev* **12:** 806–819.
Woo M, Hakem R, Furlonger C, Hakem A, Duncan GS, Sasaki T, Bouchard D, Lu L, Wu GE, Paige CJ, et al. 2003. Caspase-3 regulates cell cycle in B cells: A consequence of substrate specificity. *Nat Immunol* **4:** 1016–1022.
Xu HL, Xu WH, Cai Q, Feng M, Long J, Zheng W, Xiang YB, Shu XO. 2009. Polymorphisms and haplotypes in the caspase-3, caspase-7, and caspase-8 genes and risk for endometrial cancer: A population-based, case-control study in a Chinese population. *Cancer Epidemiol Biomarkers Prev* **18:** 2114–2122.
Xue Y, Daly A, Yngvadottir B, Liu M, Coop G, Kim Y, Sabeti P, Chen Y, Stalker J, Huckle E, et al. 2006. Spread of an inactive form of caspase-12 in humans is due to recent positive selection. *Am J Hum Genet* **78:** 659–670.
Yamasaki K, Muto J, Taylor KR, Cogen AL, Audish D, Bertin J, Grant EP, Coyle AJ, Misaghi A, Hoffman HM, et al. 2009. NLRP3/cryopyrin is necessary for interleukin-1β (IL-1β) release in response to hyaluronan, an endogenous trigger of inflammation in response to injury. *J Biol Chem* **284:** 12762–12771.
Yang GY, Schielke GP, Gong C, Mao Y, Ge HL, Liu XH, Betz AL. 1999. Expression of tumor necrosis factor-α and intercellular adhesion molecule-1 after focal cerebral

ischemia in interleukin-1β converting enzyme deficient mice. *J Cereb Blood Flow Metab* **19:** 1109–1117.

Yang X, Merchant MS, Romero ME, Tsokos M, Wexler LH, Kontny U, Mackall CL, Thiele CJ. 2003. Induction of caspase 8 by interferon γ renders some neuroblastoma (NB) cells sensitive to tumor necrosis factor-related apoptosis-inducing ligand (TRAIL) but reveals that a lack of membrane TR1/TR2 also contributes to TRAIL resistance in NB. *Cancer Res* **63:** 1122–1129.

Yeh WC, Pompa JL, McCurrach ME, Shu HB, Elia AJ, Shahinian A, Ng M, Wakeham A, Khoo W, Mitchell K, et al. 1998. FADD: Essential for embryo development and signaling from some, but not all, inducers of apoptosis. *Science* **279:** 1954–1958.

Yeh WC, Itie A, Elia AJ, Ng M, Shu HB, Wakeham A, Mirtsos C, Suzuki N, Bonnard M, Goeddel DV, et al. 2000. Requirement for Casper (c-FLIP) in regulation of death receptor-induced apoptosis and embryonic development. *Immunity* **12:** 633–642.

Yin M, Yan J, Wei S, Wei Q. 2010. CASP8 polymorphisms contribute to cancer susceptibility: Evidence from a meta-analysis of 23 publications with 55 individual studies. *Carcinogenesis* **31:** 850–857.

Ying Lu G-QC. 2011. Effector Caspases and Leukemia. *Int J Cell Biol* **2011.** doi: 10.1155/2011/738301.

Yoo NJ, Lee JW, Kim YJ, Soung YH, Kim SY, Nam SW, Park WS, Lee JY, Lee SH. 2004. Loss of caspase-2, -6 and -7 expression in gastric cancers. *APMIS* **112:** 330–335.

Yoshida H, Kong YY, Yoshida R, Elia AJ, Hakem A, Hakem R, Penninger JM, Mak TW. 1998. Apaf1 is required for mitochondrial pathways of apoptosis and brain development. *Cell* **94:** 739–750.

Yu JW, Wu J, Zhang Z, Datta P, Ibrahimi I, Taniguchi S, Sagara J, Fernandes-Alnemri T, Alnemri ES. 2006. Cryopyrin and pyrin activate caspase-1, but not NF-κB, via ASC oligomerization. *Cell Death Differ* **13:** 236–249.

Zhang J, Cado D, Chen A, Kabra NH, Winoto A. 1998. Fas-mediated apoptosis and activation-induced T-cell proliferation are defective in mice lacking FADD/Mort1. *Nature* **392:** 296–300.

Zhang WH, Wang X, Narayanan M, Zhang Y, Huo C, Reed JC, Friedlander RM. 2003. Fundamental role of the Rip2/caspase-1 pathway in hypoxia and ischemia-induced neuronal cell death. *Proc Natl Acad Sci* **100:** 16012–16017.

Zhang HZ, Kasibhatla S, Wang Y, Herich J, Guastella J, Tseng B, Drewe J, Cai SX. 2004. Discovery, characterization and SAR of gambogic acid as a potent apoptosis inducer by a HTS assay. *Bioorg Med Chem* **12:** 309–317.

Zhang H, Zhou X, McQuade T, Li J, Chan FK, Zhang J. 2011. Functional complementation between FADD and RIP1 in embryos and lymphocytes. *Nature* **471:** 373–376.

Zheng TS, Hunot S, Kuida K, Flavell RA. 1999. Caspase knockouts: Matters of life and death. *Cell Death Differ* **6:** 1043–1053.

Apoptotic and Nonapoptotic Caspase Functions in Animal Development

Masayuki Miura

Department of Genetics, Graduate School of Pharmaceutical Sciences, University of Tokyo, and CREST, JST, 7-3-1 Hongo, Bunkyo-ku, Tokyo 113-0033, Japan

Correspondence: miura@mol.f.u-tokyo.ac.jp

A developing animal is exposed to both intrinsic and extrinsic stresses. One stress response is caspase activation. Caspase activation not only controls apoptosis but also proliferation, differentiation, cell shape, and cell migration. Caspase activation drives development by executing cell death or nonapoptotic functions in a cell-autonomous manner and by secreting signaling molecules or generating mechanical forces in a non-cell-autonomous manner.

Programmed cell death or apoptosis occurs widely during development. During *C. elegans* development, 131 cells die by caspase CED-3-dependent apoptosis; however, *ced-3* mutants do not show significant developmental defects (Ellis and Horvitz 1986). In contrast, studies on caspase mutants in mouse and *Drosophila* have revealed caspases' roles in development. During development, cells are exposed to extrinsic and intrinsic stresses, and caspases are activated as one of multiple stress responses that ensure developmental robustness (Fig. 1). Caspases actively regulate animal development through both apoptosis and nonapoptotic functions that involve cell–cell communication in developing cell communities (Miura 2011). This chapter focuses on the in vivo roles of caspases in development and regeneration.

INITIATOR CASPASES

Caspase-8

Caspase-8 is a death-effector domain (DED)-containing initiator caspase, initially found as a component of the Fas/Apo-1 death-inducing signaling complex (DISC) (Muzio et al. 1996) and a MORT1/FADD-binding protein (Boldin et al. 1996). Caspase-8-knockout mice are lethal at embryonic day (E)11.5 (Varfolomeev et al. 1998; Sakamaki et al. 2002; Kang et al. 2004). At E10.5, abnormal yolk sac vasculature is observed, followed by various defects in the developing heart and neural tube (Varfolomeev et al. 1998; Sakamaki et al. 2002). Mutant embryos show hyperemia in the superficial capillaries and other blood vessels, and extensive erythrocytosis in the liver. These defects result in severe

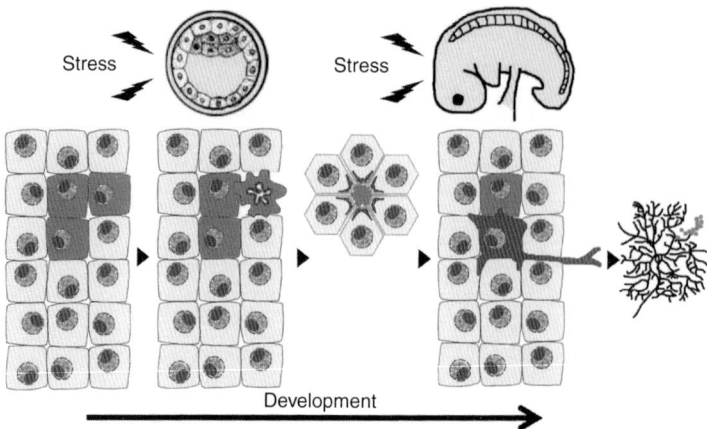

Figure 1. Caspase activation during development. An embryo undergoes intrinsic and extrinsic stress, which activates caspases to execute both apoptotic and nonapoptotic functions, including cell differentiation and dendrite pruning. Apoptotic cells affect the shape and behavior of their neighboring cells. Caspase-activated cells are shown in dark gray.

primary or secondary depletion of the hematopoietic precursor pool. Mutant embryos also show impaired heart muscle development (Varfolomeev et al. 1998).

The homozygous mutant neural and heart defects are rescued by ex vivo whole-embryo culture during E10.5–E11.5 (Sakamaki et al. 2002), suggesting that the mutant phenotypes of the neural tube and heart are caused by secondary effects of impaired angiogenesis in the yolk sac. Consistent with this idea, endothelium-specific *caspase-8*-knockout mice show the same phenotype as conventional knockout mice (Kang et al. 2004). Whereas knockout mice of Fas, TNF-R1, DR3, or TRAIL are viable (Pfeffer et al. 1993; Rothe et al. 1993; Adachi et al. 1995; Liu et al. 2001; Wang et al. 2001; Cretney et al. 2002), FADD- and cFLIPL-deficient mice show the same phenotypes as caspase-8-knockout mice (Yeh et al. 1998, 2000; Zhang et al. 1998). Thus, the caspase-8 developmental phenotype does not seem to depend on any single death ligand or receptor.

Recent studies revealed that caspase-8/FADD pathway has an important role in suppressing RIPK1/RIPK3-mediated necrosis (necroptosis). Deletion of RIPK1 or RIPK3 rescued lethality of FADD or *caspase-8* deletion (Kaiser et al. 2011; Oberst et al. 2011; Zhang et al. 2011).

The inactivation of RIPK1 or RIPK3 through caspase-8-mediated cleavage may be required for the protection of cells from necroptosis, thus *caspase-8* deletion can activate necroptosis pathway during development (Feng et al. 2007; Cho et al. 2009).

Caspase inhibitors (pan caspase-inhibitor Z-VAD, baculovirus p35, or cowpox virus CrmA) block macrophage differentiation and activation (Sordet et al. 2002) in vitro. This finding and studies in myelomonocytic-lineage-specific *caspase-8*-knockout mice suggest that the M-CSF-stimulated differentiation of macrophage precursors is caspase-8-dependent. In the absence of caspase-8, such precursors are eliminated by apoptosis, indicating that caspase-8 has a prosurvival role (Kang et al. 2004). Upon M-CSF stimulation, caspase-8 mediates cleavage of serine/threonine kinase receptor-interacting protein 1 (RIPK1), which prevents the sustained activation of NFκB and leads to macrophagic differentiation (Rebe et al. 2007).

Signal exchanges among epidermal keratinocytes, dermal fibroblasts, and leukocytic cells are critical regulatory mechanisms for skin homeostasis and wound healing. These reciprocal interactions are initiated by the paracrine signaling of interleukin 1α (IL1α), which activates both skin stem-cell proliferation and cutaneous inflammation. During wound healing, caspase-8 expression is downregulated in the epidermal granular layer, after which a p38-MAPK-mediated upregulation of NLRP3 leads to inflammasome assembly and caspase-1 activation, resulting in the noncanonical secretion of IL1α from epidermal keratinocytes (Lee et al. 2009b). Mature IL1α stimulates NFκB, which leads to growth arrest and promotes keratinocyte survival. IL1α also stimulates dermal fibroblasts to secrete secondary factors such as KGF and GM-CSF, which signal back to the epidermis to induce keratinocyte proliferation in the basal layer. Caspase-8's importance is supported by the observation that *caspase-8*-deficient skin shows epidermal hyperplasia (Lee et al. 2009b). The chronic loss of epidermal *caspase-8* in mice enhances wound-healing responses and recapitulates atopic dermatitis (Li et al. 2010a). Consistent with this, *caspase-8* is increased in diabetic mice in which wound healing is impaired (Al-Mashat et al. 2006).

Caspase-9 and Dronc

Caspase-9 and apaf-1 are key components of the intrinsic, mitochondrial apoptosis pathway. On a mixed 129/SvJ background, the knockout of *apaf-1* or *caspase-9* causes similar developmental defects, including neural-tube-closure (NTC) defects in the hindbrain and enlarged ventricular zones in the fore- and midbrain (Cecconi et al. 1998; Hakem et al. 1998; Kuida et al. 1998; Yoshida et al. 1998). Live imaging of caspase activation in cultured mouse embryo revealed the detailed apoptotic process of NTC. Two types of caspase-activated cells appeared during NTC. One is a typical apoptotic cell (classical [C]-type), and the other is an atypical apoptotic cell (dancing [D]-type), which does not show cell fragmentation and remains and dances around their original sites for a long period. Capsase-activation kinetics in D-type is slower than in C-type and caspase-7 is selectively activated in D-type. *Apaf-1*-deficiency abolished the appearance of both C- and D-type cells, and delayed the progression of neural tube closure. Thus inhibition of apoptosis could cause the NTC defect and increase the risk of exencephaly (Yamaguchi et al. 2011).

Semaphorin 7A (Sema7A) that can promote axon guidance and neuronal migration in neural development is a direct substrate for caspase-9 (Ohsawa et al. 2009). *Caspase-9* and *apaf-1* mutant mice show misrouted axons, impaired synaptic formation, and defects in olfactory sensory neuron (OSN) maturation without changes in the OSN cell number (Ohsawa et al. 2010). Thus, some developmental abnormalities resulting from caspase inhibition could be attributable to the loss of a nonapoptotic function of caspase-9.

Dronc is a *Drosophila* ortholog of caspase-9/CED-3, and the only CARD-containing caspase in *Drosophila*. The *dronc* mutation abolishes most of the programmed cell death during development and causes pupal death (Chew et al. 2004; Daish et al. 2004; Waldhuber et al. 2005; Xu et al. 2005; Kondo et al. 2006). Dronc is essential for diap1-degradation-induced cell death (Leulier et al. 2006), compensatory proliferation, and dendrite pruning, described below.

EXECUTIONER CASPASES

Caspase-3, Caspase-7, Drice, and dcp-1

Caspase-3 deficiency in the 129/SvJ mouse causes neurodevelopmental abnormalities, including an expanded ventricular zone, ectopic neural structures, and gross brain malformations, as observed in *caspase-9-* or *apaf-1*-knockout mice (Kuida et al. 1996; Leonard et al. 2002). Caspase-3 and caspase-7 are both categorized as executioner caspases, although their substrate specificity is somewhat different. Caspase-7 is a more selective protease than caspase-3 (Walsh et al. 2008). The developmental defect of executioner caspase, *apaf-1* and *caspase-9*, is strain-dependent in mice. C57BL/6-dominant background mice with a single knockout of *caspase-3* or *caspase-7* do not show severe developmental defects and have a normal life span; however, mice with double knockouts of *caspase-3* and *caspase-7 (DKO)* die shortly after birth, indicating that these caspases have redundant roles in embryonic development (Leonard et al. 2002; Lakhani et al. 2006). A small percentage (~10%) of them showed exencephaly, but the investigators assumed that this was attributable to residual genes from 129/SvJ background, as the DKO mice were backcrossed only six generations onto C57BL/6 background, which was probably insufficient to remove all of the residual genes from 129/SvJ background (Lakhani et al. 2006). Most of the DKO mice did not show any sign of exencephaly and their cause of death after birth was unclear but might be caused by defective heart development. Caspase-7's in vivo function has not been extensively studied, but evidence suggests it regulates inflammation and pathological cell death (Teixeira et al. 2008). Caspase-7, but not caspase-3, is a direct substrate for caspase-1 (Lamkanfi et al. 2008), and caspase-7 is involved in the induction of Legionella pneumophila-infected macrophages (Akhter et al. 2009).

Drice is a caspase-3 ortholog in Drosophila, in which it is the most abundantly expressed effector caspase (Fraser and Evan 1997; Kumar and Doumanis 2000). Drice mutants show pupal lethality and significantly decreased apoptosis during development, but the reduction is not as severe as in dronc mutants (Kondo et al. 2006; Muro et al. 2006; Xu et al. 2006). This is because drice shares functional redundancy with the other executioner caspase, dcp-1. dcp-1 mutants are viable and fertile (Kondo et al. 2006; Muro et al. 2006; Xu et al. 2006). drice/dcp-1 double mutants phenocopy dronc mutants, indicating that these two executioner caspases are functionally redundant, downstream of dronc (Kondo et al. 2006).

During metamorphosis, caspases are induced and activated by ecdysone, and most larval tissues including the salivary gland, midgut, and larval epidermis, are removed (Baehrecke 2000; Takemoto et al. 2007). p35 expression can only partially suppress the salivary-gland cell death; the inhibition of autophagy is required to block tissue degeneration completely (Berry and Baehrecke 2007). Despite high caspase activity in the midgut, autophagy but not caspase activity is essential for the programmed midgut cell death (Denton et al. 2009).

Caspase-Mediated Apoptosis in Fine-Tuning Neural Cell Populations

Large numbers of cells are thought to be eliminated by apoptosis in neural development (Buss et al. 2006). Live imaging enables a death-fated cell to be monitored throughout its lifetime. During sensory organ development of the Drosophila notum, about 20% of the differentiating neural cells die (Koto et al. 2011). Detailed observations of these cells indicate that the excess neural progenitors do not die in a random manner; instead, which proneural gene neuralized-positive cells die follows a rule and depends on the Notch-activation level. Thus, Notch-regulated apoptosis of aberrant, mis-specified sensory organ precursor (SOP) mediates the fine-tuning of SOP cell selection (Koto et al. 2011). It has been proposed that apoptosis-mediated cell selection eliminates harmful autoreactive lymphocytes in primary lymphoid tissues (Strasser et al. 2008). Thus, apoptosis-mediated error-correction systems may be critical for eliminating aberrant cells, achieving normal development, and maintaining tissue homeostasis.

Caspases in Morphogenesis

One of the most fundamental roles of cell death during development is in tissue sculpting. However, none of the mouse mutants of core regulators of apoptosis, except for bax/bak double-mutant mice or bid/bim/puma triple mutant mice, show severe defects in morphogenetic cell death, such as the classic example of interdigital tissue removal (Lindsten et al. 2000; Ren et al. 2010). Furthermore, although cultured cells from apaf-1, bax/bak, or bid/bim/puma mutant mice have significant defects in cell-stress-induced apoptosis, apaf-1, bax/bak, or bid/bim/puma mutant mice generally develop into mostly normal adults, depending on their genetic background (Chautan et al. 1999; Lindsten and Thompson 2006; Ren et al. 2010). These results indicate that caspase-mediated apoptosis is dispensable, or that an alternative cell-death mechanism removes cells when the apoptotic machinery is inhibited. In fact, nonapoptotic cell death is observed in the interdigital region and other areas in *apaf-1* mutant mice (Chautan et al. 1999; Cande et al. 2002; Nagasaka et al. 2010). However, the caspase functions in mammalian morphogenetic processes remain to be studied in detail.

Genetic studies of caspase signaling in *Drosophila* revealed some interesting roles of caspases in morphogenetic processes. For example, apoptotic cell death may help generate the mechanical forces that drive cell movements and cell-shape changes during epithelial morphogenesis. During *Drosophila* embryonic development, a process called dorsal closure completes the sealing of the yolk and the amnioserosa. As the dorsal epidermis spreads and amnioserosa cells constrict, multiple forces drive the closure (Hutson et al. 2003). Apoptosis of amnioserosa cells may help generate one of the forces needed to seal the dorsal epithelium over the embryo (Toyama et al. 2008).

When a discontinuity in Dpp activity is generated between neighboring cells, JNK-dependent apoptosis is triggered to restore the appropriate cell-positional information (Adachi-Yamada et al. 1999). Similar mechanism is involved in the morphogenesis of the *Drosophila* leg joint (Manjon et al. 2007). During normal segmentation of the *Drosophila* distal legs, Dpp is expressed in the most distal part of every forming segment, and only diffuses proximally; the sharp boundary in Dpp activity induces JNK-dependent apoptosis, which leads to the epithelial folding that prefigures the adult joint (Manjon et al. 2007).

During *Drosophila* embryonic development, the Hox protein Deformed helps maintain the boundary between the maxillary and mandibular head lobes, by activating the cell-death-promoting gene *reaper* to induce localized apoptosis. Abdominal-B also regulates segment boundaries through the regional activation of apoptosis. Thus, apoptosis induced by *Drosophila* Hox genes modulates segmental morphology (Lohmann et al. 2002).

Global tissue rotation is a morphogenetic movement involved in controlling tissue elongation (Haigo and Bilder 2011). *Drosophila* apoptosis mutants show an orientation defect of the male terminalia. The male terminalia normally rotates 360° during development, and incomplete rotation causes the orientation defect of male terminalia (Macias et al. 2004). During development, the male genitalia rotation accelerates to complete the full 360° rotation; this acceleration is impaired in apoptosis-defective flies (Abbott and Lengyel 1991; Benitez et al. 2010; Suzanne et al. 2010; Kuranaga et al. 2011). Genetic and live-imaging studies indicated the active roles of apoptosis in genitalia rotation (Suzanne et al. 2010; Kuranaga et al. 2011) and caspase drives the acceleration of genitalia rotation to complete the morphogenesis of male genitalia within a limited developmental window (Kuranaga et al. 2011).

Caspases in Compensatory Proliferation and Regeneration

Some animal tissues, including the limbs of urodele amphibians such as newts and salamanders, the tadpole tail of *Xenopus*, the heart and fin of fish (Johnson and Weston 1995; Akimenko et al. 2003; Poss 2007), *Drosophila* larval imaginal discs, and freshwater hydra, have a striking regenerative capacity. When such tissues are injured, many cells die by apoptosis. This apoptosis appears to be essential for the subsequent regenerative cell proliferation in hydra (Chera et al. 2009), planarians (Hwang et al. 2004; Pellettieri et al. 2010), *Drosophila* (Milan et al. 1997; Huh et al. 2004a; Ryoo et al. 2004), *Xenopus* (Tseng et al. 2007), newt (Vlaskalin et al. 2004), and mouse (Li et al. 2010c). Upon healing, a tissue's final shape and size often match those of the uninjured tissue, owing to a regenerative cell proliferation mechanism that is often called "compensatory proliferation" (Fig. 2).

The local induction of apoptosis by ectopic toxin expression in *Drosophila* wing imaginal disc results in elevated cell proliferation around the site of apoptosis, suggesting that cells can perceive apoptosis in their vicinity and undergo extra cell divisions until the original cell number is restored (Milan et al. 1997). When the *Drosophila* cell-death-inducing gene *reaper* (*rpr*) or *head involution defective* (*hid*) is expressed, or imaginal discs are subjected to X rays together with baculovirus *p35* (a potent inhibitor of

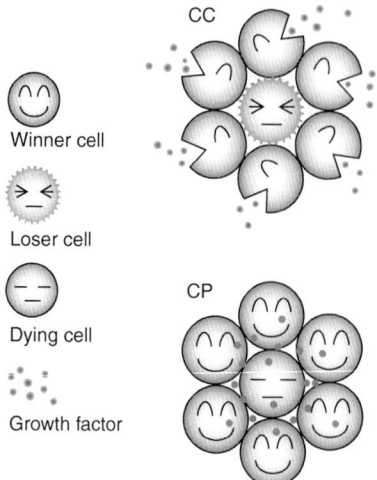

Figure 2. Cell competition (CC) and compensatory proliferation (CP). In CC, a cell that receives less growth factor is eliminated by apoptosis. Dying cells release growth factors that promote the CP of neighboring cells.

regulators Wingless (Wg) and Decapentaplegic (Dpp) are ectopically expressed (Huh et al. 2004a; Perez-Garijo et al. 2004, 2005; Ryoo et al. 2004; Kondo et al. 2006; Wells et al. 2006), and hyperplastic tissue growth occurs. In this experimental system, *dronc* was found to be required for proliferation and was suggested to be involved in the induction of mitogen expression in the proliferating cells of wing imaginal discs (Huh et al. 2004a; Kondo et al. 2006; Wells et al. 2006). In the differentiating cells of eye imaginal discs, effector caspases trigger the activation of Hedgehog (Hh) signaling for the compensatory proliferation (Fig. 3) (Fan and Bergmann 2008).

Classical tissue regeneration experiments were performed using *Drosophila* imaginal discs (Hadorn and Buck 1962; Hadorn et al. 1968). These tissue-ablation experiments required microsurgery skills and were technically difficult. Recently, nonsurgical, genetic ablation methods were used to study imaginal disc regeneration (Smith-Bolton et al. 2009; Bergantinos et al. 2010). Apoptosis induced in a portion of the wing imaginal disc by the transient expression of the cell-death-inducing gene *reaper* (diap1 antagonist) or *eiger* (TNF-α ortholog) caused caspase activation and the extrusion of dying

effector caspases such as drice and dcp-1 but not of the initiator caspase dronc [Meier et al. 2000; Xu et al. 2009]), the apoptosis-signaling cascade is activated only to the point in which dronc has been activated, and the cells fail to undergo apoptosis. Under these conditions, the proliferation

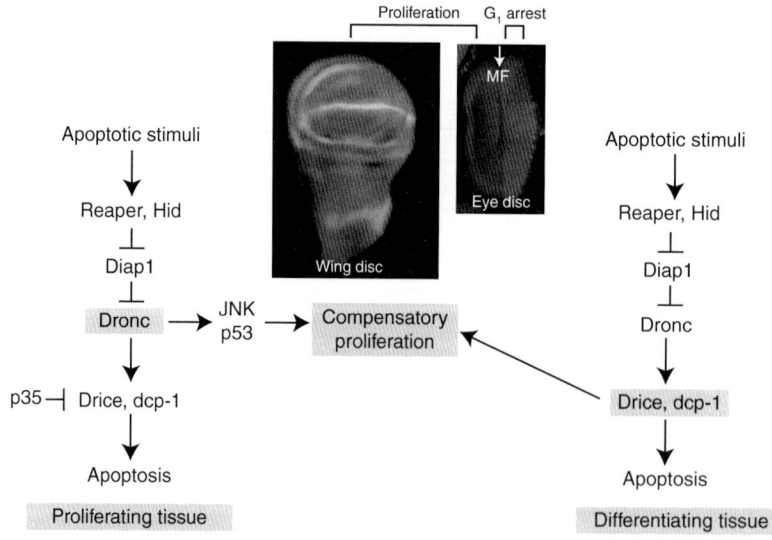

Figure 3. Genetic cascade of compensatory proliferation in proliferating or differentiating cells in *Drosophila* imaginal discs (wing and eye discs).

cells from the epithelium. As the tissue healed, cytoskeletal reorganization was observed with JNK activation at the leading edges of the healing tissue, not in the dying cells; furthermore, JNK activation was required for compensatory proliferation of the regenerating tissue (Bergantinos et al. 2010). Unlike the proliferation observed when *reaper* and caspase inhibitor *p35* are coexpressed in the wing imaginal disc, Wg is predominantly expressed in the proliferating cells, not in the dying cells during tissue regeneration (Smith-Bolton et al. 2009).

The relationship between apoptosis and compensatory proliferation was studied in Hydra during the regeneration after amputation (Chera et al. 2009). Mid-gastric bisection induces apoptosis specifically in the head, and not in the foot-regenerating region. Apoptotic cells in the head-regenerating region release Wnt3 to trigger compensatory proliferation and regeneration. The ectopic induction of apoptosis by heating in the foot-regenerating region activates the Wnt3/β-Catenin pathway, and is sufficient to cause head formation. In both cases, treatment with the pan-caspase inhibitor Z-VAD abolishes the Wnt3/β-Catenin pathway activation and head regeneration. These results showed that caspase-mediated signaling promotes compensatory proliferation in Hydra tissue regeneration.

In a mammalian wound-healing model, *caspase-3*- or *caspase-7*-deficient mice show defects in skin-wound healing and cell proliferation (Li et al. 2010c). C*aspase-3* and *caspase-7* were also important for healing in partial-hepatectomy model mice. In *caspase-3*- or *caspase-7*-knockout mice, the regenerative cell proliferation was reduced about 50%, indicating that executioner caspases are required for cell proliferation during wound healing and tissue regeneration (Li et al. 2010c). The growth signal from the caspase-activated cells might be prostaglandin E2 (PGE2). Calcium-independent phospholipase A2 (iPLA2), which produces arachidonic acid and triggers PGE2 production, is activated after cleavage by caspase-3 (Atsumi et al. 1998; Zhao et al. 2006). PGE2 is known to promote stem-cell proliferation (Feher and Gidali 1974; Liou et al. 2007; North et al. 2007), tissue regeneration (Goessling et al. 2009), and wound healing (Kuhrer et al. 1986; Paralkar et al. 2003).

Caspase in Tissue Growth Control

During regeneration, the overall size of the recovering tissue is regulated. The mechanism of tissue-growth control has been studied in *Drosophila* wing discs. The central regulator of tissue size is insulin/TOR signaling, which ultimately controls the ribosome and protein synthesis. Conditional expression of the ribosome-inactivating protein Ricin toxin A chain in the posterior wing imaginal disc causes a nonautonomous reduction in the growth and proliferation rates of adjacent cell populations (Mesquita et al. 2010). In this condition, dronc activation, detected by an antibody against activated human caspase-3 (Fan and Bergmann 2009), is observed not only in the Ricin-expressing territories but also in adjacent cells (Mesquita et al. 2010). The coexpression of baculovirus *p35* in the Ricin-expressing domain prevents cell death in that region and partially prevents it in the neighboring region. Thus, the cell proliferation is nonautonomously rescued, suggesting that effector caspases in the growth-depleted territory are required for the reduced cell proliferation in the adjacent cells. However, the nonautonomous reduction of overall tissue size is not rescued by the expression of *p35*.

p53 is activated in the growth-deficient cells, and effector caspases are activated downstream of p53. As with *p35* expression, a reduction in p53 activity in the Ricin-expressing territory nonautonomously rescues cell proliferation. However, unlike *p35* expression, the adjacent tissue growth is rescued by *p53* suppression. Targeted Ricin-expression activates p53, which induces caspase activation. The tissue size and cell proliferation are regulated by p53 nonautonomously, but caspases have a unique role in regulating the proliferation of adjacent cell populations (Mesquita et al. 2010).

Caspase in Cell Competition

Cell competition was first discovered by genetic studies in *Drosophila* (Morata and Ripoll 1975;

Simpson 1979; Simpson and Morata 1981). The genes for ribosomal proteins (RPs) are essential, and their homozygous mutations are lethal; however, their heterozygotes can result in the "*Minute*" dominant, haploinsufficient phenotypes, such as a slow growth rate and short and thin bristles (these heterozygous flies are called "*Minute*" because of their small bristles). When mosaic clones consisting of wild-type cells and *Minute* cells are generated in wing imaginal discs, the *Minute* cells are selectively eliminated by apoptosis, and wild-type cells eventually take over within a compartment (Fig. 2) (Morata and Ripoll 1975; Moreno et al. 2002). Thus, there is a competitive interaction between *Minute* cells (losers) and wild-type cells (winners). This elimination of *Minute* cells does not occur under starvation conditions (Simpson 1979), suggesting a link between the *Minute* mutation and nutrition control. Target of rapamycin complex 2 (TORC2) regulates cell survival and growth. TORC2 is activated by directly associating with a ribosome, suggesting there might be a direct link between *Minute* and growth-control mechanisms (Zinzalla et al. 2011). Cells carrying different dosages of the *myc* gene also show competitive interaction (de la Cova et al. 2004; Moreno and Basler 2004). High-*myc* cells are winners and low-*myc* cells are losers under competitive conditions. Because *myc* promotes the gene expression of ribosomal proteins, supercompetition by high *myc* might result from high ribosomal biosynthesis (Johnston 2009).

There are differences in the cell-competition processes induced by *Minute* and *myc*. The induction of cell death in *Minute* clones is initiated locally, whereas *myc*-induced competition affects up to 10 cells (de la Cova et al. 2004; Li and Baker 2007). Both *Minute* and *myc*-induced competition preserves the wing size, indicating a balance between the growth of winner and the loss of loser cells. However, cell competition itself is not a major regulatory mechanism for wing size control. Rather, growth arrest that occurred once compartment size reached the final dimension plays significant role for size control (Martin et al. 2009).

The concept of cell competition can be extended to explain the competitive nature of cell populations in the body (Baker 2011). Mosaic clones of tumor-suppressor genes that encode proteins essential for establishing epithelial apicobasal polarity, such as *scribble (scrib)*, *discs large (dlg)*, and *lethal giant larvae (lgl)*, are eliminated by cell competition (Woods and Bryant 1991; Agrawal et al. 1995; Brumby and Richardson 2003; Igaki et al. 2006). The engulfment of loser cells by winner cells is required for cell death execution and cell elimination in both the *Minute* and *scribble* models of cell competition (Li and Baker 2007; Ohsawa et al. 2011). JNK is important for elimination of *scribble* clone, but caspases may not be (Brumby and Richardson 2003; Igaki et al. 2006), suggesting that caspase-dependent and -independent mechanisms are used for cell elimination in different types of cell competition. Caspase-independent elimination of *scribble* clones by cell competition is executed by Eiger/TNF signaling (Igaki et al. 2006, 2009). Cell death signaling by Eiger/TNF has a similarity with necroptosis (Degterev et al. 2005; Christofferson and Yuan 2010) and is regulated by JNK and energy metabolic pathway (Kanda et al. 2011).

TISSUE REMODELING DURING DEVELOPMENT

Although cell competition has been extensively studied by mosaic analysis in *Drosophila* wing imaginal discs, studies on the physiological nature of cell competition are still limited. One system for studying cell-competition-like phenomena in vivo is the abdominal epithelial replacement in *Drosophila*. The adult abdominal epidermis is formed by histoblasts, which replace the polyploid larval epidermal cells (LECs) during the pupal stage. About 15 h after puparium formation, the histoblast nests expand, and the histoblasts migrate collectively over the abdomen (Madhavan and Madhavan 1980). As in cell competition, the histoblasts behave like winners and replace the LECs, which undergo apoptosis, like losers. During abdominal epithelial replacement, the coordination of LEC apoptosis and histoblast proliferation enables the orderly substitution of cell types without changing the area of the epithelium, as in cell-competition experiments in wing discs. Possible mechanisms for the replacement of abdominal epithelium include mutual signaling events, mechanical forces, and cell competition (Ninov et al. 2007, 2010). A study of caspase activation during epithelial replacement in vivo at single-cell resolution indicated that caspase activation in LECs is induced at the LEC/histoblast boundary, which expands as the LECs die. Transition from the S/G2 phases is necessary to induce nonautonomous LEC apoptosis at the LEC/histoblast boundary. This replacement boundary, formed as caspase activation is regulated locally by cell–cell communication, may drive the dynamic orchestration of cell replacement during tissue remodeling by competitive interaction (Fig. 4) (Nakajima et al. 2011).

Nonapoptotic Caspase Functions

Caspases exert nonapoptotic functions, which include roles in cell proliferation, migration, differentiation, and immunity (Kuranaga and Miura 2007; Yi and Yuan 2009). These functions are executed via three mechanisms.

Local Activation of Caspases

During neural development, some axons and dendrites degenerate or are pruned to create proper neural circuits. In *Drosophila*, dendritic pruning in class IV dendritic arborization sensory neurons occurs during metamorphosis. During this process, caspase activity is detected locally in the degenerating dendrites, and mutation of the initiator caspase *dronc* preserves most of the dendrite morphology (Kuo et al. 2006; Williams et al. 2006; Schoenmann et al. 2010). Mutations of effector caspases, such as *drice* or *dcp-1*, or expression of caspase inhibitor *p35*, only weakly reduce dendritic pruning, suggesting that dronc promotes dendritic pruning by a mechanism other than the activation of effector caspases (Schoenmann et al. 2010). The dronc activation in this system is regulated by a diap1-degrading kinase, IK2/DmIKKε (Kuranaga et al. 2006; Lee et al. 2009a). Neuromuscular degeneration caused by disruption of the spectrin/ankyrin skeleton requires *dronc*, *dcp-1*, *apaf-1 (dark)*, and *bcl-2* family gene *debcl* (Keller et al. 2011). In contrast, caspase inhibition by p35 or diap1 fails to block axon pruning during *Drosophila* metamorphosis (Awasaki et al. 2006).

Caspases are reported to be involved in axon degeneration in mammals. Caspase-6 but not caspase-3 is activated after NGF deprivation in dissociated dorsal root ganglion (DRG) culture (Nikolaev et al. 2009), and caspase-6 inhibition alone can block the axonal degeneration. In a different DRG culture system, NGF deprivation induces axonal degradation and the activation

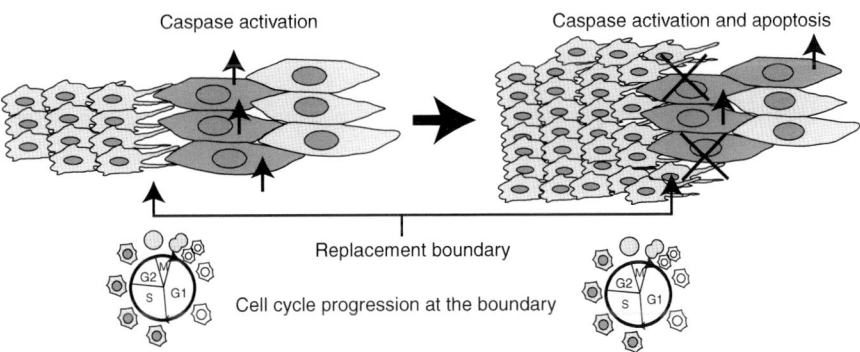

Figure 4. Ordered caspase activation during epidermal remodeling in *Drosophila* metamorphosis. Caspase-activated cells (dark grey and arrow) at the "replacement boundary" are eliminated, and histoblast (small cell) cell-cycle progression is coupled with larval epidermal cell (large cell) apoptosis. Apoptosis at the replacement boundary creates a spatiotemporal pattern of tissue removal, and maintains the epithelial integrity as histoblasts replace larval epidermal cells.

of both caspase-3 and caspase-6 in axons (Schoenmann et al. 2010). To completely protect the axons from degenerating, NAD^+ must be added to the DRG culture with caspase-6 or caspase-3 inhibitors. Because an NAD^+-sensitive pathway is thought to be involved in Wallerian degeneration (Araki et al. 2004), caspase-mediated and NAD^+-sensitive pathways may act separately or in concert to control axonal degeneration. BAX acts upstream of both NAD^+ and caspase pathways in DRG culture, and is required to eliminate misprojecting sensory axons in Plexin-A3/A4 double-knockout mice (Schoenmann et al. 2010).

In *Xenopus,* caspase-3 is required for the growth cones in cultured retinal cells to respond to the chemotropic molecules Netrin-1- or lysophosphatidic acid (LPA) (Campbell and Holt 2003).

During sperm differentiation in *Drosophila,* 64 haploid spermatids of each cyst are connected by cytoplasmic bridges. To form individual sperm, these bridges are removed, and most of the cytoplasm is expelled. This process is called "individualization." Caspases are activated in the individualization complex, and their spatially dependent activation is necessary to complete this process (Arama et al. 2003, 2006, 2007; Huh et al. 2004b; Kaplan et al. 2010).

Regulation of Endogenous Caspase Inhibitors

Three *Drosophila* genes, *reaper (rpr), head involution defective (hid),* and *grim,* are key regulators of apoptosis during *Drosophila* embryogenesis (White et al. 1994). The overexpression of one of the three genes induces apoptosis via a caspase-activation pathway. Each of these proteins binds to diap1, an endogenous caspase inhibitor, through interactions mediated by their NH_2 terminus. Genetic and biochemical data indicate that one way these proteins promote apoptosis is by inhibiting diap1's ability to prevent caspase activity (Gyrd-Hansen and Meier 2010; Sandu et al. 2010). The mammalian mitochondrial proteins Smac/DIABLO and HtrA2/Omi, whose truncated NH_2 terminus shares similarity with Rpr, Hid, and Grim, similarly inhibit the anti-apoptotic function of XIAP and enhance caspase activation (Gyrd-Hansen and Meier 2010).

A genetic modifier screen for genes that regulate caspase activation identified *Drosophila* IKK-related kinase (DmIKKε/IK2) as a diap1-degrading kinase. DmIKKε is a homolog of the noncanonical IκB kinases (IKKε/IKKi or NAK/T2K/TBK1), which regulate NF-κB activation or interferon regulatory factor (IRF)-3 and -7 activation in mammals (Kawai and Akira 2006). A low-level caspase activity is detected in the *scabrous*-expressing proneural clusters in wild-type wing discs (Kanuka et al. 2005). Similar to the phenotype of *p35* or *dronc-DN* overexpression in proneural clusters, *DmIKKε* knockdown leads to the generation of one extra bristle in the scutellar region (Kuranaga et al. 2006). Thus, DmIKKε appears to control the level of diap1, which determines the nonapoptotic caspase activity required for defining the number of bristles as two in the proneural cluster of the scutellar region (Kuranaga et al. 2006). In addition, diap2 controls the basal caspase-activity level for both apoptotic and nonapoptotic functions (Ribeiro et al. 2007).

Temporal Caspase Activation

A transient increase in caspase-3 activity is observed during neural-cell differentiation from neurospheres or PC12 cells (Rohn et al. 2004; Fernando et al. 2005). Caspase-3 and caspase-8 are transiently activated after the transduction of iPSC-inducing transcription factors, and caspase-3 or caspase-8 inhibition in human fibroblast cells prevents the iPSC induction, suggesting that transient, nonapoptotic caspase activities have critical roles in cell-fate reprogramming (Li et al. 2010b).

The diap1 expression is a critical regulator for the temporal caspase activation (Koto et al. 2009). To investigate the spatial and temporal pattern of diap1 protein expression in vivo, a reporter protein for monitoring diap1 turnover, PRAP (pre-apoptosis signal-detecting probe, based on diap1 degradation) was developed. Live-imaging analysis of PRAP revealed that diap1 degradation is regulated in a cell lineage- and stage-specific manner during *Drosophila* sensory organ development. diap1 executes two distinct functions, one in shaft-cell survival just after the final division, and one in shaft-cell morphogenesis during its maturation stage. Both functions are exerted through the regulation of caspase activity. For the maturation stage, nonapoptotic dronc activity is required and regulated by DmIKKε. Thus, precise temporal control of the endogenous caspase inhibitor diap1 is critical for maintaining a balance between cell viability and the execution of caspases' nonapoptotic functions.

CONCLUDING REMARKS

Caspase activation that occurs on intrinsic and extrinsic stress during development and tissue regeneration supports the robust cell-fate determination of the cell community through both cell-autonomous mechanisms, such as execution of cell death or nonapoptotic functions. Caspase-mediated cell death also participates in development through non-cell-autonomous mechanisms, such as secretion of signaling molecules or force generation for neighboring cells. Thus, caspases actively participate in animal development as well as tissue regeneration.

ACKNOWLEDGMENTS

We apologize to colleagues whose work could not be cited owing to space limitations. I deeply thank the members of our laboratory for stimulating discussion, especially Y. Yamaguchi for critical reading.

REFERENCES

Abbott MK, Lengyel JA. 1991. Embryonic head involution and rotation of male terminalia require the *Drosophila* locus head involution defective. *Genetics* **129**: 783–789.

Adachi M, Suematsu S, Kondo T, Ogasawara J, Tanaka T, Yoshida N, Nagata S. 1995. Targeted mutation in the Fas gene causes hyperplasia in peripheral lymphoid organs and liver. *Nat Genet* **11**: 294–300.

Adachi-Yamada T, Fujimura-Kamada K, Nishida Y, Matsumoto K. 1999. Distortion of proximodistal information causes JNK-dependent apoptosis in *Drosophila* wing. *Nature* **400**: 166–169.

Agrawal N, Kango M, Mishra A, Sinha P. 1995. Neoplastic transformation and aberrant cell-cell interactions in genetic mosaics of lethal(2)giant larvae (lgl), a tumor suppressor gene of *Drosophila*. *Dev Biol* **172**: 218–229.

Akhter A, Gavrilin MA, Frantz L, Washington S, Ditty C, Limoli D, Day C, Sarkar A, Newland C, Butchar J, et al. 2009. Caspase-7 activation by the Nlrc4/Ipaf inflammasome restricts Legionella pneumophila infection. *PLoS Pathog* **5**: e1000361.

Akimenko MA, Mari-Beffa M, Becerra J, Geraudie J. 2003. Old questions, new tools, and some answers to the mystery of fin regeneration. *Dev Dyn* **226**: 190–201.

Al-Mashat HA, Kandru S, Liu R, Behl Y, Desta T, Graves DT. 2006. Diabetes enhances mRNA levels of proapoptotic genes and caspase activity, which contribute to impaired healing. *Diabetes* **55**: 487–495.

Araki T, Sasaki Y, Milbrandt J. 2004. Increased nuclear NAD biosynthesis and SIRT1 activation prevent axonal degeneration. *Science* **305**: 1010–1013.

Arama E, Agapite J, Steller H. 2003. Caspase activity and a specific cytochrome C are required for sperm differentiation in *Drosophila*. *Dev Cell* **4**: 687–697.

Arama E, Bader M, Srivastava M, Bergmann A, Steller H. 2006. The two *Drosophila* cytochrome C proteins can function in both respiration and caspase activation. *EMBO J* **25**: 232–243.

Arama E, Bader M, Rieckhof GE, Steller H. 2007. A ubiquitin ligase complex regulates caspase activation during sperm differentiation in *Drosophila*. *PLoS Biol* **5**: e251.

Atsumi G, Tajima M, Hadano A, Nakatani Y, Murakami M, Kudo I. 1998. Fas-induced arachidonic acid release is mediated by Ca^{2+}-independent phospholipase A2 but not cytosolic phospholipase A2, which undergoes proteolytic inactivation. *J Biol Chem* **273**: 13870–13877.

Awasaki T, Tatsumi R, Takahashi K, Arai K, Nakanishi Y, Ueda R, Ito K. 2006. Essential role of the apoptotic cell engulfment genes draper and ced-6 in programmed axon pruning during *Drosophila* metamorphosis. *Neuron* **50**: 855–867.

Baehrecke EH. 2000. Steroid regulation of programmed cell death during *Drosophila* development. *Cell Death Differ* **7**: 1057–1062.

Baker NE. 2011. Cell competition. *Curr Biol* **21**: R11–15.

Benitez S, Sosa C, Tomasini N, Macias A. 2010. Both JNK and apoptosis pathways regulate growth and terminalia rotation during *Drosophila* genital disc development. *Int J Dev Biol* **54**: 643–653.

Bergantinos C, Corominas M, Serras F. 2010. Cell death-induced regeneration in wing imaginal discs requires JNK signalling. *Development* **137**: 1169–1179.

Berry DL, Baehrecke EH. 2007. Growth arrest and autophagy are required for salivary gland cell degradation in *Drosophila*. *Cell* **131**: 1137–1148.

Boldin MP, Goncharov TM, Goltsev YV, Wallach D. 1996. Involvement of MACH, a novel MORT1/FADD-interacting protease, in Fas/APO-1- and TNF receptor-induced cell death. *Cell* **85**: 803–815.

Brumby AM, Richardson HE. 2003. Scribble mutants cooperate with oncogenic Ras or Notch to cause neoplastic overgrowth in *Drosophila*. *EMBO J* **22**: 5769–5779.

Buss RR, Sun W, Oppenheim RW. 2006. Adaptive roles of programmed cell death during nervous system development. *Annu Rev Neurosci* **29**: 1–35.

Campbell DS, Holt CE. 2003. Apoptotic pathway and MAPKs differentially regulate chemotropic responses of retinal growth cones. *Neuron* **37:** 939–952.

Cande C, Cecconi F, Dessen P, Kroemer G. 2002. Apoptosis-inducing factor (AIF): Key to the conserved caspase-independent pathways of cell death? *J Cell Sci* **115**(Pt 24): 4727–4734.

Cecconi F, Alvarez-Bolado G, Meyer BI, Roth KA, Gruss P. 1998. Apaf1 (CED-4 homolog) regulates programmed cell death in mammalian development. *Cell* **94:** 727–737.

Chautan M, Chazal G, Cecconi F, Gruss P, Golstein P. 1999. Interdigital cell death can occur through a necrotic and caspase-independent pathway. *Curr Biol* **9:** 967–970.

Chera S, Ghila L, Dobretz K, Wenger Y, Bauer C, Buzgariu W, Martinou JC, Galliot B. 2009. Apoptotic cells provide an unexpected source of Wnt3 signaling to drive hydra head regeneration. *Dev Cell* **17:** 279–289.

Chew SK, Akdemir F, Chen P, Lu WJ, Mills K, Daish T, Kumar S, Rodriguez A, Abrams JM. 2004. The apical caspase dronc governs programmed and unprogrammed cell death in *Drosophila*. *Dev Cell* **7:** 897–907.

Cho YS, Challa S, Moquin D, Genga R, Ray TD, Guildford M, Chan FK. 2009. Phosphorylation-driven assembly of the RIP1-RIP3 complex regulates programmed necrosis and virus-induced inflammation. *Cell* **137:** 1112–1123.

Christofferson DE, Yuan J. 2010. Necroptosis as an alternative form of programmed cell death. *Curr Opin Cell Biol* **22:** 263–268.

Cretney E, Takeda K, Yagita H, Glaccum M, Peschon JJ, Smyth MJ. 2002. Increased susceptibility to tumor initiation and metastasis in TNF-related apoptosis-inducing ligand-deficient mice. *J Immunol* **168:** 1356–1361.

Daish TJ, Mills K, Kumar S. 2004. *Drosophila* caspase DRONC is required for specific developmental cell death pathways and stress-induced apoptosis. *Dev Cell* **7:** 909–915.

Degterev A, Huang Z, Boyce M, Li Y, Jagtap P, Mizushima N, Cuny GD, Mitchison TJ, Moskowitz MA, Yuan J. 2005. Chemical inhibitor of nonapoptotic cell death with therapeutic potential for ischemic brain injury. *Nat Chem Biol* **1:** 112–119.

de la Cova C, Abril M, Bellosta P, Gallant P, Johnston LA. 2004. *Drosophila* myc regulates organ size by inducing cell competition. *Cell* **117:** 107–116.

Denton D, Shravage B, Simin R, Mills K, Berry DL, Baehrecke EH, Kumar S. 2009. Autophagy, not apoptosis, is essential for midgut cell death in *Drosophila*. *Curr Biol* **19:** 1741–1746.

Ellis HM, Horvitz HR. 1986. Genetic control of programmed cell death in the nematode *C. elegans*. *Cell* **44:** 817–829.

Fan Y, Bergmann A. 2008. Distinct mechanisms of apoptosis-induced compensatory proliferation in proliferating and differentiating tissues in the *Drosophila* eye. *Dev Cell* **14:** 399–410.

Fan Y, Bergmann A. 2009. The cleaved-Caspase-3 antibody is a marker of Caspase-9-like DRONC activity in *Drosophila*. *Cell Death Differ* **17:** 534–539.

Feher I, Gidali J. 1974. Prostaglandin E2 as stimulator of haemopoietic stem cell proliferation. *Nature* **247:** 550–551.

Feng S, Yang Y, Mei Y, Ma L, Zhu DE, Hoti N, Castanares M, Wu M. 2007. Cleavage of RIP3 inactivates its caspase-independent apoptosis pathway by removal of kinase domain. *Cell Signal* **19:** 2056–2067.

Fernando P, Brunette S, Megeney LA. 2005. Neural stem cell differentiation is dependent upon endogenous caspase 3 activity. *FASEB J* **19:** 1671–1673.

Fraser AG, Evan GI. 1997. Identification of a *Drosophila melanogaster* ICE/CED-3-related protease, drICE. *EMBO J* **16:** 2805–2813.

Goessling W, North TE, Loewer S, Lord AM, Lee S, Stoick-Cooper CL, Weidinger G, Puder M, Daley GQ, Moon RT, et al. 2009. Genetic interaction of PGE2 and Wnt signaling regulates developmental specification of stem cells and regeneration. *Cell* **136:** 1136–1147.

Gyrd-Hansen M, Meier P. 2010. IAPs: From caspase inhibitors to modulators of NF-κB, inflammation and cancer. *Nat Rev Cancer* **10:** 561–574.

Hadorn E, Buck D. 1962. Uber entwicklungsleistungen transplantierter teilstucke von flugel-imaginalscheiben von *Drosophila* melanogaster. *Rev Suisse Zool* **69:** 302–327.

Hadorn E, Hurlimann R, Mindek G, Schubiger G, Staub M. 1968. Developmental capacity of embryonalblastema in *Drosophila* following cultivation in adult host. *Rev Suisse Zool* **75:** 557–569.

Haigo SL, Bilder D. 2011. Global tissue revolutions in a morphogenetic movement controlling elongation. *Science* **331:** 1071–1074.

Hakem R, Hakem A, Duncan GS, Henderson JT, Woo M, Soengas MS, Elia A, de la Pompa JL, Kagi D, Khoo W, et al. 1998. Differential requirement for caspase 9 in apoptotic pathways in vivo. *Cell* **94:** 339–352.

Huh JR, Guo M, Hay BA. 2004a. Compensatory proliferation induced by cell death in the *Drosophila* wing disc requires activity of the apical cell death caspase Dronc in a nonapoptotic role. *Curr Biol* **14:** 1262–1266.

Huh JR, Vernooy SY, Yu H, Yan N, Shi Y, Guo M, Hay BA. 2004b. Multiple apoptotic caspase cascades are required in nonapoptotic roles for *Drosophila* spermatid individualization. *PLoS Biol* **2:** E15.

Hutson MS, Tokutake Y, Chang MS, Bloor JW, Venakides S, Kiehart DP, Edwards GS. 2003. Forces for morphogenesis investigated with laser microsurgery and quantitative modeling. *Science* **300:** 145–149.

Hwang JS, Kobayashi C, Agata K, Ikeo K, Gojobori T. 2004. Detection of apoptosis during planarian regeneration by the expression of apoptosis-related genes and TUNEL assay. *Gene* **333:** 15–25.

Igaki T, Pagliarini RA, Xu T. 2006. Loss of cell polarity drives tumor growth and invasion through JNK activation in *Drosophila*. *Curr Biol* **16:** 1139–1146.

Igaki T, Pastor-Pareja JC, Aonuma H, Miura M, Xu T. 2009. Intrinsic tumor suppression and epithelial maintenance by endocytic activation of Eiger/TNF signaling in *Drosophila*. *Dev Cell* **16:** 458–465.

Johnson SL, Weston JA. 1995. Temperature-sensitive mutations that cause stage-specific defects in Zebrafish fin regeneration. *Genetics* **141:** 1583–1595.

Johnston LA. 2009. Competitive interactions between cells: Death, growth, and geography. *Science* **324:** 1679–1682.

Kaiser WJ, Upton JW, Long AB, Livingston-Rosanoff D, Daley-Bauer LP, Hakem R, Caspary T, Mocarski ES. 2011. RIP3 mediates the embryonic lethality of caspase-8-deficient mice. *Nature* **471:** 368–372.

Kanda H, Igaki T, Okano H, Miura M. 2011. Conserved metabolic energy production pathways govern Eiger/TNF-induced nonapoptotic cell death. *Proc Natl Acad Sci* **108:** 18977–18982.

Kang TB, Ben-Moshe T, Varfolomeev EE, Pewzner-Jung Y, Yogev N, Jurewicz A, Waisman A, Brenner O, Haffner R, Gustafsson E, et al. 2004. Caspase-8 serves both apoptotic and nonapoptotic roles. *J Immunol* **173:** 2976–2984.

Kanuka H, Kuranaga E, Takemoto K, Hiratou T, Okano H, Miura M. 2005. *Drosophila* caspase transduces Shaggy/GSK-3β kinase activity in neural precursor development. *EMBO J* **24:** 3793–3806.

Kaplan Y, Gibbs-Bar L, Kalifa Y, Feinstein-Rotkopf Y, Arama E. 2010. Gradients of a ubiquitin E3 ligase inhibitor and a caspase inhibitor determine differentiation or death in spermatids. *Dev Cell* **19:** 160–173.

Kawai T, Akira S. 2006. Innate immune recognition of viral infection. *Nat Immunol* **7:** 131–137.

Keller LC, Cheng L, Locke CJ, Muller M, Fetter RD, Davis GW. 2011. Glial-derived prodegenerative signaling in the *Drosophila* neuromuscular system. *Neuron* **72:** 760–775.

Kondo S, Senoo-Matsuda N, Hiromi Y, Miura M. 2006. DRONC coordinates cell death and compensatory proliferation. *Mol Cell Biol* **26:** 7258–7268.

Koto A, Kuranaga E, Miura M. 2009. Temporal regulation of *Drosophila* IAP1 determines caspase functions in sensory organ development. *J Cell Biol* **187:** 219–231.

Koto A, Kuranaga E, Miura M. 2011. Apoptosis ensures spacing pattern formation of *Drosophila* sensory organs. *Curr Biol* **21:** 278–287.

Kuhrer I, Kuzmits R, Linkesch W, Ludwig H. 1986. Topical PGE2 enhances healing of chemotherapy-associated mucosal lesions. *Lancet* **1:** 623.

Kuida K, Zheng TS, Na S, Kuan C, Yang D, Karasuyama H, Rakic P, Flavell RA. 1996. Decreased apoptosis in the brain and premature lethality in CPP32-deficient mice. *Nature* **384:** 368–372.

Kuida K, Haydar TF, Kuan CY, Gu Y, Taya C, Karasuyama H, Su MS, Rakic P, Flavell RA. 1998. Reduced apoptosis and cytochrome c-mediated caspase activation in mice lacking caspase 9. *Cell* **94:** 325–337.

Kumar S, Doumanis J. 2000. The fly caspases. *Cell Death Differ* **7:** 1039–1044.

Kuo CT, Zhu S, Younger S, Jan LY, Jan YN. 2006. Identification of E2/E3 ubiquitinating enzymes and caspase activity regulating *Drosophila* sensory neuron dendrite pruning. *Neuron* **51:** 283–290.

Kuranaga E, Miura M. 2007. Nonapoptotic functions of caspases: caspases as regulatory molecules for immunity and cell-fate determination. *Trends Cell Biol* **17:** 135–144.

Kuranaga E, Kanuka H, Tonoki A, Takemoto K, Tomioka T, Kobayashi M, Hayashi S, Miura M. 2006. *Drosophila* IKK-related kinase regulates nonapoptotic function of caspases via degradation of IAPs. *Cell* **126:** 583–596.

Kuranaga E, Matsunuma T, Kanuka H, Takemoto K, Koto A, Kimura K, Miura M. 2011. Apoptosis controls the speed of looping morphogenesis in *Drosophila* male terminalia. *Development* **138:** 1493–1499.

Lakhani SA, Masud A, Kuida K, Porter GA Jr, Booth CJ, Mehal WZ, Inayat I, Flavell RA. 2006. Caspases 3 and 7: Key mediators of mitochondrial events of apoptosis. *Science* **311:** 847–851.

Lamkanfi M, Kanneganti TD, Van Damme P, Vanden Berghe T, Vanoverberghe I, Vandekerckhove J, Vandenabeele P, Gevaert K, Nunez G. 2008. Targeted peptidecentric proteomics reveals caspase-7 as a substrate of the caspase-1 inflammasomes. *Mol Cell Proteomics* **7:** 2350–2363.

Lee HH, Jan LY, Jan YN. 2009a. *Drosophila* IKK-related kinase Ik2 and Katanin p60-like 1 regulate dendrite pruning of sensory neuron during metamorphosis. *Proc Natl Acad Sci* **106:** 6363–6368.

Lee P, Lee DJ, Chan C, Chen SW, Ch'en I, Jamora C. 2009b. Dynamic expression of epidermal caspase 8 simulates a wound healing response. *Nature* **458:** 519–523.

Leonard JR, Klocke BJ, D'Sa C, Flavell RA, Roth KA. 2002. Strain-dependent neurodevelopmental abnormalities in caspase-3-deficient mice. *J Neuropathol Exp Neurol* **61:** 673–677.

Leulier F, Ribeiro PS, Palmer E, Tenev T, Takahashi K, Robertson D, Zachariou A, Pichaud F, Ueda R, Meier P. 2006. Systematic in vivo RNAi analysis of putative components of the *Drosophila* cell death machinery. *Cell Death Differ* **13:** 1663–1674.

Li W, Baker NE. 2007. Engulfment is required for cell competition. *Cell* **129:** 1215–1225.

Li C, Lasse S, Lee P, Nakasaki M, Chen SW, Yamasaki K, Gallo RL, Jamora C. 2010a. Development of atopic dermatitis-like skin disease from the chronic loss of epidermal caspase-8. *Proc Natl Acad Sci* **107:** 22249–22254.

Li F, He Z, Shen J, Huang Q, Li W, Liu X, He Y, Wolf F, Li CY. 2010b. Apoptotic caspases regulate induction of iPSCs from human fibroblasts. *Cell Stem Cell* **7:** 508–520.

Li F, Huang Q, Chen J, Peng Y, Roop DR, Bedford JS, Li CY. 2010c. Apoptotic cells activate the "phoenix rising" pathway to promote wound healing and tissue regeneration. *Sci Signal* **3:** ra13.

Lindsten T, Thompson CB. 2006. Cell death in the absence of Bax and Bak. *Cell Death Differ* **13:** 1272–1276.

Lindsten T, Ross AJ, King A, Zong WX, Rathmell JC, Shiels HA, Ulrich E, Waymire KG, Mahar P, Frauwirth K, et al. 2000. The combined functions of proapoptotic Bcl-2 family members bak and bax are essential for normal development of multiple tissues. *Mol Cell* **6:** 1389–1399.

Liou JY, Ellent DP, Lee S, Goldsby J, Ko BS, Matijevic N, Huang JC, Wu KK. 2007. Cyclooxygenase-2-derived prostaglandin e2 protects mouse embryonic stem cells from apoptosis. *Stem Cells* **25:** 1096–1103.

Liu J, Na S, Glasebrook A, Fox N, Solenberg PJ, Zhang Q, Song HY, Yang DD. 2001. Enhanced $CD4^+$ T cell proliferation and Th2 cytokine production in DR6-deficient mice. *Immunity* **15:** 23–34.

Lohmann I, McGinnis N, Bodmer M, McGinnis W. 2002. The *Drosophila* Hox gene deformed sculpts head morphology via direct regulation of the apoptosis activator reaper. *Cell* **110:** 457–466.

Macias A, Romero NM, Martin F, Suarez L, Rosa AL, Morata G. 2004. PVF1/PVR signaling and apoptosis promotes the rotation and dorsal closure of the *Drosophila* male terminalia. *Int J Dev Biol* **48:** 1087–1094.

Madhavan MM, Madhavan K. 1980. Morphogenesis of the epidermis of adult abdomen of *Drosophila*. *J Embryol Exp Morphol* **60:** 1–31.

Manjon C, Sanchez-Herrero E, Suzanne M. 2007. Sharp boundaries of Dpp signalling trigger local cell death required for *Drosophila* leg morphogenesis. *Nat Cell Biol* **9:** 57–63.

Martin FA., Herrera SC., Morata G. 2009. Cell competition, growth and size control in the Drosophila wing imaginal disc. *Development* **136:** 3747–3756.

Meier P, Silke J, Leevers SJ, Evan GI. 2000. The *Drosophila* caspase DRONC is regulated by DIAP1. *EMBO J* **19:** 598–611.

Mesquita D, Dekanty A, Milan M. 2010. A dp53-dependent mechanism involved in coordinating tissue growth in *Drosophila*. *PLoS Biol* **8** 1000566.

Milan M, Campuzano S, Garcia-Bellido A. 1997. Developmental parameters of cell death in the wing disc of *Drosophila*. *Proc Natl Acad Sci* **94:** 5691–5696.

Miura M. 2011. Active participation of cell death in development and organismal homeostasis. *Dev Growth Differ* **53:** 125–136.

Morata G, Ripoll P. 1975. Minutes: Mutants of *Drosophila* autonomously affecting cell division rate. *Dev Biol* **42:** 211–221.

Moreno E, Basler K. 2004. dMyc transforms cells into supercompetitors. *Cell* **117:** 117–129.

Moreno E, Basler K, Morata G. 2002. Cells compete for decapentaplegic survival factor to prevent apoptosis in *Drosophila* wing development. *Nature* **416:** 755–759.

Muro I, Berry DL, Huh JR, Chen CH, Huang H, Yoo SJ, Guo M, Baehrecke EH, Hay BA. 2006. The *Drosophila* caspase Ice is important for many apoptotic cell deaths and for spermatid individualization, a nonapoptotic process. *Development* **133:** 3305–3315.

Muzio M, Chinnaiyan AM, Kischkel FC, O'Rourke K, Shevchenko A, Ni J, Scaffidi C, Bretz JD, Zhang M, Gentz R, et al. 1996. FLICE, a novel FADD-homologous ICE/CED-3-like protease, is recruited to the CD95 (Fas/APO-1) death–inducing signaling complex. *Cell* **85:** 817–827.

Nagasaka A, Kawane K, Yoshida H, Nagata S. 2010. Apaf-1-independent programmed cell death in mouse development. *Cell Death Differ* **17:** 931–941.

Nakajima Y, Kuranaga E, Sugimura K, Miyawaki A, Miura M. 2011. Nonautonomous apoptosis is triggered by local cell cycle progression during epithelial replacement in *Drosophila*. *Mol Cell Biol* **31:** 2499–2512.

Nikolaev A, McLaughlin T, O'Leary DD, Tessier-Lavigne M. 2009. APP binds DR6 to trigger axon pruning and neuron death via distinct caspases. *Nature* **457:** 981–989.

Ninov N, Chiarelli DA, Martin-Blanco E. 2007. Extrinsic and intrinsic mechanisms directing epithelial cell sheet replacement during *Drosophila* metamorphosis. *Development* **134:** 367–379.

Ninov N, Menezes-Cabral S, Prat-Rojo C, Manjon C, Weiss A, Pyrowolakis G, Affolter M, Martin-Blanco E. 2010. Dpp signaling directs cell motility and invasiveness during epithelial morphogenesis. *Curr Biol* **20:** 513–520.

North TE, Goessling W, Walkley CR, Lengerke C, Kopani KR, Lord AM, Weber GJ, Bowman TV, Jang IH, Grosser T, et al. 2007. Prostaglandin E2 regulates vertebrate haematopoietic stem cell homeostasis. *Nature* **447:** 1007–1011.

Oberst A, Dillon CP, Weinlich R, McCormick LL, Fitzgerald P, Pop C, Hakem R, Salvesen GS, Green DR. 2011. Catalytic activity of the caspase-8-FLIP(L) complex inhibits RIPK3-dependent necrosis. *Nature* **471:** 363–367.

Ohsawa S, Hamada S, Asou H, Kuida K, Uchiyama Y, Yoshida H, Miura M. 2009. Caspase-9 activation revealed by semaphorin 7A cleavage is independent of apoptosis in the aged olfactory bulb. *J Neurosci* **29:** 11385–11392.

Ohsawa S, Hamada S, Kuida K, Yoshida H, Igaki T, Miura M. 2010. Maturation of the olfactory sensory neurons by Apaf-1/caspase-9-mediated caspase activity. *Proc Natl Acad Sci* **107:** 13366–13371.

Ohsawa S, Sugimura K, Takino K, Xu T, Miyawaki A, Igaki T. 2011. Elimination of oncogenic neighbors by JNK-mediated engulfment in *Drosophila*. *Dev Cell* **20:** 315–328.

Paralkar VM, Borovecki F, Ke HZ, Cameron KO, Lefker B, Grasser WA, Owen TA, Li M, DaSilva-Jardine P, Zhou M, et al. 2003. An EP2 receptor-selective prostaglandin E2 agonist induces bone healing. *Proc Natl Acad Sci* **100:** 6736–6740.

Pellettieri J, Fitzgerald P, Watanabe S, Mancuso J, Green DR, Sanchez Alvarado A. 2010. Cell death and tissue remodeling in planarian regeneration. *Dev Biol* **338:** 76–85.

Perez-Garijo A, Martin FA, Morata G. 2004. Caspase inhibition during apoptosis causes abnormal signalling and developmental aberrations in *Drosophila*. *Development* **131:** 5591–5598.

Perez-Garijo A, Martin FA, Struhl G, Morata G. 2005. Dpp signaling and the induction of neoplastic tumors by caspase-inhibited apoptotic cells in *Drosophila*. *Proc Natl Acad Sci* **102:** 17664–17669.

Pfeffer K, Matsuyama T, Kundig TM, Wakeham A, Kishihara K, Shahinian A, Wiegmann K, Ohashi PS, Kronke M, Mak TW. 1993. Mice deficient for the 55 kd tumor necrosis factor receptor are resistant to endotoxic shock, yet succumb to *L. monocytogenes* infection. *Cell* **73:** 457–467.

Poss KD. 2007. Getting to the heart of regeneration in zebrafish. *Semin Cell Dev Biol* **18:** 36–45.

Rebe C, Cathelin S, Launay S, Filomenko R, Prevotat L, L'Ollivier C, Gyan E, Micheau O, Grant S, Dubart-Kupperschmitt A, et al. 2007. Caspase-8 prevents sustained activation of NF-κB in monocytes undergoing macrophagic differentiation. *Blood* **109:** 1442–1450.

Ren D, Tu HC, Kim H, Wang GX, Bean GR, Takeuchi O, Jeffers JR, Zambetti GP, Hsieh JJ, Cheng EH. 2010. BID, BIM, and PUMA are essential for activation of the BAX-

and BAK-dependent cell death program. *Science* **330**: 1390–1393.

Ribeiro PS, Kuranaga E, Tenev T, Leulier F, Miura M, Meier P. 2007. DIAP2 functions as a mechanism-based regulator of drICE that contributes to the caspase activity threshold in living cells. *J Cell Biol* **179**: 1467–1480.

Rohn TT, Cusack SM, Kessinger SR, Oxford JT. 2004. Caspase activation independent of cell death is required for proper cell dispersal and correct morphology in PC12 cells. *Exp Cell Res* **295**: 215–225.

Rothe J, Lesslauer W, Lotscher H, Lang Y, Koebel P, Kontgen F, Althage A, Zinkernagel R, Steinmetz M, Bluethmann H. 1993. Mice lacking the tumour necrosis factor receptor 1 are resistant to TNF-mediated toxicity but highly susceptible to infection by Listeria monocytogenes. *Nature* **364**: 798–802.

Ryoo HD, Gorenc T, Steller H. 2004. Apoptotic cells can induce compensatory cell proliferation through the JNK and the Wingless signaling pathways. *Dev Cell* **7**: 491–501.

Sakamaki K, Inoue T, Asano M, Sudo K, Kazama H, Sakagami J, Sakata S, Ozaki M, Nakamura S, Toyokuni S, et al. 2002. Ex vivo whole-embryo culture of caspase-8-deficient embryos normalize their aberrant phenotypes in the developing neural tube and heart. *Cell Death Differ* **9**: 1196–1206.

Sandu C, Ryoo HD, Steller H. 2010. Drosophila IAP antagonists form multimeric complexes to promote cell death. *J Cell Biol* **190**: 1039–1052.

Schoenmann Z, Assa-Kunik E, Tiomny S, Minis A, Haklai-Topper L, Arama E, Yaron A. 2010. Axonal degeneration is regulated by the apoptotic machinery or a NAD$^+$-sensitive pathway in insects and mammals. *J Neurosci* **30**: 6375–6386.

Simpson P. 1979. Parameters of cell competition in the compartments of the wing disc of *Drosophila*. *Dev Biol* **69**: 182–193.

Simpson P, Morata G. 1981. Differential mitotic rates and patterns of growth in compartments in the *Drosophila* wing. *Dev Biol* **85**: 299–308.

Smith-Bolton RK, Worley MI, Kanda H, Hariharan IK. 2009. Regenerative growth in *Drosophila* imaginal discs is regulated by Wingless and Myc. *Dev Cell* **16**: 797–809.

Sordet O, Rebe C, Plenchette S, Zermati Y, Hermine O, Vainchenker W, Garrido C, Solary E, Dubrez-Daloz L. 2002. Specific involvement of caspases in the differentiation of monocytes into macrophages. *Blood* **100**: 4446–4453.

Strasser A, Puthalakath H, O'Reilly LA, Bouillet P. 2008. What do we know about the mechanisms of elimination of autoreactive T and B cells and what challenges remain. *Immunol Cell Biol* **86**: 57–66.

Suzanne M, Petzoldt AG, Speder P, Coutelis JB, Steller H, Noselli S. 2010. Coupling of apoptosis and L/R patterning controls stepwise organ looping. *Curr Biol* **20**: 1773–1778.

Takemoto K, Kuranaga E, Tonoki A, Nagai T, Miyawaki A, Miura M. 2007. Local initiation of caspase activation in Drosophila salivary gland programmed cell death in vivo. *Proc Natl Acad Sci* **104**: 13367–13372.

Teixeira VH, Jacq L, Lasbleiz S, Hilliquin P, Oliveira CR, Cornelis F, Petit-Teixeira E. 2008. Genetic and expression analysis of CASP7 gene in a European Caucasian population with rheumatoid arthritis. *J Rheumatol* **35**: 1912–1918.

Toyama Y, Peralta XG, Wells AR, Kiehart DP, Edwards GS. 2008. Apoptotic force and tissue dynamics during *Drosophila* embryogenesis. *Science* **321**: 1683–1686.

Tseng AS, Adams DS, Qiu D, Koustubhan P, Levin M. 2007. Apoptosis is required during early stages of tail regeneration in *Xenopus laevis*. *Dev Biol* **301**: 62–69.

Varfolomeev EE, Schuchmann M, Luria V, Chiannilkulchai N, Beckmann JS, Mett IL, Rebrikov D, Brodianski VM, Kemper OC, Kollet O, et al. 1998. Targeted disruption of the mouse Caspase 8 gene ablates cell death induction by the TNF receptors, Fas/Apo1, and DR3 and is lethal prenatally. *Immunity* **9**: 267–276.

Vlaskalin T, Wong CJ, Tsilfidis C. 2004. Growth and apoptosis during larval forelimb development and adult forelimb regeneration in the newt (*Notophthalmus viridescens*). *Dev Genes Evol* **214**: 423–431.

Waldhuber M, Emoto K, Petritsch C. 2005. The *Drosophila* caspase DRONC is required for metamorphosis and cell death in response to irradiation and developmental signals. *Mech Dev* **122**: 914–927.

Walsh JG, Cullen SP, Sheridan C, Luthi AU, Gerner C, Martin SJ. 2008. Executioner caspase-3 and caspase-7 are functionally distinct proteases. *Proc Natl Acad Sci* **105**: 12815–12819.

Wang EC, Thern A, Denzel A, Kitson J, Farrow SN, Owen MJ. 2001. DR3 regulates negative selection during thymocyte development. *Mol Cell Biol* **21**: 3451–3461.

Wells BS, Yoshida E, Johnston LA. 2006. Compensatory proliferation in *Drosophila* imaginal discs requires Dronc-dependent p53 activity. *Curr Biol* **16**: 1606–1615.

White K, Grether ME, Abrams JM, Young L, Farrell K, Steller H. 1994. Genetic control of programmed cell death in *Drosophila*. *Science* **264**: 677–683.

Williams DW, Kondo S, Krzyzanowska A, Hiromi Y, Truman JW. 2006. Local caspase activity directs engulfment of dendrites during pruning. *Nat Neurosci* **9**: 1234–1236.

Woods DF, Bryant PJ. 1991. The discs-large tumor suppressor gene of *Drosophila* encodes a guanylate kinase homolog localized at septate junctions. *Cell* **66**: 451–464.

Xu D, Li Y, Arcaro M, Lackey M, Bergmann A. 2005. The CARD-carrying caspase Dronc is essential for most, but not all, developmental cell death in *Drosophila*. *Development* **132**: 2125–2134.

Xu D, Wang Y, Willecke R, Chen Z, Ding T, Bergmann A. 2006. The effector caspases drICE and dcp-1 have partially overlapping functions in the apoptotic pathway in *Drosophila*. *Cell Death Differ* **13**: 1697–1706.

Xu D, Woodfield SE, Lee TV, Fan Y, Antonio C, Bergmann A. 2009. Genetic control of programmed cell death (apoptosis) in *Drosophila*. *Fly (Austin)* **3**: 78–90.

Yamaguchi Y, Shinotsuka N, Nonomura K, Takemoto K, Kuida K, Yosida H, Miura M. 2011. Live imaging of apoptosis in a novel transgenic mouse highlights its role in neural tube closure. *J Cell Biol* **195**: 1047–1060.

Yeh WC, Pompa JL, McCurrach ME, Shu HB, Elia AJ, Shahinian A, Ng M, Wakeham A, Khoo W, Mitchell K, et al. 1998. FADD: Essential for embryo development and signaling from some, but not all, inducers of apoptosis. *Science* **279:** 1954–1958.

Yeh WC, Itie A, Elia AJ, Ng M, Shu HB, Wakeham A, Mirtsos C, Suzuki N, Bonnard M, Goeddel DV, et al. 2000. Requirement for Casper (c-FLIP) in regulation of death receptor-induced apoptosis and embryonic development. *Immunity* **12:** 633–642.

Yi CH, Yuan J. 2009. The Jekyll and Hyde functions of caspases. *Dev Cell* **16:** 21–34.

Yoshida H, Kong YY, Yoshida R, Elia AJ, Hakem A, Hakem R, Penninger JM, Mak TW. 1998. Apaf1 is required for mitochondrial pathways of apoptosis and brain development. *Cell* **94:** 739–750.

Zhang J, Cado D, Chen A, Kabra NH, Winoto A. 1998. Fas-mediated apoptosis and activation-induced T-cell proliferation are defective in mice lacking FADD/Mort1. *Nature* **392:** 296–300.

Zhang H, Zhou X, McQuade T, Li J, Chan FK, Zhang J. 2011. Functional complementation between FADD and RIP1 in embryos and lymphocytes. *Nature* **471:** 373–376.

Zhao X, Wang D, Zhao Z, Xiao Y, Sengupta S, Zhang R, Lauber K, Wesselborg S, Feng L, Rose TM, et al. 2006. Caspase-3-dependent activation of calcium-independent phospholipase A2 enhances cell migration in non-apoptotic ovarian cancer cells. *J Biol Chem* **281:** 29357–29368.

Zinzalla V, Stracka D, Oppliger W, Hall MN. 2011. Activation of mTORC2 by association with the ribosome. *Cell* **144:** 757–768.

Cellular Mechanisms Controlling Caspase Activation and Function

Amanda B. Parrish, Christopher D. Freel, and Sally Kornbluth

Department of Pharmacology and Cancer Biology, Duke University School of Medicine, Durham, North Carolina 27710

Correspondence: kornb001@mc.duke.edu

Caspases are the primary drivers of apoptotic cell death, cleaving cellular proteins that are critical for dismantling the dying cell. Initially translated as inactive zymogenic precursors, caspases are activated in response to a variety of cell death stimuli. In addition to factors required for their direct activation (e.g., dimerizing adaptor proteins in the case of initiator caspases that lie at the apex of apoptotic signaling cascades), caspases are regulated by a variety of cellular factors in a myriad of physiological and pathological settings. For example, caspases may be modified posttranslationally (e.g., by phosphorylation or ubiquitylation) or through interaction of modulatory factors with either the zymogenic or active form of a caspase, altering its activation and/or activity. These regulatory events may inhibit or enhance enzymatic activity or may affect activity toward particular cellular substrates. Finally, there is emerging literature to suggest that caspases can participate in a variety of cellular processes unrelated to apoptotic cell death. In these settings, it is particularly important that caspases are maintained under stringent control to avoid inadvertent cell death. It is likely that continued examination of these processes will reveal new mechanisms of caspase regulation with implications well beyond control of apoptotic cell death.

Apoptosis is a form of programmed cell death that eliminates individual cells within an organism while preserving the overall structure of surrounding tissue. Many of the prominent morphological features of apoptosis were first described in 1972 by Kerr, Wyllie, and Currie (Kerr et al. 1972). However, it was not until the mid-1990s that apoptosis was linked to the activation of the cysteine-dependent aspartate-driven proteases (caspases), which cleave key intracellular substrates to promote cell death (Cerretti et al. 1992; Nicholson et al. 1995; Alnemri et al. 1996; Liu et al. 1996; Thornberry and Lazebnik 1998). Given the critical role that caspases play in dismantling the cell during apoptosis, their activation and subsequent activity are highly regulated. Failure of a cell to properly modulate caspase activity can cause aberrant or untimely apoptotic cell death, potentially leading to carcinogenesis, autoimmunity, neurodegeneration, and immunodeficiency (Thompson 1995; Hanahan and Weinberg 2000; Yuan and Yankner 2000; Li and Yuan 2008).

Caspases are synthesized within the cell as inactive zymogens that lack significant protease activity. Thus, caspases are, in essence, regulated from the moment of protein synthesis in that they are not activated until receipt of specific death stimuli (Earnshaw et al. 1999). The primary structure of a caspase is an amino-terminal prodomain and a carboxy-terminal protease domain, which contains the key catalytic cysteine residue. Caspases are categorized as initiator or effector caspases, based on their position in apoptotic signaling cascades. The initiator caspases (caspase-2, -8, -9, and -10) act apically in cell death pathways and all share long, structurally similar prodomains. This group of enzymes is activated through "induced proximity" when adaptor proteins interact with the prodomains and promote caspase dimerization (Boatright et al. 2003; Baliga et al. 2004; Pop et al. 2006; Riedl and Salvesen 2007; Wachmann et al. 2010). In contrast, the effector caspases (caspase-3, -6, and -7) have shorter prodomains and exist in the cell as preformed, but inactive, homodimers. Following cleavage mediated by an initiator caspase, effector caspases act directly on specific cellular substrates to dismantle the cell. Although many individual caspase substrates have been implicated in specific aspects of cellular destruction (e.g., lamin cleavage is required for the efficient packaging of nuclei into small membrane-bound vesicles), recent proteomic approaches have greatly expanded the known repertoire of proteolytic products generated during apoptosis (Van Damme et al. 2005; Dix et al. 2008; Mahrus et al. 2008). Further work will be needed to confirm these findings and to determine how (or if) all of these substrates participate in the apoptotic process (see Poręba et al. 2013), especially as new details emerge on the relationship between posttranslational modifications, like phosphorylation, and caspase cleavage (Dix et al. 2012).

CASPASE ACTIVATION

Initiation of apoptosis occurs through either a cell-intrinsic or cell-extrinsic pathway. Extrinsic pathway cell death signals originate at the plasma membrane where an extracellular ligand (e.g., FasL) binds to its cell surface transmembrane "death receptor" (e.g., Fas receptor), inducing oligomerization of the receptor (Trauth et al. 1989; Itoh and Nagata 1993; Danial and Korsmeyer 2004). This, in turn, promotes clustering of proteins that bind to the intracellular domain of the receptor (e.g., FADD, or Fas-associated death domain-containing protein), which then binds to the prodomain of initiator caspases (e.g., caspase-8 or -10) to promote their dimerization and activation; these complexes are referred to as DISCs, or death-induced signaling complexes (Kischkel et al. 1995). As initiator caspases, caspases-8 and -10 are activated within the DISC through induced proximity dimerization (Boatright et al. 2003; Wachmann et al. 2010). Active caspase-8/-10 can then directly cleave and activate effector caspases, such as caspase-3. In some cell types, this pathway is sufficient to cause cell death (Type I cells). However, in other cells, caspase-8 must also engage the mitochondria as described below for the intrinsic pathway (Type II cells; Fig. 1) (Li et al. 1998; Luo et al. 1998; Scaffidi et al. 1998). Other death receptors that, like FasL–Fas receptor, participate in extrinsic apoptotic pathways include the TNF ligand–TNF-R1 complex and the DR4/5–Apo2 L/TRAIL ligand complexes.

The intrinsic pathway proceeds through the mitochondria and involves release of the respiratory chain component, cytochrome c, from the intermembrane space of the mitochondria into the cytoplasm. Cytochrome c interacts with the adaptor protein, Apaf-1 (apoptotic protease activating factor-1), to form the heptameric backbone of the apoptosome complex, which recruits and activates caspase-9 through dimerization (Liu et al. 1996; Zou et al. 1997; Acehan et al. 2002; Boatright et al. 2003). The remarkably similar phenotypes of the Apaf-1$^{-/-}$ and caspase-9$^{-/-}$ mice suggest that caspase-9 is indeed dependent on this Apaf-1-based complex for its activation (Ceccini et al. 1998; Hakem et al. 1998; Kuida et al. 1998). Recent data suggest that each apoptosome backbone recruits and activates only two caspase-9 molecules, creating a 7:2 ratio between Apaf-1 and caspase-9 within the apoptosome (Malladi et al. 2009). Active caspase-9 cleaves and activates downstream effector caspases, such as caspase-3 (Slee et al. 1999) (Fig. 1). Although caspase-9 can autocleave and can also be directly cleaved

Caspase Activation and Function

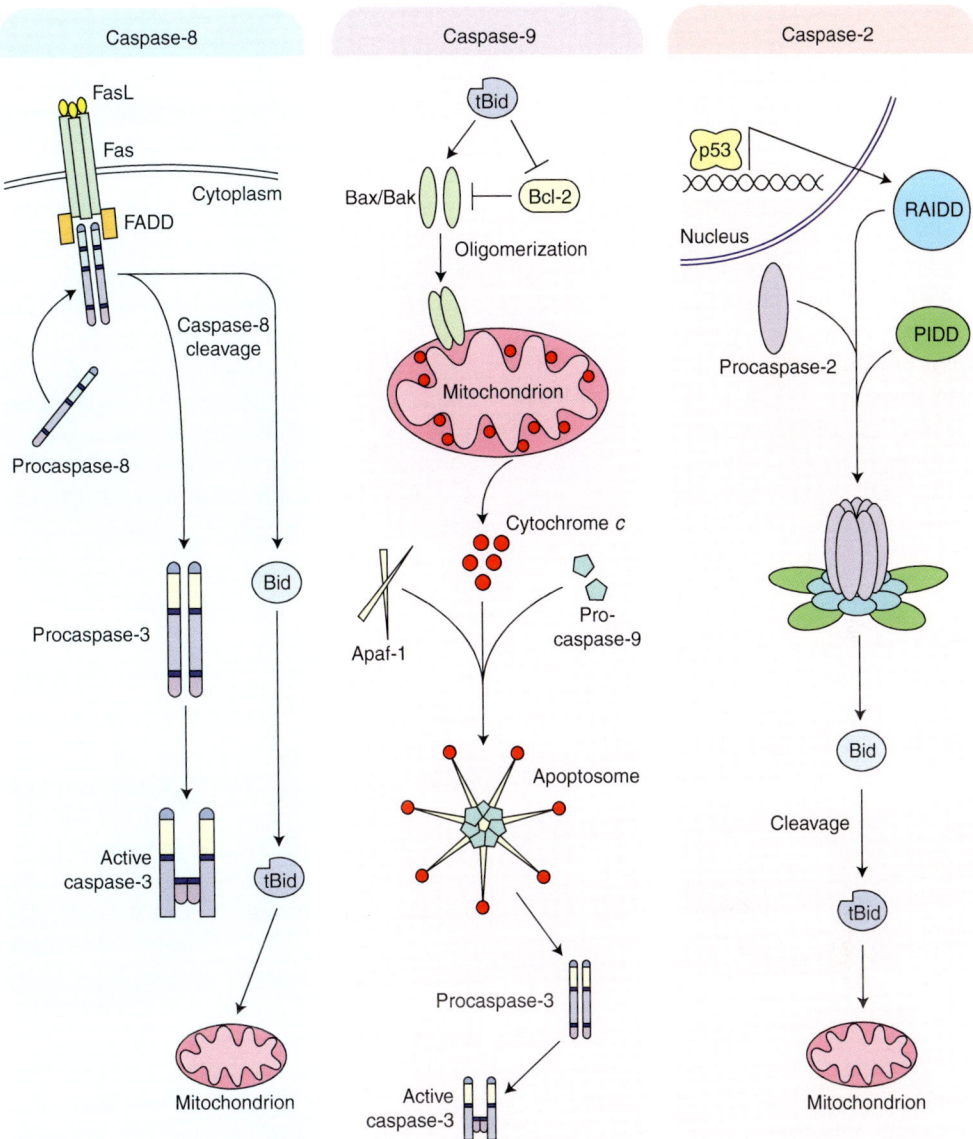

Figure 1. Activation of initiator caspases. Caspase-8: The extrinsic pathway is initiated at the plasma membrane, where a receptor (FasL) interacts with its transmembrane receptor (Fas), causing oligomerization of the receptor. The Fas receptor has an intracellular DD (death domain), which allows for the recruitment of the adaptor protein FADD from the cytoplasm through its DD. Caspase-8 is then recruited to the complex through interactions between the DED (death effector domain) on FADD and similar DED sequences on caspase-8. Active caspase-8 can directly activate caspase-3, or it can cleave Bid, which facilitates mitochondrial cytochrome *c* release. Caspase-9: Caspase-9 is activated through the mitochondria-dependent intrinsic pathway. At the mitochondria, the antiapoptotic bcl-2 family members like Bcl-2 and Bcl-xL inhibit cytochrome *c* release from the mitochondrial intermembrane space. In contrast, proapoptotic bcl-2 family members, Bax and Bak, facilitate cytochrome *c* release. BH3-only proteins help regulate the balance between the pro- and antiapoptotic bcl-2 family members. After cytochrome *c* translocates into the cytosol, it interacts with Apaf-1, which undergoes a conformational change and oligomerization into a heptameric structure known as the apoptosome. The apoptosome recruits and activates the initiator caspase, caspase-9. (*Legend continues on following page.*)

by active caspases, this modification does not appear to be necessary for its activation (Renatus et al. 2001; Boatright et al. 2003). Cleavage of caspase-9 does, however, have a variety of implications on its regulation, including the generation of a new epitope that is required for subsequent caspase-9 inhibition by XIAP. Additionally, Bratton and colleagues have suggested a model whereby cleavage of caspase-9 lowers its affinity for the apoptosome (relative to procaspase-9), promoting its replacement by new incoming procaspase-9 molecules recruited to the Apaf-1 caspase recruitment domains (CARDs) for activation (Malladi et al. 2009).

Although caspase-2 is structurally and functionally similar to the other initiator caspases, the role of caspase-2 in apoptotic processes is a bit less clear. It is activated similarly to other initiator caspases via induced proximity dimerization and autocatalytic processing within an adaptor protein complex, and it appears to function, at least under some circumstances, upstream of the mitochondria (Baliga et al. 2004). The current model for caspase-2 activation involves two adaptor proteins, PIDD (p53-induced protein with a death domain) and RAIDD (RIP-associated ICH-1/CED-3 homologous proteins with a death domain) (Duan and Dixit 1997; Read et al. 2002; Tinel and Tschopp 2004). Together, these proteins form the PIDDosome complex, which consists of five PIDDs, seven RAIDDs, and seven caspase-2 molecules (Park et al. 2007). Active caspase-2 is thought to facilitate apoptosis by cleaving the proapoptotic family member Bid to promote mitochondrial permeabilization and cytochrome c release, thereby propagating the apoptotic signal (Gao et al. 2005; Bonzon et al. 2006) (Fig. 1). Because one of its upstream activators, PIDD, is a p53 target gene, caspase-2 is thought to play a role in at least some p53-mediated cell deaths (Baptiste-Okoh et al. 2008; Sidi et al. 2008). However, caspase-2 also appears to be activated under conditions of heat shock in a p53-independent manner, which suggests that there may be PIDDosome-independent mechanisms for activating caspase-2 (Tu et al. 2006). Indeed, PIDD$^{-/-}$ mice were found to have no defects in caspase-2-initiated apoptosis following certain apoptotic stimuli, such as DNA damage and ER stress. Additionally, p53-induced caspase-2 activation following 5-FU treatment has been observed in the absence of PIDD (Vakifahmetoglu et al. 2006; Manzl et al. 2009). Conversely, cells from caspase-2 knockout mice are still susceptible to some actions of PIDD, suggesting that PIDD can use effectors other than caspase-2 to promote cell death (Berube et al. 2005).

CONSERVATION OF APOPTOTIC PATHWAYS

The caspase family of proteins is highly conserved between organisms, and, as seen in Figure 2, certain regulatory "modules" are conserved between nematodes, fruit flies, and mammals, although the precise ways in which these pathways are arranged and the individual factors involved vary in their importance/prominence in different organisms (Yan and Shi 2005). Genetic characterization of programmed cell death in the nematode identified ced-3, ced-4, ced-9, and egl-1 as central regulators of cell death in that organism (Yuan et al. 1993). EGL-1 is a Bcl-2 family member with a single block of Bcl-2 homology (a so-called BH3-only protein) that alleviates inhibition of the Apaf-1-like molecule CED-4, by CED-9, an antiapoptotic Bcl-2-like protein (Hengartner and Horvitz 1994; Conradt and Horvitz 1998). Oligomerization of the adaptor protein, CED-4 creates a high molecular weight complex that parallels those made by Apaf-1 in vertebrates (Yang et al. 1998; Qi et al.

Figure 1. (*Continued*) Active caspase-9 then directly cleaves and activates effector caspases, such as caspase-3. Caspase-2: Caspase-2 appears to function as an initiator caspase upstream of the mitochondria. The most well-understood mechanism for caspase-2 activation involves the PIDDosome. p53-dependent transcription of the adaptor protein PIDD forms the backbone of the complex with RAIDD interaction through a DD–DD interaction. Caspase-2 is recruited to RAIDD through a CARD–CARD interaction. Active caspase-2 then cleaves and activates the BH3-only protein Bid, facilitating cytochrome c release from the mitochondria.

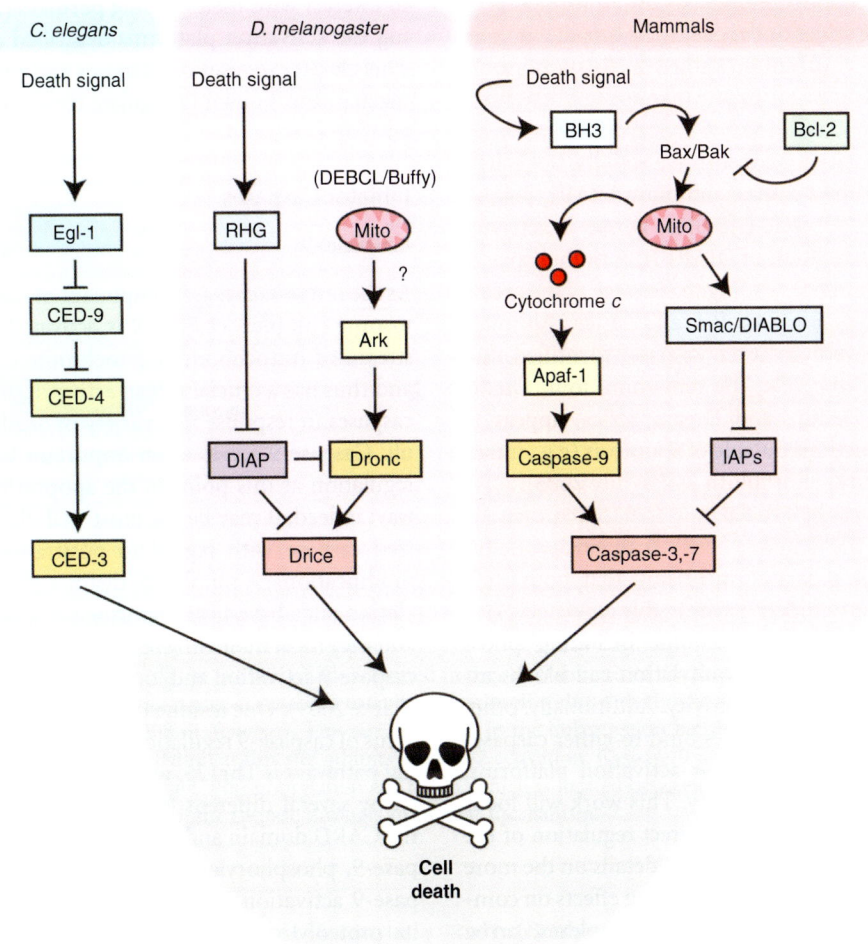

Figure 2. Similarities and differences in apoptotic machinery across evolution. Many of the core proteins of the apoptotic pathway are conserved in nematodes (*C. elegans*), fruit flies (*D. melanogaster*), and mammals, and these protein families share some functional similarities (which is noted by matching colors in the figure). As an example, Apaf-1, which forms the backbone of the caspase-9 activating apoptosome in response to cytochrome *c*, has a homolog, dApaf-1 (dArk), in the *Drosophila* pathway, which facilitates Dronc activation, although it does not appear to be activated by cytochrome *c*. In *C. elegans*, CED-4 also auto-oligomerizes to form a complex that enables activation of the CED-3 caspase. For a more complete description of each pathway, including highlights of similarities and differences between species, please see the text.

2010; Yuan et al. 2010). This CED-4 complex activates the CED-3 caspase, which functions as a prominent executioner of cell death. Although alterations in mitochondrial dynamics have been reported to modulate apoptosis in *C. elegans*, there is no evidence that the apoptosis in this organism involves mitochondrial permeabilization, and the worm apoptosome is active once EGL-1 relieves apoptosome inhibition. Similar to the mammalian apoptosome, the *C. elegans* apoptosome appears to contain two CED-3 molecules, but with a ratio of eight CED-4 to two CED-3 (Qi et al. 2010).

Drosophila cells share with their mammalian counterparts certain features of caspase activation, including activation through oligomeri-

Table 1. Caspase regulation via phosphorylation

Substrate	Kinase/phosphatase	Sites	+/−[a]	References
Initiator caspases				
Caspase-8	Src, Fyn, Lyn	Tyr380	−	Cursi et al. 2006; Senft et al. 2007; Jia et al. 2008
Caspase-8	Lyn	Tyr465	−	Jia et al. 2008
Caspase-8	p38 MAPK	Ser364	−	Alvarado-Kristensson et al. 2004
Caspase-8	CDK1	Ser387	−	Matthess et al. 2010
Caspase-8	RSK2	Thr263	+	Peng et al. 2011
Caspase-9	Erk1/2	Thr125	−	Allan et al. 2003
Caspase-9	CDK1	Thr125	−	Allan and Clarke 2007
Caspase-9	DYRK1A	Thr125	−	Seifert et al. 2008
Caspase-9	p38α	Thr125	−	Seifert and Clarke 2009
Caspase-9	PP1α	Thr125	−	Dessauge et al. 2006
Caspase-9	PKCζ	Ser144	−	Brady et al. 2005
Caspase-9	c-Abl	Tyr153	+	Raina et al. 2005
Caspase-9	PKA	Ser99, Ser183, Ser195	Unclear	Martin et al. 2005
Caspase-9	Akt	Ser196 (human)	−	Cardone et al. 1998
Caspase-9	CK2	Ser348 (mouse)	−	McDonnell et al. 2008
Caspase-2	CDK1	Ser340	−	Andersen et al. 2009
Caspase-2	CK2	Ser157	−	Shin et al. 2005
Caspase-2	CaMKII	Ser135 (*Xenopus*)	−	Nutt et al. 2005
Dronc	CaMKII	Ser130		Yang et al. 2010
Effector caspases				
Caspase-3	PKCδ	ND	+	Voss et al. 2005
Caspase-3	p38	Ser150	−	Alvarado-Kristensson et al. 2004
Caspase-3	PP2A	Ser150	−	Alvarado-Kristensson and Andersson 2005
Caspase-6	ARK5	Ser257	−	Suzuki et al. 2004
Caspase-7	PAK2	Ser30, Thr173, Ser239	−	Li et al. 2011

[a]The effect of phosphorylation at specific residues is listed as activating (+) or inactivating (−). ND, not determined.

phosphorylation sites, along with those affecting the other caspases (see further below), are summarized in Table 1.

In addition to phosphorylation, caspase-9 can be modified by nitrosylation, which occurs when intracellular nitric oxide (NO) levels are high (Torok et al. 2002). Treatment of cell lines with a pharmacological NO donor, S-nitroso-N-acetyl-D,L-penicillamine (SNAP), does not block mitochondrial cytochrome c release but reduces caspase-9 activation in a cell-free system. In vitro, SNAP was also able to negatively regulate the activity of recombinant, human capsase-9 although the relevant site was not identified (Torok et al. 2002). Mannick et al. (2001) have also showed nitrosylation of caspase-9 in different subcellular compartments, finding that a mitochondrial fraction of caspase-9 was preferentially nitrosylated.

Besides modifications that alter its activity or activation, caspase-9 can be controlled at the level of protein stability through ubiquitylation. It has been reported that caspase-9 can be ubiquitylated by XIAP (X-linked inhibitor of apoptosis protein) E3 ubiquitin ligase, a member of the IAP family of proteins, which will be discussed further below for their role in blocking caspase activity through direct interaction (see Silke and Meier 2013). XIAP contains a carboxy-terminal RING domain critical for the protein's ubiquitin ligase activity (Joazeiro and Weissman 2000). XIAP can polyubiquitylate the large subunit of active caspase-9 in vitro, but not the inactive procaspase-9 (Morizane et al. 2005). Consistent with these observations, treatment of cells with the proteasome inhibitor MG132, in conjunction with overexpression of XIAP, promotes accumulation of polyubiquitylated caspase-9 (Morizane et al. 2005). Additional studies are required to fully understand the means by which XIAP-mediated caspase-9 ubiquitylation can be controlled as well as the importance of this activity in controlling apoptotic progression.

Caspase-8

As the predominant initiator caspase in the extrinsic pathway, caspase-8 has a vital role in determining the fate of the cell following death receptor activation. Although rodents only contain caspase-8, other mammals also contain caspase-10, which appears to have at least a partially overlapping function with caspase-8 and is discussed in its own section below. Like caspase-9 in the intrinsic pathway, the activity of caspase-8 can also be altered by posttranslational modification, particularly phosphorylation.

A variety of serine/threonine kinases have been reported to directly phosphorylate caspase-8. For example, a recent report by Matthess et al. linked inhibition of the apoptotic machinery with the onset of mitosis, showing that active cdk1/cyclin B phosphorylates procaspase-8 at Ser387, which inhibits cleavage and activation of the caspase (Fig. 3). Accordingly, expression of a nonphosphorylatable caspase-8 mutant (Ser387 to Ala) rendered cells more sensitive to Fas-induced apoptosis during M phase (Matthess et al. 2010). Given the proximity of Ser387 to the caspase-8 cleavage sites, it was suggested that cdk1-mediated phosphorylation of caspase-8 blocks its autoprocessing, thereby protecting the cells from certain caspase-8-activating stimuli during mitosis.

Phosphorylation of caspase-8 by p38 MAPK at Ser364 inhibits the active caspase and protects neutrophils from Fas-induced death (Alvarado-Kristensson et al. 2004). Caspase-8 phosphorylation at Thr263 by the RSK2 kinase appears to both promote caspase-8 degradation and inhibit Fas-induced HeLa cell death (Peng et al. 2011). Interestingly, Peng and colleagues showed EGF-mediated caspase-8 ubiquitylation and degradation, which was reduced when caspase-8 phosphorylation of Thr263 was abrogated by mutation of Thr263 to Ala (Peng et al. 2011).

Caspase-8 also appears to be regulated directly by tyrosine kinases. Following EGF stimulation, the Src kinase phosphorylates caspase-8 at Tyr380, inhibiting Fas-induced caspase-8 activation and subsequent apoptosis (Cursi et al. 2006). Western blotting for Tyr380 phosphorylation revealed high levels of phosphorylated caspase-8 in colon cancer, where Src activity is often elevated (Cursi et al. 2006). This site is also phosphorylated by other kinases in this family, including Fyn and Lyn (Senft et al. 2007; Jia et al. 2008). In studies with the Lyn kinase, it was also determined that caspase-8 was phosphorylated at Tyr465, and both of these sites (Tyr380 and Tyr465) could be dephosphorylated by the Src-homology domain 2 (SH2)-containing tyrosine phosphatase 1 (SHP1) (Jia et al. 2008). In a model of neutrophil survival, this dephosphorylation rendered caspase-8 more responsive than the phosphorylated protein to apoptotic stimuli (Jia et al. 2008).

Ubiquitylation of caspase-8 also appears to have an interesting role in directly regulating the activity of the enzyme. Unlike ubiquitylation reported for other caspases, polyubiquitylation of caspase-8 by a cullin3-based E3 ligase enhances its enzymatic activity (Jin et al. 2009). This polyubiquitylation of caspase-8 occurs after recruitment of caspase-8 to the DISC, and the modification allows for the binding of active caspase-8 to the poly-Ub binding protein, p62 (Fig. 4A). Furthermore, p62 facilitates an association between this complex and other similar complexes in the cell, forming an aggregate of active caspase-8 and p62. Interestingly, caspase-8 activity appears to be enhanced within these aggregated foci, perhaps through increased stability of cleaved capase-8 (Jin et al. 2009). The deubiquitinating (DUB) enzyme A20 was reportedly involved in reversing this modification (Jin et al. 2009).

Nitrosylation appears to also play a role in regulating caspase-8 activity. Nitric oxide (NO) induces S-nitroslyation of caspase-8, which has been reported to reduce the sensitivity of hepatocytes to TNF-α/ActD-induced apoptosis (Kim et al. 2000). Specifically, elevating levels

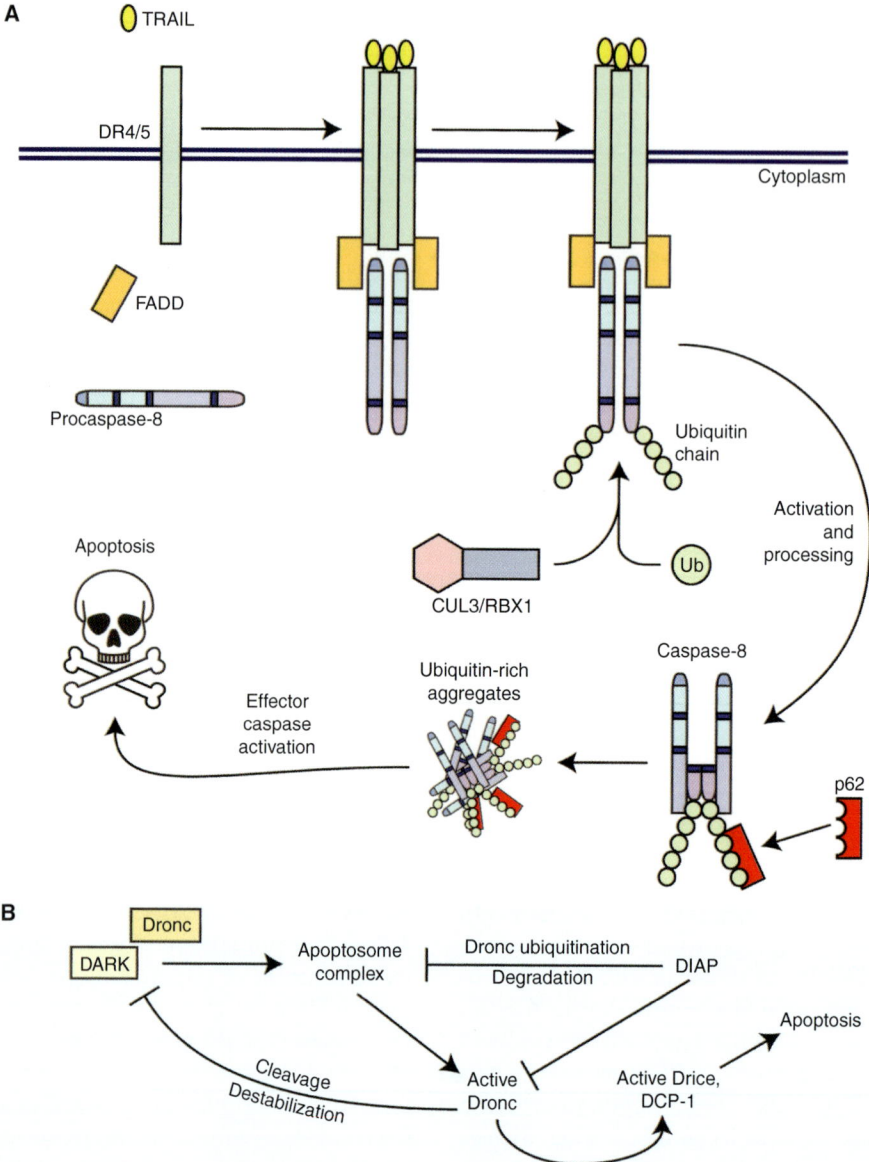

Figure 4. Novel mechanisms of ubiquitylation in caspase regulation. (A) A novel mechanism in which caspase-8 activity is positively regulated via ubiquitylation. A proapoptotic ligand, such as TRAIL, initiates the extrinsic apoptotic pathway at the plasma membrane through formation of the DISC and recruitment of procaspase-8. Caspase-8 is then activated through induced-proximity dimerization. Once active, caspase-8 can be ubiquitylated by the neddylated form of CUL3/RBX1. This posttranslational modification on caspase-8 allows for its interaction with p62. Through autoprocessing, active caspase-8 releases the catalytic domains, which remain bound to p62. This active caspase-8 is moved into cytosolic aggregates rich in ubiquitin in which the caspase-8 remains active through stabilization of the dimer. (B) A unique mechanism for a negative feedback loop between DARK (dApaf-1) and Dronc. The *Drosophila* IAP, DIAP1, inhibits Dronc and Drice/DCP-1 by interacting with the caspases. DIAP1 also indirectly reduces downstream effector caspase activity by facilitating degradation of the apoptosome complex through ubiquitylation of DARK-bound active Dronc. Active Dronc can also negatively feed back on DARK by directly cleaving the protein, leading to destabilization of DARK and reduced protein levels. Thus, Dronc and DARK are involved in a DIAP1-dependent negative feedback loop.

of NO through an inducible NO synthase inhibited caspase-8 activity, caspase-3 activity, and cytochrome c release. In vivo studies in rats have showed that a liver-specific NO donor, V-PYRRO/NO, blocked caspase-8 activity, Bid cleavage, and mitochondrial cytochrome c release in the rat livers treated with TNF-α and D-galactosamine (Kim et al. 2000).

Caspase-2

Caspase-2 is a critical initiator caspase in apoptotic pathways activated by a number of cell stresses, including nutrient depletion, heat shock, DNA damage, and spindle disruption (Robertson et al. 2002; Nutt et al. 2005; Tu et al. 2006; Rudolf et al. 2009). In some of these situations, caspase-2 is controlled through posttranslational modification.

For example, it has been shown that caspase-2 is under metabolic control in the *Xenopus* oocyte; when there are sufficient nutrients available to drive NADPH production by the pentose phosphate pathway (PPP), a suppressive phosphorylation at Ser135 (*Xenopus* numbering) within the caspase-2 prodomain is catalyzed by calcium/calmodulin-dependent protein kinase II (CaMKII) (Nutt et al. 2005). This phosphorylation appears to block its binding to the activating adaptor protein, RAIDD (Nutt et al. 2005). Ser135 phosphorylation is also controlled by metabolism via binding of the phosphoserine/phosphothreonine-binding protein, 14-3-3. 14-3-3 binding protects Ser135 from being dephosphorylated by a constitutively bound PP1 (Fig. 5). As nutrients in the oocyte are depleted, 14-3-3 is removed from caspase-2, leaving it to be dephosphorylated and subsequently activated (Nutt et al. 2009). Interestingly, caspase-2$^{-/-}$ mice have excess oocytes resulting from a failure of apoptosis, confirming the importance of caspase-2 in mammalian oocyte cell death as well (Bergeron et al. 1998).

Phosphorylation of caspase-2 appears to also be involved in modulation of the extrinsic apoptotic pathway induced by engagement of TRAIL ligand. Specifically, CK2 was reported to suppress caspase-2 through phosphorylation at Ser157 in several cancer cell lines where low levels of casein kinase 2 (CK2) correlated with sensitization to TRAIL-induced death. Because CK2-mediated phosphorylation of caspase-2 at Ser157 blocked activation of the caspase, when CK2 activity was decreased, caspase-2 activity was enhanced (Shin et al. 2005). Although TRAIL-induced death in these cells requires caspase-8, active caspase-2 appeared to cleave procaspase-8, possibly priming these cancer cells for TRAIL-induced death.

Posttranslational modification of caspase-2 appears to also provide a means to tether caspase-2 activation to cell cycle status. Cdk1/cyclin B inhibits caspase-2 during mitosis because of direct phosphorylation at Ser340 (human numbering [Andersen et al. 2009]). This phosphorylation occurs in the linker region between the large and small subunits, suggesting that this phosphorylation may play a role in blocking full cleavage and activation of the caspase. Protein phosphatase 1, PP1, may play a role in dephosphorylating this site on caspase-2. This mitotic phosphorylation of caspase-2 provides a third example of direct caspase inhibition by cdk1 during mitosis (Fig. 3). It is interesting to note that while the cdk1 phosphorylation sites on caspase-2 and caspase-9 are located in "linker regions" that are not part of the mature, active caspase, the cdk1 phosphorylation site on caspase-8 site is part of the small subunit of the mature enzyme.

Caspase-10

Although caspase-8 and caspase-10 have both been placed in the extrinsic pathway downstream of death receptor ligands, it is still a bit unclear if these caspases are functionally equivalent with regard to activation and cleavage of downstream substrates. (Kischkel et al. 2001; Wang et al. 2001; Sprick et al. 2002; Fischer et al. 2006; Bae et al. 2008; Benkova et al. 2009; Chen et al. 2009). A recent paper by Wachmann and colleagues used in vitro dimerization assays to show that caspase-10 is also activated through induced proximity dimerization, as described for other initiator caspases (Wachmann et al. 2010). In this analysis of caspase-10 activation, active caspase-10 was able to cleave

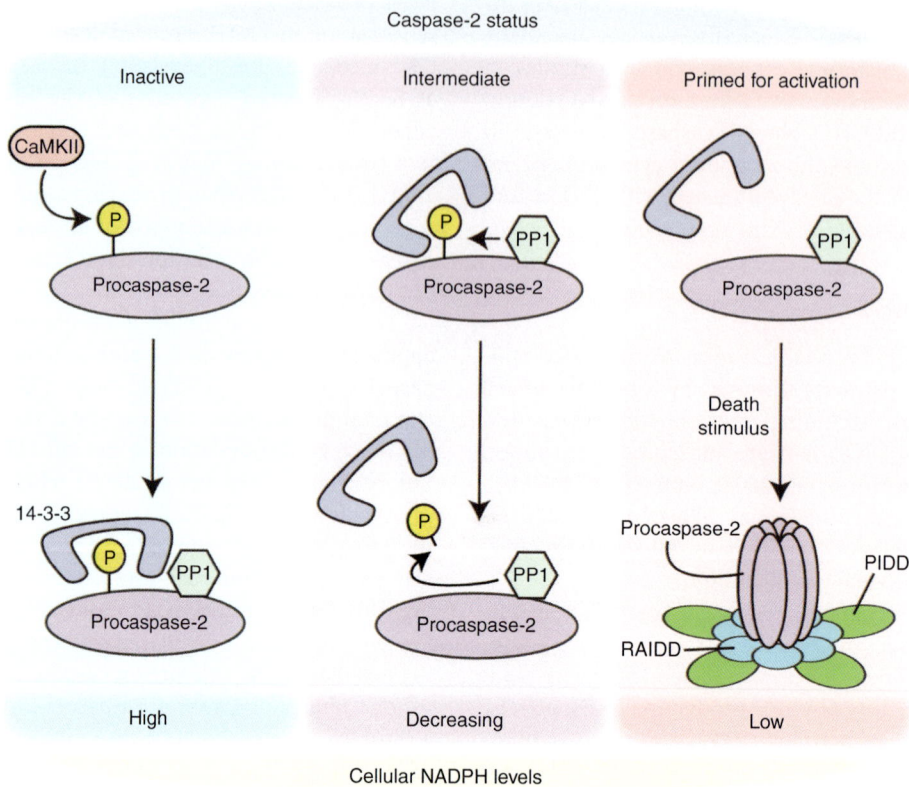

Figure 5. Metabolic regulation of caspase-2. In conditions of high intracellular NAPDH, CaMKII has been shown to phosphorylate caspase-2 at Ser135 (*Xenopus laevis* site). This phosphorylation leads to the recruitment of the phosphoserine/phosphothreonine binding protein, 14-3-3. In this state, caspase-2 is held inactive. As NADPH levels begin to drop, 14-3-3 is released from caspase-2, allowing the constitutively bound phosphatase, PP1, to dephosphorylate Ser135. This dephosphorylation leaves caspase-2 primed for activation such that it can be recruited to its activation platform (as with the PIDDosome shown here) following a prodeath stimulus.

Bid, raising the possibility that capsase-10, like caspase-8, can stimulate mitochondrial cytochrome *c* release to aid in activation of effector caspases. Further studies are necessary to more carefully differentiate between caspase-10 and caspase-8 in response to activation of death receptor signaling and to identify specific modes of caspase-10 regulation that may or may not differ from caspase-8.

Dronc

The *Drosophila* initiator caspase Dronc has both caspase-9 and caspase-2-like properties. In its activation through induced proximity by an Apaf-1-homologous adaptor protein, Dronc appears to function in the fly apoptosome in a manner similar to caspase-9. However, in a paradigm similar to caspase-2, Dronc appears to be negatively regulated by CaMKII phosphorylation in response to nutrient flux through the pentose phosphate pathway (Yang et al. 2010). Specifically, high levels of NADPH support phosphorylation at a site within the prodomain of the caspase (in this case Ser130), which reduces Dronc activation by impeding the interaction between Dronc and DARK. Decreasing NADPH levels promotes Dronc's dephosphorylation, resulting in its activation and subsequent cell death. In vivo experiments in which nonphosphorylatable Dronc was expressed in fly neurons support the observation that NADPH production attenuates *Drosophila* apoptosis and the importance of Dronc phosphorylation at Ser130 in maintaining metabolic suppression of Dronc (Yang et al. 2010).

Dronc is also regulated through ubiquitylation by the *Drosophila* IAP (DIAP1), and at least part of DIAP1's antiapoptotic activity is attributable to this function (Muro et al. 2002; Wilson et al. 2002; Shapiro et al. 2008). Recently, work by Shapiro and colleagues showed an interesting feedback mechanism between Dronc and DARK whereby active Dronc cleaves and destabilizes its adaptor (Shapiro et al. 2008) (Fig. 4B). However, DARK also appears to play a feedback role in reducing Dronc protein levels. It was shown that Dronc could be ubiquitylated (presumably by DIAP1) only when present on the apoptosome (and thus bound to DARK). Interestingly, although details are still unclear, deubiquitylases have also been suggested to play a role in *Drosophila* apoptosis, likely helping to maintain a critical balance in the levels of certain apoptotic proteins (Ryoo et al. 2002; Wing et al. 2002; Ribaya et al. 2009).

Effector Caspases

Caspase-3

Caspase-3 is the most well-characterized effector caspase. Studies directed toward ordering of the caspase cascade and characterization of caspase substrates indicates that caspase-3 has both distinct and overlapping roles with caspase-7 and caspase-6 (Slee et al. 1999; Lakhani et al. 2006; Luthi and Martin 2007; Inoue et al. 2009). Although it might appear that by the time apoptotic signals reach the point of caspase-3 activation, there is no salvaging the cell, caspase-3 is still regulated through a variety of posttranslational modifications. First, both PKCδ and p38 have been shown to phosphorylate caspase-3. The PKCδ site has not been identified, but interestingly, this modification appears to enhance caspase-3 activity (Voss et al. 2005). This isoform of PKC is sometimes referred to as a proapoptotic kinase, and thus its role here may not be surprising. Caspase-mediated cleavage of PKCδ generates a shorter, active form of the kinase, which suggests the presence of a positive proapoptotic feedback loop between these two enzymes (Emoto et al. 1995; Ghayur et al. 1996; DeVries et al. 2002). Similarly to caspase-8, p38 phosphorylation of caspase-3 at Ser150, located in the large subunit of the protein, directly inhibits caspase-3 and has been shown to impede Fas-induced apoptosis in neutrophils (Alvarado-Kristensson et al. 2004). Protein phosphatase 2A (PP2A) dephosphorylates Ser150, restoring caspase-3 activity and rendering the neutrophils sensitive to Fas-induced apoptosis (Alvarado-Kristensson and Andersson 2005).

Caspase-3 is also modified by ubiquitylation. Although the role of IAPs in regulating effector caspases such as caspase-3 is not entirely clear, at least one study has shown the ability of cIAP2 to monoubiquitylate caspase-3 in vitro (Huang et al. 2000). The physiological relevance of this modification has yet to be determined. More recent work has characterized cIAP1-dependent ubiquitylation of an intermediate processed form of caspase-3, leading to proteasome-dependent degradation of the effector caspase and increased resistance to TRAIL-induced apoptosis (Choi et al. 2009). Indeed, treatment of cells with the proteasome inhibitor lactacystin has been reported to stabilize cleaved caspase-3 (active subunits) enhancing apoptosis, while having only a small effect on caspase-8 and caspase-9 (Chen et al. 2003). Other work has shown that XIAP can polyubiquitylate active caspase-3 (not procaspase-3), leading to its proteasome-dependent degradation (Suzuki et al. 2001). Unfortunately, studies with mouse models have not provided clarity on this issue. Mice with a targeted deletion in the RING domain of XIAP have decreased levels of ubiquitylated caspase-3. They also exhibit elevated caspase-3 activity, consistent with lowered levels of caspase-3 ubiquitylation (Schile et al. 2008). However, compared with control mice, XIAP$^{-/-}$ mice do not express increased levels of caspase-3, although this may be attributable to compensation (Harlin et al. 2001).

It has been reported that caspase-3 can be nitrosylated on its active site cysteine (Mannick et al. 1999). In this same study, caspase-3 was shown to be denitrosylated in cells following treatment with Fas ligand, thus increasing caspase activity and cell death in response to Fas. Subsequent work by this group has suggested that only a subset of caspase-3 (and caspase-9, as discussed above), localized to mitochondria, is nitrosylated (Mannick et al. 2001). A recent study reported caspase-3 nitrosylation in primary adhesion fibroblasts isolated from patients, and those cells appeared to have diminished apoptotic responsiveness when caspase-3 was nitrosylated (Jiang et al. 2009). Other models have also been used to show a role for nitrosylation in regulating caspase-3, including treatment of cadriomyocytes with SNAP and doxorubicin (Dimmeler et al. 1997; Rossig et al. 1999; Maejima et al. 2005).

Caspase-7

As for caspase-8 and caspase-10, it is currently unclear the extent to which caspase-3 and caspase-7 share functions. Initial work using peptide libraries to characterize the substrate specificity of caspase-7 suggested that caspase-7 and caspase-3 have the same substrate preferences (Thornberry et al. 1997; Stennicke et al. 2000). Despite apparently overlapping substrate specificity, caspase-7 and caspase-3 have been shown to exhibit distinct subcellular distributions in mouse livers treated with anti-Fas antibody (Chandler et al. 1998). More recent studies have supported distinct roles of these caspases in cellular breakdown during apoptosis (Walsh et al. 2008). These findings are reinforced by a comparison of phenotypes from caspase-$3^{-/-}$, caspase-$7^{-/-}$, and double knockout mice (and MEFs), which suggests both separate and redundant roles for these caspases (Lakhani et al. 2006). As for caspase-3, it has been reported that cIAP1 can ubiquitylate caspase-7, although cIAP1 appears to act on the mature caspase-7 enzyme, rather than the partially processed form, as for caspase-3 (Choi et al. 2009). Although there do not appear to be parallel reports of caspases-3 and -7 phosphorylation, recent work has shown PAK2 (p21-activated kinase)-mediated phosphorylation of caspase-7 at Ser30, Thr173, and Ser239, which negatively regulate caspase-7 activity (Li et al. 2011).

Interestingly, although caspases-3 and -7 share both upstream activators and some overlapping substrate specificity, their divergent amino-terminal sequences appear to contribute to different modes of regulation. Amino-terminal sequences of caspase-7 present in the proenzyme but absent from the mature enzyme seem to negatively regulate capsase-7 function. Removal of this amino terminus potentiates caspase-7 activity, and Salvesen and colleagues have hypothesized that these residues play an as-yet undefined role in sequestering the enzyme from its upstream activators (Denault and Salvesen 2003).

Caspase-6

Although caspase-6 also functions as an effector caspase, it has been known for some time that caspase-6 has at least some unique cellular substrates. It has been reported that caspase-6 is activated only following activation of caspase-3 or caspase-7 and thus is not activated directly by an initiator caspase (Inoue et al. 2009). Once active, caspase-6 can cleave initiator caspases-2 and -8, although the functional consequences of this cleavage (away from the initiator caspase activation platform) are unclear (Slee et al. 1999; Inoue et al. 2009). In contrast, other studies have showed that caspase-6 can be activated in the absence of caspase-3 or -7 activity (LeBlanc et al. 1999; Allsopp et al. 2000; Doostzadeh-Cizeron et al. 2000). Subsequent work has provided support for the idea that caspase-6 can activate itself in vivo (Wang et al. 2010). Using the crystal structure of caspase-6, Wang and colleagues have revealed that caspase-6 can regulate its own activity through intramolecular caspase self-activation, although it is not clear how this process is engaged physiologically. Like caspase-7, the amino-terminal residues of caspase-6 appear to serve an inhibitory function (Klaiman et al. 2009; Wang et al. 2010).

Caspase-6 can be inhibited posttranslationally by kinase ARK5 (Suzuki et al. 2004). In the colon cancer cell line, SW480, which effectively evades Fas-induced apoptosis, it has been shown that caspase-6 is held inactive by the ARK5 kinase. Treatment of SW480 cells with ARK5 antisense RNA liberated caspase-6 to cleave c-FLIP to inhibit Fas signaling. Although there are two putative ARK5 sites on capase-6, the kinase appears to phosphorylate Ser257. Mutation of Ser257 to Ala overrides ARK5-mediated inhibition and allows c-FLIP cleavage and apoptosis to proceed in the SW480 cells (Suzuki et al. 2004).

DRICE AND DCP-1

The *Drosophila* effector caspases, Drice and DCP-1, are regulated through ubiquitylation, catalyzed by the IAP proteins, DIAP1 and DIAP2. Ribeiro and colleagues recently showed a role for DIAP2 in the ubiquitylation and inactivation of Drice (Ribeiro et al. 2007). Interestingly, DIAP2 acts as a pseudosubstrate of Drice; Drice's attempt to cleave DIAP2 results in trapping of DIAP2 in a covalent association with Drice, which allows proximal ubiquitylation of the associated Drice by the DIAP2 RING domain.

In a distinct mechanism, the catalytic activity of Drice also appears to be required for its ubiquitylation by DIAP1. Drice cleaves DIAP1, resulting in a conformational change, which allows DIAP1 to interact directly with Drice/DCP-1 for caspase ubiquitylation by the DIAP1 RING (Ditzel et al. 2008). In S2 cells, Drice polyubiquitylation does not appear to facilitate its degradation. Rather, ubiquitylation catalytically inactivates the protein and inhibits its ability to cleave downstream substrates. Although it is currently unclear exactly how this polyubiquitylation inactivates Drice, several possibilities exist. The polyubiquitin chains may physically block substrate access to the caspase's catalytic site, or the posttranslational modification may alter the confirmation of the catalytic domain to preclude productive enzyme–substrate interactions.

PROTEIN/PROTEIN INTERACTIONS

In addition to the changes that posttranslational modifications can have on caspase function, various proteins (by nature of their interaction with a caspase) can impact apoptotic pathways both positively and negatively. Some of the first proteins discovered as direct caspase inhibitors were of viral origin (such as p35 and CrmA) and do not appear to have direct cellular counterparts (Clem et al. 1991; Ray et al. 1992). In addition, there are many proteins that affect caspase activity indirectly by modulating, for example, the complex/platform on which the caspase is activated. For example, the backbone of the caspase-9-activating apoptosome, Apaf-1, is regulated by a complex of proteins, hsp70, CAS, and PHAPI (Kim et al. 2008). These proteins facilitate nucleotide exchange on Apaf-1, which is necessary for full apoptosome formation and subsequent caspase-9 activation. Interestingly, levels of other (nonprotein) cellular components, such nucleotides and ions, also appear to play critical roles in regulating certain steps in the activation of caspases and the apoptotic pathway (Cain et al. 2001; Chandra et al. 2006; Bao et al. 2007; Karki et al. 2007; Mei et al. 2010). Because there have been a variety of recent reviews on the proteins that regulate caspases indirectly, these circumstances are not covered here (Schafer and Kornbluth 2006; Fadeel et al. 2008; Krumschnabel et al. 2009). Instead, we focus on a few key examples of proteins that interact directly with caspases to alter either their activation or enzymatic activity.

FLIP

The FLIP (FLICE-like inhibitory protein, so-called because of the original caspase-8 name, FLICE) family of proteins was first discovered in a viral context, but homologs in other species, including mammals, were discovered nearly simultaneously (Thome et al. 1997). Interestingly, the FLIP proteins have been shown to function both as inhibitors and activators of the apoptotic process. Mammalian cellular FLIP (c-FLIP) is notably similar to procaspase-8 and procaspase-10 in primary sequence. Specifically, the long FLIP isoform contains the two death effector domains (DEDs) at the amino terminus of the protein, and at the carboxyl terminus, it has a pseudocaspase domain that lacks catalytic activity (Irmler et al. 1997). The short isoform (c-FLIP$_s$) is similar to the viral FLIP, resembling a truncated caspase-8, with two DEDs and no region homologous with the catalytic domain. It has been shown that the long form of c-FLIP is recruited to the DISC and can be partially processed, with a portion of the FLIP protein retained at the DISC. This interaction with the DISC inhibits any subsequent recruitment of caspase-8 to that site (Goltsev et al. 1997; Han et al. 1997; Irmler et al. 1997; Scaffidi et al. 1999). In addition, it appears that the short isoform can completely prevent processing of DISC-bound caspase-8, whereas the long isoform appears to allow partial processing (Krueger et al. 2001).

In contrast to the studies above, which characterized FLIP$_L$ as an inhibitor of caspase activation, some studies have shown that FLIP$_L$ can, in fact, facilitate caspase activation (Chang et al. 2002; Micheau et al. 2002). It may be that the local concentration of c-FLIP at the DISC determines its role as an inhibitor or activator of caspase-8. As an activator of caspase-8, c-FLIP$_L$ appears to heterodimerize with casapse-8, activating the caspase portion of the dimer. Indeed, this hypothesis is supported by Boatright and colleagues who used kosmotropes to induce FLIP/caspase-8 dimers. These heterodimers appeared to have a lower kinetic barrier to activation than homodimers between two wild-type caspase-8 molecules (Boatright et al. 2004). Interestingly, in contrast to the proapoptotic caspase-8 homodimer, these heterodimers of caspase-8 and FLIP$_L$ have been shown to promote cell survival (Oberst et al. 2011; van Raam and Salvesen 2012).

IAPs

The inhibitor of apoptosis proteins (IAPs) are a conserved family of proteins that are defined by the presence of at least one baculovirus IAP repeat domain (BIR), and, as discussed above, some of the IAPs also contain a RING domain that confers E3 ubiquitin ligase function (Crook et al. 1993). In addition, XIAP and DIAP proteins have been shown to act as direct stoichiometric inhibitors of caspases. XIAP can inhibit caspase-3 and caspase-7 by binding to the active site, and caspase-9 by preventing or reversing dimerization (Deveraux et al. 1997; Chai et al. 2001; Riedl et al. 2001; Srinivasula et al. 2001; Shiozaki et al. 2003). This inhibitory action of XIAP can be antagonized by mitochondrial proteins Smac/DIABLO and Omi/HtrA2, which are released during an apoptotic stimulus along with cytochrome c (Chai et al. 2000; Du et al. 2000). These proteins are similar to the Reaper, Hid, Grim (RHG) family of Drosophila proteins, which can activate apoptosis in part by displacing the DIAP1 protein from Dronc and effector caspases (Fig. 2). As a direct inhibitor of these caspases, XIAP has been detected on the apoptosome (Bratton et al. 2001, 2002). In a recent study of melanoma cell lines, XIAP was also subject to caspase cleavage in a positive feedback loop that reduces caspase inhibition and XIAP levels through proteasomal degradation (Hornle et al. 2011). Although XIAP is the best-characterized vertebrate member of the IAP family with regard to direct binding and inhibition of caspases, one recent study showed that cIAP1 can specifically block apoptosis downstream of cytochrome c release by binding to and inhibiting active caspase-9 within the apoptosome, precluding downstream activation of procaspase-3 (Burke et al. 2010).

TRANSCRIPTIONAL REGULATION OF CASPASES

Although we have focused thus far on the posttranslational regulation of caspases, some of the caspase proteins are also regulated at the level of expression through control of their mRNA transcription. Although much less is known about the mechanisms modulating mRNA expression levels, this type of regulation also appears to contribute to control of caspase activity before or during an apoptotic stimulus. Interestingly, caspase-2 transcriptional regulation appears to be coordinated with transcriptional induction of its activators, PIDD and RAIDD (Krumschnabel et al. 2009). One recent study showed a p53/p21-dependent pathway for down-regulating caspase-2 mRNA expression in resting cells and in response to DNA damage

(Baptiste-Okoh et al. 2008). Although this finding seems a bit counterintuitive, the authors postulate that this mechanism reduces unwanted extraneous cell death. Following DNA damage, the caspase-2 activator, PIDD, is transcriptionally up-regulated, potentially overcoming the reduced expression of caspase-2 and facilitating cell death.

Alternative splicing of caspase-8 plays a role in modulating apoptotic induction via the extrinsic pathway. Similarly to FLIP, the caspsase-8 L splice variant has a functional DED but lacks key residues in the catalytic domain, which allows it to interact with the DISC at the plasma membrane without being able to induce downstream caspase activation (Curtin and Cotter 2003). Thus, caspase-8 L behaves like cFLIP and acts like a caspase-8 inhibitor (Himeji et al. 2002). Methylation of the caspase-8 gene is thought to suppress caspase-8 expression in a variety of tumors (Harada et al. 2002; Curtin and Cotter 2003). Pediatric tumors, including rhabdomyosarcomas, medulloblastomas, retinoblastomas, and neuroblastomas, have all been reported to exhibit methylation of the caspase-8 gene, correlating with a decrease in caspase-8 expression. Interestingly, many of the cell lines used for this study also had down-regulated caspase-10 at a posttranscriptional level (Harada et al. 2002).

The caspase-10 gene has been characterized as a direct target of p53 following DNA damage (Rikhof et al. 2003). Treatment of both p53-wild type and p53-deficient cell lines with etopside or adriamycin-induced caspase-10 mRNA expression in a p53-dependent manner without an effect on caspase-8 expression. Caspase-10 mRNA levels are negatively regulated in Jurkat cells expressing the HIV tat gene when compared with control-treated Jurkat cells (Gibellini et al. 2005). Although the levels of caspase-8 were unchanged, the reduction of caspase-10 mRNA coupled with a concomitant increase in expression of cFLIP were sufficient to enhance the resistance of Tat-expressing T cells to TRAIL-induced apoptosis.

Two different caspase-9 transcripts can be derived from the caspase-9 gene, differing by four exons. Caspase-9a includes these four exons, whereas caspase-9b does not. Importantly, it appears that capsase-9a is proapoptotic, but the shorter isoform, caspase-9b, is antiapoptotic (Seol and Billiar 1999; Srinivasula et al. 1999). Recently, Shultz and colleagues have found that this alternative splicing of caspase-9 is defective in nonsmall cell lung carcinoma (NSCLC) (Shultz et al. 2010). Specifically, the authors showed that K-Ras12V overexpression increased the ratio of caspase-9a to capase-9b, while epidermal growth factor receptor (EGFR) overexpression lowered this ratio. Furthermore, they showed that the RNA splicing factor, SRp30a, is phosphorylated by Akt, which affects the inclusion of these four exons in the mature caspase-9 transcript. It is interesting to note that Apaf-1, the adapter protein for caspase-9, can also be transcriptionally regulated by E2F1 and p53 (Fortin et al. 2001; Moroni et al. 2001; Furukawa et al. 2002; Johnson et al. 2007).

CASPASE REGULATION IN NONAPOPTOTIC CELLULAR PROCESSES

In addition to the various mechanisms that regulate caspases during apoptosis, many studies have shown that active caspases play important roles in nonapoptotic cellular functions, including inflammation, protein secretion, and differentiation (reviewed in Kuranaga and Miura 2007; Li and Yuan 2008; Feinstein-Rotkopf and Arama 2009; Yi and Yuan 2009). Caspase activity in the absence of a cell death signal may have unintended deadly consequences in the absence of finely tuned regulation. Thus, caspase control is particularly critical in circumstances in which the proteases are activated for short, well-defined periods of time or at isolated subcellular locations. Phenotypes of the caspase knockout mice have provided additional evidence to support a role for caspases in nonapoptotic processes (reviewed in Li and Yuan 2008). An exhaustive discussion of nonapoptotic caspase activation/regulation is beyond the scope of this article, but a few recent studies are described below to illustrate the nature of such regulation and to highlight some outstanding questions.

In *Drosophila melanogaster*, tightly controlled spatial regulation of effector caspase activity is crucial during the terminal differentiation of spermatids, a process also known as individualization (Kaplan et al. 2010). Kaplan and colleagues have shown that a spermatid contains a distal-to-proximal gradient of the protein Soti, which results in a similar directional gradient of the IAP dBruce (because Soti inhibits an E3 ubiquitin ligase that ubiquitylates dBruce). During spermatid individualization, dBruce inhibits caspase activity, thus creating an opposing gradient of effector caspase activity (proximal-to-distal). This prevents unwanted death in the latest-individualizing region of the spermatid and helps to properly drive differentiation. As a result of the timing of individualization and the presence of dBruce, caspase activation is insufficient to initiate the full apoptotic program and subsequent cell death.

In hippocampal neurons, a role for transient activation of caspase-3 has been showed in synaptic long-term depression (LTD) and AMPA receptor internalization (Li et al. 2010). The authors provide evidence to support a model whereby caspase-3 is activated through a pathway similar to that which occurs during intrinsic apoptosis. Indeed, NMDA treatment facilitates mitochondrial release of cytochrome *c* and caspase-9 activation, which is necessary for the activation of caspase-3. Additional evidence suggests that cleavage and inactivation of Akt1 provides at least part of the link between caspase-3 activity and LTD; overexpression of a noncleavable Akt1 in hippocampal slice cultures inhibited LTD, confirming the importance of at least one caspase substrate during LTD and AMPA internalization. How this activity is limited to prevent neuronal death is not entirely clear.

Another recent study using olfactory sensory neurons (OSNs) has elucidated a nonapoptotic role for Apaf-1 and caspase-9 signaling in development (Ohsawa et al. 2010). This nonapoptotic caspase activity leads to cleavage of the membrane-bound protein, Semaphorin 7A, which is critical for appropriate formation of axonal projections (Pasterkamp et al. 2003). In mice lacking Apaf-1 or caspase-9 expression, a variety of problems exist, including OSN axons that are routed erroneously and OSNs that have not matured properly. However, there are no changes in the number of neurons, supporting the idea that Apaf-1 and caspase-9 are not impacting generation of OSNs, but rather specific formation of axonal projections.

CONCLUDING REMARKS

Because of the critical role that caspases play in executing the apoptotic program, to avoid unplanned cellular demise, their activation and activity must be tightly regulated. By their very structure, caspases are regulated as soon as they are expressed. They are synthesized as zymogens, which are only to be activated following the appropriate stimulus. Both before and after activation, caspases can be regulated through a variety of mechanisms including posttranslational modifications and protein/protein interactions. Although caspase activity was initially reported to occur only during apoptosis, more recent evidence suggests that caspases play critical roles in other, nonapoptosis cellular processes. Detailed analyses of these specific physiological/pathological circumstances may yield important insight into new and different modes of caspase regulation, some of which may also be important for understanding control of caspase activation in apoptosis.

ACKNOWLEDGMENTS

The authors thank Manabu Kurokawa and Reichen Yang for helpful feedback and discussion.

REFERENCES

*Reference is also in this collection.

Acehan D, Jiang X, Morgan DG, Heuser JE, Wang X, Akey CW. 2002. Three-dimensional structure of the apoptosome: Implications for assembly, procaspase-9 binding, and activation. *Mol Cell* **9:** 423–432.

Allan LA, Clarke PR. 2007. Phosphorylation of caspase-9 by CDK1/cyclin B1 protects mitotic cells against apoptosis. *Mol Cell* **26:** 301–310.

Allan LA, Clarke PR. 2009. Apoptosis and autophagy: Regulation of caspase-9 by phosphorylation. *FEBS J* **276:** 6063–6073.

Allan LA, Morrice N, Brady S, Magee G, Pathak S, Clarke PR. 2003. Inhibition of caspase-9 through phosphorylation at Thr 125 by ERK MAPK. *Nat Cell Biol* **5:** 647–654.

Allsopp TE, McLuckie J, Kerr LE, Macleod M, Sharkey J, Kelly JS. 2000. Caspase 6 activity initiates caspase 3 activation in cerebellar granule cell apoptosis. *Cell Death Differ* **7:** 984–993.

Alnemri ES, Livingston DJ, Nicholson DW, Salvesen G, Thornberry NA, Wong WW, Yuan J. 1996. Human ICE/CED-3 protease nomenclature. *Cell* **87:** 171.

Alvarado-Kristensson M, Andersson T. 2005. Protein phosphatase 2A regulates apoptosis in neutrophils by dephosphorylating both p38 MAPK and its substrate caspase 3. *J Biol Chem* **280:** 6238–6244.

Alvarado-Kristensson M, Melander F, Leandersson K, Ronnstrand L, Wernstedt C, Andersson T. 2004. p38-MAPK signals survival by phosphorylation of caspase-8 and caspase-3 in human neutrophils. *J Exp Med* **199:** 449–458.

Andersen JL, Johnson CE, Freel CD, Parrish AB, Day JL, Buchakjian MR, Nutt LK, Thompson JW, Moseley MA, Kornbluth S. 2009. Restraint of apoptosis during mitosis through interdomain phosphorylation of caspase-2. *EMBO J* **28:** 3216–3227.

Bae S, Ha TS, Yoon Y, Lee J, Cha HJ, Yoo H, Choe TB, Li S, Sohn I, Kim JY, et al. 2008. Genome-wide screening and identification of novel proteolytic cleavage targets of caspase-8 and -10 in vitro. *Int J Mol Med* **21:** 381–386.

Baliga BC, Read SH, Kumar S. 2004. The biochemical mechanism of caspase-2 activation. *Cell Death Differ* **11:** 1234–1241.

Bao Q, Lu W, Rabinowitz JD, Shi Y. 2007. Calcium blocks formation of apoptosome by preventing nucleotide exchange in Apaf-1. *Mol Cell* **25:** 181–192.

Baptiste-Okoh N, Barsotti AM, Prives C. 2008. Caspase 2 is both required for p53-mediated apoptosis and downregulated by p53 in a p21-dependent manner. *Cell Cycle* **7:** 1133–1138.

Benkova B, Lozanov V, Ivanov IP, Mitev V. 2009. Evaluation of recombinant caspase specificity by competitive substrates. *Anal Biochem* **394:** 68–74.

Bergeron L, Perez GI, Macdonald G, Shi L, Sun Y, Jurisicova A, Varmuza S, Latham KE, Flaws JA, Salter JC, et al. 1998. Defects in regulation of apoptosis in caspase-2-deficient mice. *Genes Dev* **12:** 1304–1314.

Berube C, Boucher LM, Ma W, Wakeham A, Salmena L, Hakem R, Yeh WC, Mak TW, Benchimol S. 2005. Apoptosis caused by p53-induced protein with death domain (PIDD) depends on the death adapter protein RAIDD. *Proc Natl Acad Sci* **102:** 14314–14320.

Boatright KM, Renatus M, Scott FL, Sperandio S, Shin H, Pedersen IM, Ricci JE, Edris WA, Sutherlin DP, Green DR, et al. 2003. A unified model for apical caspase activation. *Mol Cell* **11:** 529–541.

Boatright KM, Deis C, Denault JB, Sutherlin DP, Salvesen GS. 2004. Activation of caspases-8 and -10 by FLIP$_L$. *Biochem J* **382:** 651–657.

Bonzon C, Bouchier-Hayes L, Pagliari LJ, Green DR, Newmeyer DD. 2006. Caspase-2-induced apoptosis requires bid cleavage: A physiological role for bid in heat shock-induced death. *Mol Biol Cell* **17:** 2150–2157.

Brady SC, Allan LA, Clarke PR. 2005. Regulation of caspase 9 through phosphorylation by protein kinase C ζ in response to hyperosmotic stress. *Mol Cell Biol* **25:** 10543–10555.

Bratton SB, Walker G, Srinivasula SM, Sun XM, Butterworth M, Alnemri ES, Cohen GM. 2001. Recruitment, activation and retention of caspases-9 and -3 by Apaf-1 apoptosome and associated XIAP complexes. *EMBO J* **20:** 998–1009.

Bratton SB, Lewis J, Butterworth M, Duckett CS, Cohen GM. 2002. XIAP inhibition of caspase-3 preserves its association with the Apaf-1 apoptosome and prevents CD95- and Bax-induced apoptosis. *Cell Death Differ* **9:** 881–892.

Burke SP, Smith L, Smith JB. 2010. cIAP1 cooperatively inhibits procaspase-3 activation by the caspase-9 apoptosome. *J Biol Chem* **285:** 30061–30068.

Cain K, Langlais C, Sun XM, Brown DG, Cohen GM. 2001. Physiological concentrations of K^+ inhibit cytochrome c-dependent formation of the apoptosome. *J Biol Chem* **276:** 41985–41990.

Cardone MH, Roy N, Stennicke HR, Salvesen GS, Franke TF, Stanbridge E, Frisch S, Reed JC. 1998. Regulation of cell death protease caspase-9 by phosphorylation. *Science* **282:** 1318–1321.

Cecconi F, Alvarez-Bolado G, Meyer BI, Roth KA, Gruss P. 1998. Apaf1 (CED-4 homolog) regulates programmed cell death in mammalian development. *Cell* **94:** 727–737.

Cerretti DP, Kozlosky CJ, Mosley B, Nelson N, Van Ness K, Greenstreet TA, March CJ, Kronheim SR, Druck T, Cannizzaro LA, et al. 1992. Molecular cloning of the interleukin-1 β converting enzyme. *Science* **256:** 97–100.

Chai J, Du C, Wu JW, Kyin S, Wang X, Shi Y. 2000. Structural and biochemical basis of apoptotic activation by Smac/DIABLO. *Nature* **406:** 855–862.

Chai J, Shiozaki E, Srinivasula SM, Wu Q, Datta P, Alnemri ES, Shi Y. 2001. Structural basis of caspase-7 inhibition by XIAP. *Cell* **104:** 769–780.

Chandler JM, Cohen GM, MacFarlane M. 1998. Different subcellular distribution of caspase-3 and caspase-7 following Fas-induced apoptosis in mouse liver. *J Biol Chem* **273:** 10815–10818.

Chandra D, Bratton SB, Person MD, Tian Y, Martin AG, Ayres M, Fearnhead HO, Gandhi V, Tang DG. 2006. Intracellular nucleotides act as critical prosurvival factors by binding to cytochrome c and inhibiting apoptosome. *Cell* **125:** 1333–1346.

Chang DW, Xing Z, Pan Y, Algeciras-Schimnich A, Barnhart BC, Yaish-Ohad S, Peter ME, Yang X. 2002. c-FLIP$_L$ is a dual function regulator for caspase-8 activation and CD95-mediated apoptosis. *EMBO J* **21:** 3704–3714.

Chen L, Smith L, Wang Z, Smith JB. 2003. Preservation of caspase-3 subunits from degradation contributes to apoptosis evoked by lactacystin: Any single lysine or lysine pair of the small subunit is sufficient for ubiquitination. *Mol Pharmacol* **64:** 334–345.

Chen H, Xia Y, Fang D, Hawke D, Lu Z. 2009. Caspase-10-mediated heat shock protein 90 β cleavage promotes UVB irradiation-induced cell apoptosis. *Mol Cell Biol* **29:** 3657–3664.

Choi WY, Jin CY, Han MH, Kim GY, Kim ND, Lee WH, Kim SK, Choi YH. 2009. Sanguinarine sensitizes human gastric adenocarcinoma AGS cells to TRAIL-mediated apoptosis via down-regulation of AKT and activation of caspase-3. *Anticancer Res* **29:** 4457–4465.

Clem RJ, Fechheimer M, Miller LK. 1991. Prevention of apoptosis by a baculovirus gene during infection of insect cells. *Science* **254:** 1388–1390.

Conradt B, Horvitz HR. 1998. The *C. elegans* protein EGL-1 is required for programmed cell death and interacts with the Bcl-2-like protein CED-9. *Cell* **93:** 519–529.

Crook NE, Clem RJ, Miller LK. 1993. An apoptosis-inhibiting baculovirus gene with a zinc finger-like motif. *J Virol* **67:** 2168–2174.

Cursi S, Rufini A, Stagni V, Condo I, Matafora V, Bachi A, Bonifazi AP, Coppola L, Superti-Furga G, Testi R, et al. 2006. Src kinase phosphorylates Caspase-8 on Tyr380: A novel mechanism of apoptosis suppression. *EMBO J* **25:** 1895–1905.

Curtin JF, Cotter TG. 2003. Live and let die: Regulatory mechanisms in Fas-mediated apoptosis. *Cell Signal* **15:** 983–992.

Danial NN, Korsmeyer SJ. 2004. Cell death: Critical control points. *Cell* **116:** 205–219.

Denault JB, Salvesen GS. 2003. Human caspase-7 activity and regulation by its N-terminal peptide. *J Biol Chem* **278:** 34042–34050.

Dessauge F, Cayla X, Albar JP, Fleischer A, Ghadiri A, Duhamel M, Rebollo A. 2006. Identification of PP1α as a caspase-9 regulator in IL-2 deprivation-induced apoptosis. *J Immunol* **177:** 2441–2451.

Deveraux QL, Takahashi R, Salvesen GS, Reed JC. 1997. X-linked IAP is a direct inhibitor of cell-death proteases. *Nature* **388:** 300–304.

DeVries TA, Neville MC, Reyland ME. 2002. Nuclear import of PKCδ is required for apoptosis: Identification of a novel nuclear import sequence. *EMBO J* **21:** 6050–6060.

Dimmeler S, Haendeler J, Nehls M, Zeiher AM. 1997. Suppression of apoptosis by nitric oxide via inhibition of interleukin-1β-converting enzyme (ICE)-like and cysteine protease protein (CPP)-32-like proteases. *J Exp Med* **185:** 601–607.

Ditzel M, Broemer M, Tenev T, Bolduc C, Lee TV, Rigbolt KT, Elliott R, Zvelebil M, Blagoev B, Bergmann A, et al. 2008. Inactivation of effector caspases through nondegradative polyubiquitylation. *Mol Cell* **32:** 540–553.

Dix MM, Simon GM, Cravatt BF. 2008. Global mapping of the topography and magnitude of proteolytic events in apoptosis. *Cell* **134:** 679–691.

Dix MM, Simon GM, Wang C, Okerberg E, Patricelli MP, Cravatt BF. 2012. Functional interplay between caspase cleavage and phosphorylation sculpts the apoptotic proteome. *Cell* **150:** 426–440.

Doostzadeh-Cizeron J, Yin S, Goodrich DW. 2000. Apoptosis induced by the nuclear death domain protein p84N5 is associated with caspase-6 and NF-κB activation. *J Biol Chem* **275:** 25336–25341.

Du C, Fang M, Li Y, Li L, Wang X. 2000. Smac, a mitochondrial protein that promotes cytochrome *c*-dependent caspase activation by eliminating IAP inhibition. *Cell* **102:** 33–42.

Duan H, Dixit VM. 1997. RAIDD is a new "death" adaptor molecule. *Nature* **385:** 86–89.

Earnshaw WC, Martins LM, Kaufmann SH. 1999. Mammalian caspases: Structure, activation, substrates, and functions during apoptosis. *Annu Rev Biochem* **68:** 383–424.

Emoto Y, Manome Y, Meinhardt G, Kisaki H, Kharbanda S, Robertson M, Ghayur T, Wong WW, Kamen R, Weichselbaum R, et al. 1995. Proteolytic activation of protein kinase C δ by an ICE-like protease in apoptotic cells. *EMBO J* **14:** 6148–6156.

Fadeel B, Ottosson A, Pervaiz S. 2008. Big wheel keeps on turning: Apoptosome regulation and its role in chemoresistance. *Cell Death Differ* **15:** 443–452.

Feinstein-Rotkopf Y, Arama E. 2009. Can't live without them, can live with them: Roles of caspases during vital cellular processes. *Apoptosis* **14:** 980–995.

Fischer U, Stroh C, Schulze-Osthoff K. 2006. Unique and overlapping substrate specificities of caspase-8 and caspase-10. *Oncogene* **25:** 152–159.

Fortin A, Cregan SP, MacLaurin JG, Kushwaha N, Hickman ES, Thompson CS, Hakim A, Albert PR, Cecconi F, Helin K, et al. 2001. APAF1 is a key transcriptional target for p53 in the regulation of neuronal cell death. *J Cell Biol* **155:** 207–216.

Furukawa Y, Nishimura N, Satoh M, Endo H, Iwase S, Yamada H, Matsuda M, Kano Y, Nakamura M. 2002. Apaf-1 is a mediator of E2F-1-induced apoptosis. *J Biol Chem* **277:** 39760–39768.

Gao Z, Shao Y, Jiang X. 2005. Essential roles of the Bcl-2 family of proteins in caspase-2-induced apoptosis. *J Biol Chem* **280:** 38271–38275.

Ghayur T, Hugunin M, Talanian RV, Ratnofsky S, Quinlan C, Emoto Y, Pandey P, Datta R, Huang Y, Kharbanda S, et al. 1996. Proteolytic activation of protein kinase C δ by an ICE/CED 3-like protease induces characteristics of apoptosis. *J Exp Med* **184:** 2399–2404.

Gibellini D, Re MC, Ponti C, Vitone F, Bon I, Fabbri G, Grazia Di Iasio M, Zauli G. 2005. HIV-1 Tat protein concomitantly down-regulates apical caspase-10 and up-regulates c-FLIP in lymphoid T cells: A potential molecular mechanism to escape TRAIL cytotoxicity. *J Cell Physiol* **203:** 547–556.

Goltsev YV, Kovalenko AV, Arnold E, Varfolomeev EE, Brodianskii VM, Wallach D. 1997. CASH, a novel caspase homologue with death effector domains. *J Biol Chem* **272:** 19641–19644.

Hakem R, Hakem A, Duncan GS, Henderson JT, Woo M, Soengas MS, Elia A, de la Pompa JL, Kagi D, Khoo W, et al. 1998. Differential requirement for caspase 9 in apoptotic pathways in vivo. *Cell* **94:** 339–352.

Han DK, Chaudhary PM, Wright ME, Friedman C, Trask BJ, Riedel RT, Baskin DG, Schwartz SM, Hood L. 1997. MRIT, a novel death-effector domain-containing protein, interacts with caspases and BclXL and initiates cell death. *Proc Natl Acad Sci* **94:** 11333–11338.

Hanahan D, Weinberg RA. 2000. The hallmarks of cancer. *Cell* **100:** 57–70.

Harada K, Toyooka S, Shivapurkar N, Maitra A, Reddy JL, Matta H, Miyajima K, Timmons CF, Tomlinson GE, Mastrangelo D, et al. 2002. Deregulation of caspase 8 and 10 expression in pediatric tumors and cell lines. *Cancer Res* **62:** 5897–5901.

Harlin H, Reffey SB, Duckett CS, Lindsten T, Thompson CB. 2001. Characterization of XIAP-deficient mice. *Mol Cell Biol* **21:** 3604–3608.

Hengartner MO, Horvitz HR. 1994. *C. elegans* cell survival gene ced-9 encodes a functional homolog of the mammalian proto-oncogene bcl-2. *Cell* **76:** 665–676.

Himeji D, Horiuchi T, Tsukamoto H, Hayashi K, Watanabe T, Harada M. 2002. Characterization of caspase-8 L: A novel isoform of caspase-8 that behaves as an inhibitor of the caspase cascade. *Blood* **99:** 4070–4078.

Hornle M, Peters N, Thayaparasingham B, Vorsmann H, Kashkar H, Kulms D. 2011. Caspase-3 cleaves XIAP in a positive feedback loop to sensitize melanoma cells to TRAIL-induced apoptosis. *Oncogene* **30:** 575–587.

Huang H, Joazeiro CA, Bonfoco E, Kamada S, Leverson JD, Hunter T. 2000. The inhibitor of apoptosis, cIAP2, functions as a ubiquitin-protein ligase and promotes in vitro monoubiquitination of caspases 3 and 7. *J Biol Chem* **275:** 26661–26664.

Inoue S, Browne G, Melino G, Cohen GM. 2009. Ordering of caspases in cells undergoing apoptosis by the intrinsic pathway. *Cell Death Differ* **16:** 1053–1061.

Irmler M, Thome M, Hahne M, Schneider P, Hofmann K, Steiner V, Bodmer JL, Schroter M, Burns K, Mattmann C, et al. 1997. Inhibition of death receptor signals by cellular FLIP. *Nature* **388:** 190–195.

Itoh N, Nagata S. 1993. A novel protein domain required for apoptosis. Mutational analysis of human Fas antigen. *J Biol Chem* **268:** 10932–10937.

Jia SH, Parodo J, Kapus A, Rotstein OD, Marshall JC. 2008. Dynamic regulation of neutrophil survival through tyrosine phosphorylation or dephosphorylation of caspase-8. *J Biol Chem* **283:** 5402–5413.

Jiang ZL, Fletcher NM, Diamond MP, Abu-Soud HM, Saed GM. 2009. S-nitrosylation of caspase-3 is the mechanism by which adhesion fibroblasts manifest lower apoptosis. *Wound Repair Regen* **17:** 224–229.

Jin Z, Li Y, Pitti R, Lawrence D, Pham VC, Lill JR, Ashkenazi A. 2009. Cullin3-based polyubiquitination and p62-dependent aggregation of caspase-8 mediate extrinsic apoptosis signaling. *Cell* **137:** 721–735.

Joazeiro CA, Weissman AM. 2000. RING finger proteins: Mediators of ubiquitin ligase activity. *Cell* **102:** 549–552.

Johnson CE, Huang YY, Parrish AB, Smith MI, Vaughn AE, Zhang Q, Wright KM, Van Dyke T, Wechsler-Reya RJ, Kornbluth S, et al. 2007. Differential Apaf-1 levels allow cytochrome *c* to induce apoptosis in brain tumors but not in normal neural tissues. *Proc Natl Acad Sci* **104:** 20820–20825.

Kaplan Y, Gibbs-Bar L, Kalifa Y, Feinstein-Rotkopf Y, Arama E. 2010. Gradients of a ubiquitin E3 ligase inhibitor and a caspase inhibitor determine differentiation or death in spermatids. *Dev Cell* **19:** 160–173.

Karki P, Seong C, Kim JE, Hur K, Shin SY, Lee JS, Cho B, Park IS. 2007. Intracellular K^+ inhibits apoptosis by suppressing the Apaf-1 apoptosome formation and subsequent downstream pathways but not cytochrome *c* release. *Cell Death Differ* **14:** 2068–2075.

Kerr JF, Wyllie AH, Currie AR. 1972. Apoptosis: A basic biological phenomenon with wide-ranging implications in tissue kinetics. *Br J Cancer* **26:** 239–257.

Kim YM, Kim TH, Chung HT, Talanian RV, Yin XM, Billiar TR. 2000. Nitric oxide prevents tumor necrosis factor α-induced rat hepatocyte apoptosis by the interruption of mitochondrial apoptotic signaling through S-nitrosylation of caspase-8. *Hepatology* **32:** 770–778.

Kim HE, Jiang X, Du F, Wang X. 2008. PHAPI, CAS, and Hsp70 promote apoptosome formation by preventing Apaf-1 aggregation and enhancing nucleotide exchange on Apaf-1. *Mol Cell* **30:** 239–247.

Kischkel FC, Hellbardt S, Behrmann I, Germer M, Pawlita M, Krammer PH, Peter ME. 1995. Cytotoxicity-dependent APO-1 (Fas/CD95)-associated proteins form a death-inducing signaling complex (DISC) with the receptor. *EMBO J* **14:** 5579–5588.

Kischkel FC, Lawrence DA, Tinel A, LeBlanc H, Virmani A, Schow P, Gazdar A, Blenis J, Arnott D, Ashkenazi A. 2001. Death receptor recruitment of endogenous caspase-10 and apoptosis initiation in the absence of caspase-8. *J Biol Chem* **276:** 46639–46646.

Klaiman G, Champagne N, LeBlanc AC. 2009. Self-activation of Caspase-6 in vitro and in vivo: Caspase-6 activation does not induce cell death in HEK293T cells. *Biochim Biophys Acta* **1793:** 592–601.

Kornbluth S, White K. 2005. Apoptosis in *Drosophila*: Neither fish nor fowl (nor man, nor worm). *J Cell Sci* **118:** 1779–1787.

Krueger A, Schmitz I, Baumann S, Krammer PH, Kirchhoff S. 2001. Cellular FLICE-inhibitory protein splice variants inhibit different steps of caspase-8 activation at the CD95 death-inducing signaling complex. *J Biol Chem* **276:** 20633–20640.

Krumschnabel G, Sohm B, Bock F, Manzl C, Villunger A. 2009. The enigma of caspase-2: The laymen's view. *Cell Death Differ* **16:** 195–207.

Kuida K, Haydar TF, Kuan CY, Gu Y, Taya C, Karasuyama H, Su MS, Rakic P, Flavell RA. 1998. Reduced apoptosis and cytochrome *c*-mediated caspase activation in mice lacking caspase 9. *Cell* **94:** 325–337.

Kuranaga E, Miura M. 2007. Nonapoptotic functions of caspases: Caspases as regulatory molecules for immunity and cell-fate determination. *Trends Cell Biol* **17:** 135–144.

Laguna A, Aranda S, Barallobre MJ, Barhoum R, Fernandez E, Fotaki V, Delabar JM, de la Luna S, de la Villa P, Arbones ML. 2008. The protein kinase DYRK1A regulates caspase-9-mediated apoptosis during retina development. *Dev Cell* **15:** 841–853.

Lakhani SA, Masud A, Kuida K, Porter GA Jr, Booth CJ, Mehal WZ, Inayat I, Flavell RA. 2006. Caspases 3 and 7: Key mediators of mitochondrial events of apoptosis. *Science* **311:** 847–851.

LeBlanc A, Liu H, Goodyer C, Bergeron C, Hammond J. 1999. Caspase-6 role in apoptosis of human neurons, amyloidogenesis, and Alzheimer's disease. *J Biol Chem* **274:** 23426–23436.

Li J, Yuan J. 2008. Caspases in apoptosis and beyond. *Oncogene* **27:** 6194–6206.

Li H, Zhu H, Xu CJ, Yuan J. 1998. Cleavage of BID by caspase 8 mediates the mitochondrial damage in the Fas pathway of apoptosis. *Cell* **94:** 491–501.

Li Z, Jo J, Jia JM, Lo SC, Whitcomb DJ, Jiao S, Cho K, Sheng M. 2010. Caspase-3 activation via mitochondria is required for long-term depression and AMPA receptor internalization. *Cell* **141:** 859–871.

Li X, Wen W, Liu K, Zhu F, Malakhova M, Peng C, Li T, Kim HG, Ma W, Cho YY, et al. 2011. Phosphorylation of caspase-7 by P21-activated protein kinase (PAK) 2 inhibits chemotherapeutic drugs-induced apoptosis of breast cancer cell lines. *J Biol Chem* **286:** 22291–22299.

Liu X, Kim CN, Yang J, Jemmerson R, Wang X. 1996. Induction of apoptotic program in cell-free extracts: Requirement for dATP and cytochrome c. *Cell* **86:** 147–157.

Luo X, Budihardjo I, Zou H, Slaughter C, Wang X. 1998. Bid, a Bcl2 interacting protein, mediates cytochrome *c* release from mitochondria in response to activation of cell surface death receptors. *Cell* **94:** 481–490.

Luthi AU, Martin SJ. 2007. The CASBAH: A searchable database of caspase substrates. *Cell Death Differ* **14:** 641–650.

Maejima Y, Adachi S, Morikawa K, Ito H, Isobe M. 2005. Nitric oxide inhibits myocardial apoptosis by preventing caspase-3 activity via S-nitrosylation. *J Mol Cell Cardiol* **38:** 163–174.

Mahrus S, Trinidad JC, Barkan DT, Sali A, Burlingame AL, Wells JA. 2008. Global sequencing of proteolytic cleavage sites in apoptosis by specific labeling of protein N termini. *Cell* **134:** 866–876.

Malladi S, Challa-Malladi M, Fearnhead HO, Bratton SB. 2009. The Apaf-1*procaspase-9 apoptosome complex functions as a proteolytic-based molecular timer. *EMBO J* **28:** 1916–1925.

Mannick JB, Hausladen A, Liu L, Hess DT, Zeng M, Miao QX, Kane LS, Gow AJ, Stamler JS. 1999. Fas-induced caspase denitrosylation. *Science* **284:** 651–654.

Mannick JB, Schonhoff C, Papeta N, Ghafourifar P, Szibor M, Fang K, Gaston B. 2001. S-Nitrosylation of mitochondrial caspases. *J Cell Biol* **154:** 1111–1116.

Manzl C, Krumschnabel G, Bock F, Sohm B, Labi V, Baumgartner F, Logette E, Tschopp J, Villunger A. 2009. Caspase-2 activation in the absence of PIDDosome formation. *J Cell Biol* **185:** 291–303.

Martin MC, Allan LA, Lickrish M, Sampson C, Morrice N, Clarke PR. 2005. Protein kinase A regulates caspase-9 activation by Apaf-1 downstream of cytochrome c. *J Biol Chem* **280:** 15449–15455.

Matthess Y, Raab M, Sanhaji M, Lavrik IN, Strebhardt K. 2010. Cdk1/cyclin B1 controls Fas-mediated apoptosis by regulating caspase-8 activity. *Mol Cell Biol* **30:** 5726–5740.

McDonnell MA, Abedin MJ, Melendez M, Platikanova TN, Ecklund JR, Ahmed K, Kelekar A. 2008. Phosphorylation of murine caspase-9 by the protein kinase casein kinase 2 regulates its cleavage by caspase-8. *J Biol Chem* **283:** 20149–20158.

Mei Y, Stonestrom A, Hou YM, Yang X. 2010. Apoptotic regulation and tRNA. *Protein Cell* **1:** 795–801.

Micheau O, Thome M, Schneider P, Holler N, Tschopp J, Nicholson DW, Briand C, Grutter MG. 2002. The long form of FLIP is an activator of caspase-8 at the Fas death-inducing signaling complex. *J Biol Chem* **277:** 45162–45171.

Morizane Y, Honda R, Fukami K, Yasuda H. 2005. X-linked inhibitor of apoptosis functions as ubiquitin ligase toward mature caspase-9 and cytosolic Smac/DIABLO. *J Biochem* **137:** 125–132.

Moroni MC, Hickman ES, Lazzerini Denchi E, Caprara G, Colli E, Cecconi F, Muller H, Helin K. 2001. Apaf-1 is a transcriptional target for E2F and p53. *Nat Cell Biol* **3:** 552–558.

Muro I, Hay BA, Clem RJ. 2002. The *Drosophila* DIAP1 protein is required to prevent accumulation of a continuously generated, processed form of the apical caspase DRONC. *J Biol Chem* **277:** 49644–49650.

Nicholson DW, Ali A, Thornberry NA, Vaillancourt JP, Ding CK, Gallant M, Gareau Y, Griffin PR, Labelle M, Lazebnik YA, et al. 1995. Identification and inhibition of the ICE/CED-3 protease necessary for mammalian apoptosis. *Nature* **376:** 37–43.

Nutt LK, Margolis SS, Jensen M, Herman CE, Dunphy WG, Rathmell JC, Kornbluth S. 2005. Metabolic regulation of oocyte cell death through the CaMKII-mediated phosphorylation of caspase-2. *Cell* **123:** 89–103.

Nutt LK, Buchakjian MR, Gan E, Darbandi R, Yoon SY, Wu JQ, Miyamoto YJ, Gibbons JA, Andersen JL, Freel CD, et al. 2009. Metabolic control of oocyte apoptosis mediated by 14-3-3ζ-regulated dephosphorylation of caspase-2. *Dev Cell* **16:** 856–866.

Oberst A, Dillon CP, Weinlich R, McCormick LL, Fitzgerald P, Pop C, Hakem R, Salvesen GS, Green DR. 2011. Catalytic activity of the caspase-8-FLIP(L) complex inhibits RIPK3-dependent necrosis. *Nature* **471:** 363–367.

Ohsawa S, Hamada S, Kuida K, Yoshida H, Igaki T, Miura M. 2010. Maturation of the olfactory sensory neurons by Apaf-1/caspase-9-mediated caspase activity. *Proc Natl Acad Sci* **107:** 13366–13371.

Park HH, Logette E, Raunser S, Cuenin S, Walz T, Tschopp J, Wu H. 2007. Death domain assembly mechanism revealed by crystal structure of the oligomeric PIDDosome core complex. *Cell* **128:** 533–546.

Pasterkamp RJ, Peschon JJ, Spriggs MK, Kolodkin AL. 2003. Semaphorin 7A promotes axon outgrowth through integrins and MAPKs. *Nature* **424:** 398–405.

Peng C, Cho YY, Zhu F, Zhang J, Wen W, Xu Y, Yao K, Ma WY, Bode AM, Dong Z. 2011. Phosphorylation of caspase-8 (Thr-263) by ribosomal S6 kinase 2 (RSK2) mediates caspase-8 ubiquitination and stability. *J Biol Chem* **286:** 6946–6954.

Pop C, Timmer J, Sperandio S, Salvesen GS. 2006. The apoptosome activates caspase-9 by dimerization. *Mol Cell* **22:** 269–275.

* Poręba M, Stróżyk A, Salvesen GS, Drąg M. 2013. Caspase substrates and inhibitors. *Cold Spring Harb Perspect Biol* doi: 10.1101/cshperspect.a008680.

Qi S, Pang Y, Hu Q, Liu Q, Li H, Zhou Y, He T, Liang Q, Liu Y, Yuan X, et al. 2010. Crystal structure of the *Caenorhabditis elegans* apoptosome reveals an octameric assembly of CED-4. *Cell* **141:** 446–457.

Raina D, Pandey P, Ahmad R, Bharti A, Ren J, Kharbanda S, Weichselbaum R, Kufe D. 2005. c-Abl tyrosine kinase regulates caspase-9 autocleavage in the apoptotic response to DNA damage. *J Biol Chem* **280:** 11147–11151.

Ray CA, Black RA, Kronheim SR, Greenstreet TA, Sleath PR, Salvesen GS, Pickup DJ. 1992. Viral inhibition of inflammation: Cowpox virus encodes an inhibitor of the interleukin-1 β converting enzyme. *Cell* **69:** 597–604.

Read SH, Baliga BC, Ekert PG, Vaux DL, Kumar S. 2002. A novel Apaf-1-independent putative caspase-2 activation complex. *J Cell Biol* **159:** 739–745.

Renatus M, Stennicke HR, Scott FL, Liddington RC, Salvesen GS. 2001. Dimer formation drives the activation of the cell death protease caspase 9. *Proc Natl Acad Sci* **98:** 14250–14255.

Ribaya JP, Ranmuthu M, Copeland J, Boyarskiy S, Blair AP, Hay B, Laski FA. 2009. The deubiquitinase emperor's thumb is a regulator of apoptosis in *Drosophila*. *Dev Biol* **329:** 25–35.

Ribeiro PS, Kuranaga E, Tenev T, Leulier F, Miura M, Meier P. 2007. DIAP2 functions as a mechanism-based regulator of drICE that contributes to the caspase activity threshold in living cells. *J Cell Biol* **179:** 1467–1480.

Riedl SJ, Salvesen GS. 2007. The apoptosome: Signalling platform of cell death. *Nat Rev Mol Cell Biol* **8:** 405–413.

Riedl SJ, Renatus M, Schwarzenbacher R, Zhou Q, Sun C, Fesik SW, Liddington RC, Salvesen GS. 2001. Structural basis for the inhibition of caspase-3 by XIAP. *Cell* **104:** 791–800.

Rikhof B, Corn PG, El-Deiry WS. 2003. Caspase 10 levels are increased following DNA damage in a p53-dependent manner. *Cancer Biol Ther* **2:** 707–712.

Robertson JD, Enoksson M, Suomela M, Zhivotovsky B, Orrenius S. 2002. Caspase-2 acts upstream of mitochondria to promote cytochrome c release during etoposide-induced apoptosis. *J Biol Chem* **277:** 29803–29809.

Rossig L, Fichtlscherer B, Breitschopf K, Haendeler J, Zeiher AM, Mulsch A, Dimmeler S. 1999. Nitric oxide inhibits caspase-3 by S-nitrosation in vivo. *J Biol Chem* **274:** 6823–6826.

Rudolf K, Cervinka M, Rudolf E. 2009. Cytotoxicity and mitochondrial apoptosis induced by etoposide in melanoma cells. *Cancer Invest* **27:** 704–717.

Ryoo HD, Bergmann A, Gonen H, Ciechanover A, Steller H. 2002. Regulation of *Drosophila* IAP1 degradation and apoptosis by reaper and ubcD1. *Nat Cell Biol* **4:** 432–438.

Scaffidi C, Fulda S, Srinivasan A, Friesen C, Li F, Tomaselli KJ, Debatin KM, Krammer PH, Peter ME. 1998. Two CD95 (APO-1/Fas) signaling pathways. *EMBO J* **17:** 1675–1687.

Scaffidi C, Schmitz I, Krammer PH, Peter ME. 1999. The role of c-FLIP in modulation of CD95-induced apoptosis. *J Biol Chem* **274:** 1541–1548.

Schafer ZT, Kornbluth S. 2006. The apoptosome: Physiological, developmental, and pathological modes of regulation. *Dev Cell* **10:** 549–561.

Schile AJ, Garcia-Fernandez M, Steller H. 2008. Regulation of apoptosis by XIAP ubiquitin-ligase activity. *Genes Dev* **22:** 2256–2266.

Seifert A, Clarke PR. 2009. p38α- and DYRK1A-dependent phosphorylation of caspase-9 at an inhibitory site in response to hyperosmotic stress. *Cell Signal* **21:** 1626–1633.

Seifert A, Allan LA, Clarke PR. 2008. DYRK1A phosphorylates caspase 9 at an inhibitory site and is potently inhibited in human cells by harmine. *FEBS J* **275:** 6268–6280.

Senft J, Helfer B, Frisch SM. 2007. Caspase-8 interacts with the p85 subunit of phosphatidylinositol 3-kinase to regulate cell adhesion and motility. *Cancer Res* **67:** 11505–11509.

Seol DW, Billiar TR. 1999. A caspase-9 variant missing the catalytic site is an endogenous inhibitor of apoptosis. *J Biol Chem* **274:** 2072–2076.

Shapiro PJ, Hsu HH, Jung H, Robbins ES, Ryoo HD. 2008. Regulation of the *Drosophila* apoptosome through feedback inhibition. *Nat Cell Biol* **10:** 1440–1446.

Shin S, Lee Y, Kim W, Ko H, Choi H, Kim K. 2005. Caspase-2 primes cancer cells for TRAIL-mediated apoptosis by processing procaspase-8. *EMBO J* **24:** 3532–3542.

Shiozaki EN, Chai J, Rigotti DJ, Riedl SJ, Li P, Srinivasula SM, Alnemri ES, Fairman R, Shi Y. 2003. Mechanism of XIAP-mediated inhibition of caspase-9. *Mol Cell* **11:** 519–527.

Shultz JC, Goehe RW, Wijesinghe DS, Murudkar C, Hawkins AJ, Shay JW, Minna JD, Chalfant CE. 2010. Alternative splicing of caspase 9 is modulated by the phosphoinositide 3-kinase/Akt pathway via phosphorylation of SRp30a. *Cancer Res* **70:** 9185–9196.

Sidi S, Sanda T, Kennedy RD, Hagen AT, Jette CA, Hoffmans R, Pascual J, Imamura S, Kishi S, Amatruda JF, et al. 2008. Chk1 suppresses a caspase-2 apoptotic response to DNA damage that bypasses p53, Bcl-2, and caspase-3. *Cell* **133:** 864–877.

* Silke J, Meier P. 2013. Inhibitor of apoptosis (IAP) proteins—Modulators of cell death and inflammation. *Cold Spring Harb Perspect Biol* **5:** a008730.

Slee EA, Adrain C, Martin SJ. 1999. Serial killers: Ordering caspase activation events in apoptosis. *Cell Death Differ* **6:** 1067–1074.

Sprick MR, Rieser E, Stahl H, Grosse-Wilde A, Weigand MA, Walczak H. 2002. Caspase-10 is recruited to and activated at the native TRAIL and CD95 death-inducing signalling complexes in a FADD-dependent manner but can not functionally substitute caspase-8. *EMBO J* **21:** 4520–4530.

Srinivasula SM, Ahmad M, Guo Y, Zhan Y, Lazebnik Y, Fernandes-Alnemri T, Alnemri ES. 1999. Identification of an endogenous dominant-negative short isoform of caspase-9 that can regulate apoptosis. *Cancer Res* **59:** 999–1002.

Srinivasula SM, Hegde R, Saleh A, Datta P, Shiozaki E, Chai J, Lee RA, Robbins PD, Fernandes-Alnemri T, Shi Y, et al. 2001. A conserved XIAP-interaction motif in caspase-9 and Smac/DIABLO regulates caspase activity and apoptosis. *Nature* **410:** 112–116.

Stennicke HR, Renatus M, Meldal M, Salvesen GS. 2000. Internally quenched fluorescent peptide substrates disclose the subsite preferences of human caspases 1, 3, 6, 7 and 8. *Biochem J* **350** (Pt 2): 563–568.

Suzuki Y, Nakabayashi Y, Takahashi R. 2001. Ubiquitin-protein ligase activity of X-linked inhibitor of apoptosis protein promotes proteasomal degradation of caspase-3 and enhances its anti-apoptotic effect in Fas-induced cell death. *Proc Natl Acad Sci* **98:** 8662–8667.

Suzuki A, Kusakai G, Kishimoto A, Shimojo Y, Miyamoto S, Ogura T, Ochiai A, Esumi H. 2004. Regulation of caspase-6 and FLIP by the AMPK family member ARK5. *Oncogene* **23:** 7067–7075.

Thome M, Schneider P, Hofmann K, Fickenscher H, Meinl E, Neipel F, Mattmann C, Burns K, Bodmer JL, Schroter M, et al. 1997. Viral FLICE-inhibitory proteins (FLIPs) prevent apoptosis induced by death receptors. *Nature* **386:** 517–521.

Thompson CB. 1995. Apoptosis in the pathogenesis and treatment of disease. *Science* **267:** 1456–1462.

Thornberry NA, Lazebnik Y. 1998. Caspases: Enemies within. *Science* **281:** 1312–1316.

Thornberry NA, Rano TA, Peterson EP, Rasper DM, Timkey T, Garcia-Calvo M, Houtzager VM, Nordstrom PA, Roy S, Vaillancourt JP, et al. 1997. A combinatorial approach defines specificities of members of the caspase family and granzyme B. Functional relationships established for key mediators of apoptosis. *J Biol Chem* **272:** 17907–17911.

Tinel A, Tschopp J. 2004. The PIDDosome, a protein complex implicated in activation of caspase-2 in response to genotoxic stress. *Science* **304:** 843–846.

Torok NJ, Higuchi H, Bronk S, Gores GJ. 2002. Nitric oxide inhibits apoptosis downstream of cytochrome *c* release by nitrosylating caspase 9. *Cancer Res* **62:** 1648–1653.

Trauth BC, Klas C, Peters AM, Matzku S, Moller P, Falk W, Debatin KM, Krammer PH. 1989. Monoclonal antibody-mediated tumor regression by induction of apoptosis. *Science* **245:** 301–305.

Tu S, McStay GP, Boucher LM, Mak T, Beere HM, Green DR. 2006. In situ trapping of activated initiator caspases reveals a role for caspase-2 in heat shock-induced apoptosis. *Nat Cell Biol* **8:** 72–77.

Vakifahmetoglu H, Olsson M, Orrenius S, Zhivotovsky B. 2006. Functional connection between p53 and caspase-2 is essential for apoptosis induced by DNA damage. *Oncogene* **25:** 5683–5692.

Van Damme P, Martens L, Van Damme J, Hugelier K, Staes A, Vandekerckhove J, Gevaert K. 2005. Caspase-specific and nonspecific in vivo protein processing during Fas-induced apoptosis. *Nat Methods* **2:** 771–777.

van Raam BJ, Salvesen GS. 2012. Proliferative versus apoptotic functions of caspase-8 Hetero or homo: The caspase-8 dimer controls cell fate. *Biochim Biophys Acta* **1824:** 113–122.

Voss OH, Kim S, Wewers MD, Doseff AI. 2005. Regulation of monocyte apoptosis by the protein kinase Cδ-dependent phosphorylation of caspase-3. *J Biol Chem* **280:** 17371–17379.

Wachmann K, Pop C, van Raam BJ, Drag M, Mace PD, Snipas SJ, Zmasek C, Schwarzenbacher R, Salvesen GS, Riedl SJ. 2010. Activation and specificity of human caspase-10. *Biochemistry* **49:** 8307–8315.

Walsh JG, Cullen SP, Sheridan C, Luthi AU, Gerner C, Martin SJ. 2008. Executioner caspase-3 and caspase-7 are functionally distinct proteases. *Proc Natl Acad Sci* **105:** 12815–12819.

Wang J, Chun HJ, Wong W, Spencer DM, Lenardo MJ. 2001. Caspase-10 is an initiator caspase in death receptor signaling. *Proc Natl Acad Sci* **98:** 13884–13888.

Wang XJ, Cao Q, Liu X, Wang KT, Mi W, Zhang Y, Li LF, LeBlanc AC, Su XD. 2010. Crystal structures of human caspase 6 reveal a new mechanism for intramolecular cleavage self-activation. *EMBO Rep* **11:** 841–847.

Wilson R, Goyal L, Ditzel M, Zachariou A, Baker DA, Agapite J, Steller H, Meier P. 2002. The DIAP1 RING finger mediates ubiquitination of Dronc and is indispensable for regulating apoptosis. *Nat Cell Biol* **4:** 445–450.

Wing JP, Schreader BA, Yokokura T, Wang Y, Andrews PS, Huseinovic N, Dong CK, Ogdahl JL, Schwartz LM, White K, et al. 2002. *Drosophila* Morgue is an F box/ubiquitin conjugase domain protein important for grim-reaper mediated apoptosis. *Nat Cell Biol* **4:** 451–456.

Yan N, Shi Y. 2005. Mechanisms of apoptosis through structural biology. *Annu Rev Cell Dev Biol* **21:** 35–56.

Yang X, Chang HY, Baltimore D. 1998. Essential role of CED-4 oligomerization in CED-3 activation and apoptosis. *Science* **281:** 1355–1357.

Yang CS, Thomenius MJ, Gan EC, Tang W, Freel CD, Merritt TJ, Nutt LK, Kornbluth S. 2010. Metabolic regulation of *Drosophila* apoptosis through inhibitory phosphorylation of Dronc. *EMBO J* **29:** 3196–3207.

Yi CH, Yuan J. 2009. The Jekyll and Hyde functions of caspases. *Dev Cell* **16:** 21–34.

Yuan J, Yankner BA. 2000. Apoptosis in the nervous system. *Nature* **407:** 802–809.

Yuan J, Shaham S, Ledoux S, Ellis HM, Horvitz HR. 1993. The *C. elegans* cell death gene ced-3 encodes a protein similar to mammalian interleukin-1 β-converting enzyme. *Cell* **75:** 641–652.

Yuan S, Yu X, Topf M, Ludtke SJ, Wang X, Akey CW. 2010. Structure of an apoptosome-procaspase-9 CARD complex. *Structure* **18:** 571–583.

Zou H, Henzel WJ, Liu X, Lutschg A, Wang X. 1997. Apaf-1, a human protein homologous to *C. elegans* CED-4, participates in cytochrome *c*-dependent activation of caspase-3. *Cell* **90:** 405–413.

and colleagues at Genentech in San Francisco, independently found an expressed sequence tag (EST) in the public database with homology to CD95L, which was even annotated as such. The TNF-related apoptosis-inducing ligand (TRAIL), or Apo2L, as these groups named it, respectively, killed many cancer, but not normal, cells in vitro (Wiley et al. 1995; Pitti et al. 1996). So, was TRAIL going to finally fulfill TNF's promise?

In 1999, we at Immunex and Ashkenazi's team at Genentech found that systemic treatment of tumor-bearing mice with recombinant TRAIL killed tumor cells in vivo (Ashkenazi et al. 1999; Walczak et al. 1999). Importantly, a highly active form of TRAIL, capable of inducing apoptosis in mouse cells, neither induced a systemic shock syndrome, as seen with TNF, nor fulminant toxic effects owing to apoptosis induction, as observed with CD95L (Walczak et al. 1999). This meant that with TRAIL a TNF-like cytokine could be used systemically to kill tumor cells without being toxic.

Given TRAIL's potential to induce apoptosis in cancer but not normal cells, the receptor responsible for mediating this activity appeared to be an attractive target for the development of agonists capable of cross-linking it. The result of the race for the cloning of this receptor was quite surprising as TRAIL has five receptors of which two can mediate apoptosis.

At the time, in 1996 and 1997, many new human genes with homology to already known members of interesting protein families, like TRAIL itself, were found in EST databases. The EST "data mining" approach of a privately owned database led to the discovery of the first apoptosis-inducing receptor for TRAIL, now referred to as TRAIL-R1 or DR4, by the Dixit team (Pan et al. 1997a). Using a biochemical purification approach with TRAIL as bait we identified a different apoptosis-inducing receptor for TRAIL, TRAIL-R2 (Walczak et al. 1997). Shortly thereafter, TRAIL-R2 was also discovered by a number of other groups when its sequence appeared in public and private EST databases, and it received several other names including DR5, TRICK2, and KILLER (Pan et al. 1997a; Screaton et al. 1997a; Sheridan et al. 1997; Wu et al. 1997). In the following weeks and months, work by a number of groups led to the identification of two additional receptors for TRAIL, TRAIL-R3 (DcR1) and TRAIL-R4 (DcR2), which cannot induce apoptosis because they lack an intracellular DD (reviewed in Walczak and Krammer 2000). At first, these receptors were thought to exert a decoy function for TRAIL (hence the name "decoy receptor" [DcR]) by being expressed on normal but not cancer cells. When the expression patterns of the TRAIL-Rs were analyzed, it became clear, however, that this concept did not hold. To this day we know little about the physiological function of these nondeath receptors for TRAIL. Finally, osteoprotegerin (OPG), a soluble, high-affinity receptor for RANKL, another member of the TNFSF, was found to interact with TRAIL, yet with lower affinity. It is rather unlikely that this interaction is relevant in vivo because mice overexpressing TRAIL do not show any bone-related phenotype, which would have been expected if TRAIL were capable of interacting with OPG in vivo, because in mice deficient for OPG, osteoclastogenesis is not inhibited and these mice therefore develop a strong bone-related phenotype (Simonet et al. 1997). In summary, TRAIL turned out to be the most promiscuous TNFSF member.

TRAIL-R1 and TRAIL-R2 share 58% sequence homology and so far it has not been possible to identify significant distinct functions of one receptor versus the other. They both trigger apoptosis via the same pathway and, at present, even no differences to the CD95-engaged pathway are known. The big quest, therefore, is to identify the biochemical basis for the diametrically different outcome of the triggering of the TRAIL-Rs versus CD95 in normal cells, as the unraveling of this may lead the way to more targeted and patient-tailored applications of TRAIL in cancer treatment.

Based on the encouraging preclinical results with recombinant TRAIL in vivo (Ashkenazi et al. 1999; Walczak et al. 1999), Immunex and Genentech joined forces to explore the clinical potential of a first recombinant form of TRAIL in cancer treatment. Even though some clinical benefit has been detected in the various clinical trials in which this form of TRAIL has now been tested, the results of these trials were not nearly as striking as initially hoped. Apart from this form of TRAIL, "agonistic" antibodies to TRAIL-R1 and TRAIL-R2 have now been tested clinically (reviewed in Newsom-Davis et al. 2009).

Overall, the results of these studies are, however, disappointing. One reason for this is the absence of biomarkers to select patients who are likely to respond to TRAIL receptor agonist therapy, either alone or in combination with a particular TRAIL-sensitizing drug. To effectively use TRAIL-comprising cancer therapies, it will be crucial, however, to identify such biomarkers.

A perhaps more important reason for the lack of clear clinical activity of the currently developed TRAIL receptor agonists likely lies in their very nature, which simply renders them rather weak inducers of apoptosis. The recombinant form of TRAIL that has been developed clinically displays rather low apoptosis-inducing activity when compared with higher activity forms, perhaps also because the formulation of this ligand may not exactly match the biological moiety. Even though TRAIL had been suggested to be hepatotoxic (Lawrence et al. 2001), which may explain the hesitation to develop a more active form of TRAIL, we found that highly active forms of TRAIL are neither toxic to mice (Walczak et al. 1999) nor to primary human hepatocytes, even in combination with various TRAIL-sensitizing drugs (Ganten et al. 2005, 2006), suggesting that development and clinical application of high-activity forms of TRAIL and other TRAIL receptor agonists should be possible.

With respect to the antibodies, the failure likely has a different cause. Antibodies of the immunoglobulin G (IgG) type contain two antigen-binding sites. Triggering of TNFRSF members, however, requires cross-linking of at least three receptors for agonistic activity. Hence, IgG antibodies cannot do this without a cross-linking platform. This was first shown by the Krammer team for antibodies to CD95 (Dhein et al. 1992) and recently by the Genentech team for their clinically developed TRAIL-R2-specific antibody (Wilson et al. 2011).

As disappointing as it is that more than a decade after the demonstration of the tumor-specific apoptosis-inducing potential of a highly active form of TRAIL in vivo (Walczak et al. 1999) we still do not have a high-activity TRAIL receptor agonist in clinical use, it seems that this issue could soon be overcome with second-generation, high-activity TRAIL receptor agonists currently being bound for the clinic. It will be exciting to see how they fare, and whether TRAIL will finally be able to hold the promise that TNF and CD95L could not keep.

CD95- AND TRAIL-INDUCED APOPTOSIS

A prime goal for scientists working on CD95 was to elucidate how receptor cross-linking led to apoptosis induction in target cells. Nagata and colleagues made the first step by showing that apoptosis induction by CD95 required a part of the cytoplasmic domain, the DD (Itoh and Nagata 1993). They, as well as Goeddel and colleagues (Tartaglia et al. 1993), identified a similar DD in TNF-R1.

Discovery of FADD, Caspase-8, and the Death-Inducing Signaling Complex

Using yeast two-hybrid (Y2H), Dixit's team in Ann Arbor and Wallach's team in Rehovot discovered a protein they called FADD and MORT1, respectively, as a CD95-DD-interacting factor (Boldin et al. 1995a,b; Chinnaiyan et al. 1995). Apart from a DD that interacted with the CD95-DD, FADD also contained a second, DD-like domain, termed death-effector domain (DED). In the meantime, Kischkel and colleagues from the team led by Krammer and Peter biochemically studied the signaling complex that forms when CD95 is cross-linked to induce apoptosis. They dubbed this complex death-inducing signaling complex (DISC) and found that FADD and one additional, unidentified factor formed part of the CD95 DISC (Kischkel et al. 1995). Collaborating with Mann's team in Heidelberg, they determined the identity of the elusive second factor and, together with Dixit's team, were able to show how it works (Muzio et al. 1996). The identified factor was homologous to IL-1-converting enzyme (ICE) (Cerretti et al. 1992; Thornberry et al. 1992). CED-3, one of the key factors identified by Horvitz and colleagues in what later became a Nobel Prize-winning effort to define the genes required for cell death in *Caenorhabditis elegans*, was also homologous to ICE (Yuan and Horvitz 1990; Yuan et al. 1993). Consequently, the protein was named FLICE for FADD-like ICE. FLICE is FADD-like because it contains two DEDs that are homologous to the DED of FADD. The same protein was discovered by Wallach's team in a Y2H screen with the DED of MORT1 (FADD). They named the protein MACH (Boldin et al. 1996). The protein formerly known as FLICE or MACH is now referred to as caspase-8. Caspases are cysteine-dependent aspartate-directed proteases (i.e., proteases that cleave target proteins at aspartic acid residues and use cysteine in their active center). There are 14 mammalian caspases (11 in humans and 10 in rodents) which, interestingly, have been classified as "apoptotic" or "inflammatory" (Creagh and Martin 2001; Salvesen and Riedl 2008; Pop and Salvesen 2009).

The mechanism of DISC formation and CD95-induced apoptosis identified is as follows: cross-linking of CD95 results in recruitment of FADD, which in turn recruits caspase-8 to form the DISC. Caspase-8 is then activated at the DISC, resulting in cleavage of downstream substrates. One of them is caspase-3, the mammalian homolog of CED-3 and crucial effector caspase in the apoptosis pathway (Creagh and Martin 2001). Once caspase-3 is activated it cleaves many vital cellular proteins responsible for the characteristic biochemical and morphological hallmarks of apoptosis. Thus, the discovery of caspase-8 and its function at the DISC provided the missing link between CD95 cross-linking and the known part of the apoptosis pathway.

In the following years the process of DISC formation and downstream as well as regulatory events of CD95-induced apoptosis were further refined. Besides FADD and caspase-8, two related proteins, the cellular FLICE-like inhibitory protein (cFLIP) and caspase-10 (Irmler et al. 1997; Thome et al. 1997; Kischkel et al. 2001; Sprick et al. 2002), which both also contain DEDs, are recruited to the CD95 DISC. Although cFLIP lacks a catalytic cysteine in its active center and therefore blocks CD95-induced apoptosis, caspase-10 is highly homologous to caspase-8. To this day, however, the function of caspase-10 remains unclear (Kischkel et al. 2001; Sprick et al. 2002). Recently, the stoichiometry of the CD95 DISC was addressed. Surprisingly, these studies proposed that three receptors recruit one FADD protein and this FADD molecule recruits six to 10 DED-containing proteins (Fig. 2) (Dickens et al. 2012; Schleich et al. 2012).

Connecting the Death Receptor and the Mitochondrial Apoptosis Pathways

The proapoptotic BH3-only family member Bid was identified as a second critical substrate of caspase-8 for engagement of the cellular apoptosis machinery (Gross et al. 1999; Yin et al. 1999). Caspase-8-cleaved, truncated Bid (tBid) translocates from the cytosol to mitochondria where it induces mitochondrial outer membrane permeabilization (MOMP), thereby activating the mitochondrial apoptosis pathway. These processes are discussed in detail in Tait and Green (2013). In the context of this work, however, there are two crucial points about these processes: (1) cleavage of Bid by caspase-8 provides the missing link between the death receptor and mitochondrial apoptosis pathways, and (2) MOMP does not only induce release of cytochrome c but also of the second mitochondrial activator of caspases (SMAC, also known as DIABLO). Whereas cytochrome c release triggers apoptosome formation and activation of caspase-9, cytosolic SMAC binds to, and thereby neutralizes, the X-linked inhibitor of apoptosis protein (XIAP). XIAP, in turn, normally binds to, and thereby inhibits, caspase-8-cleaved caspases 3 and 9. Hence, SMAC releases these caspases from their XIAP-imposed inhibition and cell death can finally ensue (Fig. 2).

Requirement for tBid-induced MOMP distinguishes so-called type I from type II cells. Type I cells do not require MOMP to undergo

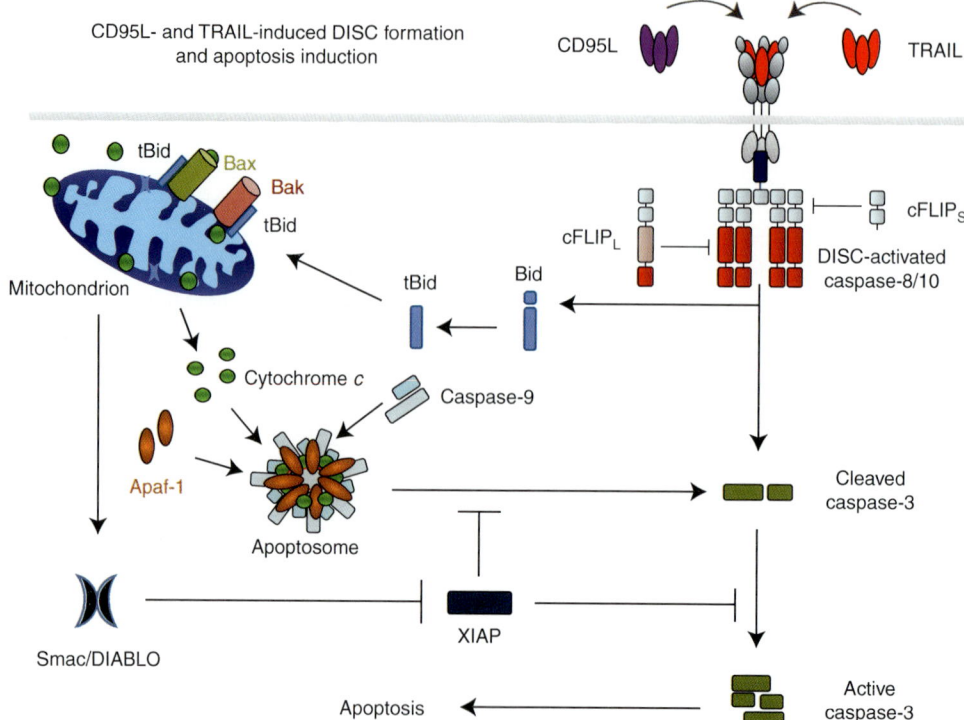

Figure 2. CD95L- and TRAIL-induced DISC formation and apoptosis induction. Binding of CD95L or TRAIL to their cognate receptors leads to receptor trimerization and formation of the death-inducing signaling complex (DISC). The DD of one FADD molecule interacts with the DDs of the three cross-linked receptors. Subsequently, procaspases 8 and 10 are recruited by interaction of their DED with that of FADD. Heterodimers between cFLIP$_L$ and caspase-8 or -10 are not inactive as proteolytic enzymes but their proteolytic activity is edited as compared with caspase-8 or -10 homodimers. cFLIP$_S$ in turn inhibits caspase activity at the DISC by preventing dimerization of caspase-8/10/cFLIP$_L$. Interestingly, one FADD molecule recruits six to 10 DED-containing caspase-8, -10, and cFLIP proteins. DISC-activated caspase-8/10 cleaves caspase-3, enabling autoactivation of caspase-3, which renders the enzyme fully active. This latter step can, however, be blocked by X-linked inhibitor of apoptosis protein (XIAP). DISC-activated caspase-8/10 also cleaves Bid (tBID). Interaction of tBid with Bak and Bax in the mitochondrial outer membrane induces Bax/Bak oligomerization, resulting in mitochondrial outer-membrane permeabilization (MOMP) so that cytochrome c and Smac/DIABLO are released from the mitochondrial intermembrane space to the cytosol. Cytochrome c, together with Apaf-1 and caspase-9, forms the apoptosome, the activation platform for caspase-9, which can, however, also be inhibited by XIAP. Smac/DIABLO binds to XIAP, which releases caspase-3 and -9 from XIAP-imposed inhibition. Activation of these caspases enables execution of apoptotic cell death.

CD95-induced apoptosis—even though it may nevertheless occur—whereas type II cells do (reviewed in Peter and Krammer 1998). In cells expressing a high XIAP/caspase-3 ratio, full activation of caspase-3 following caspase-8-mediated cleavage is blocked. Therefore, XIAP is a crucial factor for categorizing cells as type I or type II regarding CD95-mediated apoptosis (Jost et al. 2009). Differences in CD95 DISC formation, first thought to be solely responsible for the type I/type II distinction (reviewed in Peter and Krammer 1998), likely contribute to this categorization through different caspase activity outputs of the DISC in type I versus type II cells. Hence, the balance between DISC-generated caspase activity and XIAP/caspase-3 ratio is decisive for categorization of a given cell as type I or type II.

TRAIL-Induced Apoptosis

Once the TRAIL receptors were cloned, the pathway of TRAIL-induced apoptosis was investigated. Despite the initial claim that TRAIL would not use FADD as adaptor (Marsters et al. 1996; Pan et al. 1997b), we and others showed that both TRAIL-R1 and TRAIL-R2 use FADD and caspase-8 (Walczak et al. 1997; Kischkel et al. 2000, 2001; Sprick et al. 2000, 2002), just like CD95 (Fig. 2). It has been a conundrum that agonists of CD95 are so toxic, whereas even highly active TRAIL receptor agonists are not (Walczak et al. 1999), even when combined with TRAIL-apoptosis-sensitizing drugs (Ganten et al. 2005, 2006; Koschny et al. 2007). Solving this conundrum will be an important step forward in understanding the biology of death receptors in general but particularly for the use of TRAIL receptor agonists in cancer therapy.

TNF-/TNF-R1-INDUCED GENE ACTIVATION AND CELL DEATH

TNF has two cellular receptors. Whereas TNF-R1 is expressed in many tissues, TNF-R2 is almost exclusively present on lymphoid and endothelial cells. TNF-R1 contains a DD, whereas TNF-R2 does not. TNF-R1 initiates the majority of TNF-induced biological activities, including induction of cell death (Wajant et al. 2003).

The TNF-R1 Signaling Complex (TNF–RSC)

Binding of TNF to TNF-R1 triggers a series of intracellular events, but the primary output is activation of NF-κB and the mitogen-activated protein kinases (MAPKs) c-Jun amino-terminal kinase (JNK) and p38. The resulting gene activation is crucial for the inflammatory response to infection. Induction of cell death by TNF is a secondary signal that is only triggered when the gene-activatory signals are too weak or otherwise perturbed.

Binding of TNF to TNF-R1 induces receptor oligomerization and formation of the TNF-R1 signaling complex (TNF–RSC), which is initiated by recruitment of the DD-containing proteins TRADD and RIP1 to the DD of the receptor (Fig. 3). TRADD then recruits TRAF2, which in turn serves as a recruitment platform for cIAP1 and cIAP2. cIAPs are E3 ubiquitin ligases and, following their recruitment to the TNF–RSC, they polyubiquitinate RIP1, which enables activation of NF-κB and MAPKs. Ubiquitin chains can be formed via linkages of the ubiquitin subunits on different ε-amino groups of the seven lysines present in ubiquitin or via the α-amino group at the amino terminus of ubiquitin, the latter creating linear ubiquitin chains (Rieser et al. 2012). Until recently, it was thought that polyubiquitin chains involved in TNF signaling are either linked via the ε-amino groups of lysine 63 (K63) or K48 of ubiquitin. Recently, however, we showed that linear ubiquitin chains are also present on components of the TNF–RSC and are crucial in providing the physiological TNF signaling output (Gerlach et al. 2011). A protein complex termed LUBAC, for linear ubiquitin chain assembly complex, is recruited to the TNF–RSC in a cIAP-activity-dependent manner and required for activation of NF-κB and MAPKs to the full physiologically required extent. This is most likely owing to the fact that NEMO binds more strongly to linear than K63-linked ubiquitin chains (Lo et al. 2009; Rahighi et al. 2009). Besides linear and K63 ubiquitin linkages, we also found K11 and K48 linkages on RIP1 in the native TNF–RSC (Gerlach et al. 2011). Hence, different forms of ubiquitin linkages cooperate to achieve the physiological signaling output of the TNF–RSC. This suggested the following model: Rather than one long K63-linked chain on RIP1, differently linked—possibly rather short—chains, which are placed in exact positions on various TNF–RSC components, enable precise construction of this protein complex, regarding both positioning of recruited functional units and timing of complex construction and deconstruction (Walczak 2011). Thereby, the functional units of the NEMO/IKK and TAB/TAK complexes are recruited at exact times and into predefined positions within the complex so that the TNF–RSC works as designed and gene expression by NF-κB and

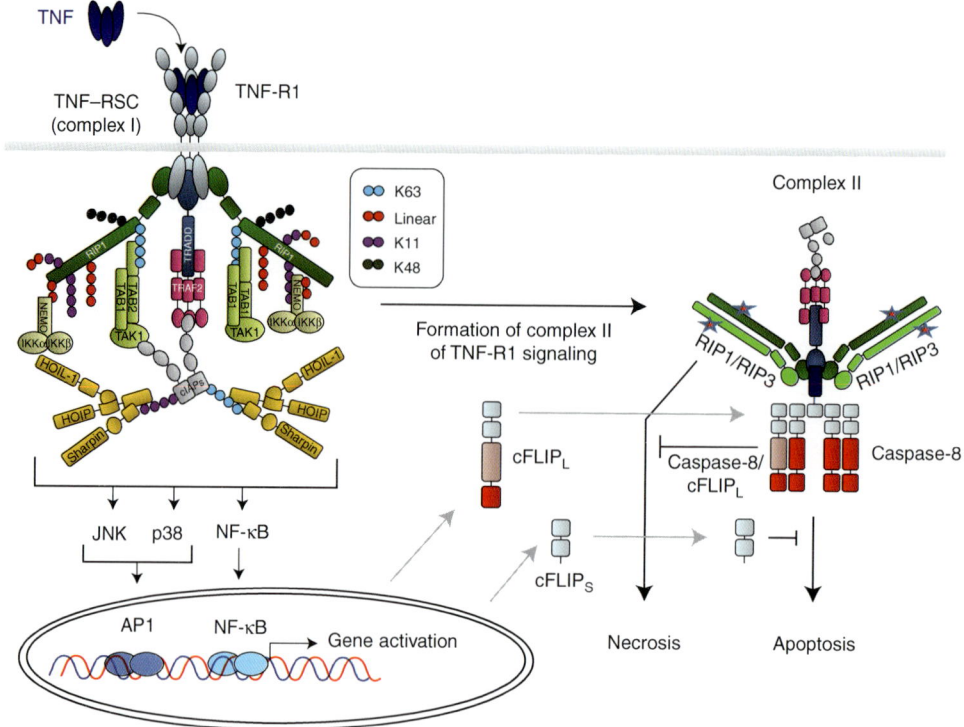

Figure 3. TNF-R1-induced gene activation and cell death signaling. Cross-linking of TNF-R1 by TNF results in formation of the TNF-R1 signaling complex (TNF–RSC). Cross-linked TNF-R1 recruits TRADD and RIP1 to the DD of the receptor. Subsequently, TRADD recruits TRAF2, which in turn provides the platform for cIAP1/2. cIAPs then place ubiquitin chains, linked via different interubiquitin linkages, on various TNF–RSC components. cIAP-Mediated ubiquitination is required to recruit LUBAC. Once recruited, LUBAC places linearly linked ubiquitin linkages on RIP1 and NEMO. Together, the different cIAP- and LUBAC-generated ubiquitin chains, placed in defined positions and sequences on specific components of the TNF–RSC, enable the physiologically required gene-activatory capacity of this complex by mediating the exact positioning of both the IKK and TAB/TAK complexes in the TNF–RSC. The different ubiquitin linkages are indicated in different colors. The depicted chain lengths, the sequence of the different linkages in them, and their exact positioning on different TNF–RSC components are only shown as examples in this model of the TNF–RSC as they are currently mostly unknown. Most likely involving the action of deubiquitinases (DUBs; not depicted here), the TNF–RSC releases TRADD, together with RIP1 and other cytoplasmic constituents of the complex, into the cytosol. This secondary complex, complex II, recruits FADD, caspase-8/10, and, when expressed, the different isoforms of cFLIP and RIP3. RIP1/3-induced necrosis from complex II is counteracted by the activity of the caspase-8/cFLIP$_L$ heteromer, and FADD/caspase-8-mediated apoptosis by cFLIP$_S$ and possibly also by cFLIP$_L$. Depending on the relative presence of the components in complex II, it can therefore either initiate FADD/caspase-8-dependent apoptosis or RIP1/RIP3-kinase-activity-dependent necrosis, or, when cFLIP and perhaps other, currently unknown inhibitory factors are present in the complex at sufficiently high levels, its cell-death-inducing capacity may be entirely inhibited.

AP-1 is activated exactly as intended and physiologically required (Fig. 3).

Disassembly of the complex is mainly mediated by so-called deubiquitinases (DUBs), enzymes that cleave ubiquitin from target proteins. It appears that certain DUBs, including the TNF–RSC-recruited A20, Cezanne, and CYLD, act in a ubiquitin-linkage-specific manner (i.e., they cleave certain types of interubiquitin linkages but not others) (Komander et al. 2009b). Their recruitment occurs later than that of the E3s and the functional units responsible for activating gene induction in response to TNF. At least with respect to A20, TNF-induced gene activation is decisive for the appearance of this protein in the complex (Gerlach et al. 2011). We are, however, only beginning to understand which DUBs cleave which interubiquitin linkages (Komander et al. 2009a), let alone in the context of particular RSCs or on specific components thereof. Likely, exact positioning of DUBs within RSCs and the timing thereof will be crucial.

Complex II of TNF-R1 Signaling: Cell Death in Two Flavors

When the TNF–RSC is deconstructed it does not disintegrate into its individual constituents. Micheau and Tschopp discovered that, following formation of the TNF–RSC, a second complex appears intracellularly that contains TRADD and some, but not all, of the other cytoplasmic components of the TNF–RSC. Additionally the cytoplasmic DISC components FADD, caspase-8, caspase-10, and cFLIP are also recruited to complex II of TNF-R1 signaling (Micheau and Tschopp 2003). It is, hence, an "intracellular DISC" whose formation temporally follows TNF–RSC (complex I) formation, and which generates the cell-death-inducing signals of TNF. However, when complex I works properly, complex II does not get to induce cell death because its signaling output is counteracted by the gene-activatory signals generated at complex I as these increase expression of prosurvival proteins, including the two isoforms of cFLIP, $cFLIP_L$, and $cFLIP_S$. Thus, if complex I is not decomposed prematurely and its output reaches physiological levels, complex II-induced cell death is prevented (Fig. 3).

It has long been known that TNF cannot only induce apoptotic but also necrotic cell death (reviewed in Vandenabeele et al. 2010). It emerged only recently, however, that TNF-induced necrosis is the result of a regulatable process that requires the kinase activities of both RIP1 and RIP3 (Holler et al. 2000; Hitomi et al. 2008; Cho et al. 2009; He et al. 2009; Zhang et al. 2009). For a long time it was a mystery why deficiency in caspase-8 or FADD is embryonically lethal if all they did was induce apoptosis. Recently, however, a plausible explanation was offered: Concomitant deletion of RIP3 or RIP1 rescued mice deficient for caspase-8 or FADD from embryonic lethality (Kaiser et al. 2011; Oberst et al. 2011; Zhang et al. 2011). This means, on the one hand, that RIP1 and RIP3 mediate lethality induced by deficiency in caspase-8 or FADD, but on the other, that the presence of caspase-8 and FADD prevents this. Thus, besides their role in apoptosis induction, FADD and caspase-8 also act together to prevent RIP1/RIP3-induced lethality. Interestingly, cFLIP deficiency is also lethal (Yeh et al. 2000). This lethality could only be reversed by concomitant deficiency in RIP3 and FADD but not when either factor was absent individually (Dillon et al. 2012). Thus, cFLIP is required for caspase-8 and FADD to block RIP1/RIP3-induced lethality (van Raam and Salvesen 2012), but also to interfere with caspase-8-induced lethality. It appears that the different isoforms of cFLIP may play distinct roles in these processes as $cFLIP_S$ interferes with apoptosis and $cFLIP_L$ with necrosis and, at least at high expression levels, also with apoptosis (Kavuri et al. 2011). Hence, RIP1/RIP3- and caspase-8/FADD/cFLIP-mediated signals control each other at complex II. Interestingly, recent studies have shown that inflammation in mice deficient for caspase-8 or FADD in either skin or gut (Kovalenko et al. 2009), does not occur when RIP3 is absent (Welz et al. 2011; Gunther et al. 2012). Whether this is solely because of prevention of RIP3-induced necrosis, or whether other RIP3-mediated signals may also play a role in preventing inflammation induced by deficiency in DISC components (Vince et al. 2012; Wallach et al. 2012), remains to be resolved.

In summary, our current model of the cell-death-inducing activities at complex II is as follows: FADD and caspase-8 induce apoptosis, which is prevented in the presence of sufficient amounts of cFLIP that are also required, together with FADD as adaptor and caspase-8 as an enzymatic partner, to interfere with RIP1/3-induced necrosis. There must be a fine balance between the two types of cell death induced by this complex, and the expression levels of cFLIP isoforms likely play an important part in tipping it one way or the other (Fig. 3).

INTEGRATING CELL DEATH AND INFLAMMATION

TRAIL and CD95L cannot only induce apoptosis but also proinflammatory, proliferative, and promigratory signals (Ehrhardt et al. 2003; Budd et al. 2006; Peter et al. 2007; Wilson et al. 2009; Lemke et al. 2010; Roder et al. 2011). Some of these signals occur in a FADD/caspase-8-independent manner but the mechanisms are still debated (Green 2010). One suggestion has been that the Src-family kinase Yes and the p85 subunit of PI3 kinase associate with the membrane-proximal portion of CD95's cytoplasmic domain on cross-linking in glioblastoma cells and that the resulting activation of the PKB/AKT pathway could be responsible for the protumorigenic signals induced by CD95L in glioblastoma (Sancho-Martinez and Martin-Villalba 2009; Parrish et al. 2013).

More is known about the DD-requiring, nonapoptotic signals induced by CD95L or TRAIL. Interestingly, Ashkenazi and colleagues found that release of FADD from the TRAIL DISC triggers formation of an intracellular complex II by recruiting TRAF2, cIAP1/2, RIP1, NEMO, and most likely several other factors required for activation of NF-κB, MAPKs, and consequently, gene induction (Varfolomeev et al. 2005). Also in this case induction of signals from complex II does not occur, or at least only to a limited extent, when the primary signal induced by complex I, in this case the TRAIL DISC, results in cell death. This suggests the following model: Although the output of the primary complex of TNF-R1 signaling is gene activation, its second complex can induce two rather different forms of cell death. In the case of CD95 and TRAIL, the signaling output of the DISC is apoptosis, whereas that of complex II is gene activation (Fig. 4).

Why are specific tasks separated into different, sequentially acting signaling complexes? One explanation could be that, when the first signal prevails, the second one is not required or should at least be minimized. If, however, the primary signal does not reach its intended end point, signaling by the second complex is initiated. If a physiological extracellular stimulus (i.e., TRAIL or CD95L) triggers the apoptosis machinery in a cell but it does not die, this should alarm the organism because infectious agents often inhibit apoptosis. Hence, an alarm signal capable of stimulating immunity should be induced now. By activating proinflammatory gene induction, and perhaps necrosis, complex II of the CD95 and TRAIL systems can do exactly that (Fig. 4).

In the case of TNF-R1, and most likely DR3, it is the other way around. When proper gene activation cannot be achieved by complex I, complex II induces cell death. TNF-induced cell death can either be apoptotic or necrotic. It is unclear which type of TNF-induced cell death prevails when its gene-activatory pathway is perturbed, and it is likely that this differs with the type of perturbation. It is, however, likely that there will be an inflammatory component in it. Thereby, a biological outcome similar to the originally intended one could be achieved (i.e., the creation of a proinflammatory environment), yet by going down a very different path than originally intended (Fig. 4).

Interestingly, inflammation occurs in both autoimmunity and cancer. TNF has been shown to be intimately involved in both processes (Balkwill 2009; Taylor and Feldmann 2009). Yet, whereas autoimmunity-associated inflammation is immunostimulatory, cancer-related inflammation suppresses immunity. It is tempting to speculate that the two cell death modalities induced by TNF and its relatives, together with their gene-activatory capacity, are decisive in determining whether inflammation will stimulate or suppress immunity. Whatever the outcome will be, it is now clear that cell death and inflammation are more closely linked than previously thought and in fact seem to represent two sides of the very same coin.

CONCLUDING REMARKS AND OUTLOOK

Studying the TNF, CD95, and TRAIL death receptor ligands has provided tremendous insight into the biochemistry and function of cell death

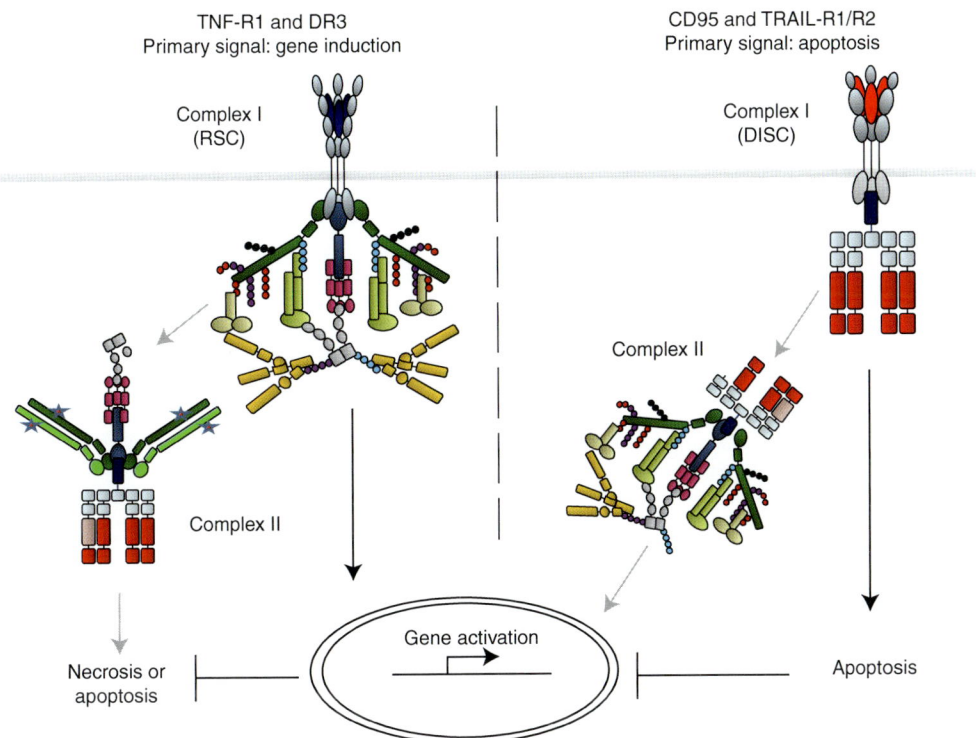

Figure 4. Comparison of CD95/TRAIL-R1/R2 and TNF-R1/DR3 signaling. For both the proapoptotic CD95 and TRAIL systems as well as the proinflammatory TNF and DR3 systems, the complex defined as complex I is the protein complex that forms at the plasma membrane and exerts the primary function of the respective receptor (i.e., apoptosis for CD95 and TRAIL-R1/R2 and gene activation via NF-κB and MAPK by TNF-R1 and DR3). The two primary complexes dissociate from the DD of the respective receptor and recruit additional proteins from the cytosol to form complex II, which triggers the respective secondary signal. In the case of CD95 and TRAIL-R1/R2 the second signal is gene activation via the NF-κB and MAPK pathways; in the case of TNF-R1/DR3 it is induction of necrosis or apoptosis. The signaling outputs of the respective secondary complexes are prevented or attenuated in case the respective primary complexes reach theirs.

and inflammation. Importantly, however, this research has also yielded three classes of drugs: TNF blockers, CD95L inhibitors, and TRAIL receptor agonists. TNF blockers have transformed the treatment of chronic inflammatory diseases. CD95L inhibitors and TRAIL receptor agonists have now entered the clinic and it will be exciting to see how their clinical potential unfolds in the future.

The understanding of death receptor-mediated signaling pathways has been at the forefront of many fascinating developments in cell biology, most importantly in the fields of cancer, inflammation, and cell death. Among the promising future areas in death receptor signaling are the specific sensitization of cancer cells to TRAIL, the elucidation of the pathways of regulated necrosis and CD95-mediated protumorigenic signaling, the deciphering of the ubiquitin code in death receptor signaling, and the reciprocal regulation of apoptosis and necrosis. It is likely that studies into these processes will, at times, provide surprising results. That, however, comes as no surprise because the journey from Coley's discovery of a tumor-necrotizing activity to today's research on the biology of death receptor–ligand systems has been packed with surprises, making this journey an exciting one that promises to continue for some time to come.

REFERENCES

*Reference is also in this collection.

Ackery A, Robins S, Fehlings MG. 2006. Inhibition of Fas-mediated apoptosis through administration of soluble Fas receptor improves functional outcome and reduces posttraumatic axonal degeneration after acute spinal cord injury. *J Neurotrauma* **23:** 604–616.

Alderson MR, Armitage RJ, Maraskovsky E, Tough TW, Roux E, Schooley K, Ramsdell F, Lynch DH. 1993. Fas transduces activation signals in normal human T lymphocytes. *J Exp Med* **178:** 2231–2235.

Alderson MR, Tough TW, Braddy S, Davis-Smith T, Roux E, Schooley K, Miller RE, Lynch DH. 1994. Regulation of apoptosis and T cell activation by Fas-specific mAb. *Int Immunol* **6:** 1799–1806.

Alderson MR, Tough TW, Davis-Smith T, Braddy S, Falk B, Schooley KA, Goodwin RG, Smith CA, Ramsdell F, Lynch DH. 1995. Fas ligand mediates activation-induced cell death in human T lymphocytes. *J Exp Med* **181:** 71–77.

Ashkenazi A, Pai RC, Fong S, Leung S, Lawrence DA, Marsters SA, Blackie C, Chang L, McMurtrey AE, Hebert A, et al. 1999. Safety and antitumor activity of recombinant soluble Apo2 ligand. *J Clin Invest* **104:** 155–162.

Baker MB, Altman NH, Podack ER, Levy RB. 1996. The role of cell-mediated cytotoxicity in acute GVHD after MHC-matched allogeneic bone marrow transplantation in mice. *J Exp Med* **183:** 2645–2656.

Balkwill F. 2009. Tumour necrosis factor and cancer. *Nat Rev Cancer* **9:** 361–371.

Barnhart BC, Legembre P, Pietras E, Bubici C, Franzoso G, Peter ME. 2004. CD95 ligand induces motility and invasiveness of apoptosis-resistant tumor cells. *EMBO J* **23:** 3175–3185.

Berke G. 1995. The CTL's kiss of death. *Cell* **81:** 9–12.

Bodmer JL, Burns K, Schneider P, Hofmann K, Steiner V, Thome M, Bornand T, Hahne M, Schroter M, Becker K, et al. 1997. TRAMP, a novel apoptosis-mediating receptor with sequence homology to tumor necrosis factor receptor 1 and Fas(Apo-1/CD95). *Immunity* **6:** 79–88.

Boldin MP, Mett IL, Varfolomeev EE, Chumakov I, Shemer-Avni Y, Camonis JH, Wallach D. 1995a. Self-association of the "death domains" of the p55 tumor necrosis factor (TNF) receptor and Fas/APO1 prompts signaling for TNF and Fas/APO1 effects. *J Biol Chem* **270:** 387–391.

Boldin MP, Varfolomeev EE, Pancer Z, Mett IL, Camonis JH, Wallach D. 1995b. A novel protein that interacts with the death domain of Fas/APO1 contains a sequence motif related to the death domain. *J Biol Chem* **270:** 7795–7798.

Boldin MP, Goncharov TM, Goltsev YV, Wallach D. 1996. Involvement of MACH, a novel MORT1/FADD-interacting protease, in Fas/APO-1- and TNF receptor-induced cell death. *Cell* **85:** 803–815.

Braun MY, Lowin B, French L, Acha-Orbea H, Tschopp J. 1996. Cytotoxic T cells deficient in both functional fas ligand and perforin show residual cytolytic activity yet lose their capacity to induce lethal acute graft-versus-host disease. *J Exp Med* **183:** 657–661.

Brunner T, Mogil RJ, LaFace D, Yoo NJ, Mahboubi A, Echeverri F, Martin SJ, Force WR, Lynch DH, Ware CF, et al. 1995. Cell-autonomous Fas (CD95)/Fas-ligand interaction mediates activation-induced apoptosis in T-cell hybridomas. *Nature* **373:** 441–444.

Budd RC, Yeh WC, Tschopp J. 2006. cFLIP regulation of lymphocyte activation and development. *Nat Rev Immunol* **6:** 196–204.

Carswell EA, Old LJ, Kassel RL, Green S, Fiore N, Williamson B. 1975. An endotoxin-induced serum factor that causes necrosis of tumors. *Proc Natl Acad Sci* **72:** 3666–3670.

Cerretti DP, Kozlosky CJ, Mosley B, Nelson N, Van Ness K, Greenstreet TA, March CJ, Kronheim SR, Druck T, Cannizzaro LA, et al. 1992. Molecular cloning of the interleukin-1β converting enzyme. *Science* **256:** 97–100.

Chen L, Park SM, Tumanov AV, Hau A, Sawada K, Feig C, Turner JR, Fu YX, Romero IL, Lengyel E, et al. 2010. CD95 promotes tumour growth. *Nature* **465:** 492–496.

Chinnaiyan AM, O'Rourke K, Tewari M, Dixit VM. 1995. FADD, a novel death domain-containing protein, interacts with the death domain of Fas and initiates apoptosis. *Cell* **81:** 505–512.

Chinnaiyan AM, O'Rourke K, Yu GL, Lyons RH, Garg M, Duan DR, Xing L, Gentz R, Ni J, Dixit VM. 1996. Signal transduction by DR3, a death domain-containing receptor related to TNFR-1 and CD95. *Science* **274:** 990–992.

Cho YS, Challa S, Moquin D, Genga R, Ray TD, Guildford M, Chan FK. 2009. Phosphorylation-driven assembly of the RIP1-RIP3 complex regulates programmed necrosis and virus-induced inflammation. *Cell* **137:** 1112–1123.

Creagh EM, Martin SJ. 2001. Caspases: Cellular demolition experts. *Biochem Soc Trans* **29:** 696–702.

Demjen D, Klussmann S, Kleber S, Zuliani C, Stieltjes B, Metzger C, Hirt UA, Walczak H, Falk W, Essig M, et al. 2004. Neutralization of CD95 ligand promotes regeneration and functional recovery after spinal cord injury. *Nat Med* **10:** 389–395.

Desbarats J, Newell MK. 2000. Fas engagement accelerates liver regeneration after partial hepatectomy. *Nat Med* **6:** 920–923.

Desbarats J, Wade T, Wade WF, Newell MK. 1999. Dichotomy between naive and memory $CD4^+$ T cell responses to Fas engagement. *Proc Natl Acad Sci* **96:** 8104–8109.

Dhein J, Daniel PT, Trauth BC, Oehm A, Moller P, Krammer PH. 1992. Induction of apoptosis by monoclonal antibody anti-APO-1 class switch variants is dependent on cross-linking of APO-1 cell surface antigens. *J Immunol* **149:** 3166–3173.

Dhein J, Walczak H, Baumler C, Debatin KM, Krammer PH. 1995. Autocrine T-cell suicide mediated by APO-1/(Fas/CD95). *Nature* **373:** 438–441.

Dickens LS, Boyd RS, Jukes-Jones R, Hughes MA, Robinson GL, Fairall L, Schwabe JW, Cain K, Macfarlane M. 2012. A death effector domain chain DISC model reveals a crucial role for Caspase-8 chain assembly in mediating apoptotic cell death. *Mol Cell* **47:** 291–305.

Dillon CP, Oberst A, Weinlich R, Janke LJ, Kang TB, Ben-Moshe T, Mak TW, Wallach D, Green DR. 2012. Survival function of the FADD-CASPASE-8-cFLIP(L) complex. *Cell Rep* **1**: 401–407.

Ehl S, Hoffmann-Rohrer U, Nagata S, Hengartner H, Zinkernagel R. 1996. Different susceptibility of cytotoxic T cells to CD95 (Fas/Apo-1) ligand-mediated cell death after activation in vitro versus in vivo. *J Immunol* **156**: 2357–2360.

Ehrhardt H, Fulda S, Schmid I, Hiscott J, Debatin KM, Jeremias I. 2003. TRAIL induced survival and proliferation in cancer cells resistant towards TRAIL-induced apoptosis mediated by NF-κB. *Oncogene* **22**: 3842–3852.

Freiberg RA, Spencer DM, Choate KA, Duh HJ, Schreiber SL, Crabtree GR, Khavari PA. 1997. Fas signal transduction triggers either proliferation or apoptosis in human fibroblasts. *J Invest Dermatol* **108**: 215–219.

Galle PR, Hofmann WJ, Walczak H, Schaller H, Otto G, Stremmel W, Krammer PH, Runkel L. 1995. Involvement of the CD95 (APO-1/Fas) receptor and ligand in liver damage. *J Exp Med* **182**: 1223–1230.

Ganten TM, Koschny R, Haas TL, Sykora J, Li-Weber M, Herzer K, Walczak H. 2005. Proteasome inhibition sensitizes hepatocellular carcinoma cells, but not human hepatocytes, to TRAIL. *Hepatology* **42**: 588–597.

Ganten TM, Koschny R, Sykora J, Schulze-Bergkamen H, Buchler P, Haas TL, Schader MB, Untergasser A, Stremmel W, Walczak H. 2006. Preclinical differentiation between apparently safe and potentially hepatotoxic applications of TRAIL either alone or in combination with chemotherapeutic drugs. *Clin Cancer Res* **12**: 2640–2646.

Gerlach B, Cordier SM, Schmukle AC, Emmerich CH, Rieser E, Haas TL, Webb AI, Rickard JA, Anderton H, Wong WW, et al. 2011. Linear ubiquitination prevents inflammation and regulates immune signalling. *Nature* **471**: 591–596.

Green DR. 2010. Cancer: A wolf in wolf's clothing. *Nature* **465**: 433.

Gross A, Yin XM, Wang K, Wei MC, Jockel J, Milliman C, Erdjument-Bromage H, Tempst P, Korsmeyer SJ. 1999. Caspase cleaved BID targets mitochondria and is required for cytochrome *c* release, while BCL-XL prevents this release but not tumor necrosis factor-R1/Fas death. *J Biol Chem* **274**: 1156–1163.

Gunther C, Martini E, Wittkopf N, Amann K, Weigmann B, Neumann H, Waldner MJ, Hedrick SM, Tenzer S, Neurath MF, et al. 2012. Caspase-8 regulates TNF-α-induced epithelial necroptosis and terminal ileitis. *Nature* **477**: 335–339.

Hall SS. 1997. *A commotion in the blood: Life, death, and the immune system*. Holt, New York.

He S, Wang L, Miao L, Wang T, Du F, Zhao L, Wang X. 2009. Receptor interacting protein kinase-3 determines cellular necrotic response to TNF-α. *Cell* **137**: 1100–1111.

Hitomi J, Christofferson DE, Ng A, Yao J, Degterev A, Xavier RJ, Yuan J. 2008. Identification of a molecular signaling network that regulates a cellular necrotic cell death pathway. *Cell* **135**: 1311–1323.

Holler N, Zaru R, Micheau O, Thome M, Attinger A, Valitutti S, Bodmer JL, Schneider P, Seed B, Tschopp J. 2000. Fas triggers an alternative, caspase-8-independent cell death pathway using the kinase RIP as effector molecule. *Nat Immunol* **1**: 489–495.

Irmler M, Thome M, Hahne M, Schneider P, Hofmann K, Steiner V, Bodmer JL, Schroter M, Burns K, Mattmann C, et al. 1997. Inhibition of death receptor signals by cellular FLIP. *Nature* **388**: 190–195.

Itoh N, Nagata S. 1993. A novel protein domain required for apoptosis. Mutational analysis of human Fas antigen. *J Biol Chem* **268**: 10932–10937.

Itoh N, Yonehara S, Ishii A, Yonehara M, Mizushima S, Sameshima M, Hase A, Seto Y, Nagata S. 1991. The polypeptide encoded by the cDNA for human cell surface antigen Fas can mediate apoptosis. *Cell* **66**: 233–243.

Jeremias I, Kupatt C, Martin-Villalba A, Habazettl H, Schenkel J, Boekstegers P, Debatin KM. 2000. Involvement of CD95/Apo1/Fas in cell death after myocardial ischemia. *Circulation* **102**: 915–920.

Jost PJ, Grabow S, Gray D, McKenzie MD, Nachbur U, Huang DC, Bouillet P, Thomas HE, Borner C, Silke J, et al. 2009. XIAP discriminates between type I and type II FAS-induced apoptosis. *Nature* **460**: 1035–1039.

Ju ST, Panka DJ, Cui H, Ettinger R, el-Khatib M, Sherr DH, Stanger BZ, Marshak-Rothstein A. 1995. Fas(CD95)/FasL interactions required for programmed cell death after T-cell activation. *Nature* **373**: 444–448.

Kaiser WJ, Upton JW, Long AB, Livingston-Rosanoff D, Daley-Bauer LP, Hakem R, Caspary T, Mocarski ES. 2011. RIP3 mediates the embryonic lethality of caspase-8-deficient mice. *Nature* **471**: 368–372.

Kavuri SM, Geserick P, Berg D, Dimitrova DP, Feoktistova M, Siegmund D, Gollnick H, Neumann M, Wajant H, Leverkus M. 2011. Cellular FLICE-inhibitory protein (cFLIP) isoforms block CD95- and TRAIL death receptor-induced gene induction irrespective of processing of caspase-8 or cFLIP in the death-inducing signaling complex. *J Biol Chem* **286**: 16631–16646.

Kennedy NJ, Kataoka T, Tschopp J, Budd RC. 1999. Caspase activation is required for T cell proliferation. *J Exp Med* **190**: 1891–1896.

Kischkel FC, Hellbardt S, Behrmann I, Germer M, Pawlita M, Krammer PH, Peter ME. 1995. Cytotoxicity-dependent APO-1 (Fas/CD95)-associated proteins form a death-inducing signaling complex (DISC) with the receptor. *EMBO J* **14**: 5579–5588.

Kischkel FC, Lawrence DA, Chuntharapai A, Schow P, Kim KJ, Ashkenazi A. 2000. Apo2L/TRAIL-dependent recruitment of endogenous FADD and caspase 8 to death receptors 4 and 5. *Immunity* **12**: 611–620.

Kischkel FC, Lawrence DA, Tinel A, LeBlanc H, Virmani A, Schow P, Gazdar A, Blenis J, Arnott D, Ashkenazi A. 2001. Death receptor recruitment of endogenous caspase-10 and apoptosis initiation in the absence of caspase-8. *J Biol Chem* **276**: 46639–46646.

Kitson J, Raven T, Jiang YP, Goeddel DV, Giles KM, Pun KT, Grinham CJ, Brown R, Farrow SN. 1996. A death-domain-containing receptor that mediates apoptosis. *Nature* **384**: 372–375.

Kleber S, Sancho-Martinez I, Wiestler B, Beisel A, Gieffers C, Hill O, Thiemann M, Mueller W, Sykora J, Kuhn A, et al. 2008. Yes and PI3K bind CD95 to signal invasion of glioblastoma. *Cancer Cell* **13**: 235–248.

Komander D, Clague MJ, Urbe S. 2009a. Breaking the chains: Structure and function of the deubiquitinases. *Nat Rev Mol Cell Biol* **10**: 550–563.

Komander D, Reyes-Turcu F, Licchesi JD, Odenwaelder P, Wilkinson KD, Barford D. 2009b. Molecular discrimination of structurally equivalent Lys 63-linked and linear polyubiquitin chains. *EMBO Rep* **10:** 466–473.

Kondo T, Suda T, Fukuyama H, Adachi M, Nagata S. 1997. Essential roles of the Fas ligand in the development of hepatitis. *Nat Med* **3:** 409–413.

Koschny R, Ganten TM, Sykora J, Haas TL, Sprick MR, Kolb A, Stremmel W, Walczak H. 2007. TRAIL/bortezomib cotreatment is potentially hepatotoxic but induces cancer-specific apoptosis within a therapeutic window. *Hepatology* **45:** 649–658.

Kovalenko A, Kim JC, Kang TB, Rajput A, Bogdanov K, Dittrich-Breiholz O, Kracht M, Brenner O, Wallach D. 2009. Caspase-8 deficiency in epidermal keratinocytes triggers an inflammatory skin disease. *J Exp Med* **206:** 2161–2177.

Lawrence D, Shahrokh Z, Marsters S, Achilles K, Shih D, Mounho B, Hillan K, Totpal K, DeForge L, Schow P, et al. 2001. Differential hepatocyte toxicity of recombinant Apo2L/TRAIL versions. *Nat Med* **7:** 383–385.

Lee P, Sata M, Lefer DJ, Factor SM, Walsh K, Kitsis RN. 2003. Fas pathway is a critical mediator of cardiac myocyte death and MI during ischemia-reperfusion in vivo. *Am J Physiol Heart Circ Physiol* **284:** H456–H463.

Lemke J, Noack A, Adam D, Tchikov V, Bertsch U, Roder C, Schutze S, Wajant H, Kalthoff H, Trauzold A. 2010. TRAIL signaling is mediated by DR4 in pancreatic tumor cells despite the expression of functional DR5. *J Mol Med (Berl)* **88:** 729–740.

Lo YC, Lin SC, Rospigliosi CC, Conze DB, Wu CJ, Ashwell JD, Eliezer D, Wu H. 2009. Structural basis for recognition of diubiquitins by NEMO. *Mol Cell* **33:** 602–615.

Loetscher H, Pan YC, Lahm HW, Gentz R, Brockhaus M, Tabuchi H, Lesslauer W. 1990. Molecular cloning and expression of the human 55 kd tumor necrosis factor receptor. *Cell* **61:** 351–359.

Lowin B, Beermann F, Schmidt A, Tschopp J. 1994a. A null mutation in the perforin gene impairs cytolytic T lymphocyte- and natural killer cell-mediated cytotoxicity. *Proc Natl Acad Sci* **91:** 11571–11575.

Lowin B, Hahne M, Mattmann C, Tschopp J. 1994b. Cytolytic T-cell cytotoxicity is mediated through perforin and Fas lytic pathways. *Nature* **370:** 650–652.

Marsters SA, Pitti RM, Donahue CJ, Ruppert S, Bauer KD, Ashkenazi A. 1996. Activation of apoptosis by Apo-2 ligand is independent of FADD but blocked by CrmA. *Curr Biol* **6:** 750–752.

Martin-Villalba A, Herr I, Jeremias I, Hahne M, Brandt R, Vogel J, Schenkel J, Herdegen T, Debatin KM. 1999. CD95 ligand (Fas-L/APO-1L) and tumor necrosis factor-related apoptosis-inducing ligand mediate ischemia-induced apoptosis in neurons. *J Neurosci* **19:** 3809–3817.

Micheau O, Tschopp J. 2003. Induction of TNF receptor I-mediated apoptosis via two sequential signaling complexes. *Cell* **114:** 181–190.

Migone TS, Zhang J, Luo X, Zhuang L, Chen C, Hu B, Hong JS, Perry JW, Chen SF, Zhou JX, et al. 2002. TL1A is a TNF-like ligand for DR3 and TR6/DcR3 and functions as a T cell costimulator. *Immunity* **16:** 479–492.

Miwa K, Hashimoto H, Yatomi T, Nakamura N, Nagata S, Suda T. 1999. Therapeutic effect of an anti-Fas ligand mAb on lethal graft-versus-host disease. *Int Immunol* **11:** 925–931.

Muzio M, Chinnaiyan AM, Kischkel FC, O'Rourke K, Shevchenko A, Ni J, Scaffidi C, Bretz JD, Zhang M, Gentz R, et al. 1996. FLICE, a novel FADD-homologous ICE/CED-3-like protease, is recruited to the CD95 (Fas/APO-1) death-inducing signaling complex. *Cell* **85:** 817–827.

Nauts HC, Swift WE, Coley BL. 1946. The treatment of malignant tumors by bacterial toxins as developed by the late William B. Coley, M.D., reviewed in the light of modern research. *Cancer Res* **6:** 205–216.

Nauts HC, Fowler GA, Bogatko FH. 1953. A review of the influence of bacterial infection and of bacterial products (Coley's toxins) on malignant tumors in man; a critical analysis of 30 inoperable cases treated by Coley's mixed toxins, in which diagnosis was confirmed by microscopic examination selected for special study. *Acta Med Scand Suppl* **276:** 1–103.

Newsom-Davis T, Prieske S, Walczak H. 2009. Is TRAIL the holy grail of cancer therapy? *Apoptosis* **14:** 607–623.

Nikolaev A, McLaughlin T, O'Leary DD, Tessier-Lavigne M. 2009. APP binds DR6 to trigger axon pruning and neuron death via distinct caspases. *Nature* **457:** 981–989.

Oberst A, Dillon CP, Weinlich R, McCormick LL, Fitzgerald P, Pop C, Hakem R, Salvesen GS, Green DR. 2011. Catalytic activity of the caspase-8-FLIP(L) complex inhibits RIPK3-dependent necrosis. *Nature* **471:** 363–367.

Oehm A, Behrmann I, Falk W, Pawlita M, Maier G, Klas C, Li-Weber M, Richards S, Dhein J, Trauth BC, et al. 1992. Purification and molecular cloning of the APO-1 cell surface antigen, a member of the tumor necrosis factor/nerve growth factor receptor superfamily. Sequence identity with the Fas antigen. *J Biol Chem* **267:** 10709–10715.

Ogasawara J, Watanabe-Fukunaga R, Adachi M, Matsuzawa A, Kasugai T, Kitamura Y, Itoh N, Suda T, Nagata S. 1993. Lethal effect of the anti-Fas antibody in mice. *Nature* **364:** 806–809.

Owen-Schaub LB, Meterissian S, Ford RJ. 1993. Fas/APO-1 expression and function on malignant cells of hematologic and nonhematologic origin. *J Immunother Emphasis Tumor Immunol* **14:** 234–241.

Oyaizu T, Shikata N, Senzaki H, Matsuzawa A, Tsubura A. 1997. Studies on the mechanism of dimethylnitrosamine-induced acute liver injury in mice. *Exp Toxicol Pathol* **49:** 375–380.

Pan G, Ni J, Wei YF, Yu G, Gentz R, Dixit VM. 1997a. An antagonist decoy receptor and a death domain-containing receptor for TRAIL. *Science* **277:** 815–818.

Pan G, O'Rourke K, Chinnaiyan AM, Gentz R, Ebner R, Ni J, Dixit VM. 1997b. The receptor for the cytotoxic ligand TRAIL. *Science* **276:** 111–113.

Pan G, Bauer JH, Haridas V, Wang S, Liu D, Yu G, Vincenz C, Aggarwal BB, Ni J, Dixit VM. 1998. Identification and functional characterization of DR6, a novel death domain-containing TNF receptor. *FEBS Lett* **431:** 351–356.

* Parrish AB, Freel CD, Kornbluth S. 2013. Cellular mechanisms controlling caspase activation and function. *Cold Spring Harb Perspect Biol* **5**: a008672.

Peter ME, Krammer PH. 1998. Mechanisms of CD95 (APO-1/Fas)-mediated apoptosis. *Curr Opin Immunol* **10**: 545–551.

Peter ME, Budd RC, Desbarats J, Hedrick SM, Hueber AO, Newell MK, Owen LB, Pope RM, Tschopp J, Wajant H, et al. 2007. The CD95 receptor: Apoptosis revisited. *Cell* **129**: 447–450.

Pitti RM, Marsters SA, Ruppert S, Donahue CJ, Moore A, Ashkenazi A. 1996. Induction of apoptosis by Apo-2 ligand, a new member of the tumor necrosis factor cytokine family. *J Biol Chem* **271**: 12687–12690.

Pop C, Salvesen GS. 2009. Human caspases: Activation, specificity, and regulation. *J Biol Chem* **284**: 21777–21781.

Rahighi S, Ikeda F, Kawasaki M, Akutsu M, Suzuki N, Kato R, Kensche T, Uejima T, Bloor S, Komander D, et al. 2009. Specific recognition of linear ubiquitin chains by NEMO is important for NF-κB activation. *Cell* **136**: 1098–1109.

Reap EA, Sobel ES, Cohen PL, Eisenberg RA. 1993. Conventional B cells, not B-1 cells, are responsible for producing autoantibodies in lpr mice. *J Exp Med* **177**: 69–78.

Rieser E, Schmukle AC, Walczak H. 2012. Linear ubiquitination: A newly discovered regulator of cell signalling. *Trends Biochem Sci* **24**: 229–231.

Roder C, Trauzold A, Kalthoff H. 2011. Impact of death receptor signaling on the malignancy of pancreatic ductal adenocarcinoma. *Eur J Cell Biol* **90**: 450–455.

Roths JB, Murphy ED, Eicher EM. 1984. A new mutation, gld, that produces lymphoproliferation and autoimmunity in C3H/HeJ mice. *J Exp Med* **159**: 1–20.

Salvesen GS, Riedl SJ. 2008. Caspase mechanisms. *Adv Exp Med Biol* **615**: 13–23.

Sancho-Martinez I, Martin-Villalba A. 2009. Tyrosine phosphorylation and CD95: A FAScinating switch. *Cell Cycle* **8**: 838–842.

Schall TJ, Lewis M, Koller KJ, Lee A, Rice GC, Wong GH, Gatanaga T, Granger GA, Lentz R, Raab H, et al. 1990. Molecular cloning and expression of a receptor for human tumor necrosis factor. *Cell* **61**: 361–370.

Schleich K, Warnken U, Fricker N, Ozturk S, Richter P, Kammerer K, Schnolzer M, Krammer PH, Lavrik IN. 2012. Stoichiometry of the CD95 death-inducing signaling complex: Experimental and modeling evidence for a death effector domain chain model. *Mol Cell* **47**: 306–319.

Screaton GR, Mongkolsapaya J, Xu XN, Cowper AE, McMichael AJ, Bell JI. 1997a. TRICK2, a new alternatively spliced receptor that transduces the cytotoxic signal from TRAIL. *Curr Biol* **7**: 693–696.

Screaton GR, Xu XN, Olsen AL, Cowper AE, Tan R, McMichael AJ, Bell JI. 1997b. LARD: A new lymphoid-specific death domain containing receptor regulated by alternative pre-mRNA splicing. *Proc Natl Acad Sci* **94**: 4615–4619.

Shear MJ, Perrault A. 1944. A chemical treatment of tumors. IX. Reactions of mice with primary subcutaneous tumors to injection of a hemorrhage-producing bacterial polysaccharide. *J Natl Cancer Inst* **44**: 461–476.

Sheridan JP, Marsters SA, Pitti RM, Gurney A, Skubatch M, Baldwin D, Ramakrishnan L, Gray CL, Baker K, Wood WI, et al. 1997. Control of TRAIL-induced apoptosis by a family of signaling and decoy receptors. *Science* **277**: 818–821.

Simonet WS, Lacey DL, Dunstan CR, Kelley M, Chang MS, Luthy R, Nguyen HQ, Wooden S, Bennett L, Boone T, et al. 1997. Osteoprotegerin: A novel secreted protein involved in the regulation of bone density. *Cell* **89**: 309–319.

Smith CA, Davis T, Anderson D, Solam L, Beckmann MP, Jerzy R, Dower SK, Cosman D, Goodwin RG. 1990. A receptor for tumor necrosis factor defines an unusual family of cellular and viral proteins. *Science* **248**: 1019–1023.

Sprick MR, Weigand MA, Rieser E, Rauch CT, Juo P, Blenis J, Krammer PH, Walczak H. 2000. FADD/MORT1 and caspase-8 are recruited to TRAIL receptors 1 and 2 and are essential for apoptosis mediated by TRAIL receptor 2. *Immunity* **12**: 599–609.

Sprick MR, Rieser E, Stahl H, Grosse-Wilde A, Weigand MA, Walczak H. 2002. Caspase-10 is recruited to and activated at the native TRAIL and CD95 death-inducing signalling complexes in a FADD-dependent manner but can not functionally substitute caspase-8. *EMBO J* **21**: 4520–4530.

Strand S, Hofmann WJ, Grambihler A, Hug H, Volkmann M, Otto G, Wesch H, Mariani SM, Hack V, Stremmel W, et al. 1998. Hepatic failure and liver cell damage in acute Wilson's disease involve CD95 (APO-1/Fas) mediated apoptosis. *Nat Med* **4**: 588–593.

Strand S, Strand D, Seufert R, Mann A, Lotz J, Blessing M, Lahn M, Wunsch A, Broering DC, Hahn U, et al. 2004. Placenta-derived CD95 ligand causes liver damage in hemolysis, elevated liver enzymes, and low platelet count syndrome. *Gastroenterology* **126**: 849–858.

Suda T, Nagata S. 1994. Purification and characterization of the Fas-ligand that induces apoptosis. *J Exp Med* **179**: 873–879.

Suda T, Takahashi T, Golstein P, Nagata S. 1993. Molecular cloning and expression of the Fas ligand, a novel member of the tumor necrosis factor family. *Cell* **75**: 1169–1178.

* Tait SWG, Green DR. 2013. Mitochondrial regulation of cell death. *Cold Spring Harb Perspect Biol* doi: 10.1101/cshperspect.a008706.

Takahashi T, Tanaka M, Brannan CI, Jenkins NA, Copeland NG, Suda T, Nagata S. 1994. Generalized lymphoproliferative disease in mice, caused by a point mutation in the Fas ligand. *Cell* **76**: 969–976.

Tartaglia LA, Ayres TM, Wong GH, Goeddel DV. 1993. A novel domain within the 55 kd TNF receptor signals cell death. *Cell* **74**: 845–853.

Taylor PC, Feldmann M. 2009. Anti-TNF biologic agents: Still the therapy of choice for rheumatoid arthritis. *Nat Rev Rheumatol* **5**: 578–582.

Thome M, Schneider P, Hofmann K, Fickenscher H, Meinl E, Neipel F, Mattmann C, Burns K, Bodmer JL, Schroter M, et al. 1997. Viral FLICE-inhibitory proteins (FLIPs) prevent apoptosis induced by death receptors. *Nature* **386**: 517–521.

Thornberry NA, Bull HG, Calaycay JR, Chapman KT, Howard AD, Kostura MJ, Miller DK, Molineaux SM,

Weidner JR, Aunins J, et al. 1992. A novel heterodimeric cysteine protease is required for interleukin-1β processing in monocytes. *Nature* **356:** 768–774.

Tracey KJ, Lowry SF, Cerami A. 1988. Cachetin/TNF-α in septic shock and septic adult respiratory distress syndrome. *Am Rev Respir Dis* **138:** 1377–1379.

Trauth BC, Klas C, Peters AM, Matzku S, Moller P, Falk W, Debatin KM, Krammer PH. 1989. Monoclonal antibody-mediated tumor regression by induction of apoptosis. *Science* **245:** 301–305.

Vandenabeele P, Galluzzi L, Vanden Berghe T, Kroemer G. 2010. Molecular mechanisms of necroptosis: An ordered cellular explosion. *Nat Rev Mol Cell Biol* **11:** 700–714.

van Raam BJ, Salvesen GS. 2012. Proliferative versus apoptotic functions of caspase-8 Hetero or homo: The caspase-8 dimer controls cell fate. *Biochim Biophys Acta* **1824:** 113–122.

Varfolomeev E, Maecker H, Sharp D, Lawrence D, Renz M, Vucic D, Ashkenazi A. 2005. Molecular determinants of kinase pathway activation by Apo2 ligand/tumor necrosis factor-related apoptosis-inducing ligand. *J Biol Chem* **280:** 40599–40608.

Via CS, Nguyen P, Shustov A, Drappa J, Elkon KB. 1996a. A major role for the Fas pathway in acute graft-versus-host disease. *J Immunol* **157:** 5387–5393.

Via CS, Rus V, Nguyen P, Linsley P, Gause WC. 1996b. Differential effect of CTLA4Ig on murine graft-versus-host disease (GVHD) development: CTLA4Ig prevents both acute and chronic GVHD development but reverses only chronic GVHD. *J Immunol* **157:** 4258–4267.

Vince JE, Wong WW, Gentle I, Lawlor KE, Allam R, O'Reilly L, Mason K, Gross O, Ma S, Guarda G, et al. 2012. Inhibitor of apoptosis proteins limit RIP3 kinase-dependent interleukin-1 activation. *Immunity* **36:** 215–227.

Vogt M, Bauer MK, Ferrari D, Schulze-Osthoff K. 1998. Oxidative stress and hypoxia/reoxygenation trigger CD95 (APO-1/Fas) ligand expression in microglial cells. *FEBS Lett* **429:** 67–72.

Wajant H, Pfizenmaier K, Scheurich P. 2003. Tumor necrosis factor signaling. *Cell Death Differ* **10:** 45–65.

Walczak H. 2011. TNF and ubiquitin at the crossroads of gene activation, cell death, inflammation, and cancer. *Immunol Rev* **244:** 9–28.

Walczak H, Krammer PH. 2000. The CD95 (APO-1/Fas) and the TRAIL (APO-2L) apoptosis systems. *Exp Cell Res* **256:** 58–66.

Walczak H, Degli-Esposti MA, Johnson RS, Smolak PJ, Waugh JY, Boiani N, Timour MS, Gerhart MJ, Schooley KA, Smith CA, et al. 1997. TRAIL-R2: A novel apoptosis-mediating receptor for TRAIL. *EMBO J* **16:** 5386–5397.

Walczak H, Miller RE, Ariail K, Gliniak B, Griffith TS, Kubin M, Chin W, Jones J, Woodward A, Le T, et al. 1999. Tumoricidal activity of tumor necrosis factor-related apoptosis-inducing ligand in vivo. *Nat Med* **5:** 157–163.

Wallach D, Kovalenko A, Kang TB. 2012. ""Necrosome"-induced inflammation: Must cells die for it? *Trends Immunol* **32:** 505–509.

Watanabe-Fukunaga R, Brannan CI, Copeland NG, Jenkins NA, Nagata S. 1992. Lymphoproliferation disorder in mice explained by defects in Fas antigen that mediates apoptosis. *Nature* **356:** 314–317.

Welz PS, Wullaert A, Vlantis K, Kondylis V, Fernandez-Majada V, Ermolaeva M, Kirsch P, Sterner-Kock A, van Loo G, Pasparakis M. 2011. FADD prevents RIP3-mediated epithelial cell necrosis and chronic intestinal inflammation. *Nature* **477:** 330–334.

Wiley SR, Schooley K, Smolak PJ, Din WS, Huang CP, Nicholl JK, Sutherland GR, Smith TD, Rauch C, Smith CA, et al. 1995. Identification and characterization of a new member of the TNF family that induces apoptosis. *Immunity* **3:** 673–682.

Williams RO, Feldmann M, Maini RN. 1992. Anti-tumor necrosis factor ameliorates joint disease in murine collagen-induced arthritis. *Proc Natl Acad Sci* **89:** 9784–9788.

Wilson NS, Dixit V, Ashkenazi A. 2009. Death receptor signal transducers: Nodes of coordination in immune signaling networks. *Nat Immunol* **10:** 348–355.

Wilson NS, Yang B, Yang A, Loeser S, Marsters S, Lawrence D, Li Y, Pitti R, Totpal K, Yee S, et al. 2011. An Fcγ receptor-dependent mechanism drives antibody-mediated target-receptor signaling in cancer cells. *Cancer Cell* **19:** 101–113.

Wu GS, Burns TF, McDonald ER 3rd, Jiang W, Meng R, Krantz ID, Kao G, Gan DD, Zhou JY, Muschel R, et al. 1997. KILLER/DR5 is a DNA damage-inducible p53–regulated death receptor gene. *Nat Genet* **17:** 141–143.

Yeh WC, Itie A, Elia AJ, Ng M, Shu HB, Wakeham A, Mirtsos C, Suzuki N, Bonnard M, Goeddel DV, et al. 2000. Requirement for Casper (c-FLIP) in regulation of death receptor-induced apoptosis and embryonic development. *Immunity* **12:** 633–642.

Yin XM, Wang K, Gross A, Zhao Y, Zinkel S, Klocke B, Roth KA, Korsmeyer SJ. 1999. Bid-deficient mice are resistant to Fas-induced hepatocellular apoptosis. *Nature* **400:** 886–891.

Yonehara S, Ishii A, Yonehara M. 1989. A cell-killing monoclonal antibody (anti-Fas) to a cell surface antigen co-downregulated with the receptor of tumor necrosis factor. *J Exp Med* **169:** 1747–1756.

Yuan JY, Horvitz HR. 1990. The *Caenorhabditis elegans* genes ced-3 and ced-4 act cell autonomously to cause programmed cell death. *Dev Biol* **138:** 33–41.

Yuan J, Shaham S, Ledoux S, Ellis HM, Horvitz HR. 1993. The *C. elegans* cell death gene ced-3 encodes a protein similar to mammalian interleukin-1 β-converting enzyme. *Cell* **75:** 641–652.

Zhang DW, Shao J, Lin J, Zhang N, Lu BJ, Lin SC, Dong MQ, Han J. 2009. RIP3, an energy metabolism regulator that switches TNF-induced cell death from apoptosis to necrosis. *Science* **325:** 332–336.

Zhang H, Zhou X, McQuade T, Li J, Chan FK, Zhang J. 2011. Functional complementation between FADD and RIP1 in embryos and lymphocytes. *Nature* **471:** 373–376.

Mitochondrial Regulation of Cell Death

Stephen W.G. Tait[1] and Douglas R. Green[2]

[1]Beatson Institute, Institute of Cancer Sciences, University of Glasgow, Glasgow G61 1BD, United Kingdom
[2]Department of Immunology, St. Jude Children's Hospital, Memphis, Tennessee 38105

Correspondence: stephen.tait@glasgow.ac.uk; douglas.green@stjude.org

Although required for life, paradoxically, mitochondria are often essential for initiating apoptotic cell death. Mitochondria regulate caspase activation and cell death through an event termed mitochondrial outer membrane permeabilization (MOMP); this leads to the release of various mitochondrial intermembrane space proteins that activate caspases, resulting in apoptosis. MOMP is often considered a point of no return because it typically leads to cell death, even in the absence of caspase activity. Because of this pivotal role in deciding cell fate, deregulation of MOMP impacts on many diseases and represents a fruitful site for therapeutic intervention. Here we discuss the mechanisms underlying mitochondrial permeabilization and how this key event leads to cell death through caspase-dependent and -independent means. We then proceed to explore how the release of mitochondrial proteins may be regulated following MOMP. Finally, we discuss mechanisms that enable cells sometimes to survive MOMP, allowing them, in essence, to return from the point of no return.

In most organisms, mitochondria play an essential role in activating caspase proteases through a pathway termed the mitochondrial or intrinsic pathway of apoptosis. Mitochondria regulate caspase activation by a process called mitochondrial outer membrane permeabilization (MOMP). Selective permeabilization of the mitochondrial outer membrane releases intermembrane space (IMS) proteins that drive robust caspase activity leading to rapid cell death. However, even in the absence of caspase activity, MOMP typically commits a cell to death and is therefore considered a point of no return (Fig. 1). Because of this pivotal role in dictating cell fate, MOMP is highly regulated, mainly through interactions between pro- and antiapoptotic members of the Bcl-2 family. In this article, we begin by discussing how mitochondria may have evolved to become central players in apoptotic cell death. We then provide an overview of current models addressing the mechanics of MOMP, outlining how this crucial event leads to cell death through both caspase-dependent or -independent mechanisms. Finally, we discuss how caspase activity may be regulated post-MOMP and define other processes that allow cells to survive MOMP and, in effect, return from the point of no return.

MITOCHONDRIA—NATURAL-BORN KILLERS?

The endosymbiosis theory of evolution posits that mitochondria are modern-day descendants

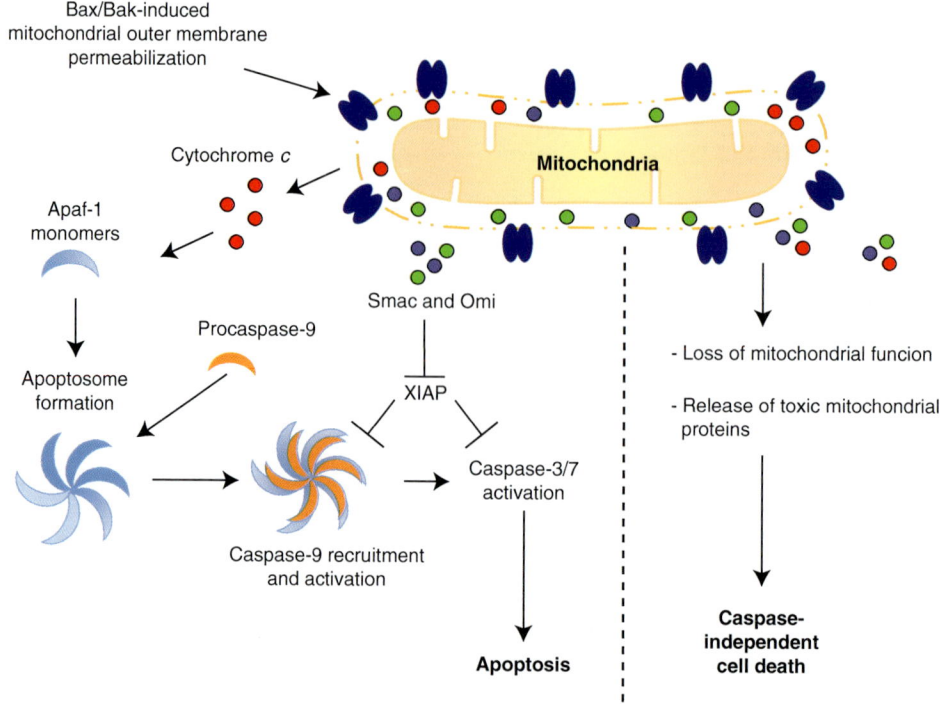

Figure 1. Mitochondrial regulation of cell death. Bax/Bak-mediated mitochondrial outer membrane permeabilization (MOMP) can lead to caspase-dependent apoptosis (*left*) or caspase-independent cell death (*right*). Following MOMP, soluble proteins are released from the mitochondrial intermembrane space into the cytoplasm. Cytochrome *c* binds to monomeric Apaf-1 leading to its conformational change and oligomerization. Procaspase-9 is recruited to heptameric Apaf-1 complexes forming the apoptosome. This leads to activation of caspase-9 and, through caspase-9-mediated cleavage, activation of the executioner caspases-3 and -7. Release of Smac and Omi from the mitochondrial intermembrane space facilitates caspase activation by neutralizing the caspase inhibitor XIAP. MOMP can also lead to nonapoptotic cell death through a gradual loss of mitochondrial function and/or release of mitochondrial proteins that kill the cell in a caspase-independent manner.

of α-proteobacteria that invaded archeon cells more than 2 billion years ago (Gray 2012). This invasion, ultimately forming the original eukaryotic cell, may have simultaneously forged a role for mitochondria in cell death. One possibility is that, following bacterial invasion, the archeon underwent altruistic cell death in order to protect the clonal population (James and Green 2002; Green 2011). Over time, some bacteria may have been able to prevent cell death, forming an endosymbiotic relationship with the archeon and eventually giving rise to mitochondria as we know them today. It may be that Bcl-2 proteins are modern-day descendants of toxins expressed by bacteria to kill one another that were initially co-opted to enable permeabilization of the mitochondrial outer membrane (which is likely host cell-derived, based on composition) while sparing the mitochondrial inner membrane (which resembles bacterial membrane composition). Accordingly, Bcl-2 proteins display structural similarities to certain bacterial toxins including diphtheria toxin β-chain and the colicins (Muchmore et al. 1996; Suzuki et al. 2000). Over time, as with most mitochondrial functions, genetic control of the proteins that regulate cell death may have transferred to the nucleus, whereas the mitochondrial outer membrane remains the battlefield.

Mitochondria play a role in apoptosis in most animals; however, the extent and importance of their contribution differ greatly between organisms (Oberst et al. 2008). In mammals, the essential requirement for MOMP as an initiating event in caspase activation and apoptosis is best evidenced in mice lacking Bax and Bak (Lindsten et al. 2000; Wei et al. 2001). Cells derived from these mice are profoundly resistant to all intrinsic apoptotic stimuli, and Bax/Bak double-knockout mice display developmental defects consistent with inhibition of cell death. In stark contrast, in the nematode *Caenorhabditis elegans* or the fly *Drosophila melanogaster*, two organisms that have been used extensively in cell death research, mitochondria do not appear to play a major role in the activation and execution of apoptosis. In *Caenorhabditis elegans*, although the proteins that control caspase activation are located on the mitochondria, this localization is not required for the regulation of apoptosis (Tan et al. 2007). In *D. melanogaster*, neither mitochondria nor Bcl-2 homologs regulate caspase activation. Instead, caspase activity is regulated primarily through interactions between caspases and inhibitor of apoptosis (IAP) proteins (Ryoo and Baehrecke 2010). Importantly, MOMP does not occur in *C. elegans* apoptotic cell death, and although MOMP has been observed during apoptosis in *D. melanogaster*, this is a consequence rather than a cause of caspase activation (Abdelwahid et al. 2007). This has led to the prevalent opinion that MOMP-dependent regulation of apoptosis evolved in higher eukaryotes. However, recent findings challenge this view; in the lophotrochozoan invertebrate *Planaria* (phylum Platyhelminthes), proapoptotic stimuli induce MOMP, and planarian caspases can be activated in cytosols by cytochrome *c* (unlike *D. melanogaster* or *C. elegans* caspases) (Bender et al. 2012). *Planaria* also encode a proapoptotic Bak homolog that can directly induce MOMP. Similarly, schistosomes (phylum Helminthes) also encode Bcl-2 proteins that can regulate MOMP (Lee et al. 2011). Cytochrome *c* can also activate caspases from an invertebrate deuterostome, the purple sea urchin, *Strongylocentrotus purpuratus* (phylum Echinodermata) (Bender et al. 2012). Collectively, these findings argue that, in cell death terms, *D. melanogaster* and *C. elegans* may be evolutionary outliers and that MOMP may be the primordial and predominant means of caspase activation in animals.

UNLEASHING THE DEATH SQUAD: MOLECULAR MECHANISMS OF MOMP

Because MOMP dictates cells fate, it is highly regulated, largely through interactions between pro- and antiapoptotic Bcl-2 family members (Youle and Strasser 2008). How antiapoptotic Bcl-2 proteins regulate MOMP is discussed elsewhere—here we review how the proteins that are required for MOMP, Bax and Bak, are activated and how, upon activation, they permeabilize the mitochondrial outer membrane.

Following activation by direct interaction with BH3-only Bcl-2 proteins, Bax and Bak undergo dramatic structural changes leading to mitochondrial targeting of Bax (which is predominantly cytosolic when inactive) and homo-oligomerization of Bax and Bak (Hsu et al. 1997; Eskes et al. 2000; Wei et al. 2000). Oligomerization of Bax and Bak is essential for MOMP because mutants that fail to oligomerize are completely inactive (George et al. 2007; Dewson et al. 2008). Given their pivotal role in deciding whether a cell dies or not, the mechanisms underlying Bax and Bak activation have been intensively investigated; however, it remains contentious how these drive MOMP (Fig. 2). One model proposes that Bax is activated by BH3-only proteins, not by binding in the hydrophobic BH3-binding pocket of Bax (which might be expected) but rather by interacting on the opposite side of Bax (Gavathiotis et al. 2008, 2010). Activated Bax then self-propagates further activation through its own, newly exposed BH3-only domain. This leads to the formation of asymmetric Bax oligomers that ultimately cause MOMP. Alternatively, BH3 proteins can activate Bax and Bak by binding in their hydrophobic BH3-binding pockets (Czabotar et al. 2013; Leshchiner et al. 2013; Moldoveanu et al. 2013). Upon activation, Bax and Bak homodimerize in a head-to-head manner (Dewson et al. 2008, 2012). Dimerization unveils a cryptic dimer–dimer binding site that allows oligomers of homodimers to form and drive MOMP (Dewson et al. 2009).

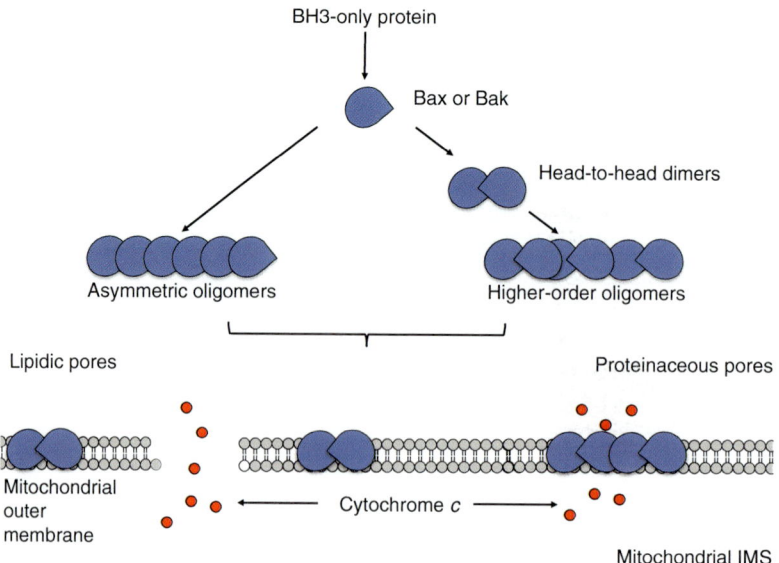

Figure 2. Mechanism of Bax/Bak activation and MOMP. BH3 domain-only proteins directly bind and activate Bax and Bak. Activated Bax and Bak form higher-order oligomers, either through asymmetric oligomers (Bax) or through the formation of higher-order oligomers formed by head-to-head Bax or Bak dimers. How oligomeric Bax and Bak permeabilize the mitochondrial outer membrane is unclear. Two prominent models argue that Bax and Bak do this either by inducing lipidic pores (*left*) or by directly forming proteinaceous pores (*right*).

Initial live-cell imaging studies, using cytochrome *c* GFP to report mitochondrial permeabilization, showed that, although the onset of MOMP is highly variable, following its initiation, permeabilization of mitochondria occurs in a rapid (<5 min) and complete manner (Goldstein et al. 2000). More recently, several studies have found that MOMP can occur at a defined point or points within a cell and propagate in a wave-like fashion over the whole cell (Lartigue et al. 2008; Bhola et al. 2009; Rehm et al. 2009). Exactly how these waves are propagated is unclear, but existing data argue against involvement of either caspases or the mitochondrial permeability transition, a change in the inner mitochondrial membrane permeability to small solutes (Crompton 1999). As discussed previously, the self-propagating nature of Bax and Bak activation might be expected to facilitate the occurrence of MOMP in a wave-like manner. Chemical inhibitors of casein kinase II inhibit wave formation, arguing that substrate(s) of this kinase (perhaps BH3-only proteins) are relevant for wave formation (Bhola et al. 2009). Alternatively, mitochondrial-derived reactive oxygen species (ROS) may promote wave formation because inhibition of ROS or addition of ROS scavengers prevents wave-like MOMP from occurring (Garcia-Perez et al. 2012). It remains unclear how permeabilization of individual mitochondria generates ROS, or, indeed, what the targets of ROS are that facilitate wave propagation.

Much interest has focused on whether MOMP permits selective or nonselective release of mitochondrial intermembrane space (IMS) proteins. At least in vitro, Bax-mediated permeabilization of liposomes leads to release of 10-kDa and 2-MDa dextrans with similar kinetics (Kuwana et al. 2002). In cells, proteins >100 kDa (predicted molecular weight of Smac-GFP dimers) are released with kinetics similar to cytochrome *c*; however, a Smac dsRed tetrameric fusion protein (predicted size 190 kDa) failed to be released from mitochondria upon MOMP (Rehm et al. 2003). Furthermore, ectopic expression of XIAP delays the kinetics of Smac release following MOMP from mitochondria dependent on the ability of XIAP to enter the mitochondrial IMS and complex with Smac (Flanagan et al. 2010). Although these results suggest that the release of IMS proteins following MOMP may have size limitations in vivo, the onset of IMS protein release from mitochondria is the same irrespective of size, thus arguing that all soluble IMS proteins exit the mitochondria through a similar mechanism (Munoz-Pinedo et al. 2006). In some settings, selective release of mitochondrial IMS proteins can be observed; for example, cells deficient in Drp-1, a dynamin-like protein required for mitochondrial fission, preferentially release Smac but not cytochrome c following MOMP (Parone et al. 2006; Estaquier and Arnoult 2007; Ishihara et al. 2009). Why loss of Drp-1 selectively inhibits cytochrome c egress from the mitochondria remains unclear, but this can inhibit the kinetics of caspase activation and apoptosis. Interestingly, Drp-1 can also act as a positive regulator of Bax-mediated MOMP (Montessuit et al. 2010).

The requirement for Bax and Bak in MOMP is clear, but how these proteins actually permeabilize the mitochondrial outer membrane remains elusive. Two prominent models propose that activated Bax and Bak cause MOMP either by forming proteinaceous pores themselves or, alternatively, by causing the formation of lipidic pores in the mitochondrial outer membrane. As discussed above, pro- and antiapoptotic Bcl-2 proteins are structurally similar to bacterial pore-forming toxins, implying that Bax and Bak themselves might directly form pores in the mitochondrial outer membrane (Muchmore et al. 1996; Suzuki et al. 2000). Along these lines, several studies have found that Bax can induce ion channels in artificial membranes; however, somewhat confusingly, antiapoptotic Bcl-2 proteins can also form membrane pores (Antonsson et al. 1997). Patch-clamp studies of isolated mitochondria have discovered that during MOMP (initiated by the addition of the BH3-only protein tBid), a mitochondrial outer membrane channel forms that increases with size over time and displays kinetics similar to MOMP (Martinez-Caballero et al. 2009). This implies that the channel (termed the mitochondrial apoptosis-induced channel [MAC]) as the perpetrator of MOMP. In support of this, inhibitors that block MAC block MOMP and apoptosis in cells (Peixoto et al. 2009). However, it remains possible that these inhibitors block the initial activation of Bax and Bak. Furthermore, in the majority of studies, the size of the MAC channels detected have only been large enough to accommodate cytochrome c release, but, as discussed above, MOMP clearly allows for the release of much larger proteins.

An alternative model proposes that activated Bax and Bak cause MOMP by inducing lipidic pores. This model would account for various characteristics of MOMP including the release of large IMS proteins and a consistent inability to detect proteinaceous pores in the mitochondrial outer membrane. Activated Bax can induce liposome permeabilization in vitro, leading to the release of encapsulated material in a size-independent manner, thereby recapitulating a key characteristic of MOMP (Basanez et al. 1999, 2002; Hardwick and Polster 2002). Moreover, cryo-EM analysis of Bax-permeabilized liposomes revealed large openings (up to 100 nm). These appeared concurrently with permeabilization and could be inhibited in a Bcl-X_L-dependent manner (Schafer et al. 2009). In further support of the lipidic pore model, Bax-induced pores were variable in size and lacked proteinaceous material—this contrasts with protein pores formed by the bacterial toxin pneumolysin that are uniform in nature and proteinaceous in composition. However, whether activated Bax and Bak induce MOMP by forming lipid pores in mitochondrial outer membranes remains unclear because similar pore-like structures have not been observed in mitochondria.

APPETITE FOR DESTRUCTION: HOW MOMP KILLS CELLS

Irrespective of mechanism, MOMP wreaks havoc on the cell. Normally, MOMP leads to the release of proteins that activate caspases leading to rapid, apoptotic cell death. However, even in the absence of caspase activity, cells generally succumb to cell death through an ill-defined process termed caspase-independent cell death

(CICD) (Tait and Green 2008) (Fig. 1). Therefore, MOMP is often considered a point of no return. Here we review how MOMP triggers cell death through caspase-dependent and -independent means.

Mitochondrial-Dependent Caspase Activation

Although the onset of MOMP is highly variable, following mitochondrial permeabilization, caspases are activated in a robust manner leading to apoptosis typically within a few minutes (Goldstein et al. 2000; Albeck et al. 2008). Of the many mitochondrial intermembrane space proteins released following MOMP, cytochrome *c* is the most important. Once in the cytoplasm, cytochrome *c* transiently binds the key caspase adaptor molecule Apaf-1. This interaction triggers extensive conformational changes in Apaf-1 leading to its oligomerization into a heptameric wheel-like structure and exposure of caspase activation and recruitment domains (CARD) (Bratton and Salvesen 2010). The Apaf-1 CARD domains bind to CARD domains of the initiator caspase procaspase-9, forming the apoptosome. At the apoptosome, dimerization of caspase-9 leads to its activation, which, in turn, cleaves and activates the executioner caspases-3 and -7, leading to rapid cell death. Cytochrome *c* is essential for mitochondrial-dependent caspase activation; cells that lack cytochrome *c* or express a mutant that poorly activates Apaf-1 (but retains respiratory function) fail to activate caspases following MOMP (Li et al. 2000; Hao et al. 2005; Matapurkar and Lazebnik 2006). Moreover, mice expressing this mutated form of cytochrome *c* phenocopy the neurological defects observed in Apaf-1- and caspase-9-deficient mice.

Besides cytochrome *c*, other mitochondrial IMS proteins facilitate caspase activation. These include Smac (also called Diablo) and Omi (also called HtrA2) (Du et al. 2000; Verhagen et al. 2000; Suzuki et al. 2001). Both proteins reside in the mitochondrial intermembrane space and are released following MOMP. In healthy cells, Omi functions as a mitochondrial chaperone, whereas the nonapoptotic function for Smac is not known. Smac and Omi promote caspase activation by binding to and neutralizing the caspase inhibitor XIAP. However, in contrast to cytochrome *c*, loss of either Omi or Smac either individually or together does not impart resistance to caspase activation and apoptosis (Okada et al. 2002; Jones et al. 2003; Martins et al. 2004). Indeed, likely because of its chaperone function, cells and mice lacking Omi are rendered more sensitive to mitochondrial damage and cell death. Although these results argue that XIAP neutralization may facilitate rather than be essential for caspase activation, recent data argue that in death-receptor-triggered apoptosis, neutralization of XIAP is essential for effective caspase activation in type II cells (cells that require MOMP for death-receptor-induced apoptosis) (Jost et al. 2009). Moreover, there may be significant redundancy with respect to XIAP inhibition given the identification of various other mitochondrial proteins that can inhibit XIAP (Zhuang et al. 2013).

Other mitochondrial IMS proteins that have been proposed to facilitate caspase activation include apoptosis-inducing factor (AIF). In contrast to cytochrome *c*, the release of AIF from the mitochondrial IMS following MOMP is slow and, in some circumstances, caspase-dependent (Arnoult et al. 2003; Munoz-Pinedo et al. 2006). As such, AIF likely does not seem to play a major role in apoptosis induction.

Even in the absence of caspase activity, cells typically succumb to a slower, ill-defined form of death termed caspase-independent cell death (CICD). CICD may serve primarily as a failsafe mechanism to ensure that cell death occurs even if caspases are inhibited (e.g., by a viral caspase inhibitor). Careful morphological analysis revealed that under physiological conditions, CICD may account for up to 10% of cell death—if this is, indeed, the case, it represents a major cell death modality (Chautan et al. 1999). Furthermore, comparison of early embryonic lethality (typically embryonic day 7 [E7], although some survive and can mature to adulthood) observed with Bax/Bak-deficient mice (unable to undergo MOMP) with the postnatal lethality of Apaf-1-deficient mice (can only undergo CICD) argues that, at the gross level,

CICD can effectively substitute for apoptosis, at least during development (Yoshida et al. 1998; Lindsten et al. 2000). That said, the ~15% of Bax/Bak-deficient animals that survive embryogenesis and mature, showing some neurological defects and expansion of lymphoid cells, represents an ongoing puzzle for the role of MOMP in development.

How CICD occurs following MOMP is unclear. Indeed, the mechanism of CICD may vary in a cell-type-dependent manner—unlike the canonical, mitochondrial pathway of caspase activity. One model supports an active role for mitochondria in mediating cell death, for example, through the release of proteins following MOMP such as AIF that can actively induce CICD. AIF may contribute to caspase-independent cell death (CICD) in some settings (Cheung et al. 2006). Alternatively, CICD may be mediated primarily by mitochondrial dysfunction that ensues following MOMP, ultimately leading to metabolic catastrophe and cell death. Along these lines, analysis of cells undergoing CICD found a rapid reduction in mitochondrial respiratory complex I and IV function (Lartigue et al. 2009). At subsequent time points post-MOMP, cytochrome c can be targeted for proteasome-dependent degradation, again promoting respiratory dysfunction (Ferraro et al. 2008). In addition to breakdown of mitochondrial respiratory function, mitochondrial proteins including TIM23 (an essential component of the mitochondrial inner membrane translocase complex) can be cleaved and inactivated following MOMP, in doing so contributing to mitochondrial dysfunction (Goemans et al. 2008). Moreover, given the important role that AIF has in maintaining respiratory complex I function (Vahsen et al. 2004), loss of AIF from the mitochondria should also promote mitochondrial dysfunction. Collectively, these findings argue that loss of mitochondrial function may be the principle reason that cells die through CICD following MOMP. However, because cells can survive complete removal of mitochondria for at least 4 d, which is typically longer than the kinetics of CICD, this still suggests that permeabilized mitochondria may also play an active role in CICD (Narendra et al. 2008). One such role may be as "ATP-sinks" because maintenance of the transmembrane potential is sustained by reversal of the F_0F_1 ATPase.

POST-MOMP REGULATION OF CASPASE ACTIVITY

Under some circumstances, MOMP need not be a death sentence. However, in order to evade cell death post-MOMP, cells must limit caspase activation. Here we review mechanisms of caspase activity regulation after MOMP, focusing on regulation of IMS protein release following MOMP and direct means of inhibiting caspase activation following mitochondrial permeabilization.

Post-MOMP Regulation of IMS Protein Release

MOMP itself does not appear to afford any specificity over which IMS proteins are released from the mitochondria. However, various studies implicate mechanisms that govern selective release of IMS proteins following MOMP; principally, these mechanisms center on IMS protein interaction with the mitochondrial membranes or by remodeling of the mitochondrial inner membrane (Fig. 3).

AIF is tethered to the mitochondrial inner membrane; consequently, its release following MOMP requires proteolytic cleavage either by caspase or calpain proteases (Arnoult et al. 2003; Polster et al. 2005). In the case of cytochrome c, electrostatic interactions with inner membrane lipids and the oxidative state of these lipids (where oxidized lipids bind cytochrome c less) have been proposed to regulate its release following MOMP (Ott et al. 2002).

The mitochondrial inner membrane is largely composed of cristae, involutions that greatly expand the mitochondrial surface area for oxidative phosphorylation and ATP generation. Far from being static, cristae are highly dynamic structures, and their accessibility to the IMS is regulated through cristae junctions. Interestingly, most cytochrome c resides in mitochondrial cristae, leading various studies to

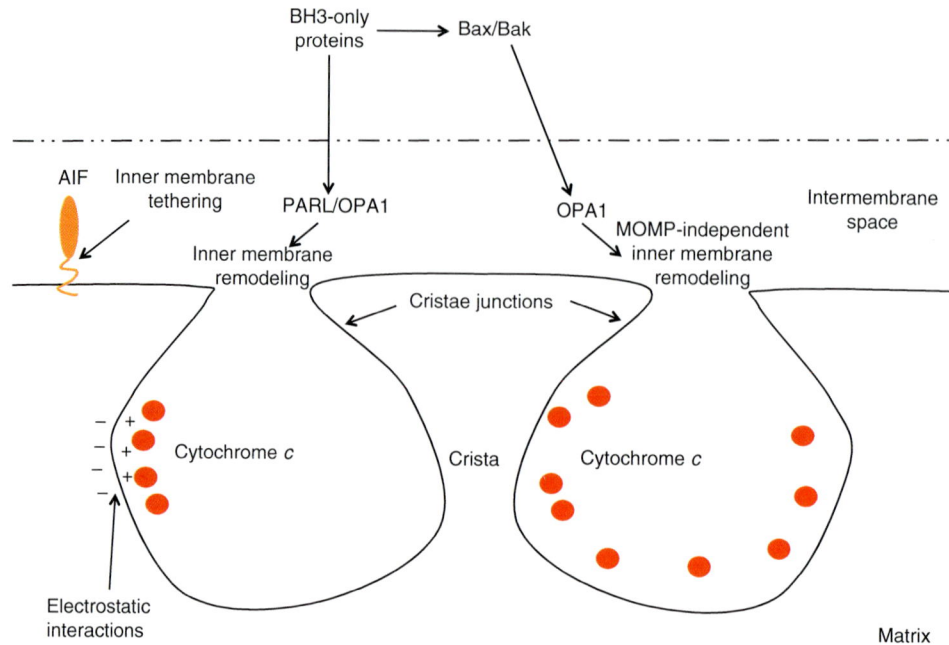

Figure 3. Post-MOMP regulation of mitochondrial intermembrane space protein release. The intermembrane space protein AIF is tethered to the mitochondrial inner membrane and requires cleavage to liberate it from the mitochondria upon MOMP. The majority of cytochrome c is sequestered within mitochondrial cristae; electrostatic interactions facilitate its association with the inner membrane. Some studies argue that cristae remodeling must occur to allow cytochrome c egress from the mitochondrial cristae following MOMP. Cristae remodeling can occur in a MOMP-independent manner by BH3 proteins (in a Bax/Bak-independent manner) or by activated Bax and Bak. Remodeling is dependent upon the intermembrane space rhomboid protease PARL and the dynamin-like GTPase OPA1.

address whether cristae remodeling provides an additional layer of regulating cytochrome c release from the mitochondria. Accordingly, several BH3-only proteins including Bid, Bim, BNIP3, and Bik have been found to regulate cristae remodeling (Scorrano et al. 2002; Germain et al. 2005; Yamaguchi et al. 2008). In vitro treatment of mitochondria with the BH3 protein tBid leads to extensive remodeling, interconnected cristae, and cytochrome c mobilization from the cristae into the IMS. Interestingly, this effect of tBid on mitochondrial inner membrane dynamics did not require the tBid BH3 domain (Scorrano et al. 2002). Other studies have found that membrane remodeling requires active Bax and Bak but does not necessitate MOMP, because pharmacological inhibitors of MOMP still allow remodeling (Yamaguchi et al. 2008). Two IMS proteins, OPA1 (a dynamin-like GTPase) and PARL (a rhomboid protease), are essential for regulating cristae dynamics. Upon MOMP, disruption of OPA1 oligomers widens cristae junctions, whereas PARL cleavage of OPA1 generates a cleavage product that maintains tight junctions (Frezza et al. 2006). However, other studies have found no gross changes in mitochondrial morphology or cristae junction size upon MOMP or only detected them following executioner caspase activity—this argues that remodeling may be consequential rather than causative in promoting IMS protein release (Sun et al. 2007). Moreover, even in a closed state, cytochrome c should be able to exit cristae junctions, arguing that cristae width is not a key determinant of release in itself (Gillick and Crompton 2008). Possibly, cristae remodeling may support IMS protein release in a cell-type-specific manner, or OPA1 and PARL may facilitate IMS protein release independently of cristae remodeling.

Besides regulating IMS protein release post-MOMP, a plethora of mechanisms have been described that can limit caspase activity. The physiological role of these mechanisms is uncertain, but perhaps they serve to restrain caspase activity and allow viability should MOMP occur in a limited number of mitochondria. As discussed above, through a well-described mechanism, XIAP can limit caspase activation by binding active caspases-9, -3, and -7. However, additional direct and indirect means of regulating caspase activity also exist that center on the formation and activation of the apoptosome. Importantly, various means of inhibiting apoptosome activation have been described in cancer, implying that this may facilitate cancer cell survival (Schafer and Kornbluth 2006).

Apoptosome Formation: Regulating the Wheel of Misfortune

Formation of the apoptosome is essential for efficient caspase-9 activation and mitochondrial-dependent apoptosis. APAF1 must bind dATP for apoptosome formation; however, paradoxically, physiological levels of nucleotides inhibit apoptosis by directly binding cytochrome c, preventing it from binding APAF1 (Chandra et al. 2006) (Fig. 4). Similarly, transfer RNA (tRNA) has also been found to bind cytochrome c, blocking its interaction with APAF1 and thereby preventing apoptosome formation (Mei et al. 2010). Physiological levels of potassium and calcium also inhibit cytochrome c-induced apoptosome formation (Cain et al. 2001; Bao et al. 2007). These inhibitory mechanisms may primarily exist to suppress accidental MOMP-induced caspase activity but are overwhelmed following rapid and extensive mitochondrial release of cytochrome c during apoptosis.

The redox status of a cell may also affect the proapoptotic activity of cytochrome c where oxidation promotes its proapoptotic activity and reduction inhibits it (Pan et al. 1999; Borutaite and Brown 2007). Mechanistically, how redox status would affect the ability of cytochrome c to induce apoptosome formation remains unclear, and some studies have found that reduced cytochrome c can still effectively activate caspases in vitro (Kluck et al. 1997). Various other proteins including HSP70, HSP90, and Cdc6 have been found to inhibit apoptosome function either by blocking its assembly or by inhibiting binding and activation of procaspase-9 at the apoptosome (Beere et al. 2000; Pandey et al. 2000; Saleh et al. 2000; Niimi et al. 2012).

Apoptosome function can also be positively regulated. The protein PHAP1 (also known as pp32) enhances apoptosome function by inhibiting aggregation of APAF1 and promoting nucleotide exchange (Jiang et al 2003; Kim et al. 2008). Importantly, reduced levels of PHAP1 inhibit apoptosis and allow clonogenic survival following chemotherapy—this finding may be relevant in small cell lung cancer because reduced PHAP expression correlates with poor clinical response to chemotherapy (Hoffarth et al. 2008).

Regulating Caspase-9 Activation

In addition to regulation of apoptosome assembly, caspase-9 activity can also be regulated. Several kinases can phosphorylate caspase-9 and inhibit its enzymatic activity. These include the MAP kinases ERK1 and ERK2 and CDK1-cyclin B1 (Allan et al. 2003; Allan and Clarke 2007). Although it is clear that phosphorylation can inhibit caspase-9 activity, how it achieves this is not understood. Because recruitment of procaspase-9 to the apoptosome does not appear to be affected by phosphorylation, perhaps phosphorylation of caspase-9 blocks its ability to dimerize. Interestingly, Rsk kinase (also a member of the MAPK family) has been found to inhibit Apaf-1 function by direct phosphorylation (Kim et al. 2012). This enables the adaptor protein 14-3-3ε; to bind Apaf-1 and prevent apoptosome assembly. At the apoptosome, autoprocessing of caspase-9 leads to a dramatic reduction in its affinity for the apoptosome, resulting in loss of caspase-9 activity. This mechanism acts as a "molecular timer" of which its activity (and ability to drive executioner caspase activity) is dictated by intracellular caspase-9

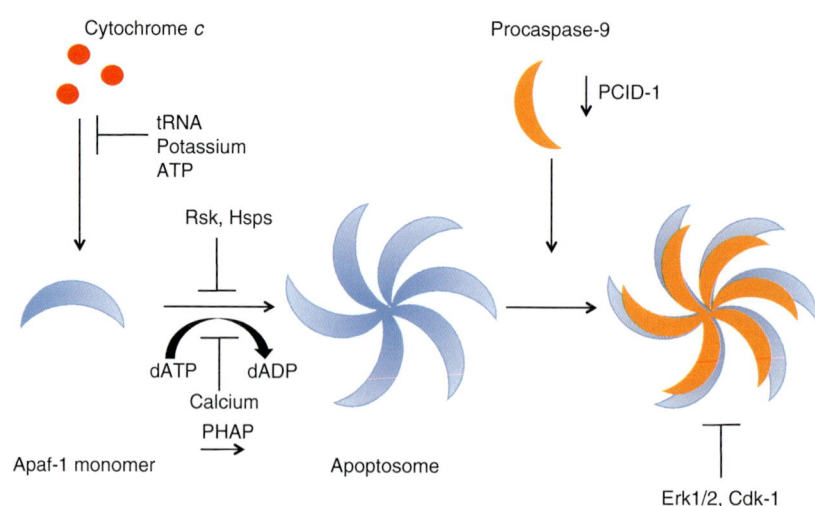

Figure 4. Regulation of apoptosome activity. Various molecules, including tRNA, potassium, and ATP, can competitively inhibit cytochrome c–Apaf-1 interactions blocking apoptosome formation. Apaf-1 oligomerization can be positively affected by proteins such as PHAP that facilitate nucleotide exchange, whereas intracellular calcium levels inhibit this event. Various proteins, including heat shock proteins (Hsps) and kinases such as Rsk can directly inhibit Apaf-1 oligomerization through interaction with Apaf-1 or by inhibitory phosphorylation. The activity of the apoptosome can also be inhibited by the kinase activity of Erk1/2 and Cdk-1. Finally, proteins such as PCID1 can regulate the intracellular levels of procaspase-9, thereby regulating apoptosome activity.

levels (Malladi et al. 2009). Consequently, regulation of caspase-9 expression can also control caspase activity post-MOMP. PCID1 is the human ortholog of Tango7, a *D. melanogaster* protein that regulates expression of the initiator caspase pro-Dronc (Chew et al. 2009). In an analogous manner, down-regulation of PCID1 reduces expression of procaspase-9. This may be clinically relevant because PCID1 is frequently down-regulated in pancreatic cancer (Jones et al. 2008).

DODGING THE BULLET—CELL SURVIVAL FOLLOWING MOMP

Although MOMP often represents a point of no return, this is not always the case. Cell survival following MOMP likely has important pathophysiological functions by facilitating long-term survival of postmitotic cells and enabling tumor cell survival. Moreover, MOMP itself may have noncytotoxic signaling functions, thereby requiring cells to survive this process. Here we discuss how cells survive MOMP and the roles, both good and bad, that survival post-MOMP can have.

Surviving "Accidental" MOMP

Live-cell imaging studies led to the initial view that MOMP is an all-or-nothing event (Goldstein et al. 2000). However, subsequent work has found that MOMP can sometimes be incomplete, leaving a minority of mitochondria intact (Tait et al. 2010). This suggests that the converse could also occur; limited mitochondria may undergo permeabilization without leading to cell death. Such accidental MOMP would necessitate that a threshold extent of MOMP must be crossed in order to trigger apoptotic caspase activity. Indeed, laser irradiation of neuronal mitochondria leading to MOMP of 15% of a cell's mitochondria was insufficient to trigger MOMP (Khodjakov et al. 2004). As already discussed, there are a plethora of mechanisms that can restrain caspase activity post-MOMP, but whether MOMP does occur in a few mitochondria without triggering cell death remains unknown.

Postmitotic Cell Survival

The life-long requirement of postmitotic cells necessitates robust prosurvival mechanisms. Both sympathetic neurons and cardiomyocytes can survive MOMP, at least in part, because they express insufficient levels of APAF-1 to activate caspases efficiently (Wright et al. 2004; Potts et al. 2005). XIAP is also a major player in conferring nonresponsiveness to MOMP in these cell types because addition of SMAC or deletion of XIAP can restore apoptotic sensitivity (Potts et al. 2003). In the case of neurons, NGF deprivation induces a so-called competence to die because it leads to XIAP down-regulation (Deshmukh and Johnson 1998; Martinou et al. 1999). Besides XIAP, the high glycolytic levels of neurons also facilitate inhibition of caspase activity (Vaughn and Deshmukh 2008). Glycolysis leads to increased glutathione synthase levels through the pentose phosphate shunt. As discussed above, reduction of cytochrome c can impair its ability to induce apoptosome activation. Comparable inhibitory mechanisms may also play a role in tumor cells given that they too are highly glycolytic.

Recovery from MOMP in Dividing Cells

In some situations, proliferating cells can survive MOMP provided that caspase function is inhibited. This has the potential to have an impact on both tumor development and therapeutic responses because cancer cells often inhibit caspase activity downstream from MOMP by a variety of mechanisms. Through a retroviral-based cDNA screen, GAPDH was found to protect cells from caspase-independent cell death downstream from MOMP (Colell et al. 2007). This protective role of GAPDH was due both to its well-established role as a key glycolytic enzyme and a newly described function by up-regulating autophagy. The ability of GAPDH to promote cell survival may be important in BCR-ABL-dependent chronic myeloid leukemia because GAPDH can promote resistance to cell death induced by BCR-ABL inhibitors (Lavallard et al. 2009).

Numerous events must occur in order for a cell to survive MOMP. Permeabilized mitochondria must be repaired or removed, and "new" mitochondria must be generated. Mitochondrial repopulation requires a cohort of mitochondria that fail to permeabilize following MOMP. The ability of certain mitochondria to evade MOMP relates to increased levels of antiapoptotic Bcl-2 proteins on their outer membrane; accordingly, Bcl-2 antagonist drugs can effectively permeabilize these mitochondria. Together with the strong correlation observed between the presence of intact mitochondria and cell survival, this suggests that the intact mitochondria provide a seed population of healthy mitochondria that ultimately repopulate the cell (Tait et al. 2010).

SUMMARY

Our understanding of MOMP and how it triggers cell death has advanced to the stage that drugs have now been developed to target this process. Nevertheless, significant gaps in our knowledge exist. For example, how activated Bax and Bak permeabilize the mitochondrial outer membrane is unknown. Secondly, although we understand how MOMP drives caspase activation, we have little mechanistic insight as to how it leads to CICD. The extent to which cells undergo CICD in vivo is difficult to gauge, mainly because of the lack of tools to detect and quantify this form of cell death accurately. Furthermore, although poorly understood, much greater attention is now being paid to how the mode of cell death influences the way the immune system perceives and reacts to a dying cell. Last, as we have discussed, MOMP need not be a death sentence. However, the mechanisms that allow cells to recover from MOMP remain poorly defined, as do its in vivo occurrence and pathophysiological importance. Ultimately, further understanding of how MOMP dictates life and death will facilitate its therapeutic targeting in a variety of diseases.

ACKNOWLEDGMENTS

S.W.G.T. is a Royal Society University Research Fellow.

REFERENCES

Abdelwahid E, Yokokura T, Krieser RJ, Balasundaram S, Fowle WH, White K. 2007. Mitochondrial disruption in *Drosophila* apoptosis. *Dev Cell* **12:** 793–806.

Albeck JG, Burke JM, Aldridge BB, Zhang M, Lauffenburger DA, Sorger PK. 2008. Quantitative analysis of pathways controlling extrinsic apoptosis in single cells. *Mol Cell* **30:** 11–25.

Allan LA, Clarke PR. 2007. Phosphorylation of caspase-9 by CDK1/cyclin B1 protects mitotic cells against apoptosis. *Mol Cell* **26:** 301–310.

Allan LA, Morrice N, Brady S, Magee G, Pathak S, Clarke PR. 2003. Inhibition of caspase-9 through phosphorylation at Thr 125 by ERK MAPK. *Nat Cell Biol* **5:** 647–654.

Antonsson B, Conti F, Ciavatta A, Montessuit S, Lewis S, Martinou I, Bernasconi L, Bernard A, Mermod JJ, Mazzei G, et al. 1997. Inhibition of Bax channel-forming activity by Bcl-2. *Science* **277:** 370–372.

Arnoult D, Gaume B, Karbowski M, Sharpe JC, Cecconi F, Youle RJ. 2003. Mitochondrial release of AIF and EndoG requires caspase activation downstream of Bax/Bak-mediated permeabilization. *EMBO J* **22:** 4385–4399.

Bao Q, Lu W, Rabinowitz JD, Shi Y. 2007. Calcium blocks formation of apoptosome by preventing nucleotide exchange in Apaf-1. *Mol Cell* **25:** 181–192.

Basanez G, Nechushtan A, Drozhinin O, Chanturiya A, Choe E, Tutt S, Wood KA, Hsu Y, Zimmberberg J, Youle RJ. 1999. Bax, but not Bcl-X_L, decreases the lifetime of planar phospholipid bilayer membranes at subnanomolar concentrations. *Proc Natl Acad Sci* **96:** 5492–5497.

Basanez G, Sharpe JC, Galanis J, Brandt TB, Hardwick JM, Zimmerberg J. 2002. Bax-type apoptotic proteins porate pure lipid bilayers through a mechanism sensitive to intrinsic monolayer curvature. *J Biol Chem* **277:** 49360–49365.

Beere HM, Wolf BB, Cain K, Mosser DD, Mahboubi A, Kuwana T, Tailor P, Morimoto RI, Cohen GM, Green DR. 2000. Heat-shock protein 70 inhibits apoptosis by preventing recruitment of procaspase-9 to the Apaf-1 apoptosome. *Nat Cell Biol* **2:** 469–475.

Bender CE, Fitzgerald P, Tait SW, Llambi F, McStay GP, Tupper DO, Pellettieri J, Sanchez Alvarado A, Salvesen GS, Green DR. 2012. Mitochondrial pathway of apoptosis is ancestral in metazoans. *Proc Natl Acad Sci* **109:** 4904–4909.

Bhola PD, Mattheyses AL, Simon SM. 2009. Spatial and temporal dynamics of mitochondrial membrane permeability waves during apoptosis. *Biophys J* **97:** 2222–2231.

Borutaite V, Brown GC. 2007. Mitochondrial regulation of caspase activation by cytochrome oxidase and tetramethylphenylenediamine via cytosolic cytochrome c redox state. *J Biol Chem* **282:** 31124–31130.

Bratton SB, Salvesen GS. 2010. Regulation of the Apaf-1-caspase-9 apoptosome. *J Cell Sci* **123:** 3209–3214.

Cain K, Langlais C, Sun XM, Brown DG, Cohen GM. 2001. Physiological concentrations of K^+ inhibit cytochrome c-dependent formation of the apoptosome. *J Biol Chem* **276:** 41985–41990.

Chandra D, Bratton SB, Person MD, Tian Y, Martin AG, Ayres M, Fearnhead HO, Gandhi V, Tang DG. 2006. Intracellular nucleotides act as critical prosurvival factors by binding to cytochrome c and inhibiting apoptosome. *Cell* **125:** 1333–1346.

Chautan M, Chazal G, Cecconi F, Gruss P, Golstein P. 1999. Interdigital cell death can occur through a necrotic and caspase-independent pathway. *Curr Biol* **9:** 967–970.

Cheung EC, Joza N, Steenaart NA, McClellan KA, Neuspiel M, McNamara S, MacLaurin JG, Rippstein P, Park DS, Shore GC, et al. 2006. Dissociating the dual roles of apoptosis-inducing factor in maintaining mitochondrial structure and apoptosis. *EMBO J* **25:** 4061–4073.

Chew SK, Chen P, Link N, Galindo KA, Pogue K, Abrams JM. 2009. Genome-wide silencing in *Drosophila* captures conserved apoptotic effectors. *Nature* **460:** 123–127.

Colell A, Ricci JE, Tait S, Milasta S, Maurer U, Bouchier-Hayes L, Fitzgerald P, Guio-Carrion A, Waterhouse NJ, Li CW, et al. 2007. GAPDH and autophagy preserve survival after apoptotic cytochrome c release in the absence of caspase activation. *Cell* **129:** 983–997.

Crompton M. 1999. The mitochondrial permeability transition pore and its role in cell death. *Biochem J* **341:** 233–249.

Czabotar PE, Westphal D, Dewson G, Ma S, Hockings C, Fairlie WD, Lee EF, Yao S, Robin AY, Smith BJ, et al. 2013. Bax crystal structures reveal how BH3 domains activate Bax and nucleate its oligomerization to induce apoptosis. *Cell* **152:** 519–531.

Deshmukh M, Johnson EM Jr. 1998. Evidence of a novel event during neuronal death: Development of competence-to-die in response to cytoplasmic cytochrome c. *Neuron* **21:** 695–705.

Dewson G, Kratina T, Sim HW, Puthalakath H, Adams JM, Colman PM, Kluck RM. 2008. To trigger apoptosis, Bak exposes its BH3 domain and homodimerizes via BH3:groove interactions. *Mol Cell* **30:** 369–380.

Dewson G, Kratina T, Czabotar P, Day CL, Adams JM, Kluck RM. 2009. Bak activation for apoptosis involves oligomerization of dimers via their α6 helices. *Mol Cell* **36:** 696–703.

Dewson G, Ma S, Frederick P, Hockings C, Tan I, Kratina T, Kluck RM. 2012. Bax dimerizes via a symmetric BH3:groove interface during apoptosis. *Cell Death Differ* **19:** 661–670.

Du C, Fang M, Li Y, Li L, Wang X. 2000. Smac, a mitochondrial protein that promotes cytochrome c-dependent caspase activation by eliminating IAP inhibition. *Cell* **102:** 33–42.

Eskes R, Desagher S, Antonsson B, Martinou JC. 2000. Bid induces the oligomerization and insertion of Bax into the outer mitochondrial membrane. *Mol Cell Biol* **20:** 929–935.

Estaquier J, Arnoult D. 2007. Inhibiting Drp1-mediated mitochondrial fission selectively prevents the release of cytochrome c during apoptosis. *Cell Death Differ* **14:** 1086–1094.

Ferraro E, Pulicati A, Cencioni MT, Cozzolino M, Navoni F, di Martino S, Nardacci R, Carri MT, Cecconi F. 2008. Apoptosome-deficient cells lose cytochrome c through proteasomal degradation but survive by autophagy-dependent glycolysis. *Mol Biol Cell* **19:** 3576–3588.

Flanagan L, Sebastia J, Tuffy LP, Spring A, Lichawska A, Devocelle M, Prehn JH, Rehm M. 2010. XIAP impairs Smac release from the mitochondria during apoptosis. *Cell Death Diff* **1**: e49.

Frezza C, Cipolat S, Martins de Brito O, Micaroni M, Beznoussenko GV, Rudka T, Bartoli D, Polishuck RS, Danial NN, De Strooper B, et al. 2006. OPA1 controls apoptotic cristae remodeling independently from mitochondrial fusion. *Cell* **126**: 177–189.

Garcia-Perez C, Roy SS, Naghdi S, Lin X, Davies E, Hajnoczky G. 2012. Bid-induced mitochondrial membrane permeabilization waves propagated by local reactive oxygen species (ROS) signaling. *Proc Natl Acad Sci* **109**: 4497–4502.

Gavathiotis E, Suzuki M, Davis ML, Pitter K, Bird GH, Katz SG, Tu HC, Kim H, Cheng EH, Tjandra N, et al. 2008. BAX activation is initiated at a novel interaction site. *Nature* **455**: 1076–1081.

Gavathiotis E, Reyna DE, Davis ML, Bird GH, Walensky LD. 2010. BH3-triggered structural reorganization drives the activation of proapoptotic BAX. *Mol Cell* **40**: 481–492.

George NM, Evans JJ, Luo X. 2007. A three-helix homo-oligomerization domain containing BH3 and BH1 is responsible for the apoptotic activity of Bax. *Genes Dev* **21**: 1937–1948.

Germain M, Mathai JP, McBride HM, Shore GC. 2005. Endoplasmic reticulum BIK initiates DRP1-regulated remodelling of mitochondrial cristae during apoptosis. *EMBO J* **24**: 1546–1556.

Gillick K, Crompton M. 2008. Evaluating cytochrome c diffusion in the intermembrane spaces of mitochondria during cytochrome c release. *J Cell Sci* **121**: 618–626.

Goemans CG, Boya P, Skirrow CJ, Tolkovsky AM. 2008. Intra-mitochondrial degradation of Tim23 curtails the survival of cells rescued from apoptosis by caspase inhibitors. *Cell Death Differ* **15**: 545–554.

Goldstein JC, Waterhouse NJ, Juin P, Evan GI, Green DR. 2000. The coordinate release of cytochrome c during apoptosis is rapid, complete and kinetically invariant. *Nat Cell Biol* **2**: 156–162.

Gray MW. 2012. Mitochondrial evolution. *Cold Spring Harb Perspect Biol* **4**: a011403.

Green DR. 2011. *Means to an end: Apoptosis and other cell death mechanisms*. Cold Spring Harbor Laboratory Press, Cold Spring Harbor, NY.

Hao Z, Duncan GS, Chang CC, Elia A, Fang M, Wakeham A, Okada H, Calzascia T, Jang Y, You-Ten A, et al. 2005. Specific ablation of the apoptotic functions of cytochrome c reveals a differential requirement for cytochrome c and Apaf-1 in apoptosis. *Cell* **121**: 579–591.

Hardwick JM, Polster BM. 2002. Bax, along with lipid conspirators, allows cytochrome c to escape mitochondria. *Mol Cell* **10**: 963–965.

Hoffarth S, Zitzer A, Wiewrodt R, Hahnel PS, Beyer V, Kreft A, Biesterfeld S, Schuler M. 2008. pp32/PHAPI determines the apoptosis response of non-small-cell lung cancer. *Cell Death Differ* **15**: 161–170.

Hsu YT, Wolter KG, Youle RJ. 1997. Cytosol-to-membrane redistribution of Bax and Bcl-X_L during apoptosis. *Proc Natl Acad Sci* **94**: 3668–3672.

Ishihara N, Nomura M, Jofuku A, Kato H, Suzuki SO, Masuda K, Otera H, Nakanishi Y, Nonaka I, Goto Y, et al. 2009. Mitochondrial fission factor Drp1 is essential for embryonic development and synapse formation in mice. *Nat Cell Biol* **11**: 958–966.

James ER, Green DR. 2002. Infection and the origins of apoptosis. *Cell Death Differ* **9**: 355–357.

Jiang X, Kim HE, Shu H, Zhao Y, Zhang H, Kofron J, Donnelly J, Burns D, Ng SC, Rosenberg S, et al. 2003. Distinctive roles of PHAP proteins and prothymosin-α in a death regulatory pathway. *Science* **299**: 223–226.

Jones JM, Datta P, Srinivasula SM, Ji W, Gupta S, Zhang Z, Davies E, Hajnoczky G, Saunders TL, Van Keuren ML, et al. 2003. Loss of Omi mitochondrial protease activity causes the neuromuscular disorder of mnd2 mutant mice. *Nature* **425**: 721–727.

Jones S, Zhang X, Parsons DW, Lin JC, Leary RJ, Angenendt P, Mankoo P, Carter, Kamiyama H, Jimeno A, et al. 2008. Core signaling pathways in human pancreatic cancers revealed by global genomic analyses. *Science* **321**: 1801–1806.

Jost PJ, Grabow S, Gray D, McKenzie MD, Nachbur U, Huang DC, Bouillet P, Thomas HE, Borner C, Silke J, et al. 2009. XIAP discriminates between type I and type II FAS-induced apoptosis. *Nature* **460**: 1035–1039.

Khodjakov A, Rieder C, Mannella CA, Kinnally KW. 2004. Laser micro-irradiation of mitochondria: Is there an amplified mitochondrial death signal in neural cells? *Mitochondrion* **3**: 217–227.

Kim HE, Jiang X, Du F, Wang X. 2008. PHAPI, CAS, and Hsp70 promote apoptosome formation by preventing Apaf-1 aggregation and enhancing nucleotide exchange on Apaf-1. *Mol Cell* **30**: 239–247.

Kim J, Parrish AB, Kurokawa M, Matsuura K, Freel CD, Andersen JL, Johnson CE, Kornbluth S. 2012. Rsk-mediated phosphorylation and 14-3-3ε binding of Apaf-1 suppresses cytochrome c–induced apoptosis. *EMBO J* **31**: 1279–1292.

Kluck RM, Martin SJ, Hoffman BM, Zhou JS, Green DR, Newmeyer DD. 1997. Cytochrome c activation of CPP32-like proteolysis plays a critical role in a Xenopus cell-free apoptosis system. *EMBO J* **16**: 4639–4649.

Kuwana T, Mackey MR, Perkins G, Ellisman MH, Latterich M, Schneiter R, Green DR, Newmeyer DD. 2002. Bid, Bax, and lipids cooperate to form supramolecular openings in the outer mitochondrial membrane. *Cell* **111**: 331–342.

Lartigue L, Medina C, Schembri L, Chabert P, Zanese M, Tomasello F, Dalibart R, Thoraval D, Crouzet M, Ichas F, et al. 2008. An intracellular wave of cytochrome c propagates and precedes Bax redistribution during apoptosis. *J Cell Sci* **121**: 3515–3523.

Lartigue L, Kushnareva Y, Seong Y, Lin H, Faustin B, Newmeyer DD. 2009. Caspase-independent mitochondrial cell death results from loss of respiration, not cytotoxic protein release. *Mol Biol Cell* **20**: 4871–4884.

Lavallard VJ, Pradelli LA, Paul A, Beneteau M, Jacquel A, Auberger P, Ricci JE. 2009. Modulation of caspase-independent cell death leads to resensitization of imatinib mesylate-resistant cells. *Cancer Res* **69**: 3013–3020.

Lee EF, Clarke OB, Evangelista M, Feng Z, Speed TP, Tchoubrieva EB, Strasser A, Kalinna BH, Colman PM, Fairlie

WD. 2011. Discovery and molecular characterization of a Bcl-2-regulated cell death pathway in schistosomes. *Proc Natl Acad Sci* **108:** 6999–7003.

Leshchiner ES, Braun CR, Bird GH, Walensky LD. 2013. Direct activation of full-length proapoptotic BAK. *Proc Natl Acad Sci* **110:** E986–E995.

Li K, Li Y, Shelton JM, Richardson JA, Spencer E, Chen ZJ, Wang X, Williams RS. 2000. Cytochrome *c* deficiency causes embryonic lethality and attenuates stress-induced apoptosis. *Cell* **101:** 389–399.

Lindsten T, Ross AJ, King A, Zong WX, Rathmell JC, Shiels HA, Ulrich E, Waymire KG, Mahar P, Frauwirth K, et al. 2000. The combined functions of proapoptotic Bcl-2 family members bak and bax are essential for normal development of multiple tissues. *Mol Cell* **6:** 1389–1399.

Malladi S, Challa-Malladi M, Fearnhead HO, Bratton SB. 2009. The Apaf-1*procaspase-9 apoptosome complex functions as a proteolytic-based molecular timer. *EMBO J* **28:** 1916–1925.

Martinez-Caballero S, Dejean LM, Kinnally MS, Oh KJ, Mannella CA, Kinnally KW. 2009. Assembly of the mitochondrial apoptosis-induced channel, MAC. *J Biol Chem* **284:** 12235–12245.

Martinou I, Desagher S, Eskes R, Antonsson B, Andre E, Fakan S, Martinou JC. 1999. The release of cytochrome *c* from mitochondria during apoptosis of NGF-deprived sympathetic neurons is a reversible event. *J Cell Biol* **144:** 883–889.

Martins LM, Morrison A, Klupsch K, Fedele V, Moisoi N, Teismann P, Albuin A, Grau E, Geppert M, Livi GP, et al. 2004. Neuroprotective role of the Reaper-related serine protease HtrA2/Omi revealed by targeted deletion in mice. *Mol Cell Biol* **24:** 9848–9862.

Matapurkar A, Lazebnik Y. 2006. Requirement of cytochrome *c* for apoptosis in human cells. *Cell Death Differ* **13:** 2062–2067.

Mei Y, Yong J, Liu H, Shi Y, Meinkoth J, Dreyfuss G, Yang X. 2010. tRNA binds to cytochrome *c* and inhibits caspase activation. *Mol Cell* **37:** 668–678.

Moldoveanu T, Grace CR, Llambi F, Nourse A, Fitzgerald P, Gehring K, Kriwacki RW, Green DR. 2013. BID-induced structural changes in BAK promote apoptosis. *Nat Struct Mol Biol* doi: 10.1038/nsmb.2563.

Montessuit S, Somasekharan SP, Terrones O, Lucken-Ardjomande S, Herzig S, Schwarzenbacher R, Manstein DJ, Bossy-Wetzel E, Basanez G, Meda P, et al. 2010. Membrane remodeling induced by the dynamin-related protein Drp1 stimulates Bax oligomerization. *Cell* **142:** 889–901.

Muchmore SW, Sattler M, Liang H, Meadows RP, Harlan JE, Yoon HS, Nettesheim D, Chang BS, Thompson CB, Wong SL, et al. 1996. X-ray and NMR structure of human Bcl-X_L, an inhibitor of programmed cell death. *Nature* **381:** 335–341.

Munoz-Pinedo C, Guio-Carrion A, Goldstein JC, Fitzgerald P, Newmeyer DD, Green DR. 2006. Different mitochondrial intermembrane space proteins are released during apoptosis in a manner that is coordinately initiated but can vary in duration. *Proc Natl Acad Sci* **103:** 11573–11578.

Narendra D, Tanaka A, Suen DF, Youle RJ. 2008. Parkin is recruited selectively to impaired mitochondria and promotes their autophagy. *J Cell Biol* **183:** 795–803.

Niimi S, Arakawa-Takeuchi S, Uranbileg B, Park JH, Jinno S, Okayama H. 2012. Cdc6 protein obstructs apoptosome assembly and consequent cell death by forming stable complexes with activated Apaf-1 molecules. *J Biol Chem* **287:** 18573–18583.

Oberst A, Bender C, Green DR. 2008. Living with death: The evolution of the mitochondrial pathway of apoptosis in animals. *Cell Death Differ* **15:** 1139–1146.

Okada H, Suh WK, Jin J, Woo M, Du C, Elia A, Duncan GS, Wakeham A, Itie A, Lowe SW, et al. 2002. Generation and characterization of Smac/DIABLO-deficient mice. *Mol Cell Biol* **22:** 3509–3517.

Ott M, Robertson JD, Gogvadze V, Zhivotovsky B, Orrenius S. 2002. Cytochrome *c* release from mitochondria proceeds by a two-step process. *Proc Natl Acad Sci* **99:** 1259–1263.

Pan Z, Voehringer DW, Meyn RE. 1999. Analysis of redox regulation of cytochrome *c*-induced apoptosis in a cell-free system. *Cell Death Differ* **6:** 683–688.

Pandey P, Saleh A, Nakazawa A, Kumar S, Srinivasula SM, Kumar V, Weichselbaum R, Nalin C, Alnemri ES, Kufe D, et al. 2000. Negative regulation of cytochrome *c*-mediated oligomerization of Apaf-1 and activation of procaspase-9 by heat shock protein 90. *EMBO J* **19:** 4310–4322.

Parone PA, James DI, Da Cruz S, Mattenberger Y, Donze O, Barja F, Martinou JC. 2006. Inhibiting the mitochondrial fission machinery does not prevent Bax/Bak-dependent apoptosis. *Mol Cell Biol* **26:** 7397–7408.

Peixoto PM, Ryu SY, Bombrun A, Antonsson B, Kinnally KW. 2009. MAC inhibitors suppress mitochondrial apoptosis. *Biochem J* **423:** 381–387.

Polster BM, Basanez G, Etxebarria A, Hardwick JM, Nicholls DG. 2005. Calpain I induces cleavage and release of apoptosis-inducing factor from isolated mitochondria. *J Biol Chem* **280:** 6447–6454.

Potts PR, Singh S, Knezek M, Thompson CB, Deshmukh M. 2003. Critical function of endogenous XIAP in regulating caspase activation during sympathetic neuronal apoptosis. *J Cell Biol* **163:** 789–799.

Potts MB, Vaughn AE, McDonough H, Patterson C, Deshmukh M. 2005. Reduced Apaf-1 levels in cardiomyocytes engage strict regulation of apoptosis by endogenous XIAP. *J Cell Biol* **171:** 925–930.

Rehm M, Dussmann H, Prehn JH. 2003. Real-time single cell analysis of Smac/DIABLO release during apoptosis. *J Cell Biol* **162:** 1031–1043.

Rehm M, Huber HJ, Hellwig CT, Anguissola S, Dussmann H, Prehn JH. 2009. Dynamics of outer mitochondrial membrane permeabilization during apoptosis. *Cell Death Differ* **16:** 613–623.

Ryoo HD, Baehrecke EH. 2010. Distinct death mechanisms in *Drosophila* development. *Curr Opin Cell Biol* **22:** 889–895.

Saleh A, Srinivasula SM, Balkir L, Robbins PD, Alnemri ES. 2000. Negative regulation of the Apaf-1 apoptosome by Hsp70. *Nat Cell Biol* **2:** 476–483.

Schafer ZT, Kornbluth S. 2006. The apoptosome: Physiological, developmental, and pathological modes of regulation. *Dev Cell* **10:** 549–561.

Schafer B, Quispe J, Choudhary V, Chipuk JE, Ajero TG, Du H, Schneiter R, Kuwana T. 2009. Mitochondrial outer membrane proteins assist Bid in Bax-mediated lipidic pore formation. *Mol Biol Cell* **20:** 2276–2285.

Scorrano L, Ashiya M, Buttle K, Weiler S, Oakes SA, Mannella CA, Korsmeyer SJ. 2002. A distinct pathway remodels mitochondrial cristae and mobilizes cytochrome *c* during apoptosis. *Dev Cell* **2:** 55–67.

Sun MG, Williams J, Munoz-Pinedo C, Perkins GA, Brown JM, Ellisman MH, Green DR, Frey TG. 2007. Correlated three-dimensional light and electron microscopy reveals transformation of mitochondria during apoptosis. *Nat Cell Biol* **9:** 1057–1065.

Suzuki M, Youle RJ, Tjandra N. 2000. Structure of Bax: Coregulation of dimer formation and intracellular localization. *Cell* **103:** 645–654.

Suzuki Y, Imai Y, Nakayama H, Takahashi K, Takio K, Takahashi R. 2001. A serine protease, HtrA2, is released from the mitochondria and interacts with XIAP, inducing cell death. *Mol Cell* **8:** 613–621.

Tait SW, Green DR. 2008. Caspase-independent cell death: Leaving the set without the final cut. *Oncogene* **27:** 6452–6461.

Tait SW, Parsons MJ, Llambi F, Bouchier-Hayes L, Connell S, Munoz-Pinedo C, Green DR. 2010. Resistance to caspase-independent cell death requires persistence of intact mitochondria. *Dev Cell* **18:** 802–813.

Tan FJ, Fire AZ, Hill RB. 2007. Regulation of apoptosis by *C. elegans* CED-9 in the absence of the C-terminal transmembrane domain. *Cell Death Differ* **14:** 1925–1935.

Vahsen N, Cande C, Briere JJ, Benit P, Joza N, Larochette N, Mastroberardino PG, Pequignot MO, Casares N, Lazar V, et al. 2004. AIF deficiency compromises oxidative phosphorylation. *EMBO J* **23:** 4679–4689.

Vaughn AE, Deshmukh M. 2008. Glucose metabolism inhibits apoptosis in neurons and cancer cells by redox inactivation of cytochrome *c*. *Nature Cell Biol* **10:** 1477–1483.

Verhagen AM, Ekert PG, Pakusch M, Silke J, Connolly LM, Reid GE, Moritz RL, Simpson RJ, Vaux DL. 2000. Identification of DIABLO, a mammalian protein that promotes apoptosis by binding to and antagonizing IAP proteins. *Cell* **102:** 43–53.

Wei MC, Lindsten T, Mootha VK, Weiler S, Gross A, Ashiya M, Thompson CB, Korsmeyer SJ. 2000. tBID, a membrane-targeted death ligand, oligomerizes BAK to release cytochrome *c*. *Genes Dev* **14:** 2060–2071.

Wei MC, Zong WX, Cheng EH, Lindsten T, Panoutsakopoulou V, Ross AJ, Roth KA, MacGregor GR, Thompson CB, Korsmeyer SJ. 2001. Proapoptotic BAX and BAK: A requisite gateway to mitochondrial dysfunction and death. *Science* **292:** 727–730.

Wright KM, Linhoff MW, Potts PR, Deshmukh M. 2004. Decreased apoptosome activity with neuronal differentiation sets the threshold for strict IAP regulation of apoptosis. *J Cell Biol* **167:** 303–313.

Yamaguchi R, Lartigue L, Perkins G, Scott RT, Dixit A, Kushnareva Y, Kuwana T, Ellisman MH, Newmeyer DD. 2008. Opa1-mediated cristae opening is Bax/Bak and BH3 dependent, required for apoptosis, and independent of Bak oligomerization. *Mol Cell* **31:** 557–569.

Yoshida H, Kong YY, Yoshida R, Elia AJ, Hakem A, Hakem R, Penninger JM, Mak TW. 1998. Apaf1 is required for mitochondrial pathways of apoptosis and brain development. *Cell* **94:** 739–750.

Youle RJ, Strasser A. 2008. The BCL-2 protein family: Opposing activities that mediate cell death. *Nat Rev Mol Cell Biol* **9:** 47–59.

Zhuang M, Guan S, Wang H, Burlingame AL, Wells JA. 2013. Substrates of IAP ubiquitin ligases identified with a designed orthogonal E3 ligase, the NEDDylator. *Mol Cell* **49:** 273–282.

Mechanisms of Action of Bcl-2 Family Proteins

Aisha Shamas-Din[1], Justin Kale[1], Brian Leber[1,2], and David W. Andrews[1]

[1]Department of Biochemistry and Biomedical Sciences, McMaster University, Hamilton, Ontario L8S4K1, Canada

[2]Department of Medicine, McMaster University, Hamilton, Ontario L8S4K1, Canada

Correspondence: andrewsd@dwalab.ca

The Bcl-2 family of proteins controls a critical step in commitment to apoptosis by regulating permeabilization of the mitochondrial outer membrane (MOM). The family is divided into three classes: multiregion proapoptotic proteins that directly permeabilize the MOM; BH3 proteins that directly or indirectly activate the pore-forming class members; and the antiapoptotic proteins that inhibit this process at several steps. Different experimental approaches have led to several models, each proposed to explain the interactions between Bcl-2 family proteins. The discovery that many of these interactions occur at or in membranes as well as in the cytoplasm, and are governed by the concentrations and relative binding affinities of the proteins, provides a new basis for rationalizing these models. Furthermore, these dynamic interactions cause conformational changes in the Bcl-2 proteins that modulate their apoptotic function, providing additional potential modes of regulation.

Apoptosis was formally described and named in 1972 as a unique morphological response to many different kinds of cell stress that was distinct from necrosis. However, despite the novelty and utility of the concept, little experimental work was performed during the following 20 years because no tools existed to manipulate the process. In the early 1990s, two seminal observations changed the landscape. First, as the complete developmental sequence of the nematode *Caenorhabditis elegans* was painstakingly elucidated at the single-cell level, it was noted that a fixed, predictable number of "intermediate" cells were destined to die, and that this process was positively and negatively regulated by specific genes. Second, a novel gene called B-cell CLL/lymphoma 2 (Bcl-2; encoded by *BCL2*) that was discovered as a partner in a reciprocal chromosomal translocation in a human tumor turned out to function not as a classic oncogene by driving cell division, but rather by preventing apoptosis. When it was discovered that the mammalian *BCL2* could substitute for *CED-9*, the *C. elegans* gene that inhibits cell death, the generality of the process was recognized, and the scientific literature exploded with now well over 10^5 publications on apoptosis. However, it is ironic to note that after a further 20 years of intensive investigation, it is clear that the mechanism of action of Bcl-2 is quite distinct from Ced-9, which sequesters the activator of the caspase protease that is the ultimate effector of apoptosis. In contrast, Bcl-2 works primarily by binding to other related proteins that regulate permeabilization of the mitochondrial outer membrane (MOM).

This review examines how apoptosis is regulated by the members of the (now very large) Bcl-2 family, composed of three groups related by structure and function (illustrated in Fig. 1): (1) the BH3 proteins that sense cellular stress and activate (either directly or indirectly); (2) the executioner proteins Bax or Bak that oligomerize in and permeabilize the MOM, thereby releasing components of the intermembrane space that activate the final, effector caspases of apoptosis; and (3) the antiapoptotic members like Bcl-2 that impede the overall process by inhibiting both the BH3 and the executioner proteins. To understand the consequence of the interactions among the three subgroups, several models have been proposed ("direct activation," "displacement," "embedded together," and "unified" models; illustrated in Fig. 2) that are briefly described here before a more detailed discussion of the Bcl-2 families.

DIRECT ACTIVATION MODEL

The distinctive feature of the direct activation model is that a BH3 protein is required to directly bind and to activate the Bcl-2 multihomology region proapoptotic proteins, Bax and Bak. The direct activation model classifies BH3 proteins as activators or sensitizers based on their affinities for binding to Bcl-2 multiregion proteins (see Table 1) (Letai et al. 2002). The activator BH3 proteins—tBid, Bim, and Puma—bind to both the proapoptotic and the antiapoptotic Bcl-2 multiregion proteins (Kim et al. 2006). The sensitizer BH3 proteins—Bad,

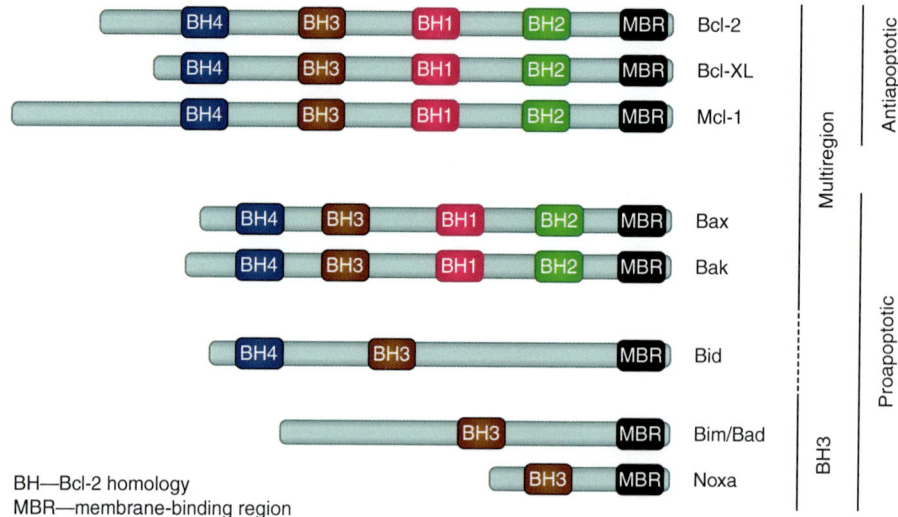

Figure 1. Schematic overview of the Bcl-2 family of proteins. The family is divided into two subgroups containing proteins that either inhibit apoptosis or promote apoptosis. The proapoptotic proteins are further subdivided functionally into those that oligomerize and permeabilize the MOM, such as Bax and Bak, or those that promote apoptosis through either activating Bax or Bak or inhibiting the antiapoptotic proteins, such as tBid, Bim, Bad, and Noxa. Proteins are included in the Bcl-2 family based on sequence homology to the founding member, Bcl-2, in one of the four Bcl-2 homology (BH) regions. All the antiapoptotic proteins, as well as Bax, Bak, and Bid, have multiple BH regions, are evolutionarily related, and share a three-dimensional (3D) structural fold. The BH3 proteins contain only the BH3 region, are evolutionarily distant from the multiregion proteins, and are intrinsically unstructured. Most members of the Bcl-2 family proteins contain a membrane-binding region (MBR) on their carboxyl termini in the form of a tail anchor, mitochondrial-targeting sequence, or as a hydrophobic amino acid sequence that facilitates binding and localization of these proteins to the MOM or to the endoplasmic reticulum (ER) membrane.

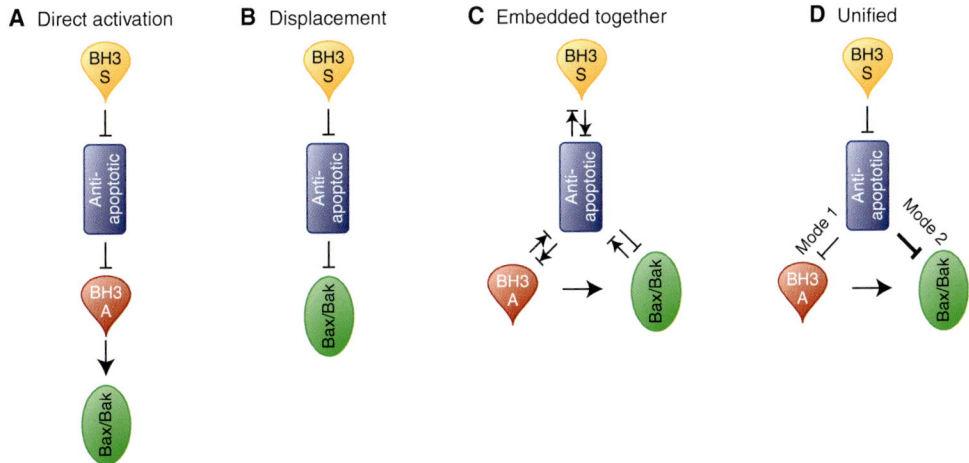

Figure 2. Schematics of the core mechanisms proposed by various models for the regulation of MOMP by Bcl-2 proteins. (↑) Activation; (⊥) inhibition; (⊥↑) mutual recruitment/sequestration. Paired forward and reverse symbols indicate the model makes explicit reference to equilibria. (A) The direct activation model divides the different BH3 proteins by qualitative differences in function. The BH3 proteins with high affinity for binding and activating Bax and Bak are termed as "activators," whereas those that only bind the antiapoptotic proteins are termed "sensitizers." The activator BH3 proteins directly interact with and activate Bax and Bak to promote MOMP. The antiapoptotic proteins inhibit MOMP by specifically sequestering the BH3 activators. The BH3 sensitizer proteins can compete for binding with the antiapoptotic proteins, thus releasing the BH3 activator proteins to avidly promote MOMP through activation and oligomerization of Bax and Bak. (B) The displacement model categorizes the BH3 proteins solely based on their affinities of binding for the antiapoptotic proteins (hence, does not recognize them as activators). In this model, Bax and Bak are constitutively active and oligomerize and induce MOMP unless held in check by the antiapoptotic proteins. Therefore, for a cell to undergo apoptosis, the correct combination of BH3 proteins must compete for binding for the different antiapoptotic proteins to liberate Bax and Bak and for MOMP to ensue. (C) The embedded together model introduces an active role for the membrane and combines the major aspects of the previous models. The interactions between members of the Bcl-2 family are governed by equlibria and therefore are contingent on the relative protein concentrations as well as their binding affinities. The latter are determined by posttranslational modifications, fraction of protein bound to the membrane, and cellular physiology. At membranes, the activator BH3 proteins directly activate Bax and Bak, which then oligomerize, inducing MOMP. Both activator and sensitizer BH3 proteins can recruit and sequester antiapoptotic proteins in the membrane. The antiapoptotic proteins inhibit apoptosis by sequestering the BH3 proteins and Bax and Bak in the membrane or by preventing their binding to membranes. At different intracellular membranes, the local concentrations of specific subsets of Bcl-2 family members alter the binding of Bcl-2 proteins to the membrane and the binding equilibria between family members. As a result, Bcl-2 family proteins have distinct but overlapping functions at different cellular locations. (D) The unified model builds on the embedded together model by proposing that the antiapoptotic proteins sequester the activator BH3 proteins (mode 1) and sequester Bax and Bak (mode 2). It differs in that in the unified model, inhibition of apoptosis through mode 1 is less efficient (smaller arrow in panel D) and therefore easier to overcome by sensitizer BH3 proteins. In addition, the unified model extends the role of Bcl-2 family proteins and the regulation of MOMP to mitochondria dynamics (not shown).

Noxa, Bik, Bmf, Hrk, and Bnip3—bind to the antiapoptotic proteins, thereby liberating activator BH3 proteins to promote mitochondrial outer membrane permeabilization (MOMP) (Letai et al. 2002; Kuwana et al. 2005; Certo et al. 2006). The antiapoptotic proteins bind to both the activator and the sensitizer BH3 proteins, but are unable to complex with Bax and Bak (Kim et al. 2006). Therefore, for a cell to evade apoptosis, antiapoptotic proteins must sequester the BH3 proteins to prevent Bax/Bak activation and apoptosis.

Table 1. Binding profiles within Bcl-2 family members

Antiapoptotic protein	Antiapoptotic protein binds to		
	Bax/Bak/Bid	BH3 proteins	
		Activator	Sensitizer
Bcl-2	Bax, Bid	Bim, Puma	Bmf, Bad
Bcl-XL	Bax, Bak, Bid	Bim, Puma	Bmf, Bad, Bik, Hrk
Bcl-w	Bax, Bak, Bid	Bim, Puma	Bmf, Bad, Bik, Hrk
Mcl-1	Bak, Bid	Bim, Puma	Noxa, Hrk
A1	Bak, Bid	Bim, Puma	Noxa, Bik, Hrk

Letai et al. (2002); Chen et al. (2005).

DISPLACEMENT MODEL

In the displacement model, BH3 proteins do not directly bind to Bax and Bak to cause their activation. Rather, Bax and Bak are constitutively active and therefore must be inhibited by the antiapoptotic proteins for the cell to survive. To initiate apoptosis, BH3 proteins displace Bax and Bak from the antiapoptotic proteins to promote Bax- or Bak-mediated MOMP. Because BH3 proteins selectively interact with a limited spectrum of antiapoptotic proteins, a combination of BH3 proteins is required to induce apoptosis in cells expressing multiple antiapoptotic Bcl-2 family members (see Table 1) (Chen et al. 2005). In support for this model, heterodimers of Bak with Mcl-1 and Bcl-XL are present in dividing cells, and overexpression of Noxa displaces Bak–Mcl-1 heterodimers, releasing Bak and forming Noxa–Mcl-1 complexes. In these cells, a combination of Bad and Noxa is required to neutralize the effects of both Bcl-XL and Mcl-1 to finally induce apoptosis (Willis et al. 2005).

EMBEDDED TOGETHER MODEL

The embedded together model incorporates the role of the membrane as the "locus of action" for most Bcl-2 family proteins because MOMP does not occur until Bax and Bak achieve their final active conformation in the membrane. The interactions with membranes result in distinct changes in conformations of the Bcl-2 family proteins that govern their affinity for the relative local concentrations of the binding partners (Leber et al. 2007, 2010; Garcia-Saez et al. 2009). For example, the cytoplasmic multiregion proteins Bax and Bcl-XL undergo large but reversible conformational changes after interacting with MOM (Edlich et al. 2011), which increase the affinity for binding to a BH3 protein, causing a further conformational change and allowing insertion in the membrane.

In this model, sensitizer BH3 proteins bind only to antiapoptotic proteins. However, the consequences of this interaction incorporate the features of both the displacement and direct activation models, because the sensitizer BH3 proteins neutralize the dual function of the antiapoptotic proteins by displacing *both* the activator BH3 proteins and Bax or Bak from the membrane-embedded conformers of the antiapoptotic proteins (Billen et al. 2008; Lovell et al. 2008). Because it is the activated forms of Bax and Bak that are bound to the membrane-embedded antiapoptotic proteins, sensitizer proteins release Bax and Bak conformers competent to oligomerize and permeabilize membranes.

Another distinguishing feature of this model is the dual role assigned to activator BH3 proteins, which directly activate proapoptotic proteins and also bind to antiapoptotic proteins. When activator BH3 proteins interact with Bax and Bak, they promote insertion into the membrane, whereupon Bax and Bak oligomerize and permeabilize cellular membranes. Similarly, interaction of activator BH3 proteins with antiapoptotic proteins promotes their insertion into membranes. However, in this case, the BH3 protein functions like a sensitizer, because the bound antiapoptotic protein is unable to bind Bax or Bak. However, sequestration goes both ways, and by binding the BH3 protein, the antiapoptotic protein inhibits it at the membrane. Moreover, because the interaction of the activator BH3 proteins with both the proapoptotic and the antiapoptotic Bcl-2 family proteins is reversible, it is therefore possible for a single BH3 protein to interact with both proapoptotic and antiapoptotic proteins (depending on their relative expression levels), thereby changing their conformation at the membranes. Recently, many of the interactions proposed by this model have been measured directly in living cells (Aranovich et al. 2012).

UNIFIED MODEL

The unified model of Bcl-2 family function builds on the embedded together model (Llambi et al. 2011). This model distinguishes the known interactions of antiapoptotic Bcl-2 proteins to sequester the activator BH3 proteins as mode 1, and to sequester the active forms of Bax and Bak as mode 2 (Fig. 2D). Although in cells both modes of inhibition take place simultaneously, in the unified model, inhibition of apoptosis through mode 1 is less efficient and is easier to overcome by BH3 sensitizers to promote MOMP than inhibition through mode 2. Importantly, the unified model also incorporates the functions of Bax and Bak in mitochondrial fission and fusion and postulates that only mode 2 repression affects this process. This model is therefore the first to explicitly link modes of MOMP regulation and mitochondrial dynamics.

The dual function assigned to antiapoptotic proteins is thus shared by both embedded together and unified models. However, in the former, the interplay between members of the Bcl-2 family are determined by competing equilibria; therefore, the abundance of proteins and specific conditions of cell physiology including posttranslational modifications will determine the prevailing interactions. As a result, the embedded together model differs from the unified model in that it predicts that either mode 1 or mode 2 can be dominant depending on circumstances such as the particular form of stress and cell type. Further work to test the different predictions of the models with full-length, wild-type proteins in different cells is required to resolve these issues.

THE MODELS: WHO BINDS TO WHOM?

One aspect of many of the models that is potentially confusing is that if an activator BH3 protein binds to an antiapoptotic family member, which is being inhibited? Whether antiapoptotic proteins sequester the BH3 proteins or the BH3 proteins sequester the antiapoptotic proteins becomes a semantic argument. A more productive way of characterizing the interaction is as a mutual sequestration that prevents their respective activation or inhibition effects on Bax and Bak. Therefore, whether MOMP ensues is determined by the relative concentrations and affinities of the proapoptotic and antiapoptotic proteins at the membrane. This recasting of the players is reminiscent of the original rheostat model proposed by the Korsmeyer group (Oltvai et al. 1993); however, it extends that model in ways not originally envisioned. For example, the rheostat model did not anticipate autoactivation. If there is sufficient cytosolic antiapoptotic Bcl-XL, then those Bcl-XL molecules recruited to the membrane by a BH3 protein can recruit additional molecules of Bcl-XL to the membrane through "autoactivation," a process also observed for Bax. Because BH3 protein binding is reversible, autoactivation ensures recruitment of sufficient Bcl-XL to provide efficient inhibition of the BH3 protein.

Another recently recognized aspect that determines the nature and fate of the binding interactions is composition of different membrane organelles. As mentioned above, the unified model provides a mechanistic link between MOMP regulation and mitochondrial fission and fusion. The importance of membranes in modifying conformations and binding partners as proposed by the embedded together model accounts for the overlapping but distinct function of the Bcl-2 family at the mitochondria and endoplasmic reticulum (ER). It also explains how other membrane sites such as the Golgi can act as a reservoir for potentially activated Bax (Dumitru et al. 2012). Therefore, the roles of Bcl-2 family proteins in cell fate decisions and other processes such as mitochondrial fusion and autophagy appear to be primarily governed by the relative concentrations and affinities of the different binding partners available in that specific subcellular membrane.

mechanism by which different BH3 proteins migrate to and insert into membranes varies. Mitochondrial-targeting and tail-anchor sequences are used to target several of the BH3 proteins to the MOM (see Table 2) (Kuwana et al. 2002; Seo et al. 2003; Hekman et al. 2006; Lovell et al. 2008). Moreover, the presence of specific lipids such as cardiolipin and cholesterol (Lutter et al. 2000; Hekman et al. 2006; Lucken-Ardjomande et al. 2008) and protein receptors such as Mtch2 at the MOM have been shown to influence the targeting of other Bcl-2 family proteins to their target membranes (Zaltsman et al. 2010).

Once at the membrane, it is likely that BH3 proteins undergo extensive conformational changes that dictate their function. For example, after cleavage by activated caspase-8, initial association of cleaved Bid with the MOM causes separation of the two fragments, with subsequent insertion and structural rearrangement of the p15 fragment (tBid) that likely orients the BH3 region to bind to Bax or Bcl-XL. Furthermore, the other BH3 proteins that are intrinsically unstructured undergo localized conformational changes upon binding membranes and antiapoptotic proteins.

Despite strong evidence for the functional interaction and activation of Bax and Bak by activator BH3 proteins, demonstration of binding of the full-length protein (as opposed to peptides from the BH3 region) has only recently been reported: Strong reversible binding of tBid to Bax was observed in liposomal MOM-like membranes (apparent K_d ~25 nM) (Lovell et al. 2008). Furthermore, when synthesized by in vitro translation, full-length BH3 proteins tBid, Bim, and Puma induced Bax- and Bak-dependent MOMP and shifted monomeric Bax and Bak to higher-order complexes in mitochondria (Kim et al. 2006).

In vitro experiments clearly show that BH3 proteins recruit and sequester the antiapoptotic Bcl-2 family proteins at the membrane. BH3 proteins bind the antiapoptotic proteins by docking on the BH3 region in the hydrophobic groove made of BH1, BH2, and BH3 regions of the antiapoptotic family proteins (Sattler et al. 1997; Liu et al. 2003; Czabotar et al. 2007). Similar to the differential binding to proapoptotic family members, experiments in vitro suggest that each BH3 protein selectively binds a defined range of antiapoptotic proteins that is determined by differences in the structure and flexibility of the hydrophobic pocket of the antiapoptotic proteins, although, to date, these interactions have been measured only with peptides from the different BH3 regions rather than the full-length proteins (see Table 1).

ANTIAPOPTOTIC MEMBERS

Structure of Family Members Alone and in Complex with BH3 Peptides

Early observations that specific mutations in Bcl-2 abrogated both antiapoptotic function and binding to Bax and the presence of BH3 regions in both classes of the proapoptotic Bcl-2 families that bind Bcl-2 as "ligands" led to the concept of a receptor surface on Bcl-2. However, it was hard to confirm the details of this binding interaction using structural studies because of difficulties with purifying recombinant full-length Bcl-2. Initial success arose from NMR studies on Bcl-XL lacking its hydrophobic carboxyl terminus (Muchmore et al. 1996), which is similar to the structure obtained for Bax (Suzuki et al. 2000), was shown to contain two hydrophobic helices (5 and 6) forming a central hairpin structure surrounded by the remaining six amphipathic helices. Thereafter, cocrystals of "tail-less" Bcl-XL with BH3 peptides derived from Bak and Bim identified the BH3-binding surface as a hydrophobic cleft formed by noncontiguous residues in BH regions 1–3 (involving parts of helices 2, 7–8, and 4–6, respectively) (Sattler et al. 1997; Liu et al. 2003). These structural observations provided the platform for measurements of the many potential interactions between the binding pockets of different antiapoptotic members and the BH3 regions of the proapoptotic family members. These differing binding affinities cluster into functional groupings (e.g., binding to multiregion vs. BH3 proteins, or activators vs. sensitizers) (Table 1) with functional consequences as elucidated below.

Multiple Mechanisms of Action of Bcl-XL: Evidence of Binding to Both Multiregion and BH3 Members

Measurements of the affinity of binding between individual pairs of antiapoptotic family members and BH3 peptides in solution provide valuable clues about functional relevance. However, in cells, most of these interactions occur at or within intracellular membranes, and, indeed, the final commitment step in apoptosis being regulated is MOMP. Thus, experiments using recombinant full-length proteins or proteins synthesized in vitro, and isolated mitochondria or liposomes, have been critical in translating these interactions into testable models. For practical reasons, such experiments are most feasible using recombinant Bcl-XL, because other antiapoptotic proteins are much more difficult to purify in sufficient quantities owing to problems with aggregation (e.g., Bcl-2) or marked protein instability (e.g., Mcl-1). Thus, details about the mechanism of action of Bcl-XL serve as a model for the other proteins, acknowledging that other members will differ in some aspects, as discussed below.

By examining membrane permeabilization in a system with recombinant Bcl-XL, Bax, and tBid (both wild type and a mutant form that is unable to bind to Bcl-XL, but still activates Bax), it was shown that Bcl-XL inhibits MOMP not only directly by binding to tBid but also by binding to membrane-bound Bax (Billen et al. 2008). Thus, both of the major interactions postulated by the competing direct activation and displacement models contribute to inhibition of apoptosis. Furthermore, other mechanisms of action of Bcl-XL independent of these binding interactions were also identified, including prevention of Bax insertion into membranes as perhaps the most potent mechanism. This initially contentious point has been recently supported by observations that Bax undergoes multiple conformational changes that ultimately lead to oligomerization and MOMP, but the first of these steps is the exposure of the amino terminus at the membrane in a reversible equilibrium (Edlich et al. 2011). Bcl-XL changes this equilibrium such that Bax is shifted out of the conformation that binds it loosely to membranes. Moreover, consistent with the postulation that dynamic conformational changes are a feature of all three Bcl-2 families, these investigators observed that Bcl-XL also undergoes reversible conformational changes that allow it to come on and off the MOM without being inserted. The structural basis of this mechanism is unclear, although it is speculated that sequestering of the opposite partner's carboxy-terminal helix 9 in the BH3-binding groove may mediate this effect. In essence, helix 9 of the other protein acts as an (inactive) BH3 mimetic.

Taken together, these observations have identified multiple mechanisms that contribute to the ultimate function of Bcl-XL. Using defined amounts of proteins with an in vitro system allows measurement of the stoichiometry of inhibition and indicates that one Bcl-XL can inhibit approximately four Bax molecules. Therefore, as a conceptual overview, the functions of Bcl-XL can be most simply summarized as a dominant-negative Bax, where it is able to undergo many of the binding interactions that Bax does but does not make the final conformational change that allows it to bind to other Bcl-XL/Bax molecules and oligomerize to form a pore. In accordance with the postulated models of oligomerization discussed above, this would imply that activated Bcl-XL cannot form a rear pocket in the analogous regions described for Bax/Bak.

Mediators of Multiple Mechanisms: Membrane Binding and Conformational Changes

Similar to Bax and Bak, there is evidence that the antiapoptotic Bcl-2 family proteins adopt multiple conformations in associating with membranes. Bcl-2 initially inserts helix 9 into the membrane, but after binding to tBid or a BH3 peptide derived from Bim, helix 5 moves to a hydrophobic environment consistent with insertion into the membrane (Kim et al. 2004). Therefore, it is plausible that Bcl-XL also adopts multiple conformations that are dictated by its interaction both with membranes and other

Bcl-2 family members that shift the dynamic equilibrium between the different forms. Specifically, the data suggest that there is a form that is loosely bound to membranes (form 1), another in which helix 9 is inserted into membranes but not other helices (form 2), and, finally, a form in which helix 9 as well as helices 5 and 6 are inserted into the membrane (form 3) (Fig. 4). It is possible that these different conformations independently mediate the different mechanisms of action of Bcl-XL in inhibiting the final process of pore formation by activated Bax. Such a scheme is also compatible with observations that mutations that do not affect the BH3-binding pocket can still enhance antiapoptotic function, either by forcing constitutive membrane insertion (into forms 2 or 3) by replacing the endogenous tail-anchor sequence (Fiebig et al. 2006), or by loosening intramolecular binding, thereby "freeing" helices 5 and 6 to insert into membranes (form 3) (Asoh et al. 2000).

Comparison of Different Antiapoptotic Members

In simpler organisms such as *Caenorhabditis elegans* and *Drosophila*, there is only one inhibitory Bcl-2 family member, whereas in vertebrates there are at least four. There are potentially multiple reasons for this redundancy. One that is firmly grounded on structural studies indicates that the different antiapoptotic family members bind to (and sequester) the multiple BH3 members differentially, including the multidomain proapoptotic members alluded to previously. Responding to multiple BH3 proteins allows fine-tuning of inhibitory responses in mammalian cells to different types of stress that "activate" specific BH3 proteins. Such a system provides multifactorial responses much more diverse than those in simpler eukaryote cells. Characterization of the differences in binding has received much attention and is conferred by the distinct sequence of each BH3 region that shares a propensity to form an amphipathic helix containing four hydrophobic residues, and the topology of the BH3-binding groove on the antiapoptotic "receptor." Peptides from certain BH3 regions like Bim bind with high affinity to all the antiapoptotic and apoptotic multiregion members, whereas others like Bad and Noxa are more selective (highly preferential binding to Bcl-2/Bcl-XL/Bcl-w or Mcl-1/Bfl-1, respectively). Some of this specificity is explained by well-defined requirements, for example, any amino acid at the fourth hydrophobic position in the BH3 region will bind to Mcl-1, which has a shallow, open pocket for this residue, as opposed to Bcl-XL, which does not accommodate charged or polar residues at this position (Lee et al. 2009; Fire et al. 2010). Other features also contribute; the higher global flexibility of Bcl-XL creates a pliable pocket for diverse BH3 mimetics compared with the deeper hydrophobic pocket with a rigid angle of entry in Mcl-1 that restricts binding to specific BH3 proteins (Lee et al. 2009).

As a consequence, no single antiapoptotic member binds to all BH3 proteins in vitro, as assessed by biophysical measurements (see Table 1). These measurements have been largely (although not entirely) confirmed by experiments in transfected cell lines where overexpression of single antiapoptotic proteins confers protection against apoptosis mediated by the BH3-binding partners identified in vitro. The discrepancy noted in a few experiments is likely because of the fact that in cells these interactions between full-length proteins occur on membranes rather than in the cytoplasm, and membrane binding may modify protein–protein interactions either allosterically or by posttranslational modifications altering the binding surfaces (Feng et al. 2009) or affecting the orientation and proximity of the binding surfaces.

Multiple antiapoptotic proteins also allow differential control of processes relevant to cell death independent of BH3-binding-pocket interactions. The BH4 region of Bcl-2 binds to the regulatory and coupling domain of the inositol 1,4,5 triphosphate (IP3) receptor that controls calcium efflux from the ER, thereby inhibiting the initiation phase of calcium-mediated apoptosis (Rong et al. 2009). A residue critical for this binding interaction in Bcl-2 (Lys17) is not conserved in the BH4 domain of Bcl-XL (Asp11), rendering the latter ineffective at

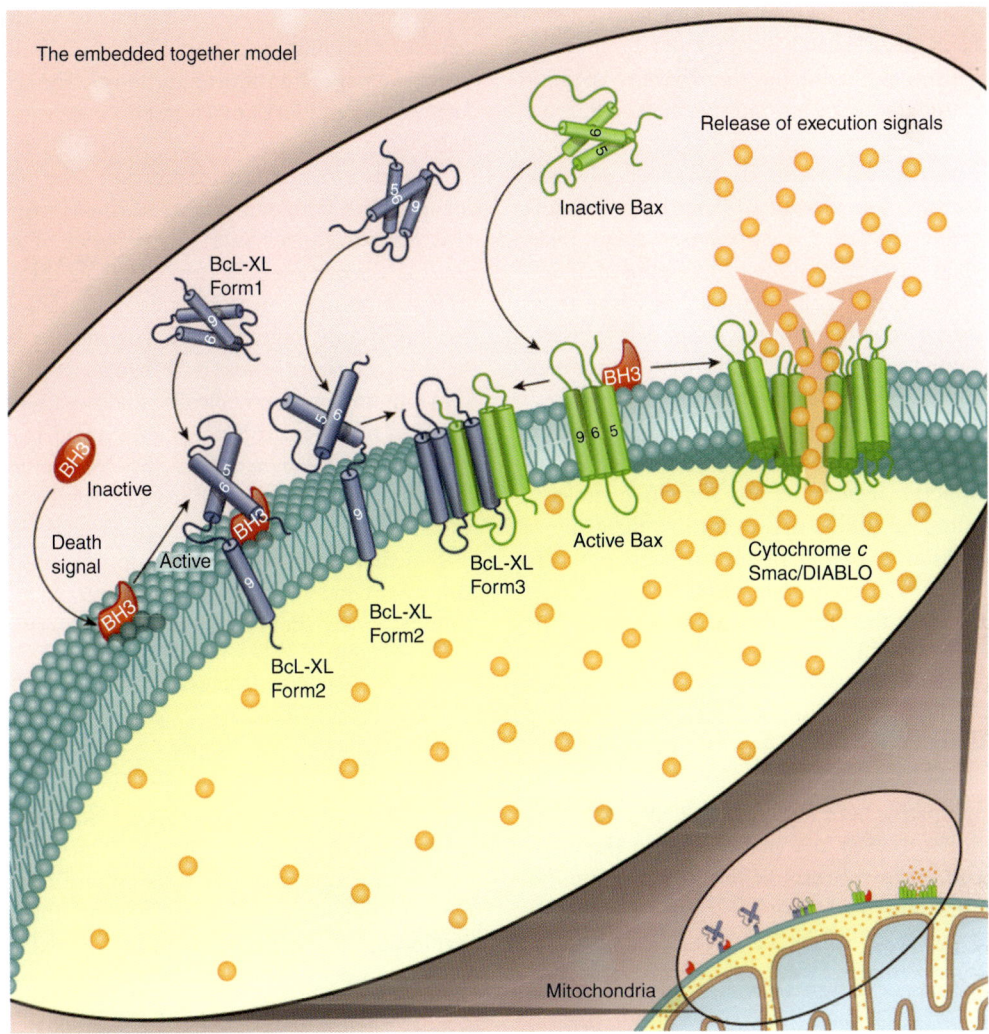

Figure 4. Schematic overview of the embedded together model. The role of the membrane is highlighted as the "locus of action" where the effects of the interactions between the Bcl-2 family members are manifest. After the cell receives a death signal, an activator BH3 protein migrates to and inserts into the MOM, where it recruits cytoplasmic Bax. Bax undergoes conformational changes at membranes that allow it to respond to chemical changes in the cell such as the generation of reactive oxygen species, ion concentration, and pH. Membrane-bound Bax or Bak changes its conformation such that they oligomerize, leading to MOMP and/or recruit other cytoplasmic Bax. Both the activator and the sensitizer BH3 proteins sequester the antiapoptotic proteins (such as Bcl-XL) by recruiting and strongly binding to them at the MOM, thereby preventing the inhibition of Bax and Bak. Bcl-XL changes its conformation depending on its binding partner. Upon binding to a BH3 protein or Bax/Bak, Bcl-XL changes from form 1 (cytoplasmic or loosely attached to the MOM) to form 2 (helix 9 inserted into MOM) or to form 3 (helices 5, 6, and 9 bound to or inserted into MOM), respectively. It is likely that form 2 binds primarily BH3 proteins but also recruits additional Bcl-XL to the membrane, whereas form 3 binds primarily Bax and Bak. No function has yet been ascribed to Bcl-XL form 1, although one is likely. Thus, by causing the proteins to adopt different conformations, the membrane regulates their function in determining the fate of the cell. Unlike other models that propose unidirectional interactions, in this model, all of the functional interactions are governed by dynamic equilibria of protein–membrane and protein–protein interactions.

inhibiting IP3-mediated calcium release (Monaco et al. 2011).

Another reason for the diversity of antiapoptotic proteins beyond specificity conferred by different binding partners is the control of subcellular localization. In particular cell types, there may be a benefit to having Bcl-2 family members constitutively present on membranes such as is the case with Bcl-2, as opposed to Bcl-XL, Mcl-1, Bfl-1, and Bcl-w, all of which must undergo a conformational change before inserting into the membrane. In Bcl-2, it is presumed that the carboxy-terminal region that is necessary and sufficient for membrane insertion (Janiak et al. 1994) is not bound to other hydrophobic regions of the protein once it is synthesized and can therefore mediate direct membrane insertion. In the other antiapoptotic proteins, the carboxy-terminal tail is sequestered until the protein is activated. Even within this group, there are different strategies that control membrane localization. Unlike the other family members, Bfl-1 does not have a hydrophobic region at the carboxyl terminus that mediates membrane insertion but has an amphipathic helix (Brien et al. 2009). Bcl-XL is thought to exist as a homodimer in the cytoplasm, where the carboxy-terminal tail is bound reciprocally to a hydrophobic groove in the dimer partner (Jeong et al. 2004). The longer carboxy-terminal helix 8 of Bcl-w binds in its own BH3-binding pocket and can be displaced by BH3 peptides to allow membrane insertion (Hinds et al. 2003; Wilson-Annan et al. 2003), a mechanism reminiscent of Bax. Before apoptosis is elicited, Mcl-1 is constitutively loosely associated with mitochondria by an EELD motif in the amino-terminal portion, which can bind to the mitochondrial import receptor Tom70 (Chou et al. 2006). For all the antiapoptotic proteins, deletion of the carboxy-terminal α helix decreases function, presumably by preventing assumption of forms 2 and 3 on the membrane where many of the relevant binding partners are localized. Furthermore, attachment of the inhibitor to the membrane increases the probability of interaction by increasing local concentration and the viscosity of the membranes restricting diffusion. A third justification for diversity of antiapoptotic proteins is the benefit of varying regulation of protein abundance as a way of fine-tuning apoptosis. The *bcl-2* gene contains two estrogen response elements controlling expression in breast tissue. Bcl-2 is a long-lived protein whose expression does not change appreciably even during advanced stages of stress, partly because of the presence of an internal ribosome entry site (IRES) in the $5'$ UTR that permits cap-independent translation (Willimott and Wagner 2010). The stability of the Bcl-2 transcript is positively regulated by the RNA-binding protein nucleolin, and negatively regulated by the microRNAs mi-R15a and 16-1 (Willimott and Wagner 2010). Bcl-XL protein levels are more variable and increase acutely in response to internal stress and extracellular signals, mediated by the Jak-STAT and rel/NF-κB pathways (Grad et al. 2000). In contrast, Mcl-1 is an extremely short-lived protein with rapid turnover tightly regulated by a complex cascade of phosphorylation-dependent deubiquitination by USPX9 (Schwickart et al. 2010) that reverses the ubiquitination and subsequent proteasomal degradation mediated by the BH3 protein E3 ubiquitin ligase MULE/ARF-BP1 (Zhong et al. 2005).

The consequences of these variations in the structure of binding pockets (control of subcellular localization and dynamic protein levels), is that despite sharing the core mechanism of inhibition, each antiapoptotic protein has a distinct personality. This is evident in the specific profile of expression of the proteins in different cell types and organs in whole animals, with the result that each protein has different physiological roles that are apparent in the phenotypes of the knockout mice with different antiapoptotic members (for review, see Hardwick and Soane 2013).

PERSPECTIVE AND FUTURE PROSPECTS

This brief overview illustrates the enormous growth in our understanding of the mechanisms behind the pivotal role that the Bcl-2 family plays in regulating apoptosis since the original identification of Bcl-2 as a chromosome translocation partner in human B-cell follicular lymphoma. We are now at a stage where this understanding is yielding practical results, as several drugs mimicking BH3 regions that bind to Bcl-2 and Bcl-XL are in late-stage clinical trials as cancer agents to elicit or enhance chemotherapy-induced apoptosis. The recognition that there are distinct binding profiles for each antiapoptotic protein that arose from fundamental studies has now motivated the search for other small molecules to expand the therapeutic tool kit (Stewart et al. 2010), so that in the future we will be able to target every antiapoptotic protein.

To date, most attention has been paid to the role of the Bcl-2 family in regulating MOMP because of the well-characterized consequences of releasing IMS proteins in activating caspases. However, it is increasingly apparent that the ER is the site of many important processes that determine cell death and survival in which the Bcl-2 family is intimately involved. Aside from controlling calcium flux (Rong et al. 2009) and regulating the activity of Beclin-1 to initiate autophagy (see Mah and Ryan 2012; Nixon and Yang 2012), other death pathways are also inhibited by Bcl-2 at the ER (Germain et al. 2002). Beyond this, there is also evidence that a portion of the antiapoptotic activity of Bcl-2/Bcl-XL does not depend on binding to and inhibiting the other two proapoptotic families (Minn et al. 1999). One recent study suggests that this mechanism involves regulation of cytoplasmic levels of acetyl-CoA as a substrate for protein α-acetylation (Yi et al. 2011). Elucidating potential binding partners that mediate this pathway is an important target of future research.

Our basic understanding of the core mechanism of the regulation of membrane permeabilization by Bcl-2 family members has passed from the stage of phenomenology to testable descriptions of mechanism. The next hurdle will be to extend quantitative measurements of the binding interactions that have been measured in vitro to what happens in organelles and in cells. This will allow further refinement and elaboration of exciting preliminary mathematical models of the control of apoptosis in whole cells (Spencer and Sorger 2011).

ACKNOWLEDGMENTS

Funding for the research in the Andrews and Leber laboratories was provided by a grant from the Center for Health Information and Research (FRN 12517). D.W.A. holds the Canada Research Chair in Membrane Biogenesis, A.S.-D. holds an Ontario Graduate Studentship, and J.K. holds a research fellowship from the Canadian Breast Cancer Foundation–Ontario Region.

REFERENCES

*Reference is also in this collection.

Annis MG, Soucie EL, Dlugosz PJ, Cruz-Aguado JA, Penn LZ, Leber B, Andrews DW. 2005. Bax forms multispanning monomers that oligomerize to permeabilize membranes during apoptosis. *EMBO J* **24:** 2096–2103.

Aouacheria A, Brunet F, Gouy M. 2005. Phylogenomics of life-or-death switches in multicellular animals: Bcl-2, BH3-Only, and BNip families of apoptotic regulators. *Mol Biol Evol* **22:** 2395–2416.

Aranovich A, Liu Q, Collins T, Geng F, Dixit S, Leber B, Andrews DW. 2012. Differences in the mechanisms of proapoptotic BH3 proteins binding to Bcl-XL and Bcl-2 quantified in live MCF-7 cells. *Mol Cell* **45:** 754–763.

Asoh S, Ohtsu T, Ohta S. 2000. The super anti-apoptotic factor Bcl-xFNK constructed by disturbing intramolecular polar interactions in rat Bcl-xL. *J Biol Chem* **275:** 37240–37245.

Billen LP, Kokoski CL, Lovell JF, Leber B, Andrews DW. 2008. Bcl-XL inhibits membrane permeabilization by competing with Bax. *PLoS Biol* **6:** e147.

Billen LP, Shamas-Din A, Andrews DW. 2009. Bid: A Bax-like BH3 protein. *Oncogene* **27** (Suppl 1): S93–S104.

Bleicken S, Classen M, Padmavathi PV, Ishikawa T, Zeth K, Steinhoff HJ, Bordignon E. 2010. Molecular details of Bax activation, oligomerization, and membrane insertion. *J Biol Chem* **285:** 6636–6647.

Brenner C, Grimm S. 2006. The permeability transition pore complex in cancer cell death. *Oncogene* **25:** 4744–4756.

Brien G, Debaud AL, Robert X, Oliver L, Trescol-Biemont MC, Cauquil N, Geneste O, Aghajari N, Vallette FM, Haser R, et al. 2009. C-terminal residues regulate localization and function of the antiapoptotic protein Bfl-1. *J Biol Chem* **284:** 30257–30263.

Brock SE, Li C, Wattenberg BW. 2010. The Bax carboxy-terminal hydrophobic helix does not determine organelle-specific targeting but is essential for maintaining Bax in an inactive state and for stable mitochondrial membrane insertion. *Apoptosis* **15:** 14–27.

Brooks C, Wei Q, Feng L, Dong G, Tao Y, Mei L, Xie ZJ, Dong Z. 2007. Bak regulates mitochondrial morphology and pathology during apoptosis by interacting with mitofusins. *Proc Natl Acad Sci* **104:** 11649–11654.

Cartron PF, Gallenne T, Bougras G, Gautier F, Manero F, Vusio P, Meflah K, Vallette FM, Juin P. 2004. The first α helix of Bax plays a necessary role in its ligand-induced activation by the BH3-only proteins Bid and PUMA. *Mol Cell* **16:** 807–818.

Certo M, Del Gaizo Moore V, Nishino M, Wei G, Korsmeyer S, Armstrong SA, Letai A. 2006. Mitochondria primed by death signals determine cellular addiction to antiapoptotic BCL-2 family members. *Cancer Cell* **9:** 351–365.

Chen L, Willis SN, Wei A, Smith BJ, Fletcher JI, Hinds MG, Colman PM, Day CL, Adams JM, Huang DC. 2005. Differential targeting of prosurvival Bcl-2 proteins by their BH3-only ligands allows complementary apoptotic function. *Mol Cell* **17:** 393–403.

Chinnadurai G, Vijayalingam S, Gibson SB. 2008. BNIP3 subfamily BH3-only proteins: Mitochondrial stress sensors in normal and pathological functions. *Oncogene* **27** (Suppl 1): S114–S127.

Chipuk JE, Fisher JC, Dillon CP, Kriwacki RW, Kuwana T, Green DR. 2008. Mechanism of apoptosis induction by inhibition of the anti-apoptotic BCL-2 proteins. *Proc Natl Acad Sci* **105:** 20327–20332.

Chittenden T, Harrington EA, O'Connor R, Flemington C, Lutz RJ, Evan GI, Guild BC. 1995. Induction of apoptosis by the Bcl-2 homologue Bak. *Nature* **374:** 733–736.

Chou JJ, Li H, Salvesen GS, Yuan J, Wagner G. 1999. Solution structure of BID, an intracellular amplifier of apoptotic signaling. *Cell* **96:** 615–624.

Chou CH, Lee RS, Yang-Yen HF. 2006. An internal EELD domain facilitates mitochondrial targeting of Mcl-1 via a Tom70-dependent pathway. *Mol Biol Cell* **17:** 3952–3963.

Czabotar PE, Lee EF, van Delft MF, Day CL, Smith BJ, Huang DC, Fairlie WD, Hinds MG, Colman PM. 2007. Structural insights into the degradation of Mcl-1 induced by BH3 domains. *Proc Natl Acad Sci* **104:** 6217–6222.

Danial NN. 2008. BAD: Undertaker by night, candyman by day. *Oncogene* **27** (Suppl 1): S53–S70.

Dejean LM, Martinez-Caballero S, Guo L, Hughes C, Teijido O, Ducret T, Ichas F, Korsmeyer SJ, Antonsson B, Jonas EA, et al. 2005. Oligomeric Bax is a component of the putative cytochrome *c* release channel MAC, mitochondrial apoptosis-induced channel. *Mol Biol Cell* **16:** 2424–2432.

Dewson G, Kratina T, Sim HW, Puthalakath H, Adams JM, Colman PM, Kluck RM. 2008. To trigger apoptosis, Bak exposes its BH3 domain and homodimerizes via BH3: groove interactions. *Mol Cell* **30:** 369–380.

Dewson G, Kratina T, Czabotar P, Day CL, Adams JM, Kluck RM. 2009. Bak activation for apoptosis involves oligomerization of dimers via their α6 helices. *Mol Cell* **36:** 696–703.

Dlugosz PJ, Billen LP, Annis MG, Zhu W, Zhang Z, Lin J, Leber B, Andrews DW. 2006. Bcl-2 changes conformation to inhibit Bax oligomerization. *EMBO J* **25:** 2287–2296.

Dumitru R, Gama V, Fagan BM, Bower JJ, Swahari V, Pevny LH, Deshmukh M. 2012. Human embryonic stem cells have constitutively active Bax at the Golgi and are primed to undergo rapid apoptosis. *Mol Cell* **46:** 573–583.

Edlich F, Banerjee S, Suzuki M, Cleland MM, Arnoult D, Wang C, Neutzner A, Tjandra N, Youle RJ. 2011. Bcl-x(L) retrotranslocates Bax from the mitochondria into the cytosol. *Cell* **145:** 104–116.

Farrow SN, White JH, Martinou I, Raven T, Pun KT, Grinham CJ, Martinou JC, Brown R. 1995. Cloning of a *bcl-2* homologue by interaction with adenovirus E1B 19 K. *Nature* **374:** 731–733.

Feng Y, Liu D, Shen X, Chen K, Jiang H. 2009. Structure assembly of Bcl-x(L) through α5–α5 and α6–α6 interhelix interactions in lipid membranes. *Biochim Biophys Acta* **1788:** 2389–2395.

Fiebig AA, Zhu W, Hollerbach C, Leber B, Andrews DW. 2006. Bcl-XL is qualitatively different from and ten times more effective than Bcl-2 when expressed in a breast cancer cell line. *BMC Cancer* **6:** 213.

Fire E, Gulla SV, Grant RA, Keating AE. 2010. Mcl-1–Bim complexes accommodate surprising point mutations via minor structural changes. *Protein Sci* **19:** 507–519.

Gallenne T, Gautier F, Oliver L, Hervouet E, Noel B, Hickman JA, Geneste O, Cartron PF, Vallette FM, Manon S, et al. 2009. Bax activation by the BH3-only protein Puma promotes cell dependence on antiapoptotic Bcl-2 family members. *J Cell Biol* **185:** 279–290.

Garcia-Saez AJ, Ries J, Orzaez M, Perez-Paya E, Schwille P. 2009. Membrane promotes tBID interaction with BCL(XL). *Nat Struct Mol Biol* **16:** 1178–1185.

Gavathiotis E, Suzuki M, Davis ML, Pitter K, Bird GH, Katz SG, Tu HC, Kim H, Cheng EH, Tjandra N, et al. 2008. BAX activation is initiated at a novel interaction site. *Nature* **455:** 1076–1081.

Gavathiotis E, Reyna DE, Davis ML, Bird GH, Walensky LD. 2010. BH3-triggered structural reorganization drives the activation of proapoptotic BAX. *Mol Cell* **40:** 481–492.

George NM, Evans JJ, Luo X. 2007. A three-helix homo-oligomerization domain containing BH3 and BH1 is responsible for the apoptotic activity of Bax. *Genes Dev* **21:** 1937–1948.

Germain M, Mathai JP, Shore GC. 2002. BH-3-only BIK functions at the endoplasmic reticulum to stimulate cytochrome *c* release from mitochondria. *J Biol Chem* **277:** 18053–18060.

Grad JM, Zeng XR, Boise LH. 2000. Regulation of Bcl-xL: A little bit of this and a little bit of STAT. *Curr Opin Oncol* **12:** 543–549.

Griffiths GJ, Dubrez L, Morgan CP, Jones NA, Whitehouse J, Corfe BM, Dive C, Hickman JA. 1999. Cell damage-induced conformational changes of the pro-apoptotic protein Bak in vivo precede the onset of apoptosis. *J Cell Biol* **144:** 903–914.

* Hardwick JM, Soane L. 2013. Multiple functions of BCL-2 family proteins. *Cold Spring Harb Perspect Biol* **5:** a008722.

Hekman M, Albert S, Galmiche A, Rennefahrt UE, Fueller J, Fischer A, Puehringer D, Wiese S, Rapp UR. 2006. Reversible membrane interaction of BAD requires two C-terminal lipid binding domains in conjunction with 14–3-3 protein binding. *J Biol Chem* **281:** 17321–17336.

Hinds MG, Lackmann M, Skea GL, Harrison PJ, Huang DC, Day CL. 2003. The structure of Bcl-w reveals a role for the

C-terminal residues in modulating biological activity. *EMBO J* **22:** 1497–1507.

Hinds MG, Smits C, Fredericks-Short R, Risk JM, Bailey M, Huang DC, Day CL. 2007. Bim, Bad and Bmf: Intrinsically unstructured BH3-only proteins that undergo a localized conformational change upon binding to prosurvival Bcl-2 targets. *Cell Death Differ* **14:** 128–136.

Hoppins S, Edlich F, Cleland MM, Banerjee S, McCaffery JM, Youle RJ, Nunnari J. 2011. The soluble form of Bax regulates mitochondrial fusion via MFN2 homotypic complexes. *Mol Cell* **41:** 150–160.

Hsu YT, Youle RJ. 1997. Nonionic detergents induce dimerization among members of the Bcl-2 family. *J Biol Chem* **272:** 13829–13834.

Hsu YT, Wolter KG, Youle RJ. 1997. Cytosol-to-membrane redistribution of Bax and Bcl-X(L) during apoptosis. *Proc Natl Acad Sci* **94:** 3668–3672.

Hu X, Han Z, Wyche JH, Hendrickson EA. 2003. Helix 6 of tBid is necessary but not sufficient for mitochondrial binding activity. *Apoptosis* **8:** 277–289.

Inohara N, Ding L, Chen S, Nunez G. 1997. *harakiri*, a novel regulator of cell death, encodes a protein that activates apoptosis and interacts selectively with survival-promoting proteins Bcl-2 and Bcl-X(L). *EMBO J* **16:** 1686–1694.

Jabbour AM, Heraud JE, Daunt CP, Kaufmann T, Sandow J, O'Reilly LA, Callus BA, Lopez A, Strasser A, Vaux DL, et al. 2009. Puma indirectly activates Bax to cause apoptosis in the absence of Bid or Bim. *Cell Death Differ* **16:** 555–563.

Janiak F, Leber B, Andrews DW. 1994. Assembly of Bcl-2 into microsomal and outer mitochondrial membranes. *J Biol Chem* **269:** 9842–9849.

Jeong SY, Gaume B, Lee YJ, Hsu YT, Ryu SW, Yoon SH, Youle RJ. 2004. Bcl-x(L) sequesters its C-terminal membrane anchor in soluble, cytosolic homodimers. *EMBO J* **23:** 2146–2155.

Kalinec GM, Fernandez-Zapico ME, Urrutia R, Esteban-Cruciani N, Chen S, Kalinec F. 2005. Pivotal role of Harakiri in the induction and prevention of gentamicin-induced hearing loss. *Proc Natl Acad Sci* **102:** 16019–16024.

Kamer I, Sarig R, Zaltsman Y, Niv H, Oberkovitz G, Regev L, Haimovich G, Lerenthal Y, Marcellus RC, Gross A. 2005. Proapoptotic BID is an ATM effector in the DNA-damage response. *Cell* **122:** 593–603.

Karbowski M, Norris KL, Cleland MM, Jeong SY, Youle RJ. 2006. Role of Bax and Bak in mitochondrial morphogenesis. *Nature* **443:** 658–662.

Kiefer MC, Brauer MJ, Powers VC, Wu JJ, Umansky SR, Tomei LD, Barr PJ. 1995. Modulation of apoptosis by the widely distributed Bcl-2 homologue Bak. *Nature* **374:** 736–739.

Kim PK, Annis MG, Dlugosz PJ, Leber B, Andrews DW. 2004. During apoptosis Bcl-2 changes membrane topology at both the endoplasmic reticulum and mitochondria. *Mol Cell* **14:** 523–529.

Kim H, Rafiuddin-Shah M, Tu HC, Jeffers JR, Zambetti GP, Hsieh JJ, Cheng EH. 2006. Hierarchical regulation of mitochondrion-dependent apoptosis by BCL-2 subfamilies. *Nat Cell Biol* **8:** 1348–1358.

Kim H, Tu HC, Ren D, Takeuchi O, Jeffers JR, Zambetti GP, Hsieh JJ, Cheng EH. 2009. Stepwise activation of BAX and BAK by tBID, BIM, and PUMA initiates mitochondrial apoptosis. *Mol Cell* **36:** 487–499.

Kutuk O, Letai A. 2008. Regulation of Bcl-2 family proteins by posttranslational modifications. *Curr Mol Med* **8:** 102–118.

Kuwana T, Newmeyer DD. 2003. Bcl-2-family proteins and the role of mitochondria in apoptosis. *Curr Opin Cell Biol* **15:** 691–699.

Kuwana T, Mackey MR, Perkins G, Ellisman MH, Latterich M, Schneiter R, Green DR, Newmeyer DD. 2002. Bid, Bax, and lipids cooperate to form supramolecular openings in the outer mitochondrial membrane. *Cell* **111:** 331–342.

Kuwana T, Bouchier-Hayes L, Chipuk JE, Bonzon C, Sullivan BA, Green DR, Newmeyer DD. 2005. BH3 domains of BH3-only proteins differentially regulate Bax-mediated mitochondrial membrane permeabilization both directly and indirectly. *Mol Cell* **17:** 525–535.

Kvansakul M, Yang H, Fairlie WD, Czabotar PE, Fischer SF, Perugini MA, Huang DC, Colman PM. 2008. Vaccinia virus anti-apoptotic F1 L is a novel Bcl-2-like domain-swapped dimer that binds a highly selective subset of BH3-containing death ligands. *Cell Death Differ* **15:** 1564–1571.

Landeta O, Landajuela A, Gil D, Taneva S, Diprimo C, Sot B, Valle M, Frolov V, Basanez G. 2011. Reconstitution of proapoptotic BAK function in liposomes reveals a dual role for mitochondrial lipids in the BAK-driven membrane permeabilization process. *J Biol Chem* **286:** 8213–8230.

Leber B, Lin J, Andrews DW. 2007. Embedded together: The life and death consequences of interaction of the Bcl-2 family with membranes. *Apoptosis* **12:** 897–911.

Leber B, Lin J, Andrews DW. 2010. Still embedded together binding to membranes regulates Bcl-2 protein interactions. *Oncogene* **29:** 5221–5230.

Lee EF, Czabotar PE, Yang H, Sleebs BE, Lessene G, Colman PM, Smith BJ, Fairlie WD. 2009. Conformational changes in Bcl-2 pro-survival proteins determine their capacity to bind ligands. *J Biol Chem* **284:** 30508–30517.

Letai A, Bassik MC, Walensky LD, Sorcinelli MD, Weiler S, Korsmeyer SJ. 2002. Distinct BH3 domains either sensitize or activate mitochondrial apoptosis, serving as prototype cancer therapeutics. *Cancer Cell* **2:** 183–192.

Li H, Zhu H, Xu CJ, Yuan J. 1998. Cleavage of BID by caspase 8 mediates the mitochondrial damage in the Fas pathway of apoptosis. *Cell* **94:** 491–501.

Liu X, Dai S, Zhu Y, Marrack P, Kappler JW. 2003. The structure of a Bcl-xL/Bim fragment complex: Implications for Bim function. *Immunity* **19:** 341–352.

Llambi F, Moldoveanu T, Tait SW, Bouchier-Hayes L, Temirov J, McCormick LL, Dillon CP, Green DR. 2011. A unified model of mammalian BCL-2 protein family interactions at the mitochondria. *Mol Cell* **44:** 517–531.

Lovell JF, Billen LP, Bindner S, Shamas-Din A, Fradin C, Leber B, Andrews DW. 2008. Membrane binding by tBid initiates an ordered series of events culminating in membrane permeabilization by Bax. *Cell* **135:** 1074–1084.

Lucken-Ardjomande S, Montessuit S, Martinou JC. 2008. Bax activation and stress-induced apoptosis delayed by the accumulation of cholesterol in mitochondrial membranes. *Cell Death Differ* **15**: 484–493.

Luo X, Budihardjo I, Zou H, Slaughter C, Wang X. 1998. Bid, a Bcl2 interacting protein, mediates cytochrome *c* release from mitochondria in response to activation of cell surface death receptors. *Cell* **94**: 481–490.

Lutter M, Fang M, Luo X, Nishijima M, Xie X, Wang X. 2000. Cardiolipin provides specificity for targeting of tBid to mitochondria. *Nat Cell Biol* **2**: 754–761.

* Mah LY, Ryan KM. 2012. Autophagy and cancer. *Cold Spring Harb Perspect Biol* **4**: a008821.

McDonnell JM, Fushman D, Milliman CL, Korsmeyer SJ, Cowburn D. 1999. Solution structure of the proapoptotic molecule BID: A structural basis for apoptotic agonists and antagonists. *Cell* **96**: 625–634.

Minn AJ, Kettlun CS, Liang H, Kelekar A, Vander Heiden MG, Chang BS, Fesik SW, Fill M, Thompson CB. 1999. Bcl-xL regulates apoptosis by heterodimerization-dependent and -independent mechanisms. *EMBO J* **18**: 632–643.

Moldoveanu T, Liu Q, Tocilj A, Watson M, Shore G, Gehring K. 2006. The X-ray structure of a BAK homodimer reveals an inhibitory zinc binding site. *Mol Cell* **24**: 677–688.

Monaco G, Decrock E, Akl H, Ponsaerts R, Vervliet T, Luyten T, De Maeyer M, Missiaen L, Distelhorst CW, De Smedt H, et al. 2011. Selective regulation of IP(3)-receptor-mediated Ca^{2+} signaling and apoptosis by the BH4 domain of Bcl-2 versus Bcl-Xl. *Cell Death Differ* **19**: 295–309.

Montessuit S, Somasekharan SP, Terrones O, Lucken-Ardjomande S, Herzig S, Schwarzenbacher R, Manstein DJ, Bossy-Wetzel E, Basanez G, Meda P, et al. 2010. Membrane remodeling induced by the dynamin-related protein Drp1 stimulates Bax oligomerization. *Cell* **142**: 889–901.

Muchmore SW, Sattler M, Liang H, Meadows RP, Harlan JE, Yoon HS, Nettesheim D, Chang BS, Thompson CB, Wong SL, et al. 1996. X-ray and NMR structure of human Bcl-xL, an inhibitor of programmed cell death. *Nature* **381**: 335–341.

Nakagawa T, Shimizu S, Watanabe T, Yamaguchi O, Otsu K, Yamagata H, Inohara H, Kubo T, Tsujimoto Y. 2005. Cyclophilin D–dependent mitochondrial permeability transition regulates some necrotic but not apoptotic cell death. *Nature* **434**: 652–658.

Nakano K, Vousden KH. 2001. PUMA, a novel proapoptotic gene, is induced by p53. *Mol Cell* **7**: 683–694.

Nechushtan A, Smith CL, Hsu YT, Youle RJ. 1999. Conformation of the Bax C-terminus regulates subcellular location and cell death. *EMBO J* **18**: 2330–2341.

Nguyen M, Millar DG, Yong VW, Korsmeyer SJ, Shore GC. 1993. Targeting of Bcl-2 to the mitochondrial outer membrane by a COOH-terminal signal anchor sequence. *J Biol Chem* **268**: 25265–25268.

* Nixon RA, Yang D-S. 2012. Autophagy and neuronal cell death in neurological disorders. *Cold Spring Harb Perspect Biol* **4**: a008839.

O'Connor L, Strasser A, O'Reilly LA, Hausmann G, Adams JM, Cory S, Huang DC. 1998. Bim: A novel member of the Bcl-2 family that promotes apoptosis. *EMBO J* **17**: 384–395.

Oda E, Ohki R, Murasawa H, Nemoto J, Shibue T, Yamashita T, Tokino T, Taniguchi T, Tanaka N. 2000. Noxa, a BH3-only member of the Bcl-2 family and candidate mediator of p53-induced apoptosis. *Science* **288**: 1053–1058.

Oh KJ, Singh P, Lee K, Foss K, Lee S, Park M, Aluvila S, Kim RS, Symersky J, Walters DE. 2010. Conformational changes in BAK, a pore-forming proapoptotic Bcl-2 family member, upon membrane insertion and direct evidence for the existence of BH3–BH3 contact interface in BAK homo-oligomers. *J Biol Chem* **285**: 28924–28937.

Oltvai ZN, Milliman CL, Korsmeyer SJ. 1993. Bcl-2 heterodimerizes in vivo with a conserved homolog, Bax, that accelerates programmed cell death. *Cell* **74**: 609–619.

Pavlov EV, Priault M, Pietkiewicz D, Cheng EH, Antonsson B, Manon S, Korsmeyer SJ, Mannella CA, Kinnally KW. 2001. A novel, high conductance channel of mitochondria linked to apoptosis in mammalian cells and Bax expression in yeast. *J Cell Biol* **155**: 725–731.

Peyerl FW, Dai S, Murphy GA, Crawford F, White J, Marrack P, Kappler JW. 2007. Elucidation of some Bax conformational changes through crystallization of an antibody–peptide complex. *Cell Death Differ* **14**: 447–452.

Ploner C, Kofler R, Villunger A. 2008. Noxa: At the tip of the balance between life and death. *Oncogene* **27**(Suppl 1): S84–S92.

Puthalakath H, Villunger A, O'Reilly LA, Beaumont JG, Coultas L, Cheney RE, Huang DC, Strasser A. 2001. Bmf: A proapoptotic BH3-only protein regulated by interaction with the myosin V actin motor complex, activated by anoikis. *Science* **293**: 1829–1832.

Qian S, Wang W, Yang L, Huang HW. 2008. Structure of transmembrane pore induced by Bax-derived peptide: evidence for lipidic pores. *Proc Natl Acad Sci* **105**: 17379–17383.

Rong YP, Bultynck G, Aromolaran AS, Zhong F, Parys JB, De Smedt H, Mignery GA, Roderick HL, Bootman MD, Distelhorst CW. 2009. The BH4 domain of Bcl-2 inhibits ER calcium release and apoptosis by binding the regulatory and coupling domain of the IP3 receptor. *Proc Natl Acad Sci* **106**: 14397–14402.

Sattler M, Liang H, Nettesheim D, Meadows RP, Harlan JE, Eberstadt M, Yoon HS, Shuker SB, Chang BS, Minn AJ, et al. 1997. Structure of Bcl-xL–Bak peptide complex: Recognition between regulators of apoptosis. *Science* **275**: 983–986.

Schafer B, Quispe J, Choudhary V, Chipuk JE, Ajero TG, Du H, Schneiter R, Kuwana T. 2009. Mitochondrial outer membrane proteins assist Bid in Bax-mediated lipidic pore formation. *Mol Biol Cell* **20**: 2276–2285.

Schwarz M, Andrade-Navarro MA, Gross A. 2007. Mitochondrial carriers and pores: Key regulators of the mitochondrial apoptotic program? *Apoptosis* **12**: 869–876.

Schwickart M, Huang X, Lill JR, Liu J, Ferrando R, French DM, Maecker H, O'Rourke K, Bazan F, Eastham-Anderson J, et al. 2010. Deubiquitinase USP9X stabilizes MCL1 and promotes tumour cell survival. *Nature* **463**: 103–107.

Seo YW, Shin JN, Ko KH, Cha JH, Park JY, Lee BR, Yun CW, Kim YM, Seol DW, Kim DW, et al. 2003. The molecular mechanism of Noxa-induced mitochondrial dysfunction in p53-mediated cell death. *J Biol Chem* **278:** 48292–48299.

Setoguchi K, Otera H, Mihara K. 2006. Cytosolic factor- and TOM-independent import of C-tail-anchored mitochondrial outer membrane proteins. *EMBO J* **25:** 5635–5647.

Shamas-Din A, Brahmbhatt H, Leber B, Andrews DW. 2011. BH3-only proteins: Orchestrators of apoptosis. *Biochim Biophys Acta* **1813:** 508–520.

Shimizu S, Narita M, Tsujimoto Y. 1999. Bcl-2 family proteins regulate the release of apoptogenic cytochrome *c* by the mitochondrial channel VDAC. *Nature* **399:** 483–487.

Simmons MJ, Fan G, Zong WX, Degenhardt K, White E, Gelinas C. 2008. Bfl-1/A1 functions, similar to Mcl-1, as a selective tBid and Bak antagonist. *Oncogene* **27:** 1421–1428.

Sinha S, Levine B. 2008. The autophagy effector Beclin 1: A novel BH3-only protein. *Oncogene* **27** (Suppl 1): S137–S148.

Spencer SL, Sorger PK. 2011. Measuring and modeling apoptosis in single cells. *Cell* **144:** 926–939.

Stewart ML, Fire E, Keating AE, Walensky LD. 2010. The MCL-1 BH3 helix is an exclusive MCL-1 inhibitor and apoptosis sensitizer. *Nat Chem Biol* **6:** 595–601.

Suzuki M, Youle RJ, Tjandra N. 2000. Structure of Bax: Coregulation of dimer formation and intracellular localization. *Cell* **103:** 645–654.

Tsujimoto Y, Shimizu S. 2007. Role of the mitochondrial membrane permeability transition in cell death. *Apoptosis* **12:** 835–840.

Wang K, Yin XM, Chao DT, Milliman CL, Korsmeyer SJ. 1996. BID: A novel BH3 domain-only death agonist. *Genes Dev* **10:** 2859–2869.

Weber A, Paschen SA, Heger K, Wilfling F, Frankenberg T, Bauerschmitt H, Seiffert BM, Kirschnek S, Wagner H, Hacker G. 2007. BimS-induced apoptosis requires mitochondrial localization but not interaction with anti-apoptotic Bcl-2 proteins. *J Cell Biol* **177:** 625–636.

Wei MC, Zong WX, Cheng EH, Lindsten T, Panoutsakopoulou V, Ross AJ, Roth KA, MacGregor GR, Thompson CB, Korsmeyer SJ. 2001. Proapoptotic BAX and BAK: A requisite gateway to mitochondrial dysfunction and death. *Science* **292:** 727–730.

Willimott S, Wagner SD. 2010. Post-transcriptional and post-translational regulation of Bcl2. *Biochem Soc Trans* **38:** 1571–1575.

Willis SN, Chen L, Dewson G, Wei A, Naik E, Fletcher JI, Adams JM, Huang DC. 2005. Proapoptotic Bak is sequestered by Mcl-1 and Bcl-xL, but not Bcl-2, until displaced by BH3-only proteins. *Genes Dev* **19:** 1294–1305.

Wilson-Annan J, O'Reilly LA, Crawford SA, Hausmann G, Beaumont JG, Parma LP, Chen L, Lackmann M, Lithgow T, Hinds MG, et al. 2003. Proapoptotic BH3-only proteins trigger membrane integration of prosurvival Bcl-w and neutralize its activity. *J Cell Biol* **162:** 877–887.

Wolter KG, Hsu YT, Smith CL, Nechushtan A, Xi XG, Youle RJ. 1997. Movement of Bax from the cytosol to mitochondria during apoptosis. *J Cell Biol* **139:** 1281–1292.

Yang L, Harroun TA, Weiss TM, Ding L, Huang HW. 2001. Barrel-stave model or toroidal model? A case study on melittin pores. *Biophys J* **81:** 1475–1485.

Yeretssian G, Correa RG, Doiron K, Fitzgerald P, Dillon CP, Green DR, Reed JC, Saleh M. 2011. Non-apoptotic role of BID in inflammation and innate immunity. *Nature* **474:** 96–99.

Yethon JA, Epand RF, Leber B, Epand RM, Andrews DW. 2003. Interaction with a membrane surface triggers a reversible conformational change in bax normally associated with induction of apoptosis. *J Biol Chem* **278:** 48935–48941.

Yi CH, Pan H, Seebacher J, Jang IH, Hyberts SG, Heffron GJ, Vander Heiden MG, Yang R, Li F, Locasale JW, et al. 2011. Metabolic regulation of protein N-α-acetylation by Bcl-xL promotes cell survival. *Cell* **146:** 607–620.

Zaltsman Y, Shachnai L, Yivgi-Ohana N, Schwarz M, Maryanovich M, Houtkooper RH, Vaz FM, De Leonardis F, Fiermonte G, Palmieri F, et al. 2010. MTCH2/MIMP is a major facilitator of tBID recruitment to mitochondria. *Nat Cell Biol* **12:** 553–562.

Zha J, Harada H, Osipov K, Jockel J, Waksman G, Korsmeyer SJ. 1997. BH3 domain of BAD is required for heterodimerization with BCL-XL and pro-apoptotic activity. *J Biol Chem* **272:** 24101–24104.

Zhang Z, Zhu W, Lapolla SM, Miao Y, Shao Y, Falcone M, Boreham D, McFarlane N, Ding J, Johnson AE, et al. 2010. Bax forms an oligomer via separate, yet interdependent, surfaces. *J Biol Chem* **285:** 17614–17627.

Zhong Q, Gao W, Du F, Wang X. 2005. Mule/ARF-BP1, a BH3-only E3 ubiquitin ligase, catalyzes the polyubiquitination of Mcl-1 and regulates apoptosis. *Cell* **121:** 1085–1095.

Zinkel SS, Hurov KE, Ong C, Abtahi FM, Gross A, Korsmeyer SJ. 2005. A role for proapoptotic BID in the DNA-damage response. *Cell* **122:** 579–591.

Multiple Functions of BCL-2 Family Proteins

J. Marie Hardwick and Lucian Soane

W. Harry Feinstone Department of Molecular Microbiology and Immunology, Johns Hopkins University Bloomberg School of Public Health, Baltimore, Maryland 21205

Correspondence: hardwick@jhu.edu

BCL-2 family proteins are the regulators of apoptosis, but also have other functions. This family of interacting partners includes inhibitors and inducers of cell death. Together they regulate and mediate the process by which mitochondria contribute to cell death known as the intrinsic apoptosis pathway. This pathway is required for normal embryonic development and for preventing cancer. However, before apoptosis is induced, BCL-2 proteins have critical roles in normal cell physiology related to neuronal activity, autophagy, calcium handling, mitochondrial dynamics and energetics, and other processes of normal healthy cells. The relative importance of these physiological functions compared to their apoptosis functions in overall organismal physiology is difficult to decipher. Apoptotic and noncanonical functions of these proteins may be intertwined to link cell growth to cell death. Disentanglement of these functions may require delineation of biochemical activities inherent to the characteristic three-dimensional shape shared by distantly related viral and cellular BCL-2 family members.

WHAT ARE BCL-2 FAMILY PROTEINS AND HOW DO THEY WORK?

Human BCL-2 was discovered as the gene located near the junction at which chromosomes 18 and 14 (t14;18) are joined anomalously in the tumor cells of follicular lymphoma patients (Tsujimoto et al. 1984). This chromosome translocation leads to misregulation of the normal BCL-2 expression pattern to contribute to cancer (Tsujimoto et al. 1985; Nunez et al. 1989). Unlike previously identified oncogenes, BCL-2 was found to promote cell survival as opposed to promoting cell proliferation (Vaux et al. 1988; Tsujimoto 1989). That is, BCL-2 increases the total cell number by preventing cell death rather than by increasing cell division rate. Given that failure of these cells to die resulted in cancer, it was logical to assume that BCL-2 blocks a form of deliberate cell death. The term apoptosis (Gk: falling off, like a tree leaf) had been coined some years earlier to refer to deliberate cell death, and thus was applied to the type of cell death blocked by BCL-2 (Kerr et al. 1972; Hockenbery et al. 1991).

Compelling genetic evidence that solidified and extended this model of apoptosis regulation came from simultaneous research on the worm *Caenorhabditis elegans*. The worm BCL-2 ortholog, CED-9, was identified as the gene responsible for preventing cell death during worm development (Hengartner et al. 1992; Hengartner and Horvitz 1994b). Further genetic studies revealed that CED-9 inhibits caspase-mediated cell death (Horvitz et al. 1983; Yuan and Horvitz 1990; Yuan et al. 1993). Although *C. elegans* has only one (multidomain) BCL-2 family member, eight additional homologs of BCL-2 ranging in size from 20 to 37 kDa (BCL-x_L, MCL-1, BCL-w, BFL-1/A1, BCL-B, BAX, BAK, and BOK) plus five less related proteins sharing significant amino acid sequence similarity [BCL2L12, BCL-Rambo (BCL2L13), BCL-G (BCL2L14), BFK (BCL2L15), and BID] have been identified in the human genome (Fig. 1) (Blaineau and Aouacheria 2009). These proteins are thought to work on membranes of mitochondria and the endoplasmic reticulum (ER) facilitated by a hydrophobic membrane anchor/targeting domain near the carboxyl terminus of most BCL-2 homologs, and by a helical hairpin (helix 5 and 6 between BH1 and BH2) suggested to insert into membranes (Muchmore et al. 1996; Minn et al. 1997; Basanez and Hardwick 2008).

Although most BCL-2 homologs inhibit cell death, a subset is classified as proapoptotic (BAX, BAK, and BID). Proapoptotic BAX was first identified as an inhibitory binding partner of BCL-2 (Oltvai et al. 1993). The pro-death function of BAX is activated in response to a range of deleterious events inside or outside the cell, causing BAX to undergo conformational changes, membrane-insertion, and oligomerization to form a channel or other structure in the mitochondrial outer membrane. This is widely assumed to be the conduit through which cytochrome *c* exits mitochondria to trigger caspase activation and cell death (Cosulich et al. 1997; Kim et al. 1997; Jurgensmeier et al. 1998; Rosse et al. 1998; Kluck et al. 1999). The role of BCL-2-like antiapoptotic proteins is to inhibit their proapoptotic partners, leading to the original rheostat model in which the balance between counteracting anti- and proapoptotic BCL-2 family proteins determines cell fate (Korsmeyer et al. 1993). Although the ratios of anti- and pro-death family proteins indeed usually correlate with cell fate, this model is oversimplified in light of multiple subsequent discoveries, including the occasional interconversion of anti- and pro-death activities (Cheng et al. 1997a; Clem et al. 1998; Lewis et al. 1999), the existence of additional BCL-2-interacting proteins (Wang et al. 1996; Kelekar et al. 1997; Strasser et al. 2000; Puthalakath et al. 2001; Shamas-Din et al. 2011), the identification of BCL-2-like proteins unable to affect cell death (Bellows et al. 2002; Peterson et al. 2007; Galindo et al. 2009; Gonzalez and Esteban 2010), and the rapidly growing list of alternative nonapoptotic functions of BCL-2 family members that may have an important impact on cell survival.

The third functional subgroup of the BCL-2 family triangle is designated BH3-only because these proteins have only one of the four different BH (BCL-2 homology) motifs (Huang and Strasser 2000; Shamas-Din et al. 2011). BH motifs (numbered in order of discovery) are 10–20 amino acid regions of greatest amino acid sequence similarity across family members, though BH sequence identity can be low, and most BCL-2 homologs lack at least one BH motif (Fig. 1). The BH3 motif of proapoptotic family members is required for their pro-death activities. Eight BH3-only proteins (BID, BAD, BIK, BIM, BMF, HRK, NOXA, and PUMA) generally range in size from ~100 to 200 amino acids and are classified as BCL-2 family members based on their ability to bind and inhibit antiapoptotic BCL-2 proteins, though they lack significant overall amino acid sequence similarity (except BID) to BCL-2 proteins or to each other. BH3-only proteins promote apoptosis using one or both of two general strategies. They bind and directly activate BAX and BAK (e.g., tBID, BIM, and PUMA), or they promote death indirectly by inserting their BH3-containing helix into a hydrophobic groove on specific antiapoptotic BCL-2 proteins (Petros et al. 2000; Strasser 2005; Deng et al. 2007; Billen et al. 2008). This triangular model further explains that antiapoptotic BCL-2 proteins, of which BCL-x_L is the best characterized, protect cells by binding and inhibiting the direct activator BH3-only proteins and the multi-BH proapoptotic proteins BAX and BAK. Other than the BH3 helix, three-dimensional structures of BH3-only proteins are unresolved except for BID, which adopts a BCL-2-like fold shared by both anti- and pro-death family members (Chou et al. 1999; McDonnell et al. 1999). Thus, BH3-only proteins are thought to

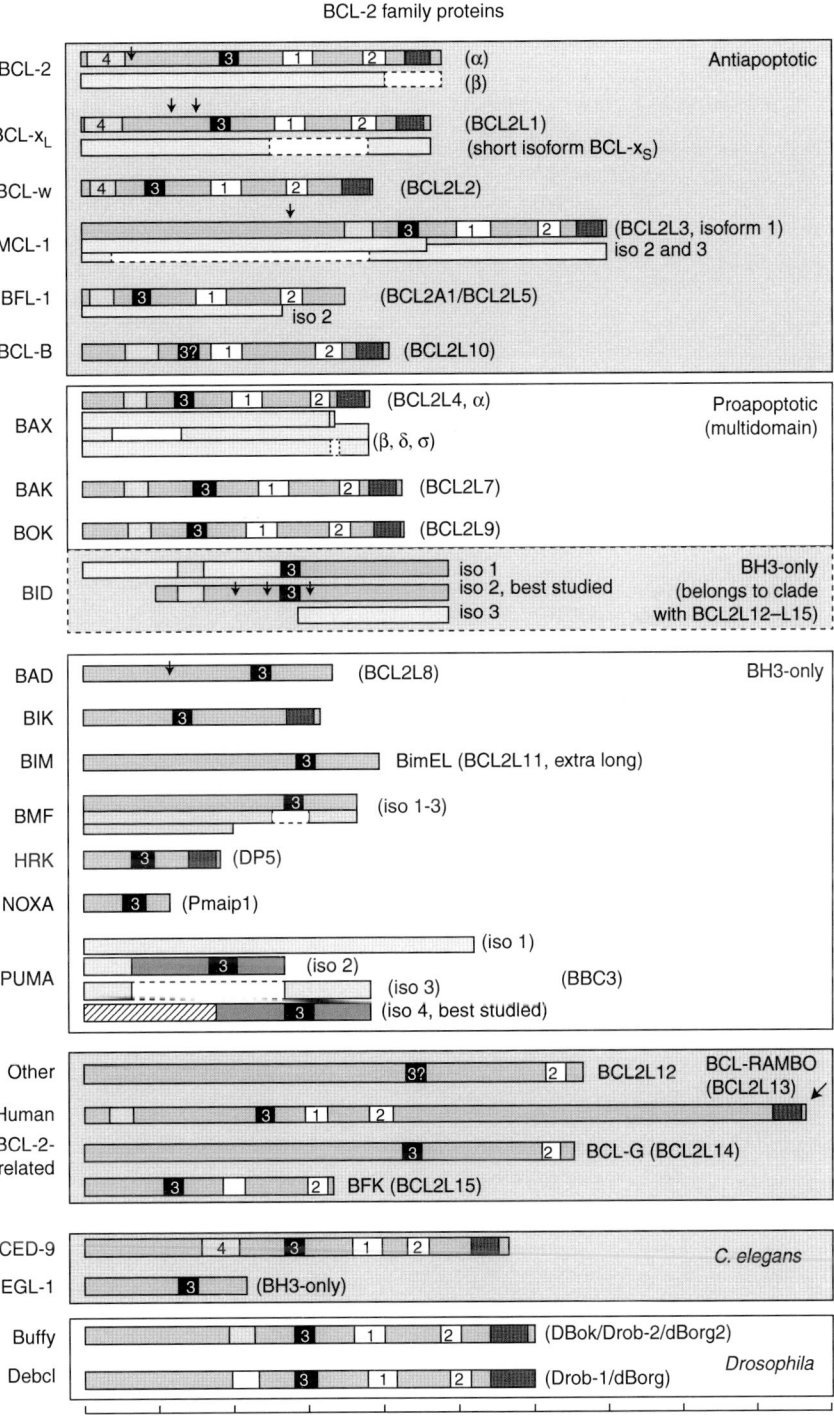

Figure 1. Human, *C. elegans*, and *Drosophila* BCL-2 family members. BH motifs are numbered, BH4 (light gray, unnumbered box indicates traditional classification without verified sequence homology), BH3 (black), BH1–BH2 (white), transmembrane (dark gray), arrows (protease cleavage sites). Splice variants (isoforms) are shown as separate or partially overlapped diagrams. Scale bar at the bottom marks increments of 50 amino acids.

be intrinsically disordered, possibly reflecting their dynamical functions and supported by extensive work (Hinds et al. 2007). Complete structures of partnered complexes would help distinguish this possibility from a case like the well ordered structures of Sgf11 and Sgf73 proteins that stretch across different domains of the SAGA deubiquitinating (DUBm) complex and would appear inherently unfolded in isolation (Samara et al. 2010).

The prevailing BCL-2 apoptosis model, in which antiapoptotic proteins are inhibited when their deep binding cleft is occupied by the BH3 helix of proapoptotic family members (e.g., BH3-only BAD), is strongly supported by the effects of a small molecule ABT-737, a BH3 mimetic designed to occupy the BCL-x_L groove (Oltersdorf et al. 2005). ABT-737 derivatives are in clinical trials as anticancer agents with promising early results, and additional therapeutics specific to antiapoptotic MCL-1, which is not targeted by ABT-737, are being pursued (Tse et al. 2008; Yecies et al. 2010; Gandhi et al. 2011).

Considerable evidence now suggests that both pro- and antiapoptotic BCL-2 family proteins have additional functions required for normal physiology of healthy cells. These noncanonical functions are unlikely to be fully explained by classical apoptosis regulatory activities in which anti-death BCL-2 proteins directly bind and inhibit proapoptotic BCL-2 family proteins to control the release of cytochrome c from mitochondria in the intrinsic apoptosis pathway. In addition, there is a growing fourth class of BCL-2 family proteins that primarily lack apoptosis regulatory activities. Viruses encode many BCL-2-shaped proteins, but most of these appear to have functions distinct from regulating cell death (see below). Similarly, the cellular BCL-2 homologs of *Drosophila* may not regulate cell death in most cell types, and the role of cytochrome c in apoptosis is not uniformly conserved through evolution (Oberst et al. 2008; Galindo et al. 2009; Tanner et al. 2011; Bender et al. 2012). Noncanonical functions of BCL-2 family proteins include their ability to alter mitochondrial shape changes and energetics, to regulate autophagy, and to modulate innate immunity during virus infections (Stack et al. 2005; Hardwick et al. 2012). Furthermore, antiapoptotic BCL-2 proteins can become proapoptotic, whereas proapoptotic proteins can promote cell survival (Bellows et al. 2002; Peterson et al. 2007; Galindo et al. 2009; Gonzalez and Esteban 2010). A major unanswered question is the relative importance of "day-job" functions versus the apoptosis-related functions of BCL-2 proteins in determining cell fate, and if the biochemical details of these functions overlap.

TURNING A NEW LEAF ON APOPTOSIS

Impressive research progress over the past 25 years has successfully driven home the fact that a subset of cells must die for an embryo to develop properly, even for the severed Planarian to regrow a new head (Gonzalez-Estevez and Salo 2010). But now it is time to let go of a few engrained assumptions to include a broader perspective. For example, we can no longer assume verbatim that the mere presence of caspase activity correlates with cell death, as caspases are also required for synaptic activity, cell growth, and inhibition of necrosis (Peter 2011; Li and Sheng 2012). We also no longer can assume that BID, BAX, and BAD are promoting cell death just because they are expressed (Fannjiang et al. 2003; Seo et al. 2004; Danial et al. 2010; Gimenez-Cassina et al. 2012). Even more complicated is the difficult task of experimentally distinguishing between noncanonical day jobs versus apoptosis functions. For example, simply evaluating the degree of cell death will not reveal whether the disrupted function of BCL-2 was an essential day job or its direct role in apoptosis, or both. One matter seems clear, most undead cells are detrimental to essentially all forms of life, and teleological reasoning suggests there is no better way to link fundamental cell functions to cell death than to use the same molecules for both processes.

The original assumption that antiapoptotic BCL-2/BCL-x_L must bind to proapoptotic BAX/BAK to inhibit cell death was first challenged by point mutants of BCL-x_L (e.g., BCL-x_L F131V/D133A, BCL-x_L Y101K) that retain significant anti-death activity but fail to interact with BAX or BAK (Cheng et al. 1996; Minn et al. 1999). However, these mutants have been reported to retain their second mode of action, the ability to bind specific BH3-only proteins, supporting the dual-strategy hypothesis whereby antiapoptotic BCL-2 proteins can suppress cell death by interfering with both subcategories of pro-death BCL-2 family proteins, the multidomain and the BH3-only (direct activator) proteins (Cheng et al. 2001; Billen et al. 2008). Elegant biochemical and computational studies have rigorously probed the classic apoptosis mechanisms with purified components and provide compelling evidence for BCL-x_L-inhibited, tBID-activated, BAX-mediated cytochrome c release. However, these findings cannot fully explain the protective effects of BCL-x_L (Billen et al. 2008). These studies however do not directly test alternative noncanonical functions of BCL-2 family proteins, as both the F131V/D133A and Y101K BCL-x_L mutants retain the ability to bind additional factors unrelated to BCL-2 family proteins that are not present or not being evaluated in these assays (Fig. 2). For example, wild type and F131V/D133A BCL-x_L can promote cell survival by binding to Aven, a regulator of both Apaf1 (apoptosome) and of ATM kinase at the G2/M checkpoint (Chau et al. 2000; Guo et al. 2008; Roelofs and Hardwick 2011; Zou et al. 2011). BCL-2 proteins also bind to the mitochondrial metabolite channel VDAC (Cheng et al. 2003), autophagy regulator Beclin 1 (Pattingre et al. 2005), the mitochondrial fission and fusion factors Drp1 and Mfn1/2 (Rolland et al. 2009), and others (Chipuk et al. 2004). At present it is difficult to incorporate all of these and other reported interactions of BCL-x_L into a unifying model for the physiological function of BCL-x_L. The field currently lacks tools equivalent to the elegant reconstituted BCL-x_L-tBID-BAX assays to probe these critical functions, though progress

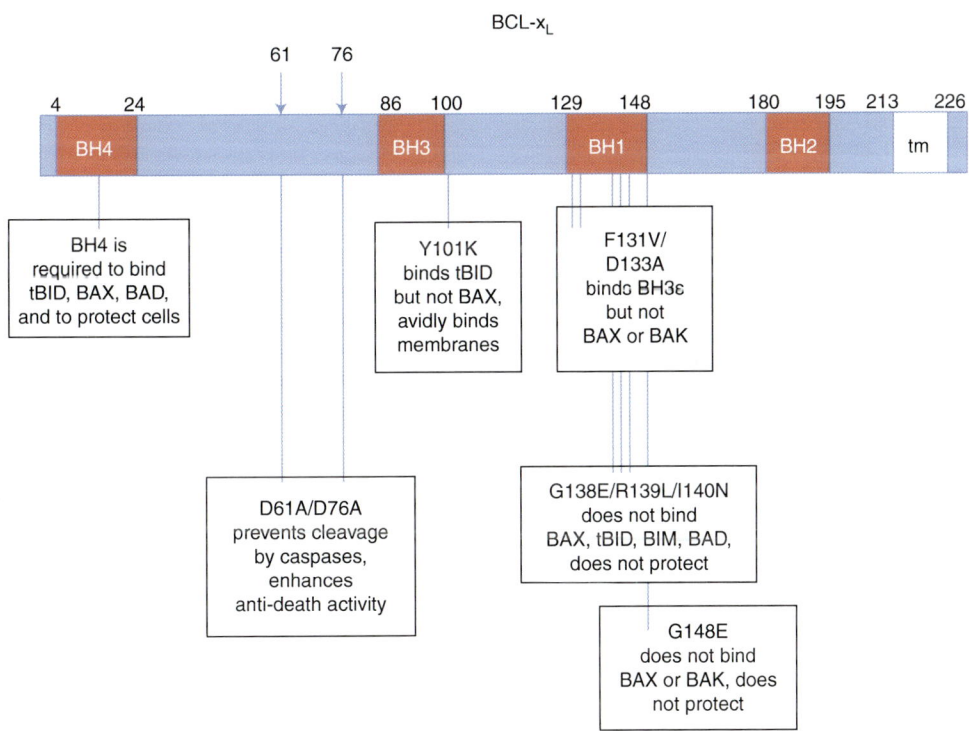

Figure 2. Map of key BCL-x_L mutants. Amino acid positions and single amino acid changes are indicated. tm, transmembrane region.

has been made (Alavian et al. 2011). We simply have insufficient knowledge at present to fully accomplish this task.

DEATH AND SURVIVAL FUNCTIONS OF PROAPOPTOTIC BCL-2 PROTEINS

BAX and BAK

Long before the discovery of BCL-2, Rita Levi-Montalcini and her colleagues were among the first to recognize programmed cell death while observing neurons in the developing chick embryo. With other colleagues, she later discovered the factor required to prevent this death of sympathetic and sensory neurons in the peripheral nervous system, nerve growth factor (NGF) (Levi-Montalcini and Angeletti 1968). Withdrawal of NGF leads to neuronal death that is mediated by the proapoptotic BCL-2 family protein BAX (Deckwerth et al. 1996). As a result, the use of BAX knockout mice has become a standard in the field for studying the functions of NGF and other factors because NGF-deprived $bax^{-/-}$ neurons can be evaluated without the complication of death of the subject under study (Glebova and Ginty 2004). In this scenario, BAX is thought to act downstream of NGF withdrawal by classical apoptosis mechanisms (see below).

As expected, BAX promotes cell death in the in the developing central nervous system, in which it is estimated that over half of the neurons born will die in a BAX-dependent manner, although this slow process can only be fully appreciated when evaluated over the long term in BAX-deficient animals (Sun et al. 2004). A role for caspase-dependent apoptosis in this process is evident from the enormous brains first observed in mice lacking caspase-3, caspase-9, or Apaf1 (Kuida et al. 1996,1998; Yoshida et al. 1998; Zheng et al. 2000). These brain phenotypes are different from the BAX/BAK double knockout, which has an obvious increase in the number of neuroprogenitors in the periventricular zone of the brain (Lindsten et al. 2000). This increase in survival of neuroprogenitors is much more pronounced in the double knockout than in the BAX single knockout and is nearly undetectable in the BAK single knockout (Lindsten et al. 2000). BAX also suppresses neurogenesis in the hippocampus and the cerebellum of adult brains (Sun et al. 2004; Garcia et al. 2012). Adult-born neurons also survive in much greater numbers in the absence of BAX even when continued neurogenesis is ablated (Sahay et al. 2011). These undead neurons do not significantly impact spatial learning and memory. However, these extra neurons apparently can confer significantly improved contextual discrimination learning, a function that normally declines with age, raising the possibility that BAX suppression could delay premature diminution of neuronal function by allowing more neurons to survive and fulfill their functions (Sahay et al. 2011).

Contrary to the pro-death developmental functions of BAX and BAK, BAX-deficiency does not rescue the death of some neuron subtypes under pathological situations (Lindsten et al. 2000; Whitmore et al. 2003; Glebova and Ginty 2004). In fact, endogenous or exogenous BAX or BAK can even protect against cell death induced by infection with Sindbis virus, which primarily infects neurons of the brain and spinal cord and causes encephalitis in mice (Lewis et al. 1999; Fannjiang et al. 2003). In an extensive analysis of BAK knockout mice, BAK was found to either inhibit or enhance neuronal death depending on the developmental stage, death stimulus, and brain region (Fannjiang et al. 2003). For example, BAK promotes death of neurons of the cortex in a stroke model, but protects hippocampal neurons following kainate-induced seizures. However, the protection by BAK in a kainate-induced seizure model appears not to be because of classical antiapoptotic function, because the degree of neuronal cell death simply correlates with more severe seizures. That is, the cell death that occurred several days later is triggered by a different process (Fannjiang et al. 2003). Because BAK knockout mice showed more seizure behaviors within minutes after kainate injection, long before neuronal death, alternative functions of BAK are implicated, such as changes in neuronal activity that give rise to a seizure. This is supported by the altered electrophysiological recordings of acute brain slices prepared from BAK knockout and control mice (Fannjiang et al. 2003).

BAK also shows bipolar effects in the spinal cord. BAK inhibits the death of virus-infected spinal cord motor neurons in young mice, but promotes motor neuron death in more mature animals (Fannjiang et al. 2003). In this scenario, BAK is expected to kill by conventional apoptosis mechanisms, but it is also possible that BAK promotes death indirectly by an alternative mechanism, such as altered neuronal activity analogous to excitotoxicity (Fannjiang et al. 2003). Perhaps this apparent switch in function is analogous to the developmental switch in excitatory to inhibitory effects of the GABA neurotransmitter (Marty and Llano 2005). Thus, BAK and BAX could contribute to both survival and death of neurons by alternative nonapoptotic mechanisms, or potentially a combination of nonapoptotic and apoptotic mechanisms. Consistent with this possibility, it is intriguing that BAX inhibits neuronal death in brain slices in which neuro-connections are preserved, but kills when these same neurons are dissociated in a culture dish. For validation of genuine anti-death activity of BAX and BAK, targeted reconstitution of BAX or BAK into the neurons of their respective knockout mice by infecting these animals with Sindbis virus encoding a copy of BAX or BAK, dramatically rescues knockout mice from Sindbis virus-induced neuronal death and mortality (Lewis et al. 1999; Fannjiang et al. 2003). This is likely not a fluke of the model system, because the same model confirmed the anti-death activity of BCL-2 and the pro-death activity of BIM_S (Seo et al. 2004).

BH3-Only Proteins

BAD is known to promote cell death by antagonizing anti-death BCL-2 proteins, but a number of studies clearly show that BAD has a normal physiological role in healthy cells. Endogenous and exogenous BAD strongly protect mice and their derived cells from Sindbis virus- and NMDA-induced neuronal death (Seo et al. 2004). The antiapoptotic functions of BAD are not limited to neurons. Overexpressed BAD in transfected cell lines can inhibit cell death similar to $BCL-x_L$ if the caspase cleavage sites in BAD are mutated to render BAD uncleavable (Condorelli et al. 2001; Kim et al. 2002; Seo et al. 2004). Interestingly, different death stimuli use distinct caspase cleavage sites to inactivate BAD. For example, mutation of the caspase cleavage site at Asp56 is required to block death following IL-3 withdrawal, whereas Asp61 must be mutated to protect against staurosporine and γ-irradiation (Seo et al. 2004). However, even proteolytically cleaved BAD (tBAD) cannot kill immature neurons in the brain, in which a further step apparently involving dephosphorylation of Ser residues (Ser112, S136, and S155) is required to turn off anti-death function and activate the death function of BAD in neonatal mice (Seo et al. 2004). This is consistent with the evidence that dephosphorylated BAD binds and inactivates antiapoptotic BCL-2 family proteins (Datta et al. 1997).

The mechanisms by which BAD inhibits cell death in these model systems is not known, but the findings are consistent with the classical view that dephosphorylation of BAD releases BAD from 14-3-3 to engage and inhibit BCL-2-like antiapoptotic proteins. However, this classical explanation of sequestering away proapoptotic BAD when phosphorylated is rather unsatisfying when attempting to explain how overexpression of phosphomimetic mutants of BAD show a gain of anti-death activity (Datta et al. 1997; Seo et al. 2004). Thus, BAD likely increases cell survival through its "day-job" mechanisms, which may involve its roles in glucose metabolism, autophagy, or cell cycle progression (Roy et al. 2009). The BH3 motif of BAD is critical for killing cells, as expected, but its BH3 is also required to promote health and well-being by activating glucokinase and increasing glucose metabolism (Danial et al. 2003, 2008). Similarly, phosphorylation is required both for its effects on glucose metabolism and for suppression of its proapoptotic activity. Hence, the cell survival and cell death functions of BAD appear to be intricately linked.

An intriguing nonapoptotic role for BAD in the control of potassium-ATP channels through its effects on glucose metabolism is also regulated by BAD phosphorylation (Gimenez-

Cassina et al. 2012). In this manner, BAD links metabolism to the control of seizure activity measured both as behavioral and electrographic seizures. Recently, BAD and BAX have also been shown to exert nonapoptotic functions in long-term depression (LTD) of synaptic transmission in CA1 hippocampal neurons (Jiao and Li 2011). BAD and BAX-mediated activation of limited caspase-3 activity is required for NMDA receptor-dependent LTD but not for mGluR-LTD. Activation of this pathway is sufficient to induce synaptic depression and is inhibited by both BAD and BAX siRNAs or knockout. Activation of BAD by dephosphorylation is limited, and apparently BAX does not translocate to mitochondria in this model (Jiao and Li 2011).

BID has a key role in cross talk between the extrinsic (extracellular ligand binding to cell surface death receptors) and intrinsic apoptosis pathways. Cleavage of BID by caspase-8 in the extrinsic pathway generates amino-terminally deleted BID known as truncated BID (tBID) (Li et al. 1998; Luo et al. 1998; Gross et al. 1999). tBID then activates the intrinsic pathway by well-studied mechanisms in which tBID transiently binds and induces BAX activation on mitochondrial membranes in vitro and in cells (Kuwana et al. 2002; Billen et al. 2008; Lovell et al. 2008). The pro-death function of the activator BH3-only proteins BID, BIM, and PUMA are apparent from studies of triple knockout mice that fail to activate BAX/BAK-dependent apoptosis in neurons and lymphocytes (Ren et al. 2010).

Although cleavage of full-length BID near its amino terminus (Fig. 1) to expose its killer BH3 domain could be analogous to the cleavage and removal of inhibitory prodomains of proteases for example, it is also possible that full-length BID has a "day job" in healthy cells. Indeed, in addition to its apoptotic function BID also has a nonapoptotic role in regulation of the DNA damage response (Zinkel et al. 2005). It has been shown that DNA damage induces translocation of BID to the nucleus in which it is phosphorylated by ATM and regulates an intra S-phase checkpoint. More recently, BID was also shown to mediate the ATR-directed DNA damage response to replicative stress by interacting with the Atrip/RPA complex (Liu et al. 2011). Interestingly, BID was recently identified in a screen for factors that facilitate innate immune responses in the gut. BID appears to be required for activation of host defense mechanisms to control bacterial infections, but may also exacerbate immune-mediated inflammatory bowel disease (Yeretssian et al. 2011). This function of BID apparently does not involve its classical apoptosis mechanisms, as knockin mice with uncleavable BID are competent for immune signaling. Conversely, mutations in the amino-terminal region of BID (not found in tBID) interfere with binding to NOD1 (member of a large family of host-defense proteins also found in plants), which forms a complex with RIP2 through their mutual CARD domains and activate IKK to mediate NF-κB activation.

NOXA, which is best known for antagonizing the anti-death function of MCL-1, has dual roles in apoptosis and metabolism, which became apparent by studying posttranslational modifications of NOXA (Lowman et al. 2010). Phosphorylation of NOXA appears to increase glucose metabolism through the pentose phosphate shunt, and this requires an intact BH3 domain, but not for binding to MCL-1 and inducing apoptosis.

MITOCHONDRIAL MEMBRANE STRUCTURE AND INTRAMITOCHONDRIAL FUNCTIONS OF BCL-2 PROTEINS

A major theme has emerged in recent years, the involvement of BCL-2 family proteins in mitochondrial shape-changes and organelle localization (Fig. 3) (Frank 2006; Detmer and Chan 2007; Knott et al. 2008). Some yet undetermined but fundamental underlying function of BCL-2 proteins, possibly in mitochondrial energetics (Vander Heiden et al. 2001), could easily influence mitochondrial morphology and function indirectly. Many studies now suggest close interactions between BCL-2 family proteins and the dynamin-like GTPases Drp1 and Mfn1/2 that physically mediate mitochondrial outer membrane fission and fusion, respectively (Li et al. 2008; Rolland et al. 2009). Here too, BCL-2 family proteins also appear to

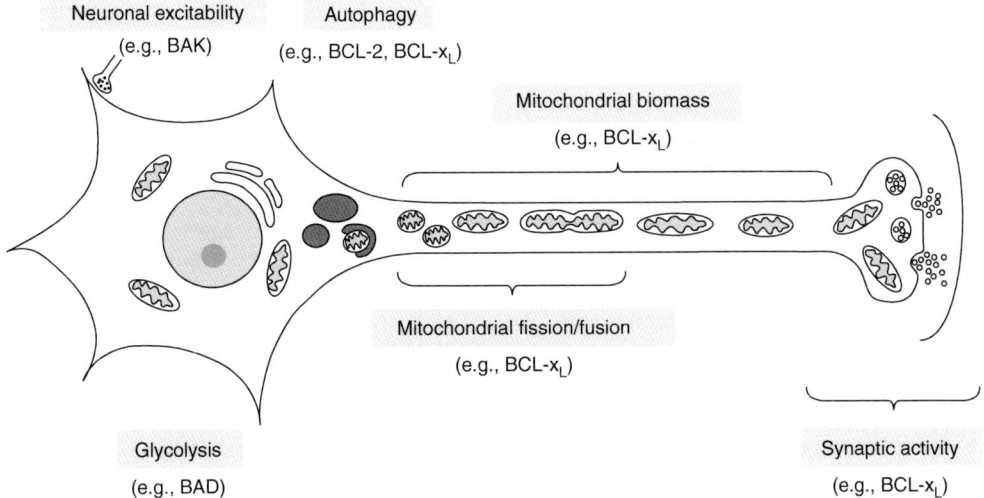

Figure 3. Neuronal functions of BCL-2 family proteins in healthy cells.

have polar opposite effects, but with a twist. Both anti- and pro-death BCL-2 family proteins functionally interact with both fission and fusion factors, but the connections between specific protein players and cell death/survival outcomes are far from clear (Jagasia et al. 2005; Cheng et al. 2006; Delivani et al. 2006; Parone et al. 2006; Li et al. 2008; Tan et al. 2008; Breckenridge et al. 2009). Nevertheless, these mechanisms may be conserved across species even when apoptosis functions of BCL-2 family proteins appear not to be conserved (Delivani et al. 2006). Genetic studies in C. elegans and biochemical strategies indicate that the fission and death functions of Drp1 are separable, and that both involve BCL-2 family members (Cassidy-Stone et al. 2008; Abdelwahid et al. 2010). A role for BAX in fusion has also been connected to the permeability transition pore opening on the mitochondrial inner membrane, leading to necrosis (Whelan et al. 2012).

The link between BCL-2 family proteins and mitochondrial dynamics was first uncovered when Drp1 (homolog of yeast Dnm1) was found to promote BAX-induced mitochondrial fission and cell death (Frank et al. 2001; Karbowski et al. 2002). In contrast to dying cells, BAX promotes mitochondrial fusion in healthy cells (Karbowski et al. 2006). This dichotomy is shared by the C. elegans BCL-2 homolog CED-9, which is an essential inhibitor of cell death but can show pro-death functions (Hengartner and Horvitz 1994a). Although CED-9 interacts with FZO-1 (human Mfn1/2) to stimulate mitochondrial fusion in healthy cells, CED-9 also is required for DRP1-dependent fragmentation of mitochondria during developmental cell death (arguably both a death and day job) (Jagasia et al. 2005). Similarly, human BCL-x_L can bind and induce Drp1-dependent mitochondrial fission, but this appears to be a day job rather than a cell death function, because BCL-x_L expression increases the number of neuronal synapses and the number of mitochondria localized to synapses, which is at least partially dependent on Drp1 (Berman et al. 2008, 2009; Li et al. 2008) (Fig. 3). These changes induced by BCL-x_L are associated with dramatically increased spontaneous synaptic activity (Li et al. 2008). More direct evidence that BCL-2 family proteins can alter synaptic activity comes from studies using the squid giant synapse. Microinjection of recombinant BCL-x_L protein into the squid nerve cell terminals dramatically increases synaptic activity within ~15 min (Jonas et al. 2003). Interestingly, BCL-x_L can have opposite effects on neuronal activity in this squid model depending on the stimulus. Microinjected

BCL-x_L delays hypoxia-induced synaptic decline, but also enhances the detrimental effects of excessive neuronal activity (Jonas et al. 2004, 2005b; Hickman et al. 2008b). These findings were confirmed using the tailor-fit small molecular inhibitor of BCL-x_L ABT-737. The effects of BCL-2 family proteins on synaptic activity correlates well with their ability to form large versus small channels on patch-clamped intracellular mitochondria, in which recombinant proteins are delivered through the inner patch pipette (Jonas et al. 2003, 2004, 2005a). The ability of caspase-cleaved BCL-x_L to promote synaptic run-down in squid under some conditions is supported by the ability of ABT-737 to suppress this effect of BCL-x_L (Hickman et al. 2008a; Ofengeim et al. 2012). This is in striking contrast to the death-promoting effects of ABT-737 in treating cancer.

Merging of day-job and apoptotic mechanisms may be most apparent in the retraction and growth of new neuronal projections and particularly dendritic spines that can be rapidly remodeled within minutes. Thus, it is conceivable that BCL-2 family proteins carry out their classical apoptotic functions except that instead of death of the entire cell, only a tiny appendage of an enormous neuron is effectively "killed" as part of normal synaptic plasticity. However, it is unlikely to be this simple. Nevertheless, recent analysis of a caspase-resistant BCL-x_L knockin mouse is consistent with this hypothesis (Ofengeim et al. 2012). A mouse in which the two caspase cleavage sites in BCL-x_L were mutated to render BCL-x_L uncleavable is strikingly resistant to transient ischemic injury. Fitting with a prodeath role of cleaved BCL-x_L, and not simply preservation of uncleaved full-length BCL-x_L, ABT-737 protects wild-type animals from neuronal loss in the hippocampus following an ischemic event, again opposite to its expected role in killing tumor cells (Ofengeim et al. 2012). However, these effects of BCL-x_L appear not to involve BAX and BAK, but may involve non-apoptotic functions of caspases.

Currently, it is not clear if the opposite effects of individual mammalian and worm (and perhaps Drosophila) BCL-2 family proteins on mitochondrial shape changes require any shared biochemical mechanisms. However, recent in vitro studies argue even stronger that BCL-2 family proteins may have an intimate role in membrane fusion reactions. Biochemical approaches suggest that Drp1 induces a hemifusion state in the mitochondrial outer membrane (Montessuit et al. 2010; Landes and Martinou 2011). In this manner, Drp1 creates a local lipid topology that promotes BAX oligomerization, consistent with colocalization of Drp1 and BAX in spots found at sites of mitochondrial fission (Karbowski et al. 2002; Montessuit et al. 2010). This function of Drp1 does not require its GTPase activity, which is required for mitochondrial fission. If BAX oligomerizes at stalled fission junctions, it is conceivable that outer membrane permeability is the result of a defective, excessive, or incomplete fission process (Montessuit et al. 2010).

On the day-job side, development of a powerful in vitro fusion assay has served to show a role for BAX in mitochondrial fusion in vitro (Meeusen and Nunnari 2007; Hoppins et al. 2011). By mixing together two populations of mitochondria decorated with different fusion protein components unique to the cells from which they were isolated, the effects of these protein components can be rigorously evaluated. These studies reveal that BAX and BCL-x_L (separately) can promote Mfn2-dependent fusion of isolated mitochondrial organelles in vitro, fitting with their functions in healthy mitochondria. These mammalian studies are built on earlier work in yeast, Drosophila, and C. elegans that first identified the mitochondrial fission, fusion, and maintenance factors and continue to significantly advance knowledge that would not otherwise be possible (Labrousse et al. 1999; Shaw and Nunnari 2002; Tan et al. 2008; Rolland et al. 2009).

Although some of the effects on mitochondrial structure are exerted at the OMM, recent findings indicate that BCL-2 proteins can also affect mitochondrial structure and function by acting intramitochondrially (Hardwick et al. 2012). Although the long-standing dogma indicates that the apoptotic functions of BCL-2 family proteins are exerted exclusively at the OMM, this topic was recently revisited as several studies have shown that antiapoptotic BCL-2 proteins can be imported into mitochondria and likely associate with the inner mitochondrial membrane in which they are suggested to carry out nonapoptotic functions (Huang and Yang-Yen 2010; Vento et al. 2010; Alavian et al. 2011; Chen et al. 2011; Warr et al. 2011; Perciavalle et al. 2012). These studies follow earlier reports of inner mitochondrial membrane localization of BCL-2 (Hardwick et al. 2012). Intramitochondrially localized BCL-x_L is suggested to regulate mitochondrial ATP production by interacting with the F_1F_0 ATP synthase complex and by stabilizing the inner membrane potential, thereby providing significant energy conservation (Alavian et al. 2011; Chen et al. 2011). MCL-1 has been shown to possess a bona-fide mitochondrial presequence that mediates $\Delta\Psi$-dependent MCL-1 import into the mitochondrial matrix. Matrix-localized MCL-1, while devoid of antiapoptotic activity, has been shown to regulate the structure of mitochondrial cristae and the ATP synthase to alter ATP production as well (Perciavalle et al. 2012).

Non-BCL-2 family apoptosis regulators have also been found to be imported into mitochondria, suggesting much broader roles for apoptosis regulators at this unexpected location. Recently, p53 was reported to accumulate in the mitochondrial matrix and trigger mitochondrial permeability transition pore (PTP) opening by interaction with the PTP regulator cyclophilin D (CypD) in response to oxidative stress (Vaseva et al. 2012). Interestingly, BCL-2, which in some studies localizes preferentially at the inner mitochondrial membrane (Gotow et al. 2000) and not OMM, was show recently to also interact with matrix-localized Cyp D (Eliseev et al. 2009). Discrepancies between studies regarding the subcellular localization of different BCL-2 family proteins may reflect different cell types and energetic states, but may also reflect hereto unappreciated conformation-dependent epitopes recognized by different antibodies. The availability of genetic knockouts/knockdowns will help clarify these issues.

REGULATORS OF CALCIUM HOMEOSTASIS

Although mitochondria are considered the primary site of action of BCL-2 family proteins, many of these proteins also localize at the ER and recent studies have shown that pro- and anti-apoptotic family members exert opposing effects on ER Ca^{2+} handling. Although early studies noted that overexpression of BCL-2 can affect Ca^{2+} signaling and redistribution of Ca^{2+} from ER to mitochondria (Baffy et al. 1993), it was subsequently found that BCL-2 reduces basal ER Ca^{2+} levels specifically through increasing Ca^{2+} leak into the cytosol (Palmer et al. 2004). BCL-x_L was later found to exert a similar effect on enhancing Ca^{2+} leak from ER and maintaining a low basal ER calcium concentration (White et al. 2005).

Although it was originally hypothesized that BCL-2 (and BAX) alter ER calcium levels through their channel-forming ability, it was later shown that this effect does not depend on their putative pore-forming domains (Chami et al. 2004). Subsequent studies suggested instead that regulation of ER calcium levels by BCL-2 proteins occurs through direct or indirect modulation of ER calcium channels. BCL-2 was found to bind and inactivate the calcium pump, SERCA (Dremina et al. 2004), and to induce a decline in SERCA2b levels following overexpression (Vanden Abeele et al. 2002). More recent studies pointed to a role of BCL-2 family proteins in regulating the IP3R function, although the mechanisms involved are still debated (Pinton and Rizzuto 2006). Several groups have shown that the prosurvival effect of BCL-2 and BCL-x_L at the ER is promoted by an increase in ER Ca^{2+} leak leading to low basal ER Ca^{2+} concentrations and thus to a reduction in stress-induced ER Ca^{2+} release (Oakes et al. 2005; White et al. 2005). BCL-2 and BCL-x_L were shown to directly bind IP3R and modulate its Ca^{2+} conductance, which in the case of BCL-2 appears to involve an effect on IP3R phosphorylation state. MCL-1 has been recently shown to bind IP3R as well and function in a similar manner at the ER (Eckenrode et al. 2010). These effects of BCL-2 proteins appear to involve activation of IP3R channel gating by an allosteric mechanism that sensitizes the channel to low inositol 1,4,5-trisphosphate concentrations and accounts for the reduced steady-state ER Ca^{2+} levels (White et al. 2005; Eckenrode et al. 2010). At least for BCL-2,

ER calcium regulation appears to be modulated by phosphorylation as phosphorylated BCL-2, which resides primarily at the ER, cannot reduce basal ER Ca^{2+} levels (Bassik et al. 2004; Oakes et al. 2006).

Other studies have found instead that interaction between IP3R and BCL-2 results in inhibition of IP3R and consequently a reduction in stress-induced IP3R-mediated Ca^{2+} release, elevation of cytosolic calcium and mitochondrial calcium overload. Consistent with this alternate model a peptide derived from IP3R has been shown to disrupt the BCL-2/IP3R interaction and reverse the inhibitory effect of BCL-2 on IP3R (Rong et al. 2008). The inhibitory effect of BCL-2 was attributed to the BH4 domain of BCL-2 that binds the regulatory and coupling domain of IP3R (Rong et al. 2009). The interaction of BCL-2 proteins with the IP3R likely involves multiple binding sites, as amino-terminal truncation of MCL-1 still binds efficiently to the IP3R (Eckenrode et al. 2010). Although this second model is also supported by substantial evidence, it does not provide an explanation for the observed effects of BCL-2 and BCL-x_L on resting ER calcium levels.

Regarding the proapoptotic BCL-2 proteins it has been found that cells from BAX and BAK double knockout (DKO) mice also have lower resting ER calcium levels and are protected from apoptotic stimuli that signal through calcium (Scorrano et al. 2003). Based on these findings it has been proposed that in contrast to BCL-2 and BCL-x_L, BAX and BAK elevate ER Ca^{2+} concentration and trigger ER Ca^{2+} release and its uptake by the mitochondria following stress (Scorrano et al. 2003; Oakes et al. 2005). Although BAX and BAK have been shown to regulate IP3R1 and Ca^{2+} leak, no direct interaction between BAX/BAK and IP3R has been observed and these effects may be mediated through modulation of BCL-2/IP3R1 interaction and IP3R1 phosphorylation state (Oakes et al. 2005). Although overexpression of BCL-2 or BCL-x_L also results in a decreased capacitative Ca^{2+} entry, reduction of calreticulin and SERCA2 levels, as well as altered IP3R levels, none of these effects were noted in BAX/BAK DKO cells (Scorrano et al. 2002, 2003).

A role in regulating ER calcium release has also been reported for several BH3-only proteins, including BIK/NBK, PUMA, and NIX/BNIP3. ER-localized BIK has been shown to be required for BAX/BAK-dependent ER Ca^{2+} release and cytochrome c release in response to genotoxic stress (Mathai et al. 2005). Similarly, PUMA has been shown to contribute to ER Ca^{2+} depletion-induced apoptosis by modulating BAX activity (Luo et al. 2005). Another study showed that ER-localized NIX/BNIP3 was required to induce Ca^{2+}-mediated PTP opening and loss of $\Delta\Psi_m$ in cardiomyocytes, although the mechanism by which NIX modulates ER Ca^{2+} levels is unknown (Diwan et al. 2008).

The release of calcium from the ER has been reported to be a control point for initiation of apoptosis in response to several stimuli, such as arachidonic acid, ceramide, and H_2O_2 (Scorrano et al. 2003). Subsequent studies have shown, however, that ER calcium regulation is also involved in nonapoptotic functions of these proteins (i.e., mitochondrial energy metabolism and T-cell activation [Jones et al. 2007]), suggesting that ER Ca^{2+} regulation represents another day-job function of these proteins. Such nonapoptotic functions of BCL-2 proteins appear to be conserved as the zebrafish BCL-2 homolog Nrz has been recently shown to control the cytoskeletal dynamics during zebrafish development by regulating ER Ca^{2+} release through direct interaction with the IP3R1 (Popgeorgiev et al. 2011).

VIRAL BCL-2 PROTEINS: AN UNEXPECTED MIXED BAG

Many viruses encode proteins that localize to mitochondria, for example, the RNA viruses HIV and influenza virus, but the functions of these proteins in virus infection and virus-host cell interactions are only partially delineated (Boya et al. 2004; Gocnikova and Russ 2007). Although BCL-2-like proteins have not been identified in the genomes of RNA viruses or small DNA viruses, in which coding capacities are preciously conserved, all three large DNA virus families infecting mammals, herpesviridae, adenoviridae, and poxviridae, encode proteins that are included in the BCL-2 family by at least one criterion. However, many of these viral BCL-2-like factors differ greatly in sequence or function from each other and from cellular BCL-2, but all appear to maintain a BCL-2-like three-dimensional structure. The open reading frames of viral BCL-2-like genes are unspliced and generally located in variable regions of their respective viral genomes, suggesting that they were acquired from their host cell to successfully establish a stable virus–host relationship. This assumption challenges some theories suggesting precellular origins for at least some viral genomes. Like the viral oncogenes of avian and other retroviruses that differ from their homologs in host cells (proto-oncogene), viral BCL-2-like proteins appear to be resistant to cellular regulatory mechanisms relative to their cellular counterparts (Bellows et al. 2000; Irusta et al. 2003).

Antiapoptotic activity has been confirmed for several unrelated poxvirus BCL-2-like proteins, including vaccinia virus F1L (Wasilenko et al. 2003) and N1L (Cooray et al. 2007), virus M11L (Su et al. 2006), parapoxvirus ORF virus ORFV125 (Westphal et al. 2007), and fowlpox virus FPV039 (Banadyga et al. 2007), plus the obvious homologs of these proteins encoded by related poxviruses (Fig. 4). Similarly, obvious BCL-2 homologs with antiapoptotic activity are found in essentially all γ herpesviruses (Henderson et al. 1993; Cheng et al. 1997b; Nava et al. 1997). Furthermore, one or both of the BCL-2 homologs of γ herpesvirus Epstein–Barr virus (BHRF1 and BALF1) is required for this virus to immortalize B cells and inhibit cell death and perhaps other functions (Altmann and Hammerschmidt 2005; Seto et al. 2010), although BALF1 lacks obvious antiapoptotic activity (Bellows et al. 2002). It has long been

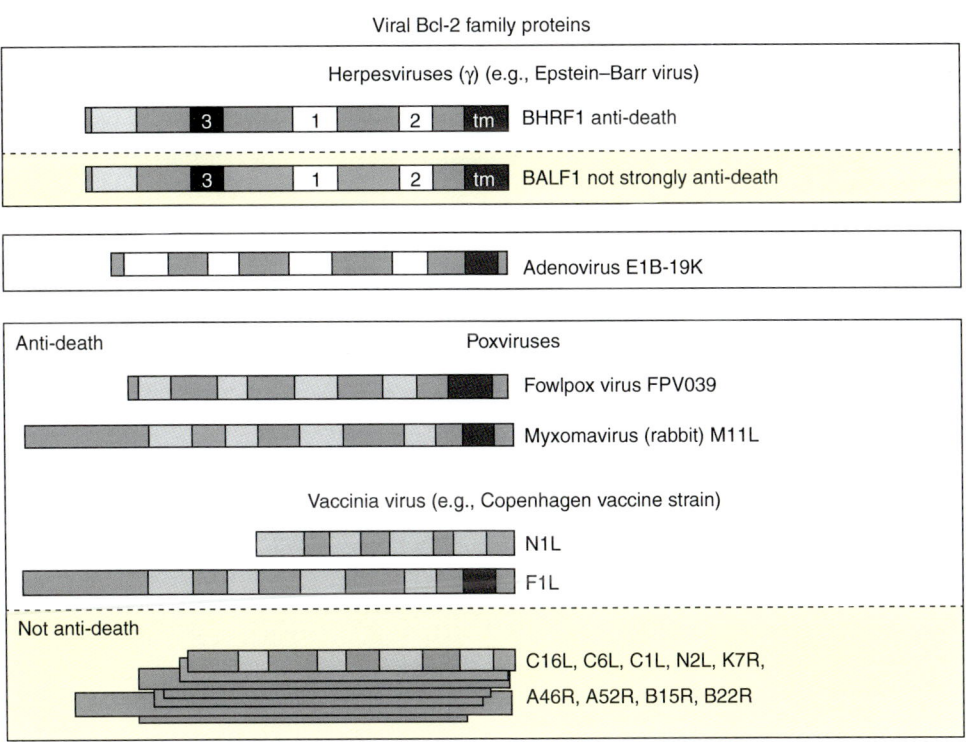

Figure 4. Viral BCL-2 family proteins with and without significant amino acid sequence similarity. BCL-2 family assignment of viral proteins are based on three-dimensional structure determinations or predicted structures. These proteins are found in the three large DNA virus families of mammals as indicated, although some lack detectable activities related to apoptosis (yellow boxes). See Figure 1 legend for BH motif key.

assumed that the antiapoptotic activity of viral BCL-2 homologs of γ herpesviruses contribute importantly to the B-cell lymphomas and many other cancers associated with these viruses. It was this transformation-promoting function of adenovirus E1B-19K that first revealed the importance of anti-death activity in E1A-induced transformation as well as mammalian virus pathogenesis (Rao et al. 1992). However, adenoviruses are not associated with any known human cancers, implying that E1B-19K serves to combat virus-induced apoptosis that is activated as a host defense-response (Degenhardt et al. 2000). The mechanisms by which viral BCL-2-like proteins inhibit BAX/BAK-dependent cell death is best characterized for E1B-19K (Perez and White 2000; Cuconati and White 2002; Shimazu et al. 2007).

Functional analyses and three-dimensional structure determination of a number of poxvirus proteins has revealed an unexpected plethora of diverse BCL-2-like proteins. Some of these proteins show antiapoptotic activity, but contain limited sequence similarity to the BH motifs that define the cellular BCL-2 family (Fig. 4; FPV039, ORFV125) (Taylor et al. 2006; Banadyga et al. 2009), whereas others have no amino acid sequence similarity to other known BCL-2 proteins, yet adopt a BCL-2-like helical structure (Fig. 4; e.g., F1L, N1L) (Taylor et al. 2006).

Three-dimensional structure determination has uncovered many other BCL-2-like proteins encoded by vaccinia virus, and thereby inferred for other related Orthopox viruses. In addition to the characterized antiapoptotic F1L and N1L proteins, there are seven additional genes in the common laboratory strain Western Reserve and 12 genes in the related Copenhagen vaccine strain (Fig. 4) that also appear to adopt a BCL-2-like fold despite the lack of any obvious sequence similarity to cellular BCL-2 proteins (Graham et al. 2008; Kalverda et al. 2009). Three-dimensional structures have been solved for a subset of these proteins referred to as the A46 subfamily (A46, A52, B14, and K7). Unlike F1L and N1L, the A46 protein family lacks detectable antiapoptotic function despite obvious BCL-2-like structural similarity. Rather, functions assigned to these factors include inhibition of TLR signaling and suppression of host immune responses (Gonzalez and Esteban 2010). Consistent with the lack of antiapoptotic activity, the three-dimensional structures of A52, B14, and K7 reveal that the BH3-binding hydrophobic groove common to antiapoptotic viral and cellular BCL-2-like proteins is occluded and unable to bind BH3-peptides. In contrast, the antiapoptotic F1L and N1L vaccinia virus proteins (and M11L of myxoma virus) have an open groove and bind with high affinity to BH3 peptides of proapoptotic proteins (Aoyagi et al. 2007; Douglas et al. 2007; Kvansakul et al. 2008). The remaining 4–5 putative BCL-2-like proteins of vaccinia viruses are more distantly related to the A46 subgroup, but sequence analysis and secondary structure predictions suggest that several members of the C6 and C16/B22 groups, as well as N2 and C1 proteins also have an all-α helical structure compatible with the 3D structural fold of BCL-2 despite lacking sequence homology to other BCL-2 proteins. Similarly, they are suggested to function in innate immunity (Gonzalez and Esteban 2010). Unlike antiapoptotic F1L, M11L, and FPV039, which are single copy genes, multiple A46-related genes are present in a single genome, raising the possibility that many more yet unidentified BCL-2-like proteins may be present in other viral and cellular genomes.

AUTOPHAGY REGULATION BY VIRAL AND CELLULAR BCL-2 PROTEINS

Recent studies indicate that in addition to their apoptotic roles, members of BCL-2 family proteins also regulate autophagy (Levine et al. 2008, 2011; He and Levine 2010; Kang et al. 2011). Antiapoptotic BCL-2 proteins including BCL-2 (Pattingre et al. 2005), BCL-x_L (Maiuri et al. 2007b), BCL-w (Erlich et al. 2007), and MCL-1 (Germain and Slack 2011; Malik et al. 2011) have been shown to interact with the autophagy regulator Beclin 1, the homolog of yeast Atg6, and to inhibit autophagy. Inhibition of autophagy and binding to Beclin 1 has also been reported for herpesvirus BCL-2 homologs, including KsBCL-2 (Pattingre et al. 2005), and M11, the to the mitochondrial matrix and couples mitochondrial fusion to respiration. *Nat Cell Biol* **14:** 575–583.

Perez D, White E. 2000. TNF-α signals apoptosis through a bid-dependent conformational change in Bax that is inhibited by E1B 19K. *Mol Cell* **6:** 53–63.

Peter ME. 2011. Programmed cell death: Apoptosis meets necrosis. *Nature* **471:** 310–312.

Peterson JS, Bass BP, Jue D, Rodriguez A, Abrams JM, McCall K. 2007. Noncanonical cell death pathways act during *Drosophila* oogenesis. *Genesis* **45:** 396–404.

Petros AM, Nettesheim DG, Wang Y, Olejniczak ET, Meadows RP, Mack J, Swift K, Matayoshi ED, Zhang H, Thompson CB, et al. 2000. Rationale for Bcl-x_L/Bad peptide complex formation from structure, mutagenesis, and biophysical studies. *Protein Sci* **9:** 2528–2534.

Pinton P, Rizzuto R. 2006. Bcl-2 and Ca^{2+} homeostasis in the endoplasmic reticulum. *Cell Death Differ* **13:** 1409–1418.

Popgeorgiev N, Bonneau B, Ferri KF, Prudent J, Thibaut J, Gillet G. 2011. The apoptotic regulator Nrz controls cytoskeletal dynamics via the regulation of Ca^{2+} trafficking in the zebrafish blastula. *Dev Cell* **20:** 663–676.

Puthalakath H, Villunger A, O'Reilly LA, Beaumont JG, Coultas L, Cheney RE, Huang DC, Strasser A. 2001. Bmf: A proapoptotic BH3-only protein regulated by interaction with the myosin V actin motor complex, activated by anoikis. *Science* **293:** 1829–1832.

Rao L, Debbas M, Sabbatini P, Hockenbery D, Korsmeyer S, White E. 1992. The adenovirus E1A proteins induce apoptosis, which is inhibited by the E1B 19-kDa and Bcl-2 proteins. *Proc Natl Acad Sci* **89:** 7742–7746.

Ren D, Tu HC, Kim H, Wang GX, Bean GR, Takeuchi O, Jeffers JR, Zambetti GP, Hsieh JJ, Cheng EH. 2010. BID, BIM, and PUMA are essential for activation of the BAX- and BAK-dependent cell death program. *Science* **330:** 1390–1393.

Roelofs BA, Hardwick JM. 2011. Flying to a halt: *Drosophila* Aven arrests the cell cycle. *Cell cycle* **10:** 1351–1352.

Rolland SG, Lu Y, David CN, Conradt B. 2009. The BCL-2-like protein CED-9 of C. elegans promotes FZO-1/Mfn1,2- and EAT-3/Opa1-dependent mitochondrial fusion. *J Cell Biol* **186:** 525–540.

Rong YP, Aromolaran AS, Bultynck G, Zhong F, Li X, McColl K, Matsuyama S, Herlitze S, Roderick HL, Bootman MD, et al. 2008. Targeting Bcl-2-IP3 receptor interaction to reverse Bcl-2's inhibition of apoptotic calcium signals. *Mol Cell* **31:** 255–265.

Rong YP, Barr P, Yee VC, Distelhorst CW. 2009. Targeting Bcl-2 based on the interaction of its BH4 domain with the inositol 1,4,5-trisphosphate receptor. *Biochim Biophys Acta* **1793:** 971–978.

Rosse T, Olivier R, Monney L, Rager M, Conus S, Fellay I, Jansen B, Borner C. 1998. Bcl-2 prolongs cell survival after Bax-induced release of cytochrome *c*. *Nature* **391:** 496–499.

Roy SS, Madesh M, Davies E, Antonsson B, Danial N, Hajnoczky G. 2009. Bad targets the permeability transition pore independent of Bax or Bak to switch between Ca^{2+}-dependent cell survival and death. *Mol Cell* **33:** 377–388.

Sahay A, Scobie KN, Hill AS, O'Carroll CM, Kheirbek MA, Burghardt NS, Fenton AA, Dranovsky A, Hen R. 2011. Increasing adult hippocampal neurogenesis is sufficient to improve pattern separation. *Nature* **472:** 466–470.

Samara NL, Datta AB, Berndsen CE, Zhang X, Yao T, Cohen RE, Wolberger C. 2010. Structural insights into the assembly and function of the SAGA deubiquitinating module. *Science* **328:** 1025–1029.

Scorrano L, Ashiya M, Buttle K, Weiler S, Oakes SA, Mannella CA, Korsmeyer SJ. 2002. A distinct pathway remodels mitochondrial cristae and mobilizes cytochrome *c* during apoptosis. *Dev Cell* **2:** 55–67.

Scorrano L, Oakes SA, Opferman JT, Cheng EH, Sorcinelli MD, Pozzan T, Korsmeyer SJ. 2003. BAX and BAK regulation of endoplasmic reticulum Ca^{2+}: A control point for apoptosis. *Science* **300:** 135–139.

Seo SY, Chen YB, Ivanovska I, Ranger AM, Hong SJ, Dawson VL, Korsmeyer SJ, Bellows DS, Fannjiang Y, Hardwick JM. 2004. BAD is a pro-survival factor prior to activation of its pro-apoptotic function. *J Biol Chem* **279:** 42240–42249.

Seto E, Moosmann A, Gromminger S, Walz N, Grundhoff A, Hammerschmidt W. 2010. Micro RNAs of Epstein–Barr virus promote cell cycle progression and prevent apoptosis of primary human B cells. *PLoS Pathog* **6**.

Shamas-Din A, Brahmbhatt H, Leber B, Andrews DW. 2011. BH3-only proteins: Orchestrators of apoptosis. *Biochim Biophysica Acta* **1813:** 508–520.

Shaw JM, Nunnari J. 2002. Mitochondrial dynamics and division in budding yeast. *Trend Cell Biol* **12:** 178–184.

Shimazu T, Degenhardt K, Nur EKA, Zhang J, Yoshida T, Zhang Y, Mathew R, White E, Inouye M. 2007. NBK/BIK antagonizes MCL-1 and BCL-x_L and activates BAK-mediated apoptosis in response to protein synthesis inhibition. *Genes Dev* **21:** 929–941.

Sinha S, Colbert CL, Becker N, Wei Y, Levine B. 2008. Molecular basis of the regulation of Beclin 1-dependent autophagy by the γ-herpesvirus 68 Bcl-2 homolog M11. *Autophagy* **4:** 989–997.

Stack J, Haga IR, Schroder M, Bartlett NW, Maloney G, Reading PC, Fitzgerald KA, Smith GL, Bowie AG. 2005. Vaccinia virus protein A46R targets multiple Toll-like-interleukin-1 receptor adaptors and contributes to virulence. *J Exp Med* **201:** 1007–1018.

Strappazzon F, Vietri-Rudan M, Campello S, Nazio F, Florenzano F, Fimia GM, Piacentini M, Levine B, Cecconi F. 2011. Mitochondrial BCL-2 inhibits AMBRA1-induced autophagy. *EMBO J* **30:** 1195–1208.

Strasser A. 2005. The role of BH3-only proteins in the immune system. *Nat Rev* **5:** 189–200.

Strasser A, Puthalakath H, Bouillet P, Huang DC, O'Connor L, O'Reilly LA, Cullen L, Cory S, Adams JM. 2000. The role of bim, a proapoptotic BH3-only member of the Bcl-2 family in cell-death control. *Ann NY Acad Sci* **917:** 541–548.

Su J, Wang G, Barrett JW, Irvine TS, Gao X, McFadden G. 2006. Myxoma virus M11L blocks apoptosis through inhibition of conformational activation of Bax at the mitochondria. *J Virol* **80:** 1140–1151.

Sun W, Winseck A, Vinsant S, Park OH, Kim H, Oppenheim RW. 2004. Programmed cell death of adult-generated hippocampal neurons is mediated by the proapoptotic gene Bax. *J Neurosci* **24**: 11205–11213.

Tan FJ, Husain M, Manlandro CM, Koppenol M, Fire AZ, Hill RB. 2008. CED-9 and mitochondrial homeostasis in *C. elegans* muscle. *J Cell Sci* **121**: 3373–3382.

Tanner EA, Blute TA, Brachmann CB, McCall K. 2011. Bcl-2 proteins and autophagy regulate mitochondrial dynamics during programmed cell death in the *Drosophila* ovary. *Development* **138**: 327–338.

Taylor JM, Quilty D, Banadyga L, Barry M. 2006. The vaccinia virus protein F1L interacts with Bim and inhibits activation of the pro-apoptotic protein Bax. *J Biol Chem* **281**: 39728–39739.

Tse C, Shoemaker AR, Adickes J, Anderson MG, Chen J, Jin S, Johnson EF, Marsh KC, Mitten MJ, Nimmer P, et al. 2008. ABT-263: A potent and orally bioavailable Bcl-2 family inhibitor. *Cancer Res* **68**: 3421–3428.

Tsujimoto Y. 1989. Overexpression of the human BCL-2 gene product results in growth enhancement of Epstein–Barr virus-immortalized B cells. *Proc Natl Acad Sci* **86**: 1958–1962.

Tsujimoto Y, Finger LR, Yunis J, Nowell PC, Croce CM. 1984. Cloning of the chromosome breakpoint of neoplastic B cells with the t(14;18) chromosome translocation. *Science* **226**: 1097–1099.

Tsujimoto Y, Cossman J, Jaffe E, Croce CM. 1985. Involvement of the bcl-2 gene in human follicular lymphoma. *Science* **228**: 1440–1443.

Twig G, Elorza A, Molina AJ, Mohamed H, Wikstrom JD, Walzer G, et al. 2008. Fission and selective fusion govern segregation and elimination by autophagy. *EMBO J* **27**: 433–446.

Vanden Abeele F, Skryma R, Shuba Y, Van Coppenolle F, Slomianny C, Roudbaraki M, Mauroy B, Wuytack F, Prevarskaya N. 2002. Bcl-2-dependent modulation of Ca^{2+} homeostasis and store-operated channels in prostate cancer cells. *Cancer Cell* **1**: 169–179.

Vander Heiden MG, Li XX, Gottleib E, Hill RB, Thompson CB, Colombini M. 2001. Bcl-x_L promotes the open configuration of the voltage-dependent anion channel and metabolite passage through the outer mitochondrial membrane. *J Biol Chem* **276**: 19414–19419.

Vaseva AV, Marchenko ND, Ji K, Tsirka SE, Holzmann S, Moll UM. 2012. p53 Opens the mitochondrial permeability transition pore to trigger necrosis. *Cell* **149**: 1536–1548.

Vaux DL, Cory S, Adams JM. 1988. Bcl-2 gene promotes haemopoietic cell survival and cooperates with c-myc to immortalize pre-B cells. *Nature* **335**: 440–442.

Vento MT, Zazzu V, Loffreda A, Cross JR, Downward J, Stoppelli MP, Iaccarino I. 2010. Praf2 is a novel Bcl-x_L/Bcl-2 interacting protein with the ability to modulate survival of cancer cells. *PLoS ONE* **5**: e15636.

Wang K, Yin XM, Chao DT, Milliman CL, Korsmeyer SJ. 1996. BID: A novel BH3 domain-only death agonist. *Genes Dev* **10**: 2859–2869.

Warr MR, Mills JR, Nguyen M, Lemaire-Ewing S, Baardsnes J, Sun KL, Malina A, Young JC, Jeyaraju DV, O'Connor-McCourt M, et al. 2011. Mitochondrion-dependent N-terminal processing of outer membrane Mcl-1 protein removes an essential Mule/Lasu1 protein-binding site. *J Biol Chem* **286**: 25098–25107.

Wasilenko ST, Stewart TL, Meyers AF, Barry M. 2003. Vaccinia virus encodes a previously uncharacterized mitochondrial-associated inhibitor of apoptosis. *Proc Natl Acad Sci* **100**: 14345–14350.

Westphal D, Ledgerwood EC, Hibma MH, Fleming SB, Whelan EM, Mercer AA. 2007. A novel Bcl-2-like inhibitor of apoptosis is encoded by the parapoxvirus ORF virus. *J Virol* **81**: 7178–7188.

Whelan RS, Konstantinidis K, Wei AC, Chen Y, Reyna DE, Jha Y, Yang Y, Calvert JW, Lindsten T, Thompson CB, et al. 2012. Bax regulates primary necrosis through mitochondrial dynamics. *Proc Natl Acad Sci* **109**: 6566–6571.

White C, Li C, Yang J, Petrenko NB, Madesh M, Thompson CB, Foskett JK. 2005. The endoplasmic reticulum gateway to apoptosis by Bcl-x_L modulation of the InsP3R. *Nat Cell Biol* **7**: 1021–1028.

Whitmore AV, Lindsten T, Raff MC, Thompson CB. 2003. The proapoptotic proteins Bax and Bak are not involved in Wallerian degeneration. *Cell Death Differ* **10**: 260–261.

Yecies D, Carlson NE, Deng J, Letai A. 2010. Acquired resistance to ABT-737 in lymphoma cells that up-regulate MCL-1 and BFL-1. *Blood* **115**: 3304–3313.

Yee KS, Wilkinson S, James J, Ryan KM, Vousden KH. 2009. PUMA- and Bax-induced autophagy contributes to apoptosis. *Cell Death Differ* **16**: 1135–1145.

Yeretssian G, Correa RG, Doiron K, Fitzgerald P, Dillon CP, Green DR, Reed JC, Saleh M. 2011. Non-apoptotic role of BID in inflammation and innate immunity. *Nature* **474**: 96–99.

Yoshida H, Kong YY, Yoshida R, Elia AJ, Hakem A, Hakem R, Penninger JM, Mak TW. 1998. Apaf1 is required for mitochondrial pathways of apoptosis and brain development. *Cell* **94**: 739–750.

Yuan JY, Horvitz HR. 1990. The *Caenorhabditis elegans* genes ced-3 and ced-4 act cell autonomously to cause programmed cell death. *Dev Biol* **138**: 33–41.

Yuan J, Shaham S, Ledoux S, Ellis HM, Horvitz HR. 1993. The *C. elegans* cell death gene ced-3 encodes a protein similar to mammalian interleukin-1 β-converting enzyme. *Cell* **75**: 641–652.

Zheng TS, Hunot S, Kuida K, Momoi T, Srinivasan A, Nicholson DW, Lazebnik Y, Flavell RA. 2000. Deficiency in caspase-9 or caspase-3 induces compensatory caspase activation. *Nat Med* **6**: 1241–1247.

Zinkel SS, Hurov KE, Ong C, Abtahi FM, Gross A, Korsmeyer SJ. 2005. A role for proapoptotic BID in the DNA-damage response. *Cell* **122**: 579–591.

Zou S, Chang J, LaFever L, Tang W, Johnson EL, Hu J, Wilk R, Krause HM, Drummond-Barbosa D, Irusta PM. 2011. Identification of dAven, a *Drosophila melanogaster* ortholog of the cell cycle regulator Aven. *Cell Cycle* **10**: 989–998.

degradation through recognition by the 26S proteasome, recent evidence indicates that other linkage types, such as M1 and K63, can regulate biological processes in a degradation-independent manner (Komander and Rape 2012).

Whether ubiquitylation targets proteins for degradation or mediates nondegradative signaling depends on protein interactions between the ubiquitylated protein and Ub-binding proteins, which can therefore be considered as Ub "receptors" (Hoeller et al. 2006). Ub receptors carry specialized Ub-binding domains (UBDs) that enable them to interact with the specific linkage types and assemble Ub-dependent signaling hubs. The generation of K63- and M1-linked Ub chains are particularly important for the generation of Ub-dependent complexes that activate NF-κB and MAPK in response to cytokine signaling (Chen 2012; Schmukle and Walczak 2012). Studies in both mammalian systems and *Drosophila* have revealed that IAPs control apoptotic and innate immune signaling pathways via degradative and nondegradative ubiquitylation. Accordingly, IAPs have been found to mediate K48-, K63-, as well as K11-linked Ub chains.

Regulation of E3 Activity

In the absence of signaling, cIAP1 resides in an inactive monomeric configuration (Fig. 2). This is achieved via an intramolecular interaction between cIAP1's BIR3 and RING finger domain (Dueber et al. 2011; Feltham et al. 2011; Lopez

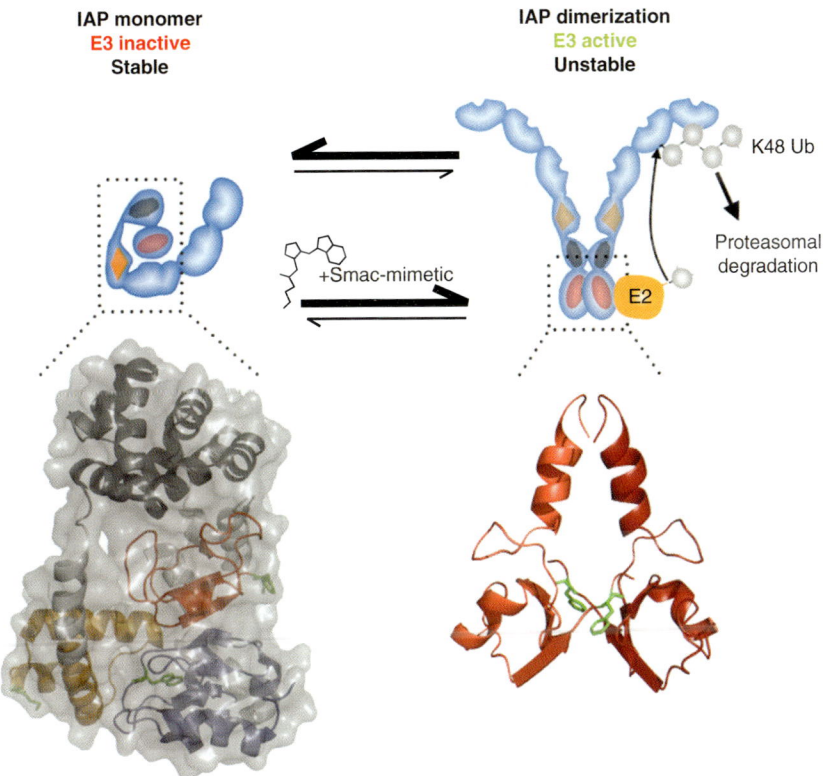

Figure 2. Mechanism of Smac-mimetic (SM)-induced activation of cIAPs. Monomeric cIAP1 BIR3, UBA, CARD RING, represented as in Fig. 1 using coordinates from Dueber and colleagues (Dueber et al. 2011), PDB: 3T6P, is an inactive E3 ligase. Binding of an SM releases the BIR3-mediated inhibition on RING dimerization (PDB:3EB5), resulting in activation of the E3 ligase function, auto-K48 ubiquitylation, and proteasomal degradation.

et al. 2011), which prevents RING-dimerization, E2 binding, and E2 activation. Autoinhibition of cIAP's E3 activity is supported by electrostatic intramolecular interactions via positively charged residues of the CARD (Dueber et al. 2011; Lopez et al. 2011). cIAP1's E3 ligase activity can be activated following binding to a substrate. Substrate-binding to the BIR3 liberates the RING from BIR3-mediated inhibition, exposing two interaction surfaces required for RING dimerization and E2 binding, respectively (Mace et al. 2008). RING dimerization is of particular importance as it is indispensable for the transfer of Ub from the E2 to a lysine residue of the target substrate. Binding of SM to the BIR3 causes autoubiquitylation and proteasomal degradation of cIAPs (Fig. 2). In effect, SMs mimic the presence of a substrate, leading to activation of cIAP's E3 activity. However, in the absence of a bona fide protein substrate, lysine residues of cIAP1 serve as acceptor lysines for Ub, resulting in polyubiquitylation and degradation of cIAP1. Regulation of the E3 activity clearly differs among IAPs. cIAP2, which shares high sequence conservation with cIAP1, readily exists in a dimeric state (Feltham et al. 2011). Although this explains how SMs activate cIAP1's auto-E3 ligase activity, it is currently unknown how the E3 activity of cIAP1, cIAP2, or XIAP are regulated in normal signaling processes.

FUNCTION OF IAP PROTEINS

IAP-Mediated Regulation of Caspases and Cell Death

The apoptotic cell death program culminates in the activation of caspases, a family of highly specific cysteine proteases essential for the destruction of the cell. Caspases are expressed as zymogens consisting of a prodomain, a large subunit (p20), and a small subunit (p10) (Riedl and Shi 2004) (see also Fig. 4). Caspases reside in proteolytic cascades that are typically started by so-called initiator caspases, such as caspase-9 and caspase-8, which cleave and activate downstream effector caspases, such as caspase-3 and caspase-7 (Berger et al. 2006; Denault et al. 2006). Following zymogen activation, caspases are regulated by certain members of the IAP protein family (Salvesen and Duckett 2002). In Drosophila, DIAP1-mediated inhibition of caspases is essential for cell survival as loss of DIAP1 function instigates spontaneous caspase-mediated apoptosis (Rodriguez et al. 1999; Wang et al. 1999; Goyal et al. 2000; Lisi et al. 2000).

DIAP1 represents an essential negative regulator of the initiator caspase DRONC and the effector caspases drICE and DCP-1 (Fig. 3). These caspases bind to distinct BIR domains: the BIR1 region of DIAP1 is essential for binding to the effector caspases drICE and DCP-1 (Kaiser et al. 1998; Wang et al. 1999; Zachariou et al. 2003), whereas the BIR2 region directly associates with DRONC (Meier et al. 2000; Chai et al. 2003). Although DIAP1-caspase association is the decisive step in the regulation of Drosophila apoptosis, physical interaction between DIAP1 and caspases alone is insufficient to regulate caspases. This is evident because DIAP1-bound effector caspases remain catalytically active under in vitro conditions (Tenev et al. 2005). Moreover, DIAP1 mutants with a dysfunctional RING finger fail to suppress caspase-mediated cell death, even though these mutants bind to caspases with the same affinity as their wild-type counterparts. Therefore, DIAP1 does not act as a classical active-site enzyme inhibitor, but rather regulates the catalytic potential of caspases. Ultimately, suppression of caspases and apoptosis results from DIAP1-mediated ubiquitylation of the zymogenic form of DRONC (Fig. 3C) and active drICE or DCP-1 (Fig. 3D,E) (Lisi et al. 2000; Wilson et al. 2002; Chai et al. 2003; Ditzel et al. 2008). The mechanism through which ubiquitylation of DRONC results in its inactivation appears to be context dependent and involves degradative as well as nondegradative ubiquitylation (Fig. 3C). Outside of the apoptosome, DIAP1-mediated ubiquitylation of DRONC neutralizes it through an unknown mechanism that operates independent of the proteasome. However, when it is part of the apoptosome, DIAP1 conjugates K48-linked poly-Ub chains to DRONC, targeting it for proteasomal destruction (Shapiro et al. 2008). Hence, only apoptosome-associated DRONC (as well as DARK itself), but not free DRONC monomer, is target-

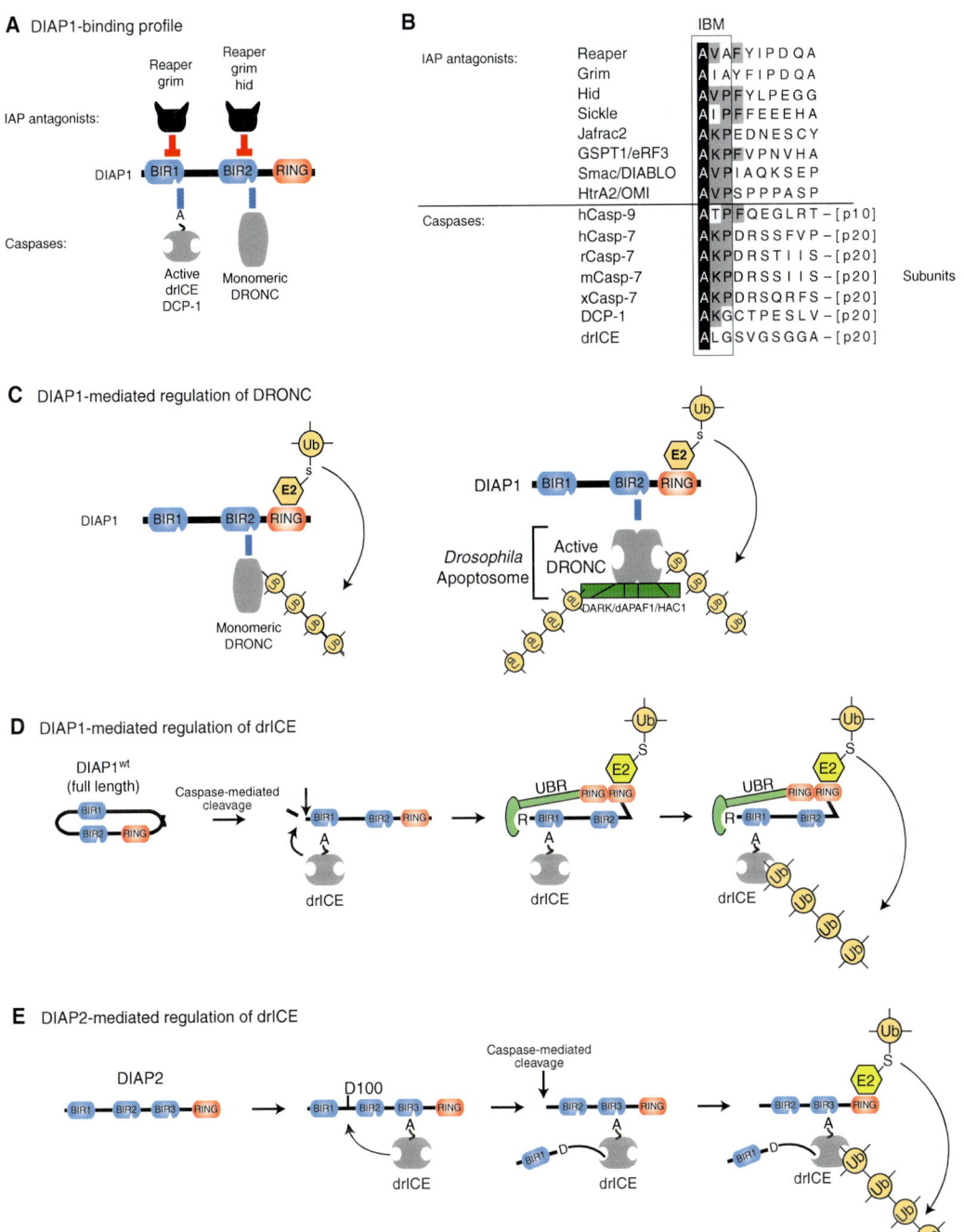

Figure 3. IAP-mediated regulation of caspases in *Drosophila*. (*A*) Binding profile of DIAP1 with caspases and IAP antagonists. Direct physical interaction with the effector caspases drICE or DCP-1 and the initiator caspase DRONC is mediated through DIAP1's BIR1 and BIR2 domains, respectively. Following their activation, drICE and DCP-1 expose an NH$_2$-terminal IBM (depicted as A), which allows their binding to BIR1. (*B*) Sequence alignment of IBM-bearing proteins. Identical residues are highlighted in black. Residues conserved in four or more IBM proteins are indicated in gray. (*C*) DIAP1's BIR2-DRONC association is essential for DIAP1 to neutralize DRONC. Following binding, DIAP's RING finger promotes Ub conjugation of DRONC, leading to its inactivation through nondegradative ubiquitylation of monomeric DRONC (*left* panel) and by targeting apoptosome-associated active DRONC for degradation (*right* panel). (*Legend continues on following page.*)

ed for proteasomal degradation. Interestingly, DRONC-mediated cleavage of DARK is required for proteasomal degradation of the DRONC/DARK complex, suggesting that the cleavage event recruits the E3 ligase (Shapiro et al. 2008).

DIAP1-mediated regulation of effector caspases is also dependent on the conjugation of Ub (Fig. 3D,E). Attachment of nondegradative (K63-linked) poly-Ub chains to the effector caspase drICE (homolog of caspase-3/-7) directly reduces its proteolytic potency, affecting kinetic parameters of the enzyme (Ditzel et al. 2008). Computational modeling of a ubiquitylated effector caspase suggests that the Ub chains sterically occlude the catalytic pocket of the caspase, thereby interfering with substrate entry. In addition to Ub, DIAP1 can also inactivate effector caspases via the covalent attachment of NEDD8. NEDD8-mediated suppression of drICE occurs via a mechanism that relies on noncompetitive inhibition, most likely through a NEDD8-induced conformational change of the caspase. Disruption of drICE ubiquitylation or NEDDylation, either by loss of DIAP1's E3 activity or generation of a nonmodifyable form of drICE, renders this effector caspase resistant to DIAP1-mediated inactivation (Ditzel et al. 2008; Broemer et al. 2010). In addition to its own RING finger domain, DIAP1 recruits a second Ub-E3 ligase that belongs to the N-end-rule pathway (UBR), to effectively block caspase activity (Ditzel et al. 2003; Herman-Bachinsky et al. 2007; Tenev et al. 2007). Recruitment of the N-end rule E3 ligase requires caspase-mediated cleavage of DIAP1 and exposure of a docking site for the UBR at the neo-NH_2 terminus of cleaved DIAP1 (Ditzel and Meier 2005). drICE, but not DRONC or DCP-1, is also regulated by the second *Drosophila* IAP, DIAP2.

DIAP2, based on its domain architecture, is the closest homolog of mammalian IAPs and directly regulates drICE and contributes to the overall caspase activity threshold in living cells. Consistently, *diap2* mutant animals harbor increased levels of drICE activity and are sensitized to apoptosis following exposure to genotoxic stress (Zimmermann et al. 2002; Ribeiro et al. 2007). Conversely, overexpression of DIAP2 suppresses developmental cell death and phenotypes caused by apoptosis inducers (Hay et al. 1995). Moreover, it rescues apoptosis triggered by RNAi-mediated depletion of DIAP1 (Leulier et al. 2006b). However, compared to DIAP1, DIAP2 exhibits a more restricted specificity for caspases as it exclusively regulates drICE and does not bind other caspases (Leulier et al. 2006b).

Intriguingly, DIAP2 functions as a mechanism-based regulator of drICE, whereby it acts as a pseudosubstrate that, following cleavage, traps the active caspase via a covalent linkage between DIAP2 and the catalytic machinery of drICE (Ribeiro et al. 2007) (Fig. 3E). The mechanism of caspase inhibition of DIAP2 is highly similar to the one used by the viral caspase inhibitor p35, which also functions as a suicide substrate that locks on to the active caspase through a covalent linkage with the catalytic cysteine of the caspase. DIAP2 mutants that cannot be cleaved fail to bind drICE and suppress drICE-mediated cell death. This method of enzyme inhibition is referred to as "mechanism based" because it relies on the activity of the enzyme. It is unusual for proteins that regulate enzymes to use a mechanism-based strategy, and most enzyme inhibitors, such as XIAP, avoid the catalytic machinery by simply blocking the substrate cleft. In addition to its mechanism-

Figure 3. (*Continued*) (*D,E*) Mechanism of effector caspase (drICE) inactivation by DIAP1 (*D*) and DIAP2 (*E*). (*D*) Full-length wild-type DIAP1 is held in an inactive conformation and requires caspase-mediated proteolytic cleavage at residue 20 for its activation. After cleavage, BIR-mediated caspase binding occurs more efficiently. Cleavage also facilitates recruitment of N-end rule UBR E3 ligases, which together with DIAP1's RING domain, promote ubiquitylation and inactivation of drICE and DCP-1. (*E*) drICE is also subject to regulation by DIAP2. drICE binds to the BIR3 of DIAP2 in an IBM-dependent manner and, following binding, cleaves DIAP2 at D100. DIAP2 cleavage results in a covalent adduct between D100 and the catalytic machinery of drICE, trapping the caspase. Full inactivation of drICE is achieved through RING-mediated ubiquitylation.

based interaction, DIAP2's E3 ligase activity is also essential for proper drICE regulation (Ribeiro et al. 2007).

The mode of caspase regulation by DIAP1 and DIAP2 differs from that of mammalian XIAP, which is a potent, classical, active site enzyme inhibitor of caspases-3 and -7 (Fig. 4A,B) (Deveraux et al. 1997). Residues within a small segment that is NH_2 terminal to XIAP's BIR2 domain directly bind to a surface found above the active-site pocket of caspase-3 and caspase-7 that is specific for these caspases, explaining the selectivity of BIR2. This prevents substrate entry and thereby results in inhibition of the caspases' catalytic activity. In this respect, XIAP acts as an inhibitor of caspases in a strict biochemical sense, blocking caspases through a "key-lock" type of mechanism (Eckelman et al. 2006). Surprisingly, the BIR2 domain itself plays little direct role in the inhibitory mechanism as almost all inhibitory contacts are made by the linker region preceding the BIR2 domain. Nevertheless, the BIR domain is functionally important as it makes additional contacts with residues outside of the catalytic pocket, thereby strengthening caspase binding. The strategy through which XIAP inhibits caspase-9 is fundamentally different (Fig. 4C,D). Here the BIR3 domain of XIAP binds to the homo-dimerization surface of caspase-9. This results in caspase-9 inactivation because caspase-9 requires a dimerization-induced conformational change to

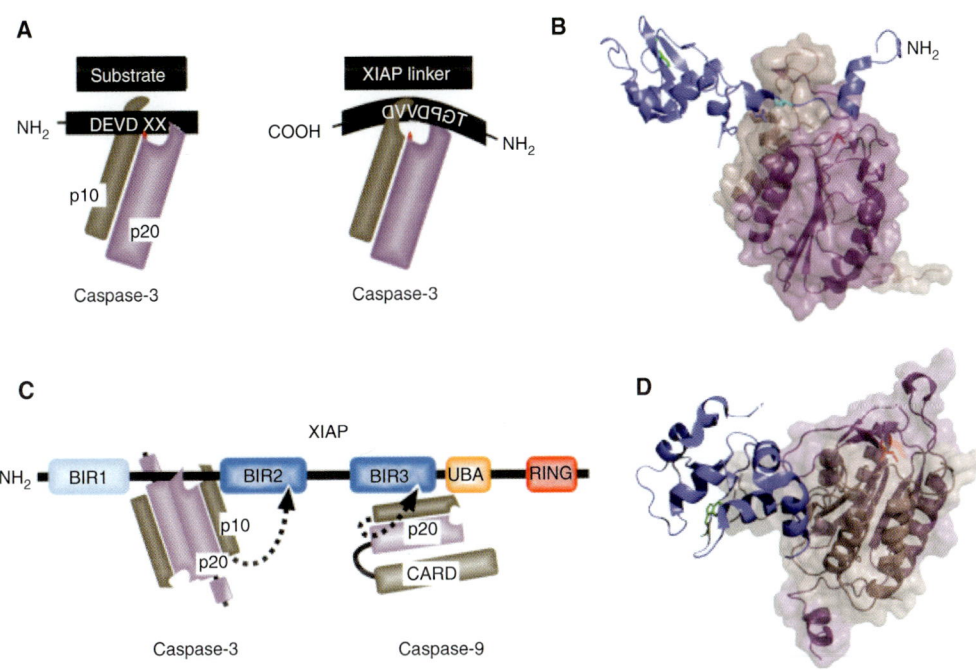

Figure 4. XIAP-mediated inhibition of caspase-3 and caspase-9. (A) Schematic comparison of substrate and XIAP linker interaction with caspase-3, catalytic cysteine, indicated in red. (B) The BIR2 of XIAP and NH_2-terminal linker (blue cartoon) embedded in the active site groove of the caspase-3 p10 (brown)/p20 (violet) heterodimer (PDB: 1I30), revealing the reverse (C-N) linker, cleavage incompatible, orientation. The reference Trp in the IBM binding groove of XIAP's BIR2 is indicated in green, Asp148 in cyan, and the position of the catalytically active cysteine in the p20 of caspase-3 in red stick format. (C) The distinct mechanisms of caspase inhibition used by XIAP represented schematically. The BIR3 binds to a dimerization surface of a caspase-9 monomer, preventing it from dimerizing and autoactivating. (D) The structure (PDB: 1NW9) of the BIR3 of XIAP (blue) bound to caspase-9 (p10 in brown, p20 in violet). The conserved reference Trp in the IBM binding groove of BIR3 is indicated in green stick format and the position of the catalytic cysteine in red stick format.

generate a productive catalytic pocket, which is no longer possible because XIAP interferes with caspase-9 dimerization. Other mammalian IAPs, such as cIAP1 and cIAP2, can also bind to caspases, particularly caspase-7, but are inefficient in inhibiting them through mere physical interactions under in vitro conditions.

Despite a large body of evidence showing that IAPs, and particularly XIAP, are important physiological regulators of caspases, $Xiap^{-/-}$, $Ciap1^{-/-}$, and $Ciap2^{-/-}$ knockout animals are surprisingly normal and display only limited cell-death related phenotypes. This seems to be due to a measure of functional redundancy among IAPs in mammals. Surprisingly, even $Xiap^{-/-}Ciap2^{-/-}$ animals are phenotypically normal (Moulin et al. 2012). However, $Xiap^{-/-}Ciap1^{-/-}$ and $Ciap1^{-/-}Ciap2^{-/-}$ are embryonic lethal. Such embryos die at around embryonic day E10.5, which is at a similar time to when $Caspase\text{-}8^{-/-}$ mice die. Likewise, animals that lack the caspase-8 adaptor FADD or FLIP, which resembles caspase-8 but lacks a catalytic site (Wilson et al. 2009), also die at E10.5. The $Ciap1^{-/-}Ciap2^{-/-}$ double knockout mice are partially rescued to birth (but not beyond) by crossing them to $Tnfr1^{-/-}$ mice (Moulin et al. 2012), demonstrating that the cIAPs regulate a developmentally important TNF-R1 signaling pathway at E10.5 and suggesting that caspase-8, FADD, and cFLIP are likewise involved in this pathway. Embryonic lethality is also partially rescued by crossing the $Xiap^{-/-}Ciap1^{-/-}$ and $Ciap1^{-/-}Ciap2^{-/-}$ mice to $Ripk1^{-/-}$ and $Ripk3^{-/-}$ mice (Moulin et al. 2012), showing that these IAPs function together as critical regulators of an embryonic decision point involving RIP kinase activity. Consistently, recent evidence indicates that XIAP, cIAP1, and cIAP2 together control the assembly of an upstream cell-death-inducing platform dubbed the "Ripoptosome" (also referred to as Complex-II and Necrosome) (Fig. 5). The Ripoptosome assembles in response to simultaneous genetic deletion of XIAP, cIAP1, and cIAP2, or in response to SM treatment. It can also form following genotoxic stress-induced depletion of XIAP, cIAP1, and cIAP2. This large ~2MDa macromolecular complex contains the core components RIPK1, FADD, and caspase-8, and can stimulate caspase-8-mediated apoptosis as well as caspase-independent necrosis. The Ripoptosome can also include additional proteins such as caspase-10, $cFLIP_L$, RIPK3, and TRIF, depending on cell type and stimulus. Assembly of the Ripoptosome depends on the kinase activity of RIPK1. It is negatively regulated by cIAP1, cIAP2, and XIAP, as well as cFLIP. Among the IAPs, cIAP1 and cIAP2 are the most critical regulators of Ripoptosome assembly (Geserick et al. 2009; Feoktistova et al. 2011; Tenev et al. 2011a). Nevertheless, XIAP also contributes to the regulation of this RIPK1-based platform because in the absence of XIAP, depletion of cIAPs results in increased assembly of this complex. The extent to which individual IAPs contribute to the inhibition of Ripoptosome assembly will, most likely, depend on cell type and stimulus. IAP-mediated inactivation of RIPK1 and/or Ripoptosome occurs in an Ub-dependent manner, most likely by targeting RIPK1 and other components of the Ripoptosome for proteasomal degradation. Caspase-8-mediated cleavage of cFLIP leads to the generation of cFLIP(p43), which allows its binding to TRAF2 and the formation of a cFLIP(p43)-caspase-8-TRAF2 tertiary complex (Micheau et al. 2002; Kataoka and Tschopp 2004). TRAF2 then recruits cIAPs, which target cleaved cFLIP and caspase-8 for ubiquitylation (Tenev et al. 2011b). This indicates that cIAP1 and cIAP2 target "active" cFLIP-caspase-8 complexes for ubiquitylation and inactivation.

Even though SM-mediated inhibition of cIAP1, cIAP2, and XIAP does not necessarily lead to immediate cell death, SM-induced assembly of the Ripoptosome can prime cells for death (Geserick et al. 2009; Feoktistova et al. 2011; Tenev et al. 2011b). For example, Fas signaling in resistant cells can be converted to a prodeath stimulus by SM treatment that leads to formation of an RIPK1-containing cell-death-inducing complex (Geserick et al. 2009). Likewise, activation of TLR3 (which normally signals for inflammatory responses through NF-κB and type-I interferon induction) in the presence of SM (or absence of cIAPs), leads to Ripoptosome-

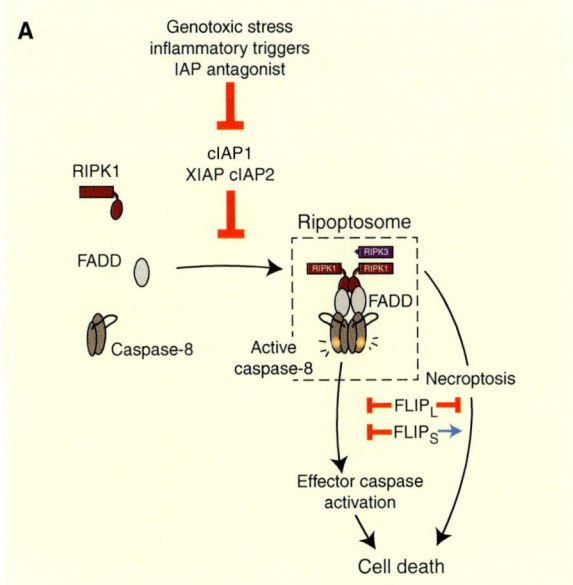
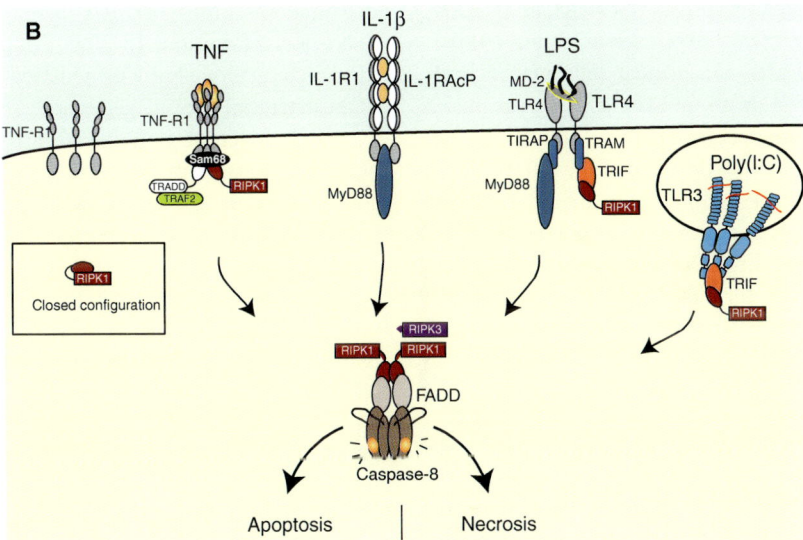

Figure 5. cIAP1, cIAP2, and XIAP prevent the formation of a RIPK1-dependent platform, dubbed the Ripoptosome, Necrosome, or Complex-II. (A) All three IAPs target RIPK1 and components of the Ripoptosome (caspase-8 and cFLIP$_L$) for Ub-mediated inactivation. Following genotoxic stress, cytokine signaling-induced depletion of cIAPs, or SM treatment, cIAP1, cIAP2, and XIAP levels rapidly decline and/or are inactivated. This allows formation and accumulation of the Ripoptosome. In the presence of high levels of RIPK3, this can lead to necroptosis. cFLIP also regulates Ripoptosome-mediated cell death. cFLIP$_L$ thereby prevents apoptosis and necroptosis, whereas FLIP$_S$ inhibits apoptosis but promotes necroptosis. (B) Under steady-state conditions, the majority of RIPK1 appears to be in a closed configuration that prevents it from binding to partner proteins. Cytokine receptor stimulation can convert a small fraction of RIPK1 into an "open," binding-competent configuration. In the presence of cIAPs and XIAP, binding-competent RIPK1 is targeted for Ub-mediated inactivation, most likely via proteasomal degradation. Under conditions where IAP levels are low, however, unmodified and binding-competent RIPK1 accumulates and can form the Ripoptosome. In the presence of high levels of cFLIP$_L$, the Ripoptosome is dissolved via caspase-8-cFLIPL-mediated cleavage of RIPK1. When cFLIP$_L$ levels are low, the Ripoptosome can promote caspase-dependent or caspase-independent cell death.

mediated cell death (Fig. 5B). Moreover, the proinflammatory cytokines TNF, TWEAK, and LIGHT trigger Ripoptosome-mediated cell death in the presence of SM or genotoxic stress-mediated depletion of IAPs. This suggests that mere inactivation of IAPs is not sufficient to induce cell death. For death to occur, an additional "RIPK1-activating" signal is required. This can be provided in the form of DNA damage or cytokine signaling (Fig. 5B).

The Ripoptosome can also mediate caspase-independent necroptosis. This form of death depends on the presence of RIPK3 and generation of reactive oxygen species (ROS) (Vandenabeele et al. 2010). Interestingly, different isoforms of cFLIP determine whether the Ripoptosome induces RIPK3-dependent necroptosis or caspase-mediated apoptosis (Geserick et al. 2009; Feoktistova et al. 2011; Tenev et al. 2011b). Whereas cFLIP$_L$ can protect cells against both forms of cell death, cFLIP$_S$ actively promotes RIP-dependent necroptosis while blocking apoptosis. This paradoxical role of the different isoforms of cFLIP is likely due to the fact that cFLIP$_L$ limits recruitment of RIPK1 into the Ripoptosome and allows localized activation of caspase-8, which results in cleavage and inactivation of RIPK1 and CYLD (O'Donnell and Ting 2011; Pop et al. 2011) and suppression of necroptotic cell death. This observation also provides a mechanistic explanation why genetic deletion of FADD, FLIP$_L$, or caspase-8 causes embryonic lethality. Although caspase-8- and FADD-deficient mice die at embryonic stage 10.5, they are rescued, at least in part, by codeletion of RIPK1 and RIPK3 (Kaiser et al. 2011; Oberst et al. 2011; Zhang et al. 2011). Therefore, caspase-8 appears to be required to suppress caspase-independent necroptosis. It is not known presently how caspase-8 is activated to regulate RIPK-dependent necrosis.

IAP-Mediated Regulation of Innate Immunity and Cell Survival

In addition to controlling the assembly and activity of caspase-activating platforms, IAPs also contribute to cell survival by regulating NF-κB signal transduction and innate immune responses (Damgaard and Gyrd-Hansen 2011). NF-κB transcription factors are important regulators of the genes necessary for innate and adaptive immune responses and for the survival and proliferation of certain cell types (Karin and Greten 2005). The realization that IAPs function as critical components of NF-κB signal transduction first came from *Drosophila* where DIAP2 was found to be essential to fend off Gram-negative bacterial infection (Gesellchen et al. 2005; Kleino et al. 2005; Leulier et al. 2006a; Huh et al. 2007). In *Drosophila*, infection by Gram-negative bacteria triggers the innate immune response by activating the immune deficiency (IMD) signaling cascade (Lemaitre and Hoffmann 2007), a Rel/NF-κB-dependent pathway that shares striking similarities with mammalian tumor necrosis factor receptor 1 (TNF-R1) (Tanji and Ip 2005). *diap2* mutant flies fail to activate NF-κB-mediated expression of antibacterial peptide genes and, consequently, rapidly succumb to bacterial infection (Leulier et al. 2006a; Huh et al. 2007). DIAP2-mediated signaling to NF-κB critically depends on its Ub-E3 ligase activity (Leulier et al. 2006a; Huh et al. 2007; Paquette et al. 2010; Meinander et al. 2012). In conjunction with the E2 Ub-conjugating enzymes Effete (UBC5) and UEV1a/Bendless (UEV1a/Ubc13) (Zhou et al. 2005; Paquette et al. 2010), DIAP2 promotes the conjugation of K63-linked Ub chains on IMD and DREDD (also referred to as DCP-2) (Paquette et al. 2010; Meinander et al. 2012), the *Drosophila* ortholog of caspase-8. Activation of the pattern-recognition receptor PGRP-LCx triggers recruitment of IMD, DREDD, and dFADD (also referred to as BG4). Recruitment of DIAP2 targets DREDD for K63-linked ubiquitylation, which allows Ub-mediated aggregation and activation of DREDD (Meinander et al. 2012). Active DREDD subsequently cleaves IMD (Paquette et al. 2010). Upon cleavage, IMD exposes an IBM at its neo-NH$_2$ terminus, which binds to the BIR2/3 of DIAP2. This provides DIAP2 with an additional docking site, reinforcing complex stability and allowing DIAP2-mediated ubiquitylation of IMD, and quite possibly other components of the signaling complex (Paquette et al. 2010). The Ub chains on IMD and DREDD appear to serve as scaffolds for the recruitment of dTAK1, IKK, and the precursor form of the NF-κB transcription factor Relish (Rutschmann et al. 2000, 2002; Silverman et al. 2000, 2003; Lu et al. 2001; Vidal et al. 2001; Kanayama et al. 2004; Kleino et al. 2005; Zhuang et al. 2006; Ferrandon et al. 2007). This brings Relish into close proximity of ubiquitylated and active DREDD, allowing DREDD-mediated proteolysis of Relish. The proximity to the signaling complex also allows phospho-mediated activation of Relish (Erturk-Hasdemir et al. 2009). Subsequently, cleaved and phosphoryated Relish translocates to the nucleus where it drives expression of antimicrobial peptide genes.

IAP-mediated activation of NF-κB also occurs in mammals. cIAP1 and cIAP2 are required for canonical activation of NF-κB and MAPK by members of the TNF-receptor family (Mahoney et al. 2008; Varfolomeev et al. 2008). For instance, binding of trimeric TNF to TNF-R1 triggers the initial recruitment of the adaptor proteins TRADD, TRAF2, the E3 ligases, cIAP1 and cIAP2, and the protein kinase RIPK1 (Fig. 6). This complex is frequently referred to as Complex-I (Micheau and Tschopp 2003). cIAP-mediated conjugation of Ub to components of Complex-I, such as RIPK1, allows subsequent recruitment of the linear Ub chain assembly complex (LUBAC, composed of HOIL/HOIP/Sharpin), the kinase complexes (composed of TAK1/TAB2/TAB3), and IKK (composed of NEMO/IKKα/IKKβ) (Silke 2011). This mechanism of Complex-I assembly is well established for some cell types, and the molecular interactions between TRADD, TRAF2, and cIAPs have been revealed by crystallography, as depicted schematically in Figure 6. It is worth remarking, however, that neither TRADD nor RIPK1 are essential components for Complex-I-mediated activation of NF-κB in all cells (Chen et al. 2008; Ermolaeva et al. 2008; Pobezinskaya et al. 2008; Wong et al. 2010). In stark contrast, cIAPs are indispensable for TNF-induced activation of NF-κB in all cell types tested so far (Mahoney et al. 2008; Varfolomeev et al. 2008; Haas et al. 2009). Ub-dependent recruitment of LUBAC, TAK1/TAB2/TAB3, and IKKs is mediated by UBDs present in TAB2, NEMO, and HOIP. Once recruited, LUBAC then modifies NEMO and RIPK1 with M1-linked Ub chains, resulting in increased stability of the TNF signaling complex. Additionally, the binding of NEMO to M1-linked Ub chains causes a conformational change of the IKK complex that is thought to facilitate its activation (Rahighi et al. 2009). Following activation, IKKβ phosphorylates IκBα, which targets it for K48-linked ubiquitylation and proteasomal degradation. Depletion of IκB liberates NF-κB dimers, which subsequently translocate to the nucleus and drive expression of target genes. Loss of LUBAC components markedly impairs, but does not abolish, Complex-I mediated signaling (Haas et al. 2009; Tokunaga et al. 2009, 2011; Gerlach et al. 2011; Ikeda et al. 2011). Therefore, it seems likely that cIAPs, and cIAP-mediated ubiquitylation, are sufficient for limited activation of genes important for inflammation and cell survival. Perhaps, the newly described K11-linked Ub chains that can be generated by cIAPs and that can recruit NEMO (Dynek et al. 2010) might contribute to this partial NF-κB activity. cIAPs are also required for JNK signaling (Matsuzawa et al. 2008; Gardam et al. 2011). This has been most clearly demonstrated for signaling that emanates from CD40, a TNF-superfamily receptor, but similar concepts likely hold true for TNF-R1 signaling too.

cIAPs are also required for constitutive suppression of the noncanonical NF-κB pathways (Varfolomeev et al. 2007; Vince et al. 2007). Activation of the noncanonical NF-κB pathway occurs in response to ligands of a subset of the TNF receptor superfamily that includes BAFF, CD40 L, and TWEAK. Under unstimulated conditions, noncanonical NF-κB signaling is normally suppressed because of the constitutive proteasomal degradation of the kinase NIK by a Ub E3 ligase complex consisting of TRAF2, TRAF3, and cIAPs (Varfolomeev et al. 2007; Vince et al. 2007; Vallabhapurapu et al. 2008; Zarnegar et al. 2008). TRAF3 binds directly to NIK and recruits it to TRAF2 through its ability to heterodimerize with TRAF2. TRAF2 in turn recruits cIAP1 or cIAP2, which are responsible for the conjugation of K48-linked Ub chains

Clearing the Dead: Apoptotic Cell Sensing, Recognition, Engulfment, and Digestion

Amelia Hochreiter-Hufford and Kodi S. Ravichandran

Department of Microbiology, Immunology and Cancer Biology, Center for Cell Clearance and Beirne Carter Center for Immunology Research, University of Virginia, Charlottesville, Virginia 22908

Correspondence: ravi@virginia.edu

Clearance of apoptotic cells is the final stage of programmed cell death. Uncleared corpses can become secondarily necrotic, promoting inflammation and autoimmunity. Remarkably, even in tissues with high cellular turnover, apoptotic cells are rarely seen because of efficient clearance mechanisms in healthy individuals. Recently, significant progress has been made in understanding the steps involved in prompt cell clearance in vivo. These include the sensing of corpses via "find me" signals, the recognition of corpses via "eat me" signals and their cognate receptors, the signaling pathways that regulate cytoskeletal rearrangement necessary for engulfment, and the responses of the phagocyte that keep cell clearance events "immunologically silent." This study focuses on our understanding of these steps.

Multicellular organisms execute the majority of unwanted cell populations in a regulated fashion via the process of apoptosis (Henson and Hume 2006; Nagata et al. 2010). Examples of unwanted cells include excess cells generated during development, cells infected with intracellular bacteria or viruses, transformed or malignant cells capable of tumorigenesis, and cells irreparably damaged by cytotoxic agents. Swift removal of these cells is necessary for maintenance of overall health and homeostasis and prevention of autoimmunity, pathogen burden, or cancer. Quick removal of dying cells is a key final step, if not the ultimate goal of the apoptotic program.

The term "phagocytosis" refers to an internalization process by which larger particles, such as bacteria and dead/dying cells, are engulfed and processed within a membrane-bound vesicle called the phagosome (Ravichandran and Lorenz 2007). A phagocyte is any cell that is capable of engulfment, including "professional" phagocytes such as macrophages, immature dendritic cells, and neutrophils. Metazoa have multiple mechanisms for clearing apoptotic cells, often depending on the tissue and apoptotic cell type (Gregory 2009). Macrophages and immature dendritic cells readily engulf dead or dying cells in tissues such as bone marrow (where a large number of new hematopoietic cells are generated), spleen (during or after an immune response), and the thymus (in young animals during T-lymphocyte development). In other tissues, neighboring "nonprofessional" phagocytes can also mediate the clearance of apoptotic targets. For example, in the mammary epithelium, viable mammary epithelial cells engulf apoptotic mammary epithelial cells after cessation of lactation (Monks et al. 2005, 2008). What distinguishes the phagocytosis of apoptotic cells from the phagocytosis of most bacteria or necrotic cells is the lack of a pro-inflammatory immune response (Henson 2005). This article discusses apoptotic cell engulfment, specifically the recruitment of phagocytes, through "find me" signals, the recognition of apoptotic cells by phagocytes via "eat me" signals, the internalization process and signaling pathways used for cytoskeletal rearrangement, and finally the digestion of apoptotic cells and phagocytic response to this process (Fig. 1).

RECRUITMENT OF PHAGOCYTES TO THEIR APOPTOTIC MEAL

Remarkably, even in tissues with high cellular turnover, apoptotic cells are rarely seen in situ, which is thought to be due to efficient clearance mechanisms. Early studies in the nematode *Caenorhabditis elegans* suggested that apoptotic cells are recognized and cleared before they are "fully dead" (Hoeppner et al. 2001; Reddien et al. 2001). This work led to the idea that apoptotic cells advertise their status to local and distant phagocytes at their earliest stages of death, perhaps via the release of "find me" signals (Ravichandran 2003).

"Find Me" Signals: Establishing a Chemotactic Gradient to Direct Phagocyte Migration

The role of "find me" signals is to establish a chemotactic gradient stimulating the migration of phagocytes to the apoptotic cell. To date, several proposed "find me" signals released by dying cells have been reported (Fig. 2). These include fractalkine, lysophosphatidylcholine (LPC), sphingosine-1-phosphate (S1P), and

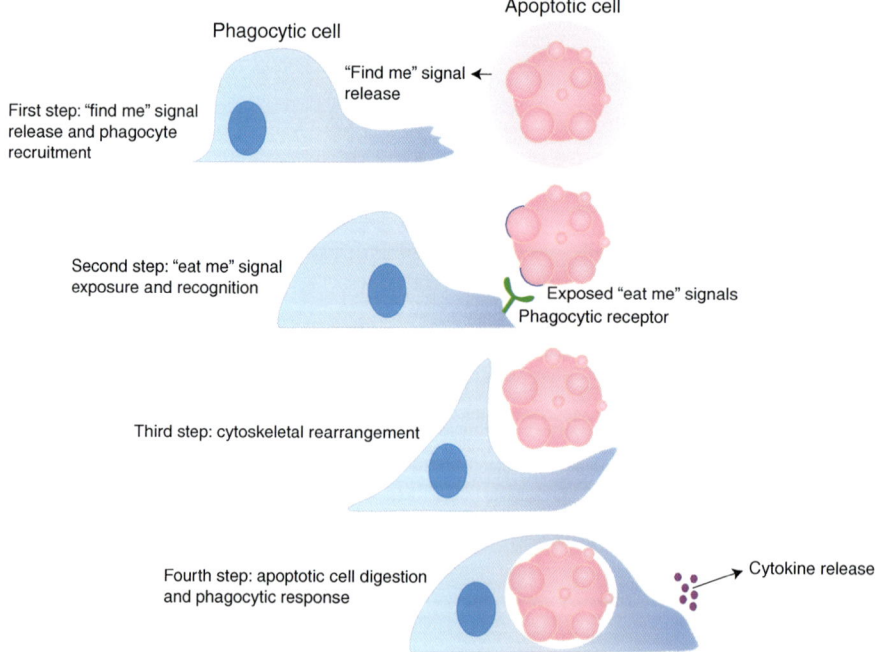

Figure 1. The steps of efficient apoptotic cell clearance. First, "find me" signals released by apoptotic cells are recognized via their cognate receptors on the surface of phagocytes. This is the sensing stage and stimulates phagocyte migration to the location of apoptotic cells. Second, phagocytes recognize exposed "eat me" signals on the surface of apoptotic cells via their phagocytic receptors, which leads to downstream signaling events culminating in Rac activation. Finally, further signaling events within the phagocyte regulate the digestion and processing of the apoptotic cell meal and the secretion of anti-inflammatory cytokines.

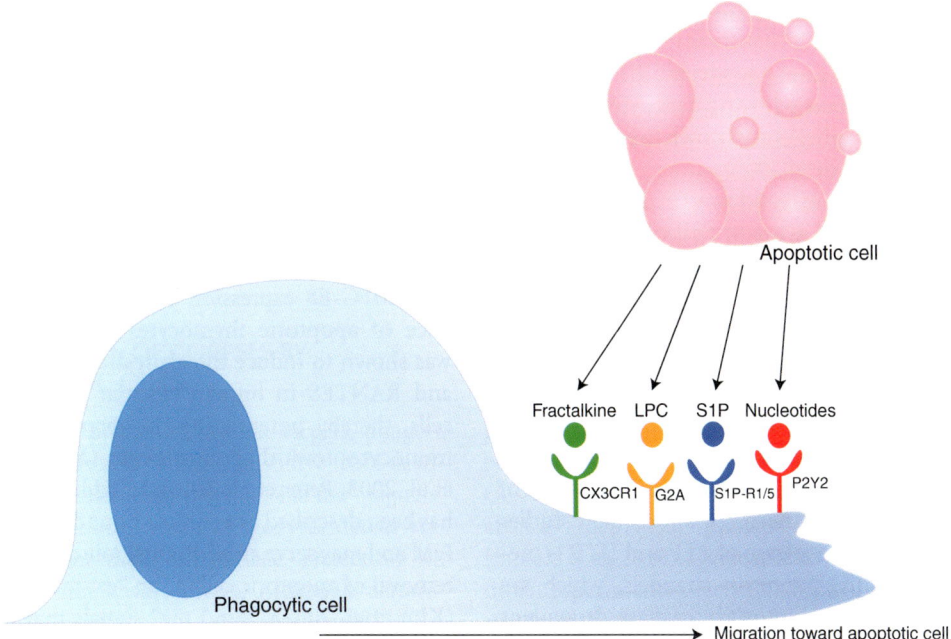

Figure 2. "Find me" signals and their receptors. Apoptotic cells release "find me" signals including fractalkine, LPC, S1P, and nucleotides. These molecules bind their cognate receptors (CX3CR1, G2A, S1P-R1/5, and P2Y2, respectively) present on the phagocyte surface. "Find me" signal recognition by the phagocyte stimulates migration toward the dying target.

the nucleotides ATP and UTP (Lauber et al. 2003; Gude et al. 2008; Truman et al. 2008; Elliott et al. 2009).

Fractalkine (i.e., CXC3CL1) is currently the only classical chemokine "find me" signal identified (Peter et al. 2010). It is a membrane-associated protein that is released from apoptotic B cells and neurons by a yet unknown protease; the released fractalkine is sensed via CX3CR1, which, in turn, directs macrophages to the dying targets (Truman et al. 2008). During affinity maturation of an antibody response, B cells in germinal centers undergo a high rate of apoptosis, and in an experiment measuring the clearance of these apoptotic B cells, fractalkine/CX3CR1 attraction was shown to influence macrophage recruitment to germinal centers in vivo (Truman et al. 2008). Interestingly, no increase in the presence of apoptotic or secondarily necrotic cells was observed in the germinal centers of CX3CR1 knockout mice, suggesting that fractalkine-mediated macrophage attraction is not required for clearance. Expression of fractalkine is limited to only a few cell types, so other mechanisms of phagocyte attraction must also exist.

LPC was the first discovered lipid "find me" signal, and it is released from apoptotic cells by the caspase-3-dependent activation of phospholipase A2, leading to the conversion of phosphatidylcholine to LPC (Lauber et al. 2003). The recognition of LPC is thought to occur via the G-protein-coupled receptor G2A and thereby stimulates macrophage chemotaxis toward apoptotic cells (Peter et al. 2008). However, the concentration of LPC reported to be required for macrophage chemotaxis appears to be quite high. In addition, the amount of LPC present in circulation is higher than LPC levels released by apoptotic cells, making LPC an unlikely candidate for a chemotactic mediator (Nagata et al. 2010). S1P, another lipid reported to function as a "find me" signal for apoptotic cells, is produced by sphingosine kinase 1 (SphK1) and secreted by apoptotic cells (Gude et al. 2008).

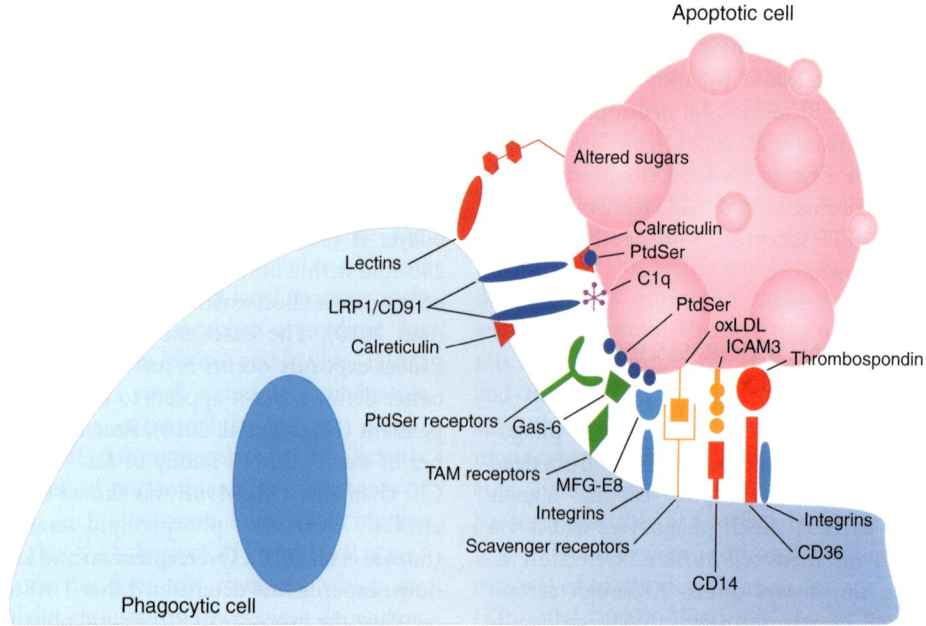

Figure 3. Apoptotic cell "eat me" signals and phagocytic receptors. As apoptotic cells undergo programmed cell death, they begin to expose "eat me" signals on their surfaces. Phosphatidylserine (PtdSer) is the best studied "eat me" signal; however, several others are also pictured here. "Eat me" signals are recognized by phagocytic engulfment receptors either directly (as with PtdSer receptors including TIM-4, BAI1, and Stabilin-2) or indirectly via bridging molecules or accessory receptors (as with Gas-6/TAM receptors, MFG-E8/$\alpha_v\beta_{3/5}$, and $\alpha_v\beta_{3/5}$ in conjunction with CD36 in the recognition of thrombospondin).

of apoptotic cells and enhance the uptake of targets by phagocytes (Arur et al. 2003; Gardai et al. 2005). Furthermore, some reports have detected oxidized PtdSer on the surface of apoptotic cells (Kagan et al. 2002), so perhaps PtdSer modification is another method by which dead cells identify themselves.

Apoptotic Cell Engulfment Receptors

A plethora of receptors are expressed on the surface of phagocytic cells that recognize "eat me" signals displayed on dying cells (Fig. 3). These include the lectins that bind altered sugars on apoptotic cells (Ezekowitz et al. 1990), CD36 (in conjunction with integrins $\alpha_v\beta_3$ and $\alpha_v\beta_5$) that binds thrombospondin (Savill et al. 1990), LRP1/CD91 (in conjunction with calreticulin) that binds complement C1q (Ogden et al. 2001), CD14 that binds ICAM3 (Gregory et al. 1998), and the scavenger receptors that bind oxidized LDL (Gordon 1999). Furthermore, PtdSer is recognized by an assortment of bridging proteins and direct recognition receptors, adding complexity to the process of dying cell recognition. For example, LRP1/CD91 has also been shown to bind calreticulin colocalized with PtdSer (Gardai et al. 2005). In addition, secreted proteins such as MFG-E8, growth-arrest-specific 6 (Gas6), and protein S have also been shown to bind PtdSer on the surface of apoptotic cells and promote engulfment via their cognate receptors on phagocytes. MFG-E8 (originally discovered in milk fat globules in the mammary gland) is expressed and secreted by professional phagocytes, associates with the $\alpha_v\beta_{3/5}$ integrins on phagocytes via its RGD motif, and binds PtdSer on apoptotic cells via its C1 and C2 domains (Hanayama et al. 2002; Nagata et al. 2010). Gas6 and protein S bridge PtdSer on apoptotic cells with the Tyro-3-Axl-Mer family of receptors (TAM receptors) on phagocytes (Nakano et al. 1997; Nagata et al. 2010).

Recently, receptors that directly recognize PtdSer on apoptotic cells and promote phagocytosis have been discovered, each with a unique expression pattern. Brain-specific angiogenesis inhibitor 1 (BAI1) is a transmembrane protein belonging to the adhesion-type G-protein-coupled receptor family that binds PtdSer via thrombospondin type 1 repeats (Park et al. 2007). BAI1 is expressed in macrophage cell lines and immunological tissues such as the bone marrow and spleen; however, BAI1 is expressed at a much higher degree in the brain glia and neuronal cells (Mori et al. 2002). Another set of recently described PtdSer receptors, T-cell immunoglobulin and mucin-domain-containing molecule 4 (TIM-4) and TIM-1, are small transmembrane proteins that bind PtdSer via their IgV domain (Kobayashi et al. 2007; Miyanishi et al. 2007; Santiago et al. 2007). TIM-4 is highly expressed by professional phagocytes; however, TIM-1 is primarily expressed in kidney cells. In contrast to the other identified PtdSer receptors, TIM-4 (and perhaps TIM-1) does not mediate direct signaling for engulfment and therefore likely functions as a tethering receptor (Park et al. 2009). Stabilin-2 (also known as hyaluronic acid receptor for endocytosis, or HARE) is a large type I membrane protein that stereospecifically recognizes PtdSer on the surface of apoptotic cells via its epidermal growth factor-like domains (Park et al. 2008a,c). It is primarily expressed on sinusoidal endothelial cells; however, Stabilin-2 is also expressed on human monocyte-derived macrophages (Park et al. 2008a). The most recent PtdSer receptor described to date is the receptor for advanced glycation end products (RAGE), which binds PtdSer in both its membrane-bound and soluble forms (He et al. 2011). The soluble form of RAGE appears to act as a decoy receptor, blocking PtdSer recognition and engulfment mediated by the membrane-bound form and other PtdSer receptors. RAGE is primarily expressed in lung tissue, including alveolar macrophages; however, its expression is induced by inflammation, perhaps suggesting a role for RAGE in the resolution of inflammation (Armstrong and Ravichandran 2011). Finally, the protein referred to as simply the "phosphatidylserine receptor" (PSR), was originally described as capable of binding PtdSer and mediating engulfment; however, the identity of PSR has turned out to be that of a Jumonji domain-containing nuclear protein, which does not directly mediate phagocytosis of apoptotic cells (Bose et al. 2004; Bratton and Henson 2008). The signaling mechanisms activated by the aforementioned receptors are discussed in the following section.

"Don't Eat Me" Signals

In addition to detecting the "eat me" signals on the surface of apoptotic cells, phagocytes can further distinguish between live and dead targets by the presence of "don't eat me" signals on the surface of living cells. CD47 (also known as integrin-associated protein) is a membrane protein expressed on the surface of healthy cells, recognized by its cognate receptor, SIRPα, which inhibits engulfment by phagocytes even in the presence of PtdSer (Gardai et al. 2005; Tsai and Discher 2008). CD47 expression is suppressed or down-modulated during apoptosis, thereby permitting clearance. Recently, mouse hematopoietic stem cells were shown to transiently increase CD47 expression during inflammation-mediated migration in order to avoid clearance; however, constitutive CD47 expression was detected in both mouse and human myeloid leukemias (Jaiswal et al. 2009). This poses a clever mechanism by which cancer cells may evade immune recognition and clearance, and new clinical therapies for the treatment of acute lymphoblastic leukemia are currently being developed that use blocking antibodies for CD47 (Chao et al. 2011). Research has shown that CD47 blocking antibodies increases engulfment in vitro and inhibits tumor engraftment in vivo; however, one large caveat to such treatment is that CD47 expression is important for the survival of many cell types including erythrocytes (Oldenborg et al. 2000), suggesting that blocking its recognition, even temporarily, may cause severe side effects, including anemia. CD31 is another example of a "don't eat me" signal (Brown et al. 2002); however, currently the CD47 literature is more advanced. Finally, the inhibitory receptor CD300a, broadly expressed on many hematopoietic cell types, has been recently shown to bind both PtdSer and PtdEtn on the surface of apoptotic cells and block their engulfment by macrophages (Simhadri et al. 2012). Interestingly, this form of regulation is intrinsic to the macrophage and independent of the apoptotic target.

ENGULFMENT SIGNALING PATHWAYS FOR CYTOSKELETON REARRANGEMENT

Once the apoptotic target is captured, the phagocyte undergoes cytoskeletal rearrangements necessary for corpse internalization. Genetic studies in *Caenorhabditis elegans* provided the initial insight into our understanding of the engulfment signaling pathways leading to the rearrangement of the phagocyte cytoskeleton necessary for corpse internalization (Kinchen 2010). The ongoing identification of the fly and mammalian homologs, as well as identification of new proteins exclusive to the mammalian contexts, are further defining signaling pathways in higher organisms (Table 1).

Rho-Family GTPases

GTPases are the molecular "switches" that turn signaling pathways on or off depending on the state of their bound guanine nucleotide. Active GTPase is the GTP-bound form, which is formed by the action of specific guanine-nucleotide-exchange factors (GEFs) that catalyze the GDP-to-GTP exchange. Conversely, GTPase-activating proteins (GAPs) stimulate the irreversible hydrolysis of GTP to GDP, effectively turning off the GTPase switch. The Rho family GTPases are members of the Ras superfamily of small signaling proteins with known functions in regulation of cellular movement (Ridley 2001). Members of this family include RhoA, Cdc42, and Rac. RhoA loss or suppression is associated with increased engulfment of apoptotic cells, and, conversely, forced activation is associated with inhibited engulfment (Tosello-Trampont et al. 2003; Nakaya et al. 2006). Regulation of clearance by RhoA is executed via the Rho-associated coiled-coil-containing protein kinase (ROCK). Active, or GTP-bound, Rho increases the kinase activity of ROCK, which, in turn, mediates phosphorylation of the myosin light chain (MLC) and promotes cell contraction (Riento and Ridley 2003). Contractility

Table 1. Evolutionarily conserved apoptotic cell engulfment signaling proteins

C. elegans	Drosophila	Mammal	Function
CED-1	DRPR	LRP1 (CD91)/MEGF10	Surface receptors on phagocytes that recognize ligands on the surface of apoptotic cells
CED-2	DCrk	CrkII	Adaptor protein that is proposed to localize the CED-5/CED-12 complex to the membrane; the precise role in apoptotic cell engulfment signaling is currently unclear
CED-5	Myoblast City	Dock180	RAC GEF containing a Docker (DHR2) domain that functions as a DH domain
CED-6	Dced-6	GULP	Adaptor proteins that bind CED-1; the precise role in apoptotic cell engulfment signaling is currently unclear
CED-7	Unknown	ABCA1/ABCA7	ABC transporter involved in cholesterol efflux downstream from apoptotic cell engulfment; the precise role is unknown, but its function is required in both the phagocyte and the apoptotic cell
CED-10	DRac	Rac	Rho family GTPase; regulates Arp2/3 activation, actin polymerization, and cytoskeletal rearrangement via the Scar/WAVE complex
CED-12	Dced-12	ELMO	Adaptor protein containing a PH domain; binds to CED-5 and enhances its RAC GEF activity

likely inhibits the extension of pseudopods and phagocytic cup formation, necessary for the early stages of engulfment. RhoA activation at subsequent stages of engulfment, however, is thought to promote apoptotic cell digestion by regulating the acidification of phagosomes (Erwig and Henson 2008). In contrast to RhoA, the current function of Cdc42 in the clearance of apoptotic cells is unclear; however, some reports suggest that it may serve a beneficial role in this process (Leverrier and Ridley 2001). Finally, GTP-bound Rac has an evolutionarily conserved positive effect on engulfment, and Rac activation at sites of apoptotic cell recognition subsequently leads to Arp2/3 activation/actin polymerization/cytoskeletal rearrangement via the Scar/WAVE complex (Miki et al. 1998; Castellano et al. 2000). Presently, at least three distinct signaling pathways governing apoptotic cell clearance have been described, which ultimately lead to downstream activation of Rac.

Rac Activation via the CrkII–Dock180–ELMO Pathway

A combination of studies in *C. elegans*, *Drosophila melanogaster*, and mammalian models led to the identification of a signaling pathway for Rac activation composed of the proteins Dock180 (homologs include CED-5 in *C. elegans* and Myoblast City in the fly), ELMO (*C. elegans* CED-12 and fly DCed-12), and CrkII (*C. elegans* CED-2 and fly DCrk) (Wu and Horvitz 1998b; Reddien and Horvitz 2000; Gumienny et al. 2001; Zhou et al. 2001a). This group of proteins is involved in several cytoskeletal rearrangement processes including cell migration, neurite growth, muscle fusion, and phagocytosis of apoptotic cells (Nagata et al. 2010). Dock180 was identified as a Rac GEF that lacks the traditional Dbl-homology and pleckstrin-homology (DH-PH) domains required among GEFs to mediate GTP exchange (Brugnera et al. 2002). Although Dock180 contains a Docker (DHR2) domain that functions as a DH domain, monomeric Dock180 exists in a "closed" confirmation, thereby preventing the interaction of the DHR2 domain with Rac (Lu et al. 2005).

ELMO is an adaptor protein that binds directly to Dock180, creating an unconventional bipartite Rac GEF (Brugnera et al. 2002). There are at least three possible mechanisms by which ELMO enhances Dock180 activity. First, binding of ELMO to the carboxy-terminal region of Dock180 relieves a steric inhibition that blocks the Dock180 DHR2 domain interaction with Rac (Lu et al. 2005). Second, the PH domain of ELMO then stabilizes the bond between Rac and Dock180 (Lu et al. 2004). Finally, ELMO targets Dock180 to the phagocytic membrane via direct binding to the carboxyl terminus of the transmembrane receptor BAI1 (Park et al. 2007).

Our understanding of Dock180–ELMO regulation is currently evolving. Original reports suggested that CrkII, an adaptor protein, recruits Dock180 and ELMO to the cell membrane after cell surface receptor/ligand engagement (Gumienny et al. 2001). Further experiments showed that CrkII can be coimmunoprecipitated with the Dock180–ELMO complex and is indeed required for efficient engulfment; however, direct interaction between CrkII and Dock180 is not essential for the removal of dead cells (Tosello-Trampont et al. 2007). Moreover, it was shown that ELMO also possesses a membrane-targeting signal and that activation of RhoG by the GEF Trio leads to membrane recruitment of the Dock180–ELMO complex via binding of ELMO Armadillo (ARM) repeats to active RHOG (deBakker et al. 2004). Finally, other studies have determined that Dock180, too, has a membrane-targeting signal, raising further doubt regarding the requirement of CrkII for this purpose (Cote et al. 2005). Therefore, the exact role of CrkII in this pathway is presently unknown.

Several phagocytic receptors introduced in the previous section have been shown to use the CrkII–Dock180–ELMO pathway for clearance of apoptotic cargo. These include integrins $\alpha_v\beta_3$ and $\alpha_v\beta_5$, which recognize PtdSer via bridging molecules (Albert et al. 2000; Akakura et al. 2004; Hanayama et al. 2004), and the TAM receptor Mer, which recognizes PtdSer via Gas6 (Wu et al. 2005). Likewise, the PtdSer receptor BAI1 also binds ELMO (via its cytoplasmic tail) and activates this signaling pathway during cell clearance (Park et al. 2007).

Rac Activation via the LRP1/MEGF10–GULP–ABCA1/ABCA7 Pathway

The second evolutionarily conserved signaling pathway for the removal of apoptotic cells includes LRP1 (CD91)/MEGF10 (*C. elegans* homolog CED-1 and fly Draper), GULP (*C. elegans* CED-6 and fly dCed-6), and ABCA1/ABCA7 (*C. elegans* CED-7) (Liu and Hengartner 1998; Wu and Horvitz 1998a; Zhou et al. 2001b; Su et al. 2002; Manaka et al. 2004). Although this pathway leads to downstream Rac activation (Kinchen et al. 2005), the intermediate steps are unclear. LRP1/CD91 is a surface receptor expressed on many cells, including phagocytes that recognize "eat me" signals on the surface of the apoptotic cells. Ligands for LRP1 (including calreticulin and thrombospondin) and the fly homolog, Draper (including Pretaporter), are known (Kuraishi et al. 2009); however, the ligand(s) for the worm homolog, CED-1, remains unidentified. After apoptotic cell recognition, LRP1 directly interacts with the adaptor protein, GULP, and may lead to cytoskeletal rearrangements (Su et al. 2002). Interestingly, the PtdSer receptor Stabilin-2 was also shown to interact with GULP via its NPXY motif, and knocking down endogenous GULP significantly reduced Stabilin-2-mediated engulfment of apoptotic cells, suggesting that it uses this pathway for clearance (Park et al. 2008b). Presently, the role of ABCA1/ABCA7 in this pathway is less well characterized. ABCA1/ABCA7 are members of the ATP-binding cassette transporter (ABC-transporter) protein family. How these proteins function in this engulfment signaling pathway is currently unclear, although their function is required in both the target and phagocyte (Wu and Horvitz 1998a). Interestingly, these proteins function in the transport of cholesterol and maintenance of lipid subdomains on the plasma membrane and may thus play a role in cholesterol homeostasis in the phagocyte during apoptotic cell engulfment (Kiss et al. 2006a,b; Landry et al. 2006).

A Third Parallel Engulfment Signaling Pathway?

In addition to the two described parallel engulfment signaling pathways, the existence of a third pathway has long been proposed because of the remaining (albeit inefficient) engulfment that occurs in the *C. elegans* CED-1, CED-5, or CED-10 mutants (Kinchen 2010). In fact, a recently described pathway was proposed to participate in engulfment signaling, consisting of the tyrosine kinase Abl (*C. elegans* ABL-1) and the Abl-interacting protein Abi (*C. elegans* ABI-1) (Hurwitz et al. 2009). Using *C. elegans*, this group reported that ABI-1 is required for engulfment of apoptotic cells and proper distal tip cell migration. ABI-1 was proposed to act in parallel to the two known engulfment signaling pathways, and ABL-1 likely opposes apoptotic cell engulfment by inhibiting ABI-1 function. The molecular pathways functioning upstream and downstream of these proteins are not yet defined, and it is still unclear whether ABI-1 signals through Rac (CED-10) similar to the other signaling pathways described above.

APOPTOTIC CELL DIGESTION AND PHAGOCYTE RESPONSE

After recognition and engulfment, the process of apoptotic cell clearance is not complete. Recent studies suggest that events downstream from internalization (collectively termed phagosome maturation) influence the phagocytic capacity of a cell to internalize additional targets (Wu et al. 2000; Krieser et al. 2002; Schrijvers et al. 2005; Park et al. 2011). Proper recognition, clearance, and degradation of apoptotic cell material are vital to maintaining an environment that protects the host against unchecked inflammation and eventual autoimmunity. Unengulfed apoptotic cells have the propensity to leak their cellular contents over time (secondary necrosis), resulting in inflammation, exposure of self-antigens, and a break in tolerance (Franz et al. 2006). On the other hand, recognition of apoptotic cells can trigger the secretion of anti-inflammatory cytokines (such as TGF-β and IL-10) from the phagocyte, thereby dampening or resolving inflammation (Henson 2005; Elliott and Ravichandran 2010). Furthermore, the proper degradation and processing of apoptotic cell material by professional phagocytes, particularly dendritic cells, are imperative for self-antigen presentation necessary to establish and maintain tolerance (Delamarre et al. 2005; Erwig and Henson 2008). The steps involved in apoptotic cell digestion and the effects this event has on the phagocyte are discussed further below.

Phagosome Maturation: Preparing the Meal for Digestion

Presently, the process of phagosome maturation is actively studied, and our knowledge is evolving. Following internalization, phagosomes (membrane-bound compartments containing the phagocytosed target) become increasingly acidic, ultimately fusing with lysosomes, which contain the digestive enzymes required for degradation (Kinchen and Ravichandran 2008). This process begins with recruitment of the large GTPase, Dynamin (*C. elegans* DYN-1), to the apoptotic cell/phagocyte interface (Fig. 4) (Yu et al. 2006). Dynamin interaction with Vps34 [a phosphatidylinositol(3)-kinase] on the forming phagosome leads to recruitment of the GDP-bound small GTPase, Rab5, to the phagosome surface (Kinchen et al. 2008). Rab5 is subsequently activated by an unidentified protein (the currently known Rab5 GEFs are not required in this process). In mammalian cells, however, depletion of the GEF, Gapex-5, was shown to inhibit Rab5 activation at sites of engulfment (Kitano et al. 2008). GTP-bound, active Rab5 promotes Vps34 activation, leading to generation of PtdIns(3)P on the phagosome surface, which is subsequently removed by the lipid phosphatase MTM-1 (Zou et al. 2009; Kinchen 2010). Recent studies describe that Mon1a (*C. elegans* SAND-1) and its binding partner, Ccz1, tie Rab5 activation to Rab7 recruitment to the phagosome (Kinchen and Ravichandran 2010) and its subsequent activation (Nordmann et al. 2010). At the Rab7-positive stage, the HOPS complex is recruited to the phagosome, leading to Rab7 activation and, ultimately, fusion of the phagosome with the lysosomal network (Kinchen et al. 2008). It is at this stage that acidic proteases and nucleases get activated and the apoptotic cell targets are degraded (Lennon-Dumenil et al. 2002). Rab5 must also be inactivated, and PtdIns(3)P must be removed from

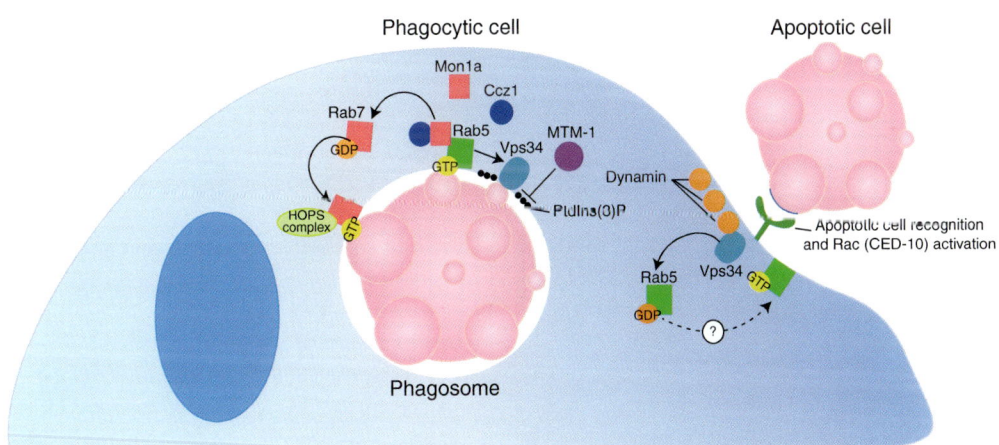

Figure 4. Phagosome maturation: The post-engulfment processing of apoptotic cells. In the first step of phagosome maturation, Dynamin is recruited to the apoptotic cell/phagocyte interface. Dynamin interacts with Vps34, leading to Rab5 recruitment and activation. Activated Rab5 further promotes Vps34 activation, leading to generation of PtdIns(3)P on the phagosome surface (subsequently removed by MTM-1). Mon1a and its binding partner, Ccz1, tie Rab5 activation and Rab7 recruitment to the phagosome. At this stage, the HOPS complex is recruited, leading to Rab7 activation and eventual fusion of the phagosome with the lysosome.

Degradation and Processing of Apoptotic Cells

After engulfment, apoptotic cells are digested into their basic cellular building blocks including nucleotides, fats, sterols, and peptides/amino acids. Phagocytes often consume multiple corpses at one time; therefore, they must efficiently process the apoptotic cell constituents to maintain homeostasis. Several mechanisms have been identified that help a phagocyte maintain homeostasis. The mitochondrial membrane protein UCP2, which uncouples oxidative phosphorylation from ATP synthesis and reduces mitochondrial membrane potential in cells, has been shown to positively regulate the engulfment capacity of phagocytes (Park et al. 2011). Overexpression of UCP2 effectively enhanced phagocytosis of apoptotic cells in vitro, and loss of UCP2 expression reduced phagocytosis both in vitro and in vivo, suggesting that mitochondrial membrane potential modulates engulfment. UCP2 may thus provide a mechanism by which the energy balance within phagocytes is maintained during the process of engulfment. DNase II is a lysosomal enzyme responsible for degrading DNA, and lack of this enzyme leads to accumulation of DNA fragments within phagocytes (Kawane et al. 2001, 2003). Furthermore, conditional deletion of DNase II in the mouse led to the development of polyarthritis and an increase of inflammatory cytokines in the joint tissues (Kawane et al. 2006). This inflammation is thought to result from the buildup of target cell DNA in macrophages, which, in turn, were shown to produce the inflammatory cytokine TNF-α (Kawane et al. 2006). Cholesterol also serves several important functions within a cell (regulates plasma membrane fluidity, plays a role in cell signaling and transport, etc.) and is therefore strictly regulated. Intriguingly, studies show that apoptotic cell-derived cholesterol becomes part of the cholesterol pool within the phagocyte, and phagocytes handle the additional sterol load by increasing their basal efflux mechanism via ABCA1 (Gerbod-Giannone et al. 2006; Kiss et al. 2006a). Conversely, increased cholesterol efflux is not seen when macrophages are fed necrotic cells or when macrophages are forced to internalize apoptotic cells via the FcR, suggesting an apoptotic-cell-specific trigger. PtdSer on the surface of apoptotic cells appears to be one such trigger; blocking PtdSer recognition was shown to inhibit enhanced cholesterol efflux, whereas providing purified PtdSer liposomes to macrophages augments it (Kiss et al. 2006a). Finally, processing of apoptotic cell proteins is linked to establishment and maintenance of self-tolerance via cross-presentation of engulfed cell peptides by the major histocompatibility complex (MHC) class I molecules (Bellone et al. 1997; Albert et al. 1998; Huang et al. 2000). Presentation of apoptotic-cell-derived peptides appears to be excluded from MHC class II presentation, perhaps reflecting the endogenous source of antigen (Blander and Medzhitov 2006).

Consequences of Apoptotic Cell Recognition and Internalization

The hallmark feature of apoptotic cell clearance is the non-inflammatory nature of this process. Many human diseases are caused or worsened by inflammation; however, the inflammatory process can also be of benefit to host survival as in the case of infection or cancer (Henson 2005). It is currently understood that a balance occurs between immunogenic and tolerogenic cell death/clearance depending on considerations such as the phagocytic cell type that clears the debris and the factors produced by the engulfing cells in response to apoptotic cell recognition.

Apoptotic cell death occurs in healthy organisms as part of normal tissue turnover (Henson and Hume 2006). The location of cell death and the means by which the cells are cleared will affect the response (or lack thereof) of the immune system. For example, organs such as the spleen and liver are considered tolerogenic, and apoptotic cells localized to these tissues will not generally elicit an immune response (Green et al. 2009). In addition, phagocytes are "programmed" to elicit specific responses after recognition of targets, dependent on their cell type. For instance, engulfment of uninfected apoptotic cells by tissue resident macrophages is usually immunologically silent; however, engulfment of apoptotic cells via infiltrating macrophages and/or dendritic cells in an inflamed environment has the potential to drive an antigen-specific immune response and break tolerance (Geissmann et al. 2010; Hochreiter-Hufford and Ravichandran 2012). Signals on the surface of dying cells can direct specific phagocytic cell types (i.e., dendritic cells or macrophages) toward them for clearance; however, in a recent study, lipid oxidation by one type of phagocyte was shown to inhibit phagocytosis of apoptotic cells via another phagocytic cell type. Briefly, expression of 12/15-lipoxygenase by resident peritoneal macrophages led to exposure of oxidized lipid on their cell surface; this oxidized lipid sequestered the MFG-E8 necessary for inflammatory monocyte recognition and engulfment of apoptotic cells (Uderhardt et al. 2012). This is the first example of a phagocyte-intrinsic mechanism for regulating apoptotic cell sorting to specific phagocytes. Annexin V is a naturally occurring PtdSer binding protein that blocks phagocytosis of apoptotic cells by macrophages but not dendritic cells (Krahling et al. 1999; Munoz et al. 2007; Frey et al. 2009). Necrotic cell immunization of Annexin V-deficient mice did not lead to elicitation of an immune response, unlike wild-type mice; macrophages isolated from Annexin V-deficient mice were shown to secrete significantly higher levels of IL-10 in response to incubation with LPS and necrotic cells than macrophages from wild-type mice (Frey et al. 2009). The investigators suggest that Annexin V blocks PtdSer-mediated anti-inflammatory cytokine release as well as phagocytosis of necrotic cells by macrophages, thereby allowing engulfment by dendritic cells and the initiation of an immune response. Further studies will elucidate whether this mechanism functions in an in vivo setting.

As mentioned above, one method by which apoptotic cells induce tolerance is by stimulating phagocyte secretion of anti-inflammatory and immunosuppressive cytokines such as TGF-β and IL-10. These cytokines, in turn, induce differentiation of regulatory T-cells and T helper 2 cells important in prevention of an inflammatory response (Green et al. 2009). Likewise, apoptotic cells have been shown to repress inflammatory cytokine secretion by monocytes in vitro (Voll et al. 1997; Kim et al. 2004). Interestingly, internalization of apoptotic cells is not required for stimulated cytokine secretion. In fact, in in vitro studies, apoptotic cell membranes were sufficient to inhibit IL-12 secretion by LPS-induced macrophages, and PtdSer vesicles, alone, stimulated TGF-β secretion by macrophages (Huynh et al. 2002; Kim et al. 2004). Work is currently underway to piece together the receptors and signaling molecules that connect PtdSer recognition with cytokine secretion.

Although apoptotic cell clearance is typically considered immunologically silent, there are exceptions in which apoptotic cells can elicit an immune response. The clearance of apoptotic tumor cells is an example in which immunogenic cell death is beneficial to the host so that residual tumor cells and metastases are maintained under control. Engulfment of oncogenic cells can give rise to the presentation of altered self-peptides, which may be presented to the immune system (Segal et al. 2008); however, the tolerogenic nature of cell clearance can dampen a subsequent immune response. Certain forms of chemotherapy and radiotherapy have been linked to the exposure or release of factors by dying cells that can initiate a robust immune response. For example, treatment of tumor cells with antracyclins led to cell-surface exposure of calreticulin early during the death program, which was essential for phagocytosis of dying tumor cells by dendritic cells and the initiation of an immune response (Obeid et al. 2007b). Additionally, calreticulin exposure was linked to immunogenic cell death caused by irradiation and UVC-light treatment (Obeid et al. 2007a). Uncleared, late apoptotic or necrotic cells with "leaky" plasma membranes can release damage-associated molecular-pattern molecules (DAMPs), such as heat shock proteins, high mobility group box 1 (HMGB-1) protein, and nucleic acids. These released intracellular molecules activate Toll-like receptors (TLRs) on phagocytes, facilitating antigen presentation and the initiation of an immune response (Peter et al. 2010; Kepp et al. 2011).

CONCLUDING REMARKS

Apoptotic cell death is an integral part of cell turnover in many tissues, and proper corpse clearance is vital to maintaining tissue homeostasis in all multicellular organisms. All living tissues have some mechanism(s) in place to handle corpse clearance, and most cell types (not just professional phagocytes) possess the ability to phagocytose apoptotic cells, underlining the relevance of this process in metazoan health.

Defects in clearance are believed to play a role in a wide variety of human pathologies including autoimmune diseases (systemic lupus erythematosus [SLE] and rheumatoid arthritis); pulmonary diseases (chronic obstructive pulmonary disease [COPD] and asthma); cardiovascular diseases (atherosclerosis); neurological diseases (Alzheimer's disease); infectious diseases (bacterial invasion by certain *Shigella*, *Yersinia*, and *Salmonella* species); and cancer (Elliott and Ravichandran 2010). Furthermore, the tolerogenic nature of apoptotic cell clearance has led to the emergence of potential new therapies using apoptotic cells for the prevention of transplant rejection and modulating autoimmune conditions such as type 1 diabetes (Castro et al. 2010; Morelli and Larregina 2010).

The field of apoptotic cell clearance is relatively young, and we are just beginning to better define the individual steps of this process and the consequences to the phagocyte, the tissue where it resides, and the organism. A better understanding of the sensing, recognition, engulfment, and processing of apoptotic cells will likely present new targets for therapy in the treatment of multiple immunological and metabolic diseases.

REFERENCES

Aderem A, Underhill DM. 1999. Mechanisms of phagocytosis in macrophages. *Annu Rev Immunol* **17**: 593–623.

Akakura S, Singh S, Spataro M, Akakura R, Kim JI, Albert ML, Birge RB. 2004. The opsonin MFG-E8 is a ligand for the $\alpha_v\beta_5$ integrin and triggers DOCK180-dependent Rac1 activation for the phagocytosis of apoptotic cells. *Exp Cell Res* **292**: 403–416.

Albert ML, Pearce SF, Francisco LM, Sauter B, Roy P, Silverstein RL, Bhardwaj N. 1998. Immature dendritic cells phagocytose apoptotic cells via $\alpha_v\beta_5$ and CD36, and cross-present antigens to cytotoxic T lymphocytes. *J Exp Med* **188**: 1359–1368.

Albert ML, Kim JI, Birge RB. 2000. $\alpha_v\beta_5$ integrin recruits the CrkII–Dock180–rac1 complex for phagocytosis of apoptotic cells. *Nat Cell Biol* **2**: 899–905.

Armstrong A, Ravichandran KS. 2011. Phosphatidylserine receptors: What is the new RAGE? *EMBO Rep* **12**: 287–288.

Arur S, Uche UE, Rezaul K, Fong M, Scranton V, Cowan AE, Mohler W, Han DK. 2003. Annexin I is an endogenous ligand that mediates apoptotic cell engulfment. *Dev Cell* **4**: 587–598.

Balasubramanian K, Schroit AJ. 2003. Aminophospholipid asymmetry: A matter of life and death. *Annu Rev Physiol* **65**: 701–734.

Bellone M, Iezzi G, Rovere P, Galati G, Ronchetti A, Protti MP, Davoust J, Rugarli C, Manfredi AA. 1997. Processing of engulfed apoptotic bodies yields T cell epitopes. *J Immunol* **159**: 5391–5399.

Blander JM, Medzhitov R. 2006. Toll-dependent selection of microbial antigens for presentation by dendritic cells. *Nature* **440**: 808–812.

Borisenko GG, Matsura T, Liu S-X, Tyurin VA, Jianfei J, Serinkan FB, Kagan VE. 2003. Macrophage recognition of externalized phosphatidylserine and phagocytosis of apoptotic Jurkat cells—existence of a threshold. *Arch Biochem Biophys* **413**: 41–52.

Bose J, Gruber AD, Helming L, Schiebe S, Wegener I, Hafner M, Beales M, Kontgen F, Lengeling A. 2004. The phosphatidylserine receptor has essential functions during embryogenesis but not in apoptotic cell removal. *J Biol* **3**: 15.

Bournazou I, Pound JD, Duffin R, Bournazos S, Melville LA, Brown SB, Rossi AG, Gregory CD. 2009. Apoptotic human cells inhibit migration of granulocytes via release of lactoferrin. *J Clin Invest* **119**: 20–32.

Bratton DL, Henson PM. 2008. Apoptotic cell recognition: Will the real phosphatidylserine receptor(s) please stand up? *Curr Biol* **18**: R76–R79.

Brown S, Heinisch I, Ross E, Shaw K, Buckley CD, Savill J. 2002. Apoptosis disables CD31-mediated cell detachment from phagocytes promoting binding and engulfment. *Nature* **418**: 200–203.

Brugnera E, Haney L, Grimsley C, Lu M, Walk SF, Tosello-Trampont AC, Macara IG, Madhani H, Fink GR, Ravichandran KS. 2002. Unconventional Rac-GEF activity is mediated through the Dock180–ELMO complex. *Nat Cell Biol* **4**: 574–582.

Castellano F, Montcourrier P, Chavrier P. 2000. Membrane recruitment of Rac1 triggers phagocytosis. *J Cell Sci* **113**: 2955–2961.

Castro CN, Barcala Tabarrozi AE, Noguerol MA, Liberman AC, Dewey RA, Arzt E, Morelli AE, Perone MJ. 2010. Disease-modifying immunotherapy for the management of autoimmune diabetes. *Neuroimmunomodulation* **17**: 173–176.

Chao MP, Alizadeh AA, Tang C, Jan M, Weissman-Tsukamoto R, Zhao F, Park CY, Weissman IL, Majeti R. 2011. Therapeutic antibody targeting of CD47 eliminates human acute lymphoblastic leukemia. *Cancer Res* **71**: 1374–1384.

Chekeni FB, Elliott MR, Sandilos JK, Walk SF, Kinchen JM, Lazarowski ER, Armstrong AJ, Penuela S, Laird DW, Salvesen GS, et al. 2010. Pannexin 1 channels mediate "find-me" signal release and membrane permeability during apoptosis. *Nature* **467**: 863–867.

Chen Y, Corriden R, Inoue Y, Yip L, Hashiguchi N, Zinkernagel A, Nizet V, Insel PA, Junger WG. 2006. ATP release guides neutrophil chemotaxis via P2Y2 and A3 receptors. *Science* **314**: 1792–1795.

Cote JF, Motoyama AB, Bush JA, Vuori K. 2005. A novel and evolutionarily conserved PtdIns(3,4,5)P3-binding domain is necessary for DOCK180 signalling. *Nat Cell Biol* **7**: 797–807.

deBakker CD, Haney LB, Kinchen JM, Grimsley C, Lu M, Klingele D, Hsu PK, Chou BK, Cheng LC, Blangy A, et al. 2004. Phagocytosis of apoptotic cells is regulated by a UNC-73/TRIO-MIG-2/RhoG signaling module and Armadillo repeats of CED-12/ELMO. *Curr Biol* **14**: 2208–2216.

Delamarre L, Pack M, Chang H, Mellman I, Trombetta ES. 2005. Differential lysosomal proteolysis in antigen-presenting cells determines antigen fate. *Science* **307**: 1630–1634.

Devitt A, Moffatt OD, Raykundalia C, Capra JD, Simmons DL, Gregory CD. 1998. Human CD14 mediates recognition and phagocytosis of apoptotic cells. *Nature* **392**: 505–509.

Elliott MR, Ravichandran KS. 2010. Clearance of apoptotic cells: Implications in health and disease. *J Cell Biol* **189**: 1059–1070.

Elliott MR, Chekeni FB, Trampont PC, Lazarowski ER, Kadl A, Walk SF, Park D, Woodson RI, Ostankovich M, Sharma P, et al. 2009. Nucleotides released by apoptotic cells act as a find-me signal to promote phagocytic clearance. *Nature* **461**: 282–286.

Erwig LP, Henson PM. 2008. Clearance of apoptotic cells by phagocytes. *Cell Death Differ* **15**: 243–250.

Ezekowitz RA, Sastry K, Bailly P, Warner A. 1990. Molecular characterization of the human macrophage mannose receptor: Demonstration of multiple carbohydrate recognition-like domains and phagocytosis of yeasts in Cos-1 cells. *J Exp Med* **172**: 1785–1794.

Fadok VA, Voelker DR, Campbell PA, Cohen JJ, Bratton DL, Henson PM. 1992. Exposure of phosphatidylserine on the surface of apoptotic lymphocytes triggers specific recognition and removal by macrophages. *J Immunol* **148**: 2207–2216.

Fadok VA, Bratton DL, Frasch SC, Warner ML, Henson PM. 1998a. The role of phosphatidylserine in recognition of apoptotic cells by phagocytes. *Cell Death Differ* **5**: 551–562.

Fadok VA, Warner ML, Bratton DL, Henson PM. 1998b. CD36 is required for phagocytosis of apoptotic cells by human macrophages that use either a phosphatidylserine receptor or the vitronectin receptor ($\alpha_v\beta_3$). *J Immunol* **161**: 6250–6257.

Fadok VA, de Cathelineau A, Daleke DL, Henson PM, Bratton DL. 2001. Loss of phospholipid asymmetry and surface exposure of phosphatidylserine is required for phagocytosis of apoptotic cells by macrophages and fibroblasts. *J Biol Chem* **276**: 1071–1077.

Florey O, Haskard DO. 2009. Sphingosine 1-phosphate enhances Fc γ receptor-mediated neutrophil activation and recruitment under flow conditions. *J Immunol* **183**: 2330–2336.

Franz S, Gaipl US, Munoz LE, Sheriff A, Beer A, Kalden JR, Herrmann M. 2006. Apoptosis and autoimmunity: When apoptotic cells break their silence. *Curr Rheumatol Rep* **8**: 245–247.

Frey B, Munoz LE, Pausch F, Sieber R, Franz S, Brachvogel B, Poschl E, Schneider H, Rodel F, Sauer R, et al. 2009. The immune reaction against allogeneic necrotic cells is reduced in Annexin A5 knock out mice whose macrophages display an anti-inflammatory phenotype. *J Cell Mol Med* **13**: 1391–1399.

Gardai SJ, McPhillips KA, Frasch SC, Janssen WJ, Starefeldt A, Murphy-Ullrich JE, Bratton DL, Oldenborg PA, Michalak M, Henson PM. 2005. Cell-surface calreticulin initiates clearance of viable or apoptotic cells through trans-activation of LRP on the phagocyte. *Cell* **123**: 321–334.

Geissmann F, Manz MG, Jung S, Sieweke MH, Merad M, Ley K. 2010. Development of monocytes, macrophages, and dendritic cells. *Science* **327**: 656–661.

Gerbod-Giannone MC, Li Y, Holleboom A, Han S, Hsu LC, Tabas I, Tall AR. 2006. TNFα induces ABCA1 through NF-κB in macrophages and in phagocytes ingesting apoptotic cells. *Proc Natl Acad Sci* **103**: 3112–3117.

Gordon S. 1999. Macrophage-restricted molecules: Role in differentiation and activation. *Immunol Lett* **65**: 5–8.

Green DR, Ferguson T, Zitvogel L, Kroemer G. 2009. Immunogenic and tolerogenic cell death. *Nat Rev Immunol* **9**: 353–363.

Gregory C. 2009. Cell biology: Sent by the scent of death. *Nature* **461**: 181–182.

Gregory CD, Devitt A, Moffatt O. 1998. Roles of ICAM-3 and CD14 in the recognition and phagocytosis of apoptotic cells by macrophages. *Biochem Soc Trans* **26**: 644–649.

Gude DR, Alvarez SE, Paugh SW, Mitra P, Yu J, Griffiths R, Barbour SE, Milstien S, Spiegel S. 2008. Apoptosis induces expression of sphingosine kinase 1 to release sphingosine-1-phosphate as a "come-and-get-me" signal. *FASEB J* **22**: 2629–2638.

Gumienny TL, Brugnera E, Tosello-Trampont AC, Kinchen JM, Haney LB, Nishiwaki K, Walk SF, Nemergut ME, Macara IG, Francis R, et al. 2001. CED-12/ELMO, a novel member of the CrkII/Dock180/Rac pathway, is required for phagocytosis and cell migration. *Cell* **107**: 27–41.

Hanayama R, Tanaka M, Miwa K, Shinohara A, Iwamatsu A, Nagata S. 2002. Identification of a factor that links apoptotic cells to phagocytes. *Nature* **417**: 182–187.

Hanayama R, Tanaka M, Miyasaka K, Aozasa K, Koike M, Uchiyama Y, Nagata S. 2004. Autoimmune disease and impaired uptake of apoptotic cells in MFG-E8-deficient mice. *Science* **304**: 1147–1150.

He M, Kubo H, Morimoto K, Fujino N, Suzuki T, Takahasi T, Yamada M, Yamaya M, Maekawa T, Yamamoto Y, et al. 2011. Receptor for advanced glycation end products binds to phosphatidylserine and assists in the clearance of apoptotic cells. *EMBO Rep* **12:** 358–364.

Helming L, Gordon S. 2009. Molecular mediators of macrophage fusion. *Trends Cell Biol* **19:** 514–522.

Henson PM. 2005. Dampening inflammation. *Nat Immunol* **6:** 1179–1181.

Henson PM, Hume DA. 2006. Apoptotic cell removal in development and tissue homeostasis. *Trends Immunol* **27:** 244–250.

Hochreiter-Hufford AE, Ravichandran KS. 2012. Oxygenated lipids: A mode to WiPE out inflammation? *Immunity* **36:** 699–701.

Hoeppner DJ, Hengartner MO, Schnabel R. 2001. Engulfment genes cooperate with ced-3 to promote cell death in *Caenorhabditis elegans*. *Nature* **412:** 202–206.

Huang FP, Platt N, Wykes M, Major JR, Powell TJ, Jenkins CD, MacPherson GG. 2000. A discrete subpopulation of dendritic cells transports apoptotic intestinal epithelial cells to T cell areas of mesenteric lymph nodes. *J Exp Med* **191:** 435–444.

Huppertz B, Bartz C, Kokozidou M. 2006. Trophoblast fusion: Fusogenic proteins, syncytins and ADAMs, and other prerequisites for syncytial fusion. *Micron* **37:** 509–517.

Hurwitz ME, Vanderzalm PJ, Bloom L, Goldman J, Garriga G, Horvitz HR. 2009. Abl kinase inhibits the engulfment of apoptotic [corrected] cells in *Caenorhabditis elegans*. *PLoS Biol* **7:** e99.

Huynh ML, Fadok VA, Henson PM. 2002. Phosphatidylserine-dependent ingestion of apoptotic cells promotes TGF-β1 secretion and the resolution of inflammation. *J Clin Invest* **109:** 41–50.

Jaiswal S, Jamieson CH, Pang WW, Park CY, Chao MP, Majeti R, Traver D, van Rooijen N, Weissman IL. 2009. CD47 is upregulated on circulating hematopoietic stem cells and leukemia cells to avoid phagocytosis. *Cell* **138:** 271–285.

Johann AM, Weigert A, Eberhardt W, Kuhn AM, Barra V, von Knethen A, Pfeilschifter JM, Brune B. 2008. Apoptotic cell-derived sphingosine-1-phosphate promotes HuR-dependent cyclooxygenase-2 mRNA stabilization and protein expression. *J Immunol* **180:** 1239–1248.

Kagan VE, Gleiss B, Tyurina YY, Tyurin VA, Elenstrom-Magnusson C, Liu SX, Serinkan FB, Arroyo A, Chandra J, Orrenius S, et al. 2002. A role for oxidative stress in apoptosis: Oxidation and externalization of phosphatidylserine is required for macrophage clearance of cells undergoing Fas-mediated apoptosis. *J Immunol* **169:** 487–499.

Kawane K, Fukuyama H, Kondoh G, Takeda J, Ohsawa Y, Uchiyama Y, Nagata S. 2001. Requirement of DNase II for definitive erythropoiesis in the mouse fetal liver. *Science* **292:** 1546–1549.

Kawane K, Fukuyama H, Yoshida H, Nagase H, Ohsawa Y, Uchiyama Y, Okada K, Iida T, Nagata S. 2003. Impaired thymic development in mouse embryos deficient in apoptotic DNA degradation. *Nat Immunol* **4:** 138–144.

Kawane K, Ohtani M, Miwa K, Kizawa T, Kanbara Y, Yoshioka Y, Yoshikawa H, Nagata S. 2006. Chronic polyarthritis caused by mammalian DNA that escapes from degradation in macrophages. *Nature* **443:** 998–1002.

Kepp O, Galluzzi L, Martins I, Schlemmer F, Adjemian S, Michaud M, Sukkurwala AQ, Menger L, Zitvogel L, Kroemer G. 2011. Molecular determinants of immunogenic cell death elicited by anticancer chemotherapy. *Cancer Metastasis Rev* **30:** 61–69.

Kim SJ, Gershov D, Ma X, Brot N, Elkon KB. 2002. I-PLA(2) activation during apoptosis promotes the exposure of membrane lysophosphatidylcholine leading to binding by natural immunoglobulin M antibodies and complement activation. *J Exp Med* **196:** 655–665.

Kim S, Elkon KB, Ma X. 2004. Transcriptional suppression of interleukin-12 gene expression following phagocytosis of apoptotic cells. *Immunity* **21:** 643–653.

Kinchen JM. 2010. A model to die for: Signaling to apoptotic cell removal in worm, fly and mouse. *Apoptosis* **15:** 998–1006.

Kinchen JM, Ravichandran KS. 2007. Journey to the grave: Signaling events regulating removal of apoptotic cells. *J Cell Sci* **120:** 2143–2149.

Kinchen JM, Ravichandran KS. 2008. Phagosome maturation: Going through the acid test. *Nat Rev Mol Cell Biol* **9:** 781–795.

Kinchen JM, Ravichandran KS. 2010. Identification of two evolutionarily conserved genes regulating processing of engulfed apoptotic cells. *Nature* **464:** 778–782.

Kinchen JM, Cabello J, Klingele D, Wong K, Feichtinger R, Schnabel H, Schnabel R, Hengartner MO. 2005. Two pathways converge at CED-10 to mediate actin rearrangement and corpse removal in *C. elegans*. *Nature* **434:** 93–99.

Kinchen JM, Doukoumetzidis K, Almendinger J, Stergiou L, Tosello-Trampont A, Sifri CD, Hengartner MO, Ravichandran KS. 2008. A pathway for phagosome maturation during engulfment of apoptotic cells. *Nat Cell Biol* **10:** 556–566.

Kiss RS, Elliott MR, Ma Z, Marcel YL, Ravichandran KS. 2006a. Apoptotic cells induce a phosphatidylserine-dependent homeostatic response from phagocytes. *Curr Biol* **16:** 2252–2258.

Kiss RS, Ma Z, Nakada-Tsukui K, Brugnera E, Vassiliou G, McBride HM, Ravichandran KS, Marcel YL. 2006b. The lipoprotein receptor-related protein-1 (LRP) adapter protein GULP mediates trafficking of the LRP ligand prosaposin, leading to sphingolipid and free cholesterol accumulation in late endosomes and impaired efflux. *J Biol Chem* **281:** 12081–12092.

Kitano M, Nakaya M, Nakamura T, Nagata S, Matsuda M. 2008. Imaging of Rab5 activity identifies essential regulators for phagosome maturation. *Nature* **453:** 241–245.

Kobayashi N, Karisola P, Pena-Cruz V, Dorfman DM, Jinushi M, Umetsu SE, Butte MJ, Nagumo H, Chernova I, Zhu B, et al. 2007. TIM-1 and TIM-4 glycoproteins bind phosphatidylserine and mediate uptake of apoptotic cells. *Immunity* **27:** 927–940.

Krahling S, Callahan MK, Williamson P, Schlegel RA. 1999. Exposure of phosphatidylserine is a general feature in the phagocytosis of apoptotic lymphocytes by macrophages. *Cell Death Differ* **6:** 183–189.

Krieser RJ, MacLea KS, Longnecker DS, Fields JL, Fiering S, Eastman A. 2002. Deoxyribonuclease IIα is required during the phagocytic phase of apoptosis and its loss causes perinatal lethality. *Cell Death Differ* **9:** 956–962.

Kuraishi T, Nakagawa Y, Nagaosa K, Hashimoto Y, Ishimoto T, Moki T, Fujita Y, Nakayama H, Dohmae N, Shiratsuchi A, et al. 2009. Pretaporter, a *Drosophila* protein serving as a ligand for Draper in the phagocytosis of apoptotic cells. *EMBO J* **28:** 3868–3878.

Landry YD, Denis M, Nandi S, Bell S, Vaughan AM, Zha X. 2006. ATP-binding cassette transporter A1 expression disrupts raft membrane microdomains through its ATPase-related functions. *J Biol Chem* **281:** 36091–36101.

Lauber K, Bohn E, Krober SM, Xiao YJ, Blumenthal SG, Lindemann RK, Marini P, Wiedig C, Zobywalski A, Baksh S, et al. 2003. Apoptotic cells induce migration of phagocytes via caspase-3-mediated release of a lipid attraction signal. *Cell* **113:** 717–730.

Lennon-Dumenil AM, Bakker AH, Maehr R, Fiebiger E, Overkleeft HS, Rosemblatt M, Ploegh HL, Lagaudriere-Gesbert C. 2002. Analysis of protease activity in live antigen-presenting cells shows regulation of the phagosomal proteolytic contents during dendritic cell activation. *J Exp Med* **196:** 529–540.

Leverrier Y, Ridley AJ. 2001. Requirement for Rho GTPases and PI 3-kinases during apoptotic cell phagocytosis by macrophages. *Curr Biol* **11:** 195–199.

Li W, Zou W, Zhao D, Yan J, Zhu Z, Lu J, Wang X. 2009. *C. elegans* Rab GTPase activating protein TBC-2 promotes cell corpse degradation by regulating the small GTPase RAB-5. *Development* **136:** 2445–2455.

Liu QA, Hengartner MO. 1998. Candidate adaptor protein CED-6 promotes the engulfment of apoptotic cells in *C. elegans*. *Cell* **93:** 961–972.

Lu M, Kinchen JM, Rossman KL, Grimsley C, deBakker C, Brugnera E, Tosello-Trampont AC, Haney LB, Klingele D, Sondek J, et al. 2004. PH domain of ELMO functions in *trans* to regulate Rac activation via Dock180. *Nat Struct Mol Biol* **11:** 756–762.

Lu M, Kinchen JM, Rossman KL, Grimsley C, Hall M, Sondek J, Hengartner MO, Yajnik V, Ravichandran KS. 2005. A steric-inhibition model for regulation of nucleotide exchange via the Dock180 family of GEFs. *Curr Biol* **15:** 371–377.

Manaka J, Kuraishi T, Shiratsuchi A, Nakai Y, Higashida H, Henson P, Nakanishi Y. 2004. Draper-mediated and phosphatidylserine-independent phagocytosis of apoptotic cells by *Drosophila* hemocytes/macrophages. *J Biol Chem* **279:** 48466–48476.

Miki H, Suetsugu S, Takenawa T. 1998. WAVE, a novel WASP-family protein involved in actin reorganization induced by Rac. *EMBO J* **17:** 6932–6941.

Miksa M, Amin D, Wu R, Ravikumar TS, Wang P. 2007. Fractalkine-induced MFG-E8 leads to enhanced apoptotic cell clearance by macrophages. *Mol Med* **13:** 553–560.

Miyanishi M, Tada K, Koike M, Uchiyama Y, Kitamura T, Nagata S. 2007. Identification of Tim4 as a phosphatidylserine receptor. *Nature* **450:** 435–439.

Monks J, Rosner D, Geske FJ, Lehman L, Hanson L, Neville MC, Fadok VA. 2005. Epithelial cells as phagocytes: Apoptotic epithelial cells are engulfed by mammary alveolar epithelial cells and repress inflammatory mediator release. *Cell Death Differ* **12:** 107–114.

Monks J, Smith-Steinhart C, Kruk ER, Fadok VA, Henson PM. 2008. Epithelial cells remove apoptotic epithelial cells during post-lactation involution of the mouse mammary gland. *Biol Reprod* **78:** 586–594.

Morelli AE, Larregina AT. 2010. Apoptotic cell-based therapies against transplant rejection: Role of recipient's dendritic cells. *Apoptosis* **15:** 1083–1097.

Mori K, Kanemura Y, Fujikawa H, Nakano A, Ikemoto H, Ozaki I, Matsumoto T, Tamura K, Yokota M, Arita N. 2002. Brain-specific angiogenesis inhibitor 1 (BAI1) is expressed in human cerebral neuronal cells. *Neurosci Res* **43:** 69–74.

Munoz LE, Franz S, Pausch F, Furnrohr B, Sheriff A, Vogt B, Kern PM, Baum W, Stach C, von Laer D, et al. 2007. The influence on the immunomodulatory effects of dying and dead cells of Annexin V. *J Leukoc Biol* **81:** 6–14.

Murugesan G, Sandhya Rani MR, Gerber CE, Mukhopadhyay C, Ransohoff RM, Chisolm GM, Kottke-Marchant K. 2003. Lysophosphatidylcholine regulates human microvascular endothelial cell expression of chemokines. *J Mol Cell Cardiol* **35:** 1375–1384.

Nagata S, Hanayama R, Kawane K. 2010. Autoimmunity and the clearance of dead cells. *Cell* **140:** 619–630.

Nakano T, Ishimoto Y, Kishino J, Umeda M, Inoue K, Nagata K, Ohashi K, Mizuno K, Arita H. 1997. Cell adhesion to phosphatidylserine mediated by a product of growth arrest-specific gene 6. *J Biol Chem* **272:** 29411–29414.

Nakaya M, Tanaka M, Okabe Y, Hanayama R, Nagata S. 2006. Opposite effects of Rho family GTPases on engulfment of apoptotic cells by macrophages. *J Biol Chem* **281:** 8836–8842.

Nordmann M, Cabrera M, Perz A, Brocker C, Ostrowicz C, Engelbrecht-Vandre S, Ungermann C. 2010. The Mon1–Ccz1 complex is the GEF of the late endosomal Rab7 homolog Ypt7. *Curr Biol* **20:** 1654–1659.

Obeid M, Panaretakis T, Joza N, Tufi R, Tesniere A, van Endert P, Zitvogel L, Kroemer G. 2007a. Calreticulin exposure is required for the immunogenicity of γ-irradiation and UVC light-induced apoptosis. *Cell Death Differ* **14:** 1848–1850.

Obeid M, Tesniere A, Ghiringhelli F, Fimia GM, Apetoh L, Perfettini JL, Castedo M, Mignot G, Panaretakis T, Casares N, et al. 2007b. Calreticulin exposure dictates the immunogenicity of cancer cell death. *Nat Med* **13:** 54–61.

Ogden CA, deCathelineau A, Hoffmann PR, Bratton D, Ghebrehiwet B, Fadok VA, Henson PM. 2001. C1q and mannose binding lectin engagement of cell surface calreticulin and CD91 initiates macropinocytosis and uptake of apoptotic cells. *J Exp Med* **194:** 781–795.

Oldenborg PA, Zheleznyak A, Fang YF, Lagenaur CF, Gresham HD, Lindberg FP. 2000. Role of CD47 as a marker of self on red blood cells. *Science* **288:** 2051–2054.

Park D, Tosello-Trampont AC, Elliott MR, Lu M, Haney LB, Ma Z, Klibanov AL, Mandell JW, Ravichandran KS. 2007. BAI1 is an engulfment receptor for apoptotic cells upstream of the ELMO/Dock180/Rac module. *Nature* **450:** 430–434.

Park SY, Jung MY, Kim HJ, Lee SJ, Kim SY, Lee BH, Kwon TH, Park RW, Kim IS. 2008a. Rapid cell corpse clearance by stabilin-2, a membrane phosphatidylserine receptor. *Cell Death Differ* **15:** 192–201.

Park SY, Kang KB, Thapa N, Kim SY, Lee SJ, Kim IS. 2008b. Requirement of adaptor protein GULP during Stabilin-2-mediated cell corpse engulfment. *J Biol Chem* **283:** 10593–10600.

Park SY, Kim SY, Jung MY, Bae DJ, Kim IS. 2008c. Epidermal growth factor-like domain repeat of Stabilin-2 recognizes phosphatidylserine during cell corpse clearance. *Mol Cell Biol* **28:** 5288–5298.

Park D, Hochreiter-Hufford A, Ravichandran KS. 2009. The phosphatidylserine receptor TIM-4 does not mediate direct signaling. *Curr Biol* **19:** 346–351.

Park D, Han CZ, Elliott MR, Kinchen JM, Trampont PC, Das S, Collins S, Lysiak JJ, Hoehn KL, Ravichandran KS. 2011. Continued clearance of apoptotic cells critically depends on the phagocyte Ucp2 protein. *Nature* **477:** 220–224.

Peter C, Waibel M, Radu CG, Yang LV, Witte ON, Schulze-Osthoff K, Wesselborg S, Lauber K. 2008. Migration to apoptotic "find-me" signals is mediated via the phagocyte receptor G2A. *J Biol Chem* **283:** 5296–5305.

Peter C, Wesselborg S, Herrmann M, Lauber K. 2010. Dangerous attraction: Phagocyte recruitment and danger signals of apoptotic and necrotic cells. *Apoptosis* **15:** 1007–1028.

Ravichandran KS. 2003. "Recruitment signals" from apoptotic cells: Invitation to a quiet meal. *Cell* **113:** 817–820.

Ravichandran KS. 2010. Find-me and eat-me signals in apoptotic cell clearance: Progress and conundrums. *J Exp Med* **207:** 1807–1817.

Ravichandran KS, Lorenz U. 2007. Engulfment of apoptotic cells: Signals for a good meal. *Nat Rev Immunol* **7:** 964–974.

Reddien PW, Horvitz HR. 2000. CED-2/CrkII and CED-10/Rac control phagocytosis and cell migration in *Caenorhabditis elegans*. *Nat Cell Biol* **2:** 131–136.

Reddien PW, Cameron S, Horvitz HR. 2001. Phagocytosis promotes programmed cell death in *C. elegans*. *Nature* **412:** 198–202.

Ridley AJ. 2001. Rho family proteins: Coordinating cell responses. *Trends Cell Biol* **11:** 471–477.

Riento K, Ridley AJ. 2003. Rocks: Multifunctional kinases in cell behaviour. *Nat Rev Mol Cell Biol* **4:** 446–456.

Rosen H, Goetzl EJ. 2005. Sphingosine 1-phosphate and its receptors: An autocrine and paracrine network. *Nat Rev Immunol* **5:** 560–570.

Santiago C, Ballesteros A, Martinez-Munoz L, Mellado M, Kaplan GG, Freeman GJ, Casasnovas JM. 2007. Structures of T cell immunoglobulin mucin protein 4 show a metal-ion-dependent ligand binding site where phosphatidylserine binds. *Immunity* **27:** 941–951.

Savill J. 1997. Apoptosis in resolution of inflammation. *J Leukoc Biol* **61:** 375–380.

Savill J, Dransfield I, Hogg N, Haslett C. 1990. Vitronectin receptor-mediated phagocytosis of cells undergoing apoptosis. *Nature* **343:** 170–173.

Savill J, Dransfield I, Gregory C, Haslett C. 2002. A blast from the past: Clearance of apoptotic cells regulates immune responses. *Nat Rev Immunol* **2:** 965–975.

Schlegel RA, Krahling S, Callahan MK, Williamson P. 1999. CD14 is a component of multiple recognition systems used by macrophages to phagocytose apoptotic lymphocytes. *Cell Death Differ* **6:** 583–592.

Schrijvers DM, De Meyer GR, Kockx MM, Herman AG, Martinet W. 2005. Phagocytosis of apoptotic cells by macrophages is impaired in atherosclerosis. *Arterioscler Thromb Vasc Biol* **25:** 1256–1261.

Segal NH, Parsons DW, Peggs KS, Velculescu V, Kinzler KW, Vogelstein B, Allison JP. 2008. Epitope landscape in breast and colorectal cancer. *Cancer Res* **68:** 889–892.

Segawa K, Suzuki J, Nagata S. 2011. Constitutive exposure of phosphatidylserine on viable cells. *Proc Natl Acad Sci* **108:** 19246–19251.

Simhadri VR, Andersen JF, Calvo E, Choi SC, Coligan JE, Borrego F. 2012. Human CD300a binds to phosphatidylethanolamine and phosphatidylserine, and modulates the phagocytosis of dead cells. *Blood* **119:** 2799–2809.

Su HP, Nakada-Tsukui K, Tosello-Trampont AC, Li Y, Bu G, Henson PM, Ravichandran KS. 2002. Interaction of CED-6/GULP, an adapter protein involved in engulfment of apoptotic cells with CED-1 and CD91/low density lipoprotein receptor-related protein (LRP). *J Biol Chem* **277:** 11772–11779.

Suzuki J, Umeda M, Sims PJ, Nagata S. 2010. Calcium-dependent phospholipid scrambling by TMEM16F. *Nature* **468:** 834–838.

Tosello-Trampont AC, Nakada-Tsukui K, Ravichandran KS. 2003. Engulfment of apoptotic cells is negatively regulated by Rho-mediated signaling. *J Biol Chem* **278:** 49911–49919.

Tosello-Trampont AC, Kinchen JM, Brugnera E, Haney LB, Hengartner MO, Ravichandran KS. 2007. Identification of two signaling submodules within the CrkII/ELMO/Dock180 pathway regulating engulfment of apoptotic cells. *Cell Death Differ* **14:** 963–972.

Truman LA, Ford CA, Pasikowska M, Pound JD, Wilkinson SJ, Dumitriu IE, Melville L, Melrose LA, Ogden CA, Nibbs R, et al. 2008. CX3CL1/fractalkine is released from apoptotic lymphocytes to stimulate macrophage chemotaxis. *Blood* **112:** 5026–5036.

Tsai RK, Discher DE. 2008. Inhibition of "self" engulfment through deactivation of myosin-II at the phagocytic synapse between human cells. *J Cell Biol* **180:** 989–1003.

Uderhardt S, Herrmann M, Oskolkova OV, Aschermann S, Bicker W, Ipseiz N, Sarter K, Frey B, Rothe T, Voll R, et al. 2012. 12/15-lipoxygenase orchestrates the clearance of apoptotic cells and maintains immunologic tolerance. *Immunity* **36:** 834–846.

van den Eijnde SM, van den Hoff MJ, Reutelingsperger CP, van Heerde WL, Henfling ME, Vermeij-Keers C, Schutte B, Borgers M, Ramaekers FC. 2001. Transient expression of phosphatidylserine at cell–cell contact areas is required for myotube formation. *J Cell Sci* **114:** 3631–3642.

Voll RE, Herrmann M, Roth EA, Stach C, Kalden JR, Girkontaite I. 1997. Immunosuppressive effects of apoptotic cells. *Nature* **390:** 350–351.

Weigert A, Tzieply N, von Knethen A, Johann AM, Schmidt H, Geisslinger G, Brune B. 2007. Tumor cell apoptosis polarizes macrophages role of sphingosine-1-phosphate. *Mol Biol Cell* **18:** 3810–3819.

Wu YC, Horvitz HR. 1998a. The *C. elegans* cell corpse engulfment gene *ced-7* encodes a protein similar to ABC transporters. *Cell* **93:** 951–960.

Wu YC, Horvitz HR. 1998b. *C. elegans* phagocytosis and cell-migration protein CED-5 is similar to human DOCK180. *Nature* **392:** 501–504.

Wu YC, Stanfield GM, Horvitz HR. 2000. NUC-1, a *Caenorhabditis elegans* DNase II homolog, functions in an intermediate step of DNA degradation during apoptosis. *Genes Dev* **14:** 536–548.

Wu Y, Singh S, Georgescu MM, Birge RB. 2005. A role for Mer tyrosine kinase in $\alpha_v\beta_5$ integrin-mediated phagocytosis of apoptotic cells. *J Cell Sci* **118:** 539–553.

Yu X, Odera S, Chuang CH, Lu N, Zhou Z. 2006. *C. elegans* Dynamin mediates the signaling of phagocytic receptor CED-1 for the engulfment and degradation of apoptotic cells. *Dev Cell* **10:** 743–757.

Zhou Z, Caron E, Hartwieg E, Hall A, Horvitz HR. 2001a. The *C. elegans* PH domain protein CED-12 regulates cytoskeletal reorganization via a Rho/Rac GTPase signaling pathway. *Dev Cell* **1:** 477–489.

Zhou Z, Hartwieg E, Horvitz HR. 2001b. CED-1 is a transmembrane receptor that mediates cell corpse engulfment in *C. elegans*. *Cell* **104:** 43–56.

Zou W, Lu Q, Zhao D, Li W, Mapes J, Xie Y, Wang X. 2009. *Caenorhabditis elegans* myotubularin MTM-1 negatively regulates the engulfment of apoptotic cells. *PLoS Genet* **5:** e1000679.

The Endolysosomal System in Cell Death and Survival

Urška Repnik[1], Maruša Hafner Česen[1], and Boris Turk[1,2]

[1]Department of Biochemistry and Molecular and Structural Biology, J. Stefan Institute, Jamova 39, SI-1000 Ljubljana, Slovenia

[2]Center of Excellence CIPKEBIP, Ljubljana, Slovenia

Correspondence: boris.turk@ijs.si

The endocytic pathway is a system specialized for the uptake of compounds from the cell microenvironment for their degradation. It contains an arsenal of hydrolases, including proteases, which are normally enclosed in membrane-bound organelles, but if released to the cytosol can initiate apoptosis signaling pathways. Endogenous and exogenous compounds have been identified that can mediate destabilization of lysosomal membranes, and it was shown that lysosomal proteases are not only able to initiate apoptotic signaling but can also amplify the apoptotic pathways initiated in other cellular compartments. The endocytic pathway also receives cargo destined for degradation via the autophagic pathway. By recycling energy and biosynthetic substrates, and by degrading damaged organelles and molecules, the endocytic system assists the autophagic system in resisting apoptotic stimuli. Steps leading to lysosomal membrane permeabilization and subsequent triggering of cell death as well as the therapeutic potential of intervention in lysosomal membrane permeabilization will be discussed.

Since the discovery of lysosomes in 1950s (de Duve et al. 1955), the concept of the endocytic pathway has changed. Although there has been huge progress in understanding the molecular mechanisms of targeting and fusion of organelles, several conceptual dilemmas have not been completely resolved. The primary function of the endocytic pathway is bulk degradation and recycling of the internalized material and redundant cellular components. Over the years, additional functions have been associated with it. Endosomes and lysosomes can fuse with the plasma membrane to repair it and to release the accumulated nondegradable material (Medina et al. 2011). Intraluminal vesicles are the source of exosomes, which have multiple functions, especially for the immune system (Ludwig and Giebel 2012). Endosomes have numerous functions in fighting infections: they can signal the presence of pathogens through Toll-like receptors, they are the site of antigenic peptide generation and their assembly with major histocompatibility complex class II molecules, and they can also kill residing pathogens (Gruenberg and van der Goot 2006). Because of a high content of proteases, de Duve (1959) coined the figurative term "suicide bags" for lysosomes, a concept since supported by a wealth of experimental reports (de Duve 1959). Perhaps the best examples of this concept are natural killer cells and cytotoxic T cells. Both have specialized lysosome-related organelles, secretory granules, that contain perforin and granzyme B, which can mediate apoptosis in the target cell (Blott and Griffiths 2002; Trapani and Smyth 2002). However, every cell can potentially become a victim of its own lysosomal hydrolases, especially if lysosomal membranes are destabilized so that the enzymes can escape into the cytosol. These offer great potential to exploit scenarios for therapy for certain diseases, most importantly cancer. On the other hand, by enabling degradation of the material sequestered by autophagy, the endocytic pathway can assist autophagy in counteracting apoptosis when cells are challenged with an apoptotic stimulus (Repnik and Turk 2010; Hafner Česen et al. 2012; Repnik et al. 2012).

ENDOCYTIC PATHWAY

Whenever we talk about lysosomes in the context of cell death and survival, we should more accurately refer also to other organelles of the endocytic pathway (see below for details; Fig. 1). Understanding the biology of the endocytic pathway is a prerequisite for the understanding of its role in the processes of cell death and survival.

We favor the concept proposed by Griffiths in 1996 that the endocytic pathway involves stable compartments connected by vesicular traffic as opposed to the maturation model, which implies that organelles are formed de novo and mature into the next organelle along the endocytic pathway. Griffiths defines a compartment as a complex, multifunctional membrane organelle that is specialized for a particular set of essential functions for the cell. According to this hypothesis, compartments cannot be assembled de novo, but arise only once during evolution and must be transferred from the mother cell to daughter cells during mitosis. In contrast, vesicles are considered to be transient organelles, simpler in composition, and are defined as membrane-enclosed containers that form de novo by budding from a preexisting compartment. Compartments can undergo homotypic fusion with themselves, whereas vesicles cannot fuse with themselves but only with a preexisting compartment. In contrast to compartments, vesicles can undergo maturation, which is a physiologically irreversible series of biochemical changes such that the particle develops into a biochemically different end stage. According to this hypothesis, early endosomes and late endosomes represent stable compartments in the endocytic pathway, while primary endocytic vesicles, phagosomes, multivesicular bodies (MVBs) or endosome carrier vesicles (ECVs), secretory granules, and even lysosomes represent vesicles (Griffiths 1996).

Endocytosis is an entrance route to the cell and is important for the uptake of nutrients, recycling of the plasma membrane, and surveillance of the extracellular environment. From the functional point of view, the endocytic pathway can be divided into *the recycling circuit* for the plasma membrane and its components, *the degradative system* for the degradation of molecules, and *the connecting system* for the transport of degradation-destined cargo from the early recycling circuit to the degradative system (Huotari and Helenius 2011). In Figure 1, the endocytic pathway is schematically presented (Griffiths 1996; Luzio et al. 2000; Eskelinen et al. 2011). The endocytic vesicle, which arises at the plasma membrane, most prominently from clathrin-coated pits and vesicles, first fuses with the early endosome, which is a major sorting compartment. A large part of the cargo and membranes internalized are recycled back to the plasma membrane through the recycling vesicles. Components that should be degraded are transported to the late endosome via MVBs, also referred to as ECVs. They are spherical vesicles and contain many small, membrane-enclosed intraluminal vesicles (ILVs). The formation of ILVs involves inward budding of the limiting membrane containing transmembrane ubiquitinated proteins and is regulated by ESCRT (endosomal sorting complex required for transport) machinery (Mayers and Audhya 2012). ILVs are present already in early endosomes and become particularly numerous in MVB/ECV, late endosomes, and hybrid organelles.

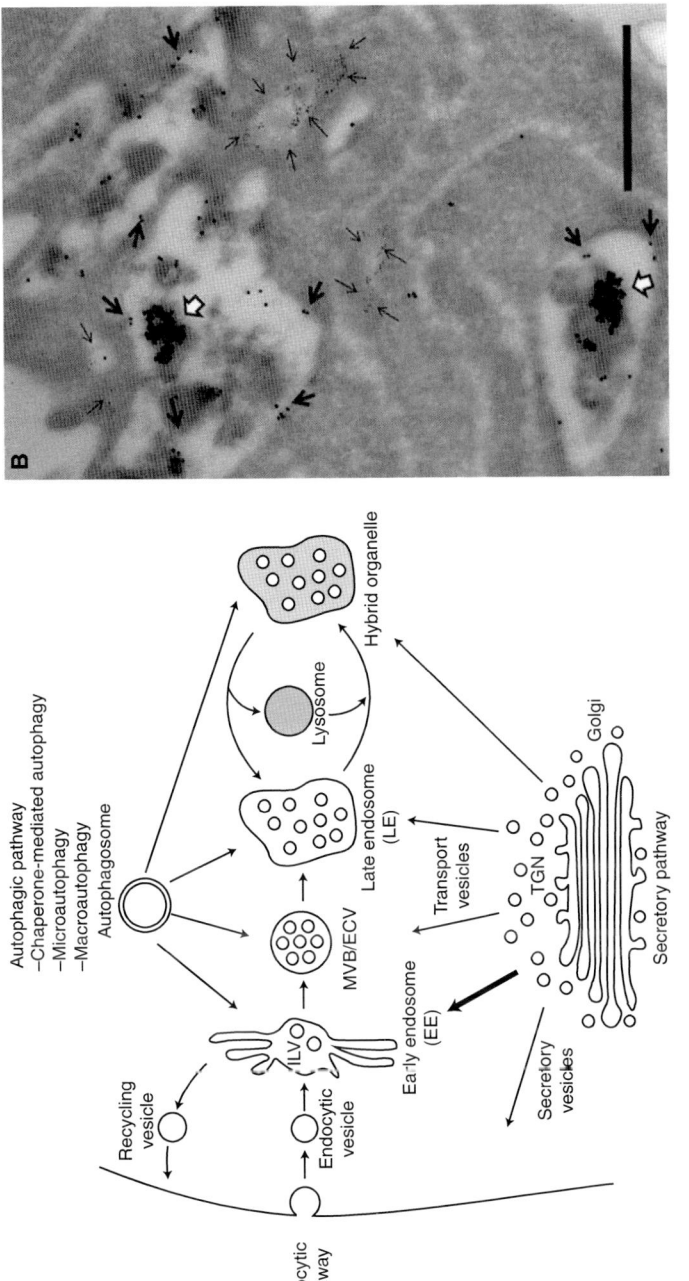

Figure 1. Endocytic pathway. (A) A schematic representation of the endocytic pathway (details are explained in the text). EE, early endosome; ILV, intraluminal vesicle; MVB, multivesicular body; ECV, endosome carrier vesicle; LE, late endosome; TGN, *trans*-Golgi network. (B) A transmission electron micrograph showing endocytic organelles in RAW cells, labeled with three different sizes of gold particles. Cells were pulsed with a 15-nm gold–bovine serum albumin (BSA) conjugate (block arrows) for 4 h and chased for 24 h to label LEs. During the last 6 min of culture, cells were fed with a 5-nm gold–BSA conjugate (thin arrows) to label endocytic vesicles and EEs. A thin Tokuyasu section was labeled with anti-LAMP2 antibody in a three-step reaction involving the primary antibody, a secondary rabbit anti-mouse antibody, and a 10-nm gold–protein A conjugate (thick arrows). Scale bar, 500 nm.

The late endosome represents the stomach of the cell. The enzymes needed for degradation are delivered by transport vesicles from the *trans*-Golgi network (TGN) to early and late endosomes. Most lysosomal hydrolases, including the proteases named cathepsins, acquire mannose 6-phospate during their transport through the Golgi complex and are recognized by mannose phosphate receptors in the TGN. In addition, some hydrolases, like sphingomyelinase, can use sortilin as an alternative sorting receptor (Saftig and Klumperman 2009). Arguably, lysosomes are vesicles that can store mature lysosomal enzymes and deliver them to a late endosomal compartment when needed. The resulting organelle is called the hybrid organelle or endolysosome. Lysosomes bud off the hybrid organelle in a process referred to as lysosome re-formation (Bright et al. 1997). It has recently been shown that lysosome re-formation depends on mTOR activation, which in turn depends on monomeric metabolites, which are produced during lysosomal degradation (Yu et al. 2010). It seems likely that lysosome re-formation serves to redistribute lysosomal enzymes (Griffiths 1996; Tjelle et al. 1996). It is also possible that for some enzymes transient escape from the harsh environment of late endosomes prolongs their stability.

Although the bulk of acid hydrolases have been localized in the lysosomes and not in the late endosomes, the latter are the main degradative compartment of the cell (Tjelle et al. 1996). More than 50 different hydrolases, including proteases, phosphatases, nucleases, glycosidases, sulphatases, lipases, etc., and some activator proteins that assist the hydrolysis of substrates, such as (pro)saposins, have been thus identified in these organelles. The luminal conditions of pH and perhaps the redox potential within the late endolysosomal system are designed to denature protein substrates and thereby make them susceptible to enzymatic degradation (Pillay et al. 2002). The pH thus decreases along the endocytic pathway: early endosomes have a pH between 6.8 and 6.1, late endosomes have a pH in the range of 6.0–4.8, and in lysosomes the pH can decrease to 4.5 (Maxfield and Yamashiro 1987). Because some lysosomal enzymes have pH optima in the neutral to alkaline range, it was suggested that the pH in lysosomes may actually fluctuate over a wider range, with a preference for a more acidic state (Butor et al. 1995). The role of redox potential inside lysosomes is more controversial. Initially it was proposed that it is reducing (Pillay et al. 2002); however, recent findings have shown that it is oxidizing and comparable to the endoplasmic reticulum (Austin et al. 2005).

In the limiting membrane there are transport proteins, which enable the translocation of ions or small metabolites into the cytosol. The integral membrane proteins also include structural glycoproteins LAMP (lysosome-associated membrane protein)-1, LAMP-2, CD63, and LIMP-2 (Schroder et al. 2010). LAMP isoforms are among the most abundant lysosomal proteins. They have large luminal domains, which are heavily glycosylated to protect the protein scaffold from the lysosomal enzymes. Although it was suggested that they provide structural integrity to the lysosomal membrane by forming a continuous protective carbohydrate lining along its inner leaflet, this idea lacks experimental support (Kundra and Kornfeld 1999). LAMPs have only short cytoplasmic tails, which are involved in vesicle trafficking along the microtubules (Huynh et al. 2007) and serve as receptors for chaperone-mediated autophagy (Cuervo and Dice 1996).

Late endosomes, lysosomes, and hybrid organelles are extremely dynamic organelles, and distinction between them is often difficult, even with electron microscopy analysis of immunolabeled thin sections, in particular in certain cell types including RAW and J774 macrophage cell lines and HeLa cells (Griffiths 1996). So far, specific markers for lysosomes that are absent in late endosomes have not been identified. LAMP molecules and acid hydrolases are found in both late endosomes and lysosomes. However, cation-independent mannose phosphate receptor (CI-MPR), Rab7, and the regulatory (RII) domain of the cAMP-dependent protein kinase are present only on late endosomes but not on lysosomes, offering some possibility for discrimination (Griffiths et al. 1988, 1990a,b; Tjelle et al. 1996). In electron micrographs late endosomes contain many ILV and membrane fragments, whereas lysosomes appear electron dense, rather homogeneous, and with fewer membrane fragments (Tjelle et al. 1996; Huotari and Helenius 2011); however, their discrimination is still not an easy task.

The degradative system of the endocytic pathway receives input for degradation not only from the cell microenvironment via endocytosis and phagocytosis, but also from the cell itself via the autophagic pathway. There are three main forms of autophagy: macroautophagy, characterized by the formation of double-membrane autophagosomes; chaperone-mediated autophagy, which depends on transmembrane receptors in the endosome/lysosome limiting membrane; and microautophagy, which includes internalization by invagination of the endosomal/lysosomal membrane (Cuervo 2004). Although it is often described that autophagosomes fuse with lysosomes, it has actually been shown that autophagosomes fuse with a LAMP-positive compartment before they acquire lysosomal proteases. It therefore seems likely that autophagosomes can enter the endocytic pathway at multiple stages, including early endosome, MVB/ECV, and, in particular, late endosomes and hybrid organelles (Fig. 1) (Eskelinen et al. 2011).

Of all lysosomal components, lysosomal proteases seem to be the main effector molecules that interface with the apoptotic machinery, if released into the cytosol. To initiate the lysosomal pathway of apoptosis, the degradative part of the endocytic pathway must therefore be targeted, which, however, includes not only lysosomes but also late endosomes and hybrid organelles, endolysosomes. Some hydrolases, particularly those with pH optima close to neutral, may be active already in the early endosomes. Most of the compounds that have lysosomal membrane–destabilizing properties, for example, lysosomotropic detergents, cannot specifically target a particular organelle, because differences in pH are too small (Fig. 2). Consequently, targeting the complex endolysosomal system will result in complex and often diverse cell responses. Therefore, we would like to emphasize that the use of the term "lysosome" is too narrow, as the complete degradative system of the endocytic pathway, which includes late endosomes, lysosomes, and hybrid organelles, harbors an arsenal of proteases that can initiate apoptotic signaling pathways if released into the cytosol. A more accurate alternative for the term "lysosome" is the term "endolysosomal system"; it also implies the inherent complexity of the system.

THE ROLE OF LYSOSOMES IN CELL DEATH

Apoptosis is the most investigated form of cell death and is characterized by caspase activation, chromatin condensation, nuclear fragmentation, and formation of apoptotic bodies. Alternative forms of cell death include necrosis, necroptosis, and secondary necrosis, which manifest with cytoplasmic swelling (oncosis), dilated organelles, loss of plasma membrane integrity, and the absence of caspase activation and chromatin condensation (Vanlangenakker et al. 2008; Vanden Berghe et al. 2010). Although necrosis has long been considered accidental and uncontrolled, it is now considered to be a well-orchestrated form of cell demise (Yamashima 2012). In necroptosis and secondary necrosis, lysosomal membrane permeabilization (LMP) is a late process, which coincides with the cellular disintegration phase and may, through proteolysis, contribute to the generation of modified damage-associated molecular patterns, which activate the immune system (Vanden Berghe et al. 2010). At least in some forms of necrosis, LMP is an early event that determines the necrotic cell death as a result of massive release of hydrolases (Yamashima 2012). However, LMP can also initiate or amplify apoptotic signaling. A quantitative relationship between the amount of lysosomal rupture and the mode of cell death has been suggested to explain different cell death pathways following LMP, so that moderate insults trigger a limited release of lysosomal contents into the cytosol and lead to apoptosis, whereas strong insults result in a complete release of lysosomal contents and lead to necrosis (Kagedal et al. 2001; Turk and Turk 2009).

Apoptotic cell death is mediated through two main signaling pathways: the extrinsic or de Duve C. 1959. *Subcellular particles* (ed. Hayashi T), pp. 128–159. The Ronald Press, New York.

de Duve C, Pressman BC, Gianetto R, Wattiaux R, Appelmans F. 1955. Tissue fractionation studies. 6. Intracellular distribution patterns of enzymes in rat-liver tissue. *Biochem J* **60:** 604–617.

de Duve C, de Barsy T, Poole B, Trouet A, Tulkens P, Van Hoof F. 1974. Lysosomotropic agents. *Biochem Pharmacol* **23:** 2495–2531.

Denamur S, Tyteca D, Marchand-Brynaert J, Van Bambeke F, Tulkens PM, Courtoy PJ, Mingeot-Leclercq MP. 2011. Role of oxidative stress in lysosomal membrane permeabilization and apoptosis induced by gentamicin, an aminoglycoside antibiotic. *Free Radic Biol Med* **51:** 1656–1665.

DiFranco KM, Gupta A, Galusha LE, Perez J, Nguyen TV, Fineza CD, Kachlany SC. 2012. Leukotoxin (Leukothera®) targets active leukocyte function antigen-1 (LFA-1) protein and triggers a lysosomal mediated cell death pathway. *J Biol Chem* **287:** 17618–17627.

Droga-Mazovec G, Bojic L, Petelin A, Ivanova S, Romih R, Repnik U, Salvesen GS, Stoka V, Turk V, Turk B. 2008. Cysteine cathepsins trigger caspase-dependent cell death through cleavage of Bid and antiapoptotic Bcl-2 homologues. *J Biol Chem* **283:** 19140–19150.

Eskelinen EL, Reggiori F, Baba M, Kovacs AL, Seglen PO. 2011. Seeing is believing: The impact of electron microscopy on autophagy research. *Autophagy* **7:** 935–956.

Fehrenbacher N, Bastholm L, Kirkegaard-Sorensen T, Rafn B, Bottzauw T, Nielsen C, Weber E, Shirasawa S, Kallunki T, Jaattela M. 2008. Sensitization to the lysosomal cell death pathway by oncogene-induced downregulation of lysosome-associated membrane proteins 1 and 2. *Cancer Res* **68:** 6623–6633.

Feofanov AV, Sharonov GV, Astapova MV, Rodionov DI, Utkin YN, Arseniev AS. 2005. Cancer cell injury by cytotoxins from cobra venom is mediated through lysosomal damage. *Biochem J* **390:** 11–18.

Filicko-O'Hara J, Grosso D, Flomenberg PR, Friedman TM, Brunner J, Drobyski W, Ferber A, Kakhniashvili I, Keever-Taylor C, Mookerjee B, et al. 2009. Antiviral responses following L-leucyl-L-leucine methyl ester (LLME)-treated lymphocyte infusions: Graft-versus-infection without graft-versus-host disease. *Biol Blood Marrow Transplant* **15:** 1609–1619.

Friedman TM, Filicko-O'Hara J, Mookerjee B, Wagner JL, Grosso DA, Flomenberg N, Korngold R. 2007. T cell repertoire complexity is conserved after LLME treatment of donor lymphocyte infusions. *Biol Blood Marrow Transplant* **13:** 1439–1447.

Gil-Parrado S, Fernandez-Montalvan A, Assfalg-Machleidt I, Popp O, Bestvater F, Holloschi A, Knoch TA, Auerswald EA, Welsh K, Reed JC, et al. 2002. Ionomycin-activated calpain triggers apoptosis. A probable role for Bcl-2 family members. *J Biol Chem* **277:** 27217–27226.

Gonzalez P, Mader I, Tchoghandjian A, Enzenmuller S, Cristofanon S, Basit F, Debatin KM, Fulda S. 2012. Impairment of lysosomal integrity by B10, a glycosylated derivative of betulinic acid, leads to lysosomal cell death and converts autophagy into a detrimental process. *Cell Death Differ* **19:** 1337–1346.

Griffiths G. 1996. On vesicles and membrane compartments. *Protoplasma* **195:** 37–58.

Griffiths G, Hoflack B, Simons K, Mellman I, Kornfeld S. 1988. The mannose 6-phosphate receptor and the biogenesis of lysosomes. *Cell* **52:** 329–341.

Griffiths G, Hollinshead R, Hemmings BA, Nigg EA. 1990a. Ultrastructural localization of the regulatory (RII) subunit of cyclic AMP-dependent protein kinase to subcellular compartments active in endocytosis and recycling of membrane receptors. *J Cell Sci* **96:** 691–703.

Griffiths G, Matteoni R, Back R, Hoflack B. 1990b. Characterization of the cation-independent mannose 6-phosphate receptor-enriched prelysosomal compartment in NRK cells. *J Cell Sci* **95:** 441–461.

Groth-Pedersen L, Jaattela M. 2010. Combating apoptosis and multidrug resistant cancers by targeting lysosomes. *Cancer Lett* doi: 10.1016/j.canlet.2010.05.021.

Gruenberg J, van der Goot FG. 2006. Mechanisms of pathogen entry through the endosomal compartments. *Nat Rev Mol Cell Biol* **7:** 495–504.

Gyrd-Hansen M, Farkas T, Fehrenbacher N, Bastholm L, Hoyer-Hansen M, Elling F, Wallach D, Flavell R, Kroemer G, Nylandsted J, et al. 2006. Apoptosome-independent activation of the lysosomal cell death pathway by caspase-9. *Mol Cell Biol* **26:** 7880–7891.

Hafner Česen M, Pegan K, Špes A, Turk B. 2012. Lysosomal pathways to cell death and their therapeutic applications. *Exp Cell Res* **318:** 1245–1251.

Hayashi MA, Nascimento FD, Kerkis A, Oliveira V, Oliveira EB, Pereira A, Radis-Baptista G, Nader HB, Yamane T, Kerkis I, et al. 2008. Cytotoxic effects of crotamine are mediated through lysosomal membrane permeabilization. *Toxicon* **52:** 508–517.

Heinrich M, Neumeyer J, Jakob M, Hallas C, Tchikov V, Winoto-Morbach S, Wickel M, Schneider-Brachert W, Trauzold A, Hethke A, et al. 2004. Cathepsin D links TNF-induced acid sphingomyelinase to Bid-mediated caspase-9 and -3 activation. *Cell Death Differ* **11:** 550–563.

Hornick JR, Vangveravong S, Spitzer D, Abate C, Berardi F, Goedegebuure P, Mach RH, Hawkins WG. 2012. Lysosomal membrane permeabilization is an early event in σ2 receptor ligand mediated cell death in pancreatic cancer. *J Exp Clin Cancer Res* **31:** 41.

Horvath I, Vigh L. 2010. Cell biology: Stability in times of stress. *Nature* **463:** 436–438.

Hou W, Han J, Lu C, Goldstein LA, Rabinowich H. 2010. Autophagic degradation of active caspase-8: A crosstalk mechanism between autophagy and apoptosis. *Autophagy* **6:** 891–900.

Huotari J, Helenius A. 2011. Endosome maturation. *EMBO J* **30:** 3481–3500.

Huynh KK, Eskelinen EL, Scott CC, Malevanets A, Saftig P, Grinstein S. 2007. LAMP proteins are required for fusion of lysosomes with phagosomes. *EMBO J* **26:** 313–324.

Ishisaka R, Utsumi K, Utsumi T. 2002. Involvement of lysosomal cysteine proteases in hydrogen peroxide-induced apoptosis in HL-60 cells. *Biosci Biotechnol Biochem* **66:** 1865–1872.

Kagedal K, Zhao M, Svensson I, Brunk UT. 2001. Sphingosine-induced apoptosis is dependent on lysosomal proteases. *Biochem J* **359**: 335–343.

Kagedal K, Johansson AC, Johansson U, Heimlich G, Roberg K, Wang NS, Jurgensmeier JM, Ollinger K. 2005. Lysosomal membrane permeabilization during apoptosis—Involvement of Bax? *Int J Exp Pathol* **86**: 309–321.

Kirkegaard T, Roth AG, Petersen NH, Mahalka AK, Olsen OD, Moilanen I, Zylicz A, Knudsen J, Sandhoff K, Arenz C, et al. 2010. Hsp70 stabilizes lysosomes and reverts Niemann-Pick disease-associated lysosomal pathology. *Nature* **463**: 549–553.

Kreuzaler P, Watson CJ. 2012. Killing a cancer: What are the alternatives? *Nat Rev Cancer* doi: 10.1038/nrc3264.

Kreuzaler PA, Staniszewska AD, Li W, Omidvar N, Kedjouar B, Turkson J, Poli V, Flavell RA, Clarkson RW, Watson CJ. 2011. Stat3 controls lysosomal-mediated cell death in vivo. *Nat Cell Biol* **13**: 303–309.

Kronke M. 1999. Biophysics of ceramide signaling: Interaction with proteins and phase transition of membranes. *Chem Phys Lipids* **101**: 109–121.

Kundra R, Kornfeld S. 1999. Asparagine-linked oligosaccharides protect Lamp-1 and Lamp-2 from intracellular proteolysis. *J Biol Chem* **274**: 31039–31046.

Kurz T, Terman A, Brunk UT. 2007. Autophagy, ageing and apoptosis: The role of oxidative stress and lysosomal iron. *Arch Biochem Biophys* **462**: 220–230.

Li W, Yuan X, Nordgren G, Dalen H, Dubowchik GM, Firestone RA, Brunk UT. 2000. Induction of cell death by the lysosomotropic detergent MSDH. *FEBS Lett* **470**: 35–39.

Lin Y, Epstein DL, Liton PB. 2010. Intralysosomal iron induces lysosomal membrane permeabilization and cathepsin D-mediated cell death in trabecular meshwork cells exposed to oxidative stress. *Invest Ophthalmol Vis Sci* **51**: 6483–6495.

Liu L, Zhang Z, Xing D. 2011. Cell death via mitochondrial apoptotic pathway due to activation of Bax by lysosomal photodamage. *Free Radic Biol Med* **51**: 53–68.

Ludwig AK, Giebel B. 2012. Exosomes: Small vesicles participating in intercellular communication. *Int J Biochem Cell Biol* **44**: 11–15.

Luzio JP, Rous BA, Bright NA, Pryor PR, Mullock BM, Piper RC. 2000. Lysosome-endosome fusion and lysosome biogenesis. *J Cell Sci* **113**: 1515–1524.

Maxfield FR, Yamashiro DJ. 1987. Endosome acidification and the pathways of receptor-mediated endocytosis. *Adv Exp Med Biol* **225**: 189–198.

Mayers JR, Audhya A. 2012. Vesicle formation within endosomes: An ESCRT marks the spot. *Commun Integr Biol* **5**: 50–56.

Medina DL, Fraldi A, Bouche V, Annunziata F, Mansueto G, Spampanato C, Puri C, Pignata A, Martina JA, Sardiello M, et al. 2011. Transcriptional activation of lysosomal exocytosis promotes cellular clearance. *Dev Cell* **21**: 421–430.

Melo FR, Lundequist A, Calounova G, Wernersson S, Pejler G. 2011. Lysosomal membrane permeabilization induces cell death in human mast cells. *Scand J Immunol* **74**: 354–362.

Miller DK, Griffiths E, Lenard J, Firestone RA. 1983. Cell killing by lysosomotropic detergents. *J Cell Biol* **97**: 1841–1851.

Oberle C, Huai J, Reinheckel T, Tacke M, Rassner M, Ekert PG, Buellesbach J, Borner C. 2010. Lysosomal membrane permeabilization and cathepsin release is a Bax/Bak-dependent, amplifying event of apoptosis in fibroblasts and monocytes. *Cell Death Differ* **17**: 1167–1178.

Ostenfeld MS, Fehrenbacher N, Hoyer-Hansen M, Thomsen C, Farkas T, Jaattela M. 2005. Effective tumor cell death by σ2 receptor ligand siramesine involves lysosomal leakage and oxidative stress. *Cancer Res* **65**: 8975–8983.

Pillay CS, Elliott E, Dennison C. 2002. Endolysosomal proteolysis and its regulation. *Biochem J* **363**: 417–429.

Pourahmad J, Hosseini MJ, Eskandari MR, Shekarabi SM, Daraei B. 2010. Mitochondrial/lysosomal toxic crosstalk plays a key role in cisplatin nephrotoxicity. *Xenobiotica* **40**: 763–771.

Ravikumar B, Berger Z, Vacher C, O'Kane CJ, Rubinsztein DC. 2006. Rapamycin pre-treatment protects against apoptosis. *Hum Mol Genet* **15**: 1209–1216.

Repnik U, Stoka V, Turk V, Turk B. 2012. Lysosomes and lysosomal cathepsins in cell death. *Biochim Biophys Acta* **1824**: 22–33.

Repnik U, Turk B. 2010. Lysosomal-mitochondrial crosstalk during cell death. *Mitochondrion* **10**: 662–669.

Rudolf E, Cervinka M. 2011. Sulforaphane induces cytotoxicity and lysosome- and mitochondria-dependent cell death in colon cancer cells with deleted p53. *Toxicol In Vitro* **25**: 1302–1309.

Saftig P, Klumperman J. 2009. Lysosome biogenesis and lysosomal membrane proteins: Trafficking meets function. *Nat Rev Mol Cell Biol* **10**: 623–635.

Sahara S, Yamashima T. 2010. Calpain-mediated Hsp70.1 cleavage in hippocampal CA1 neuronal death. *Biochem Biophys Res Commun* **393**: 806–811.

Schrader K, Huai J, Jockel L, Oberle C, Borner C. 2010. Non-caspase proteases: Triggers or amplifiers of apoptosis? *Cell Mol Life Sci* **67**: 1607–1618.

Schroder BA, Wrocklage C, Hasilik A, Saftig P. 2010. The proteome of lysosomes. *Proteomics* **10**: 4053–4076.

Stoka V, Turk B, Schendel SL, Kim TH, Cirman T, Snipas SJ, Ellerby LM, Bredesen D, Freeze H, Abrahamson M, et al. 2001. Lysosomal protease pathways to apoptosis. Cleavage of Bid, not pro-caspases, is the most likely route. *J Biol Chem* **276**: 3149–3157.

Terman A, Kurz T, Gustafsson B, Brunk UT. 2006. Lysosomal labilization. *IUBMB Life* **58**: 531–539.

Thelen M, Damme M, Schweizer M, Hagel C, Wong AM, Cooper JD, Braulke T, Galliciotti G. 2012. Disruption of the autophagy-lysosome pathway is involved in neuropathology of the *nclf* mouse model of neuronal ceroid lipofuscinosis. *PLoS ONE* **7**: e35493.

Thiele DL, Lipsky PE. 1990. Mechanism of L-leucyl-L-leucine methyl ester-mediated killing of cytotoxic lymphocytes: Dependence on a lysosomal thiol protease, dipeptidyl peptidase I, that is enriched in these cells. *Proc Natl Acad Sci* **87**: 83–87.

Tjelle TE, Brech A, Juvet LK, Griffiths G, Berg T. 1996. Isolation and characterization of early endosomes, late endosomes and terminal lysosomes: Their role in protein degradation. *J Cell Sci* **109**: 2905–2914.

Trapani JA, Smyth MJ. 2002. Functional significance of the perforin/granzyme cell death pathway. *Nat Rev Immunol* **2**: 735–747.

Turk B, Turk V. 2009. Lysosomes as "suicide bags" in cell death: Myth or reality? *J Biol Chem* **284**: 21783–21787.

Turk B, Stoka V, Rozman-Pungerčar J, Cirman T, Droga-Mazovec G, Orešič K, Turk V. 2002. Apoptotic pathways: Involvement of lysosomal proteases. *Biol Chem* **383**: 1035–1044.

Turk B, Turk D, Turk V. 2012. Protease signalling: The cutting edge. *EMBO J* **31**: 1630–1643.

Vanden Berghe T, Vanlangenakker N, Parthoens E, Deckers W, Devos M, Festjens N, Guerin CJ, Brunk UT, Declercq W, Vandenabeele P. 2010. Necroptosis, necrosis and secondary necrosis converge on similar cellular disintegration features. *Cell Death Differ* **17**: 922–930.

Vanlangenakker N, Vanden Berghe T, Krysko DV, Festjens N, Vandenabeele P. 2008. Molecular mechanisms and pathophysiology of necrotic cell death. *Curr Mol Med* **8**: 207–220.

Wang Y, Singh R, Massey AC, Kane SS, Kaushik S, Grant T, Xiang Y, Cuervo AM, Czaja MJ. 2008. Loss of macroautophagy promotes or prevents fibroblast apoptosis depending on the death stimulus. *J Biol Chem* **283**: 4766–4777.

Werneburg NW, Guicciardi ME, Bronk SF, Gores GJ. 2002. Tumor necrosis factor-α-associated lysosomal permeabilization is cathepsin B dependent. *Am J Physiol Gastrointest Liver Physiol* **283**: G947–G956.

Werneburg NW, Guicciardi ME, Bronk SF, Kaufmann SH, Gores GJ. 2007. Tumor necrosis factor-related apoptosis-inducing ligand activates a lysosomal pathway of apoptosis that is regulated by Bcl-2 proteins. *J Biol Chem* **282**: 28960–28970.

Werneburg NW, Bronk SF, Guicciardi ME, Thomas L, Dikeakos JD, Thomas G, Gores GJ. 2012. Tumor necrosis factor-related apoptosis inducing ligand (TRAIL) induced lysosomal translocation of proapoptotic effectors is mediated by phosphofurin acidic cluster sorting protein-2 (PACS-2). *J Biol Chem* **287**: 24427–24437.

Wilson PD, Firestone RA, Lenard J. 1987. The role of lysosomal enzymes in killing of mammalian cells by the lysosomotropic detergent N-dodecylimidazole. *J Cell Biol* **104**: 1223–1229.

Yamashima T. 2012. Hsp70.1 and related lysosomal factors for necrotic neuronal death. *J Neurochem* **120**: 477–494.

Yu L, McPhee CK, Zheng L, Mardones GA, Rong Y, Peng J, Mi N, Zhao Y, Liu Z, Wan F, et al. 2010. Termination of autophagy and reformation of lysosomes regulated by mTOR. *Nature* **465**: 942–946.

Yue XL, Lehri S, Li P, Barbier-Chassefiere V, Petit E, Huang QF, Albanese P, Barritault D, Caruelle JP, Papy-Garcia D, et al. 2009. Insights on a new path of pre-mitochondrial apoptosis regulation by a glycosaminoglycan mimetic. *Cell Death Differ* **16**: 770–781.

Zhao M, Brunk UT, Eaton JW. 2001. Delayed oxidant-induced cell death involves activation of phospholipase A2. *FEBS Lett* **509**: 399–404.

Zhao M, Antunes F, Eaton JW, Brunk UT. 2003. Lysosomal enzymes promote mitochondrial oxidant production, cytochrome c release and apoptosis. *Eur J Biochem* **270**: 3778–3786.

Metabolic Stress in Autophagy and Cell Death Pathways

Brian J. Altman and Jeffrey C. Rathmell

Department of Pharmacology and Cancer Biology, Department of Immunology, Sarah Stedman Nutrition and Metabolism Center, Duke University, Durham, North Carolina 27710

Correspondence: jeff.rathmell@duke.edu

Growth factors and oncogenic kinases play important roles in stimulating cell growth during development and transformation. These processes have significant energetic and synthetic requirements and it is apparent that a central function of growth signals is to promote glucose metabolism to support these demands. Because metabolic pathways represent a fundamental aspect of cell proliferation and survival, there is considerable interest in targeting metabolism as a means to eliminate cancer. A challenge, however, is that molecular links between metabolic stress and cell death are poorly understood. Here we review current literature on how cells cope with metabolic stress and how autophagy, apoptosis, and necrosis are tightly linked to cell metabolism. Ultimately, understanding of the interplay between nutrients, autophagy, and cell death will be a key component in development of new treatment strategies to exploit the altered metabolism of cancer cells.

Although single-celled organisms grow and proliferate based on nutrient availability, metazoan cells rely on growth factor input to promote nutrient uptake, regulate growth and proliferation, and survive (Raff 1992; Rathmell et al. 2000). Access and competition for these signals are critical in developmental patterning and to maintain homeostasis of mature tissues. Cells that do not receive proper growth factor signals typically atrophy, lose the ability to uptake and use extracellular nutrients, and instead induce the self-digestive process of autophagy as an intracellular energy source before ultimately undergoing programmed cell death. Cancer cells, in contrast, often become independent of extracellular growth signals by gaining mutations or expressing oncogenic kinases to drive intrinsic growth signals that mimic growth factor input, which can be the source of oncogene addiction. Growth factor input or oncogenic signals often drive highly elevated glucose uptake and metabolism (Rathmell et al. 2000; DeBerardinis et al. 2008; Michalek and Rathmell 2010). First described in cancer by Warburg in the 1920s, this highly glycolytic metabolic program is termed aerobic glycolysis and is a general feature of many nontransformed proliferative cells (Warburg 1956; DeBerardinis et al. 2008).

Nutrient uptake and aerobic glycolysis induced by growth signals play key roles in cell survival (Vander Heiden et al. 2001). Manipulating cell metabolism as a means to promote the death of inappropriately dividing cells, therefore, is a promising new avenue to treat disease.

Targeting the altered metabolism of cancer cells in particular is of great interest. It is still unclear at the molecular level, however, how inhibiting or modulating cell metabolism leads to apoptosis, and how these pathways may best be exploited (Dang et al. 2009; Wise and Thompson 2010).

Growth factor or oncogenic kinases promote multiple metabolic pathways that are essential to prevent metabolic stress and may be targets in efforts to link metabolism and cell death (Vander Heiden et al. 2001). Decreased glucose metabolism on loss of growth signals leads to decreased ATP generation as well as loss in generation of many biosynthetic precursor molecules, including nucleic acids, fatty acids, and acetyl-CoA for acetylation (Zhao et al. 2007; Wellen et al. 2009; Coloff et al. 2011). Glucose is also important as a precursor for the hexosamine pathway, to allow proper glycosylation and protein folding in the endoplasmic reticulum (Dennis et al. 2009; Kaufman et al. 2010). If glucose metabolism remains insufficient or disrupted, the cells can switch to rely on mitochondrial oxidation of fatty acids and amino acids, which are energy rich but do not readily support cell growth and can lead to potentially dangerous levels of reactive oxygen species (Wellen and Thompson 2010). Amino acid deficiency can directly inhibit components of the signaling pathways downstream from growth factors and activate autophagy (Lynch 2001; Beugnet et al. 2003; Byfield et al. 2005; Nobukuni et al. 2005). Finally, hypoxia induces a specific pathway to increase nutrient uptake and metabolism via the hypoxia-inducible factor (HIF1/2α) that promotes adaptation to anaerobic conditions, but may lead to apoptosis if hypoxia is severe (Saikumar et al. 1998; Suzuki et al. 2001; Fulda and Debatin 2007).

Typically a combination of metabolic stresses rather than loss of a single nutrient input occur at a given time (Degenhardt et al. 2006) and autophagy is activated to mitigate damage and provide nutrients for short-term survival (Bernales et al. 2006; Tracy et al. 2007; Altman et al. 2011; Guo et al. 2011). Autophagy is a cellular process of bulk cytoplasmic and organelle degradation common to nearly all eukaryotes. Unique double-membraned vesicles known as autophagosomes engulf cellular material and fuse with lysosomes to promote degradation of the contents (Kelekar 2005). Described in greater detail below, autophagy can reduce sources of stress, such as protein aggregates and damaged or dysfunctional intracellular organelles, and provide nutrients during times of transient and acute nutrient withdrawal.

Despite the protective effects of autophagy, cells deprived of growth signals, nutrients, or oxygen for prolonged times will eventually succumb to cell death. Apoptosis is the initial death response on metabolic stress and is regulated by Bcl-2 family proteins. In healthy cells, antiapoptotic Bcl-2 family proteins, such as Bcl-2, Bcl-xl, and Mcl-1, bind and inhibit the multidomain proapoptotic proteins Bax and Bak (van Delft and Huang 2006; Walensky 2006; Chipuk et al. 2010). In metabolic stress, proapoptotic "BH3-only" proteins of the Bcl-2 family are induced or activated and bind to and inhibit the antiapoptotic Bcl-2 family proteins to allow activation of the proapoptotic Bax and Bak (Galonek and Hardwick 2006). The BH3-only proteins Bim, Bid, and Puma can also directly bind and activate Bax and Bak (Letai et al. 2002; Ren et al. 2010). Active Bax and Bak disrupt the outer mitochondrial membrane (termed mitochondrial outer-membrane permeabilization, or MOMP) and release several proapoptotic factors including cytochrome-C that activate the apoptosome that in turn activates effector caspases to cleave a variety of cellular proteins and drive apoptosis (Schafer and Kornbluth 2006). In cases in which these apoptotic pathways are suppressed, metabolic stress can instead lead to necrotic cell death (Jin et al. 2007).

ROLE OF GROWTH FACTORS IN METABOLISM AND CELL DEATH

The phosphatidylinositol-3-kinase (PI3K)/AKT/mTOR pathway, outlined in Figure 1, is central in growth factor-stimulated control of glucose uptake and metabolism and is often altered in cancer. PI3K catalyzes the phosphorylation of phosphatidylinositol (PI) 4,5 bisphosphate (PIP2) to PI 3,4,5 triphosphate (PIP3). After

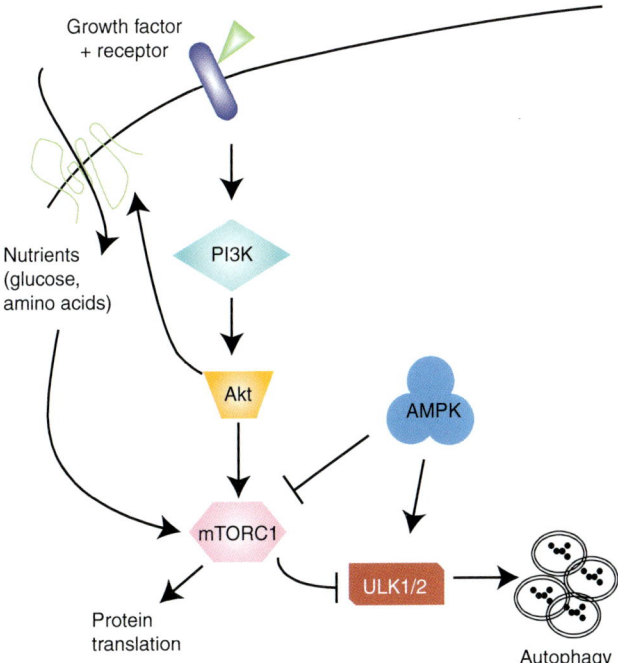

Figure 1. The PI3K-Akt-mTORC1 pathway and control of autophagy. PI3K is induced downstream from growth factor input and activates Akt and mTORC1. Akt supports the localization of nutrient transporters to the cell surface and maintains nutrient uptake. mTORC1 both supports protein translation and inhibits autophagy through phosphorylation of ULK1 and ULK2. AMPK, activated under energetic stress, inhibits mTORC1 and directly associates with ULK1/2 to activate autophagy.

growth factor signals are received, PIP3 accumulates, leading to recruitment and phosphorylation of Akt (Frech et al. 1997; Zoncu et al. 2011). Activated Akt supports metabolism in a number of ways, notably by maintaining protein translation, glucose metabolism, and inhibiting autophagy and apoptosis. Akt is essential for growth factor-stimulated glucose uptake and glycolysis by localizing Glut1 to the cell surface (Plas et al. 2001; Vander Heiden et al. 2001; Wieman et al. 2007; Wofford et al. 2008) and also increasing the activity and proper localization of hexokinases (HKs) (Gottlob et al. 2001; Robey and Hay 2006) to support the phosphorylation and downstream catabolism of glucose. Akt can also promote glucose metabolism through the pentose phosphate pathway (PPP), which generates NADPH and ribonucleotides to control cellular redox as well as lipid and nucleic acid synthesis (Rathmell et al. 2003; Duvel et al. 2010). In addition, Akt suppresses the catabolism of intracellular components such as fatty acids that would otherwise be used to support cell growth (Deberardinis et al. 2006). Thus, Akt activation promotes glucose metabolism and inhibits other metabolic pathways, leading to a glycolytic phenotype in growth factor-stimulated cells and the aerobic glycolysis characteristic of cancer cells (Elstrom et al. 2004). Importantly, Akt is highly antiapoptotic, but requires glucose to protect cells from death (Vander Heiden et al. 2001; Rathmell et al. 2003; Coloff et al. 2011), highlighting the connections between metabolism and apoptosis and enforcing a glucose addiction on stimulated or cancerous cells.

mTORC1 (mechanistic target of rapamycin complex 1) is activated downstream from Akt and is directly regulated by amino acid availability (Lynch 2001; Beugnet et al. 2003; Roccio et al. 2006; Kim et al. 2008; Sancak et al. 2010) to control translation and autophagy (Zoncu et al. 2011). mTORC1 regulates cell growth by promoting protein translation through phosphorylation of two key enzymes. Protein synthesis inhibitor eIF4E binding protein (4EBP) is inactivated by mTORC1 to allow for cap-dependent translation, and S6 kinase (S6K) is phosphorylated and activated by mTORC1 to activate the S6-ribisomal protein that appears to be important for cell metabolism and growth (Hara et al. 1998; Fingar et al. 2002; Tandon et al. 2011). mTORC1 also negatively regulates autophagy through inhibitory phosphorylation of the autophagy-essential kinases Unc-51-like kinase-1 (ULK1) and ULK2 (Ganley et al. 2009; Jung et al. 2009). In addition, mTOR signaling can be suppressed and autophagy induced by the AMP-activated protein kinase (AMPK) (Hoyer-Hansen and Jaattela 2007a; Canto and Auwerx 2010).

Many oncogenes can mimic growth signals to support metabolism in the absence of exogenous growth factor signaling (Vander Heiden et al. 2009). The PI3K/Akt/mTOR pathway plays a major role downstream from these oncogenic mutations to drive aerobic glycolysis. The BCR-Abl fusion protein, for example, promotes trafficking of Glut1 to the cell surface. Inhibition of PI3K, however, blocks this function and Glut1 is internalized thus limiting glucose uptake and glycolysis (Barnes et al. 2005). Myc can directly promote transcription of essentially all key glycolytic genes (Osthus et al. 2000; Dang et al. 2009). Indeed, recent studies have shown that c-Myc and Akt have complementary effects in promoting aerobic glycolysis (Fan et al. 2010). In addition to glucose, proliferating cells also require glutamine (Wise et al. 2008; Dang et al. 2009) and c-Myc has been shown to drive glutamine uptake and metabolism to render cells dependent on this important nutrient to support energy generation and biosynthesis (Wise et al. 2008; Gao et al. 2009).

AUTOPHAGY AND CONTROL OF METABOLIC STRESS

Decreased metabolism leads to induction of autophagy to generate nutrients from intracellular components until proper extracellular nutrient uptake is restored. Autophagy can play a prodeath role when prolonged or in certain developmental conditions, such as elimination of blastocyst inner cell mass (see review by Das et al. 2012), but in most circumstances autophagic generation of nutrients prevents or delays cell death. In autophagy, the cell packages organelles, bulk cytoplasm, and long-lived proteins in double-membraned vesicles for delivery to the lysosome and eventual degradation (Kelekar 2005). Autophagy occurs in all metazoan cells at low levels under basal conditions as a quality-control and waste-disposal mechanism (Ganley et al. 2009; Jung et al. 2009), but is induced as a protective mechanism when cells are under stress (Fig. 2).

Induction of autophagy can relieve a variety of cell stresses (Fig. 2). Autophagy can prevent overaccumulation of mitochondria and remove damaged mitochondria to prevent cell death (Colell et al. 2007; Pua et al. 2009) and to prevent an increase in damaging reactive oxygen species (ROS) (Mathew et al. 2009; Rouschop et al. 2009; Tal et al. 2009). Although some amounts of ROS are required for normal cell signaling (Hamanaka and Chandel 2010), excess ROS can lead to cellular damage, p53 activation, induction of proapoptotic proteins, and cell death in a wide variety of tissues (von Harsdorf et al. 1999; Sade and Sarin 2004; Karawajew et al. 2005; Liu et al. 2008a; Niizuma et al. 2009; Bodet et al. 2010). Autophagy is also important to engulf damaged endoplasmic reticulum (ER) in the unfolded protein response (Bernales et al. 2006; Hoyer-Hansen and Jaattela 2007b) and degrade protein aggregates that may otherwise lead to neurodegeneration (Bjorkoy et al. 2005; Komatsu et al. 2006). Finally, autophagy is essential to limit DNA damage and genomic instability, possibly through modulation of protein aggregates and the adaptor protein p62 (Mathew et al. 2007, 2009). As a consequence, autophagy-deficient cancer cells have increased genomic damage that can paradoxically promote tumor progression (Liang et al. 1999; Yue et al. 2003). This DNA damage may lead to a p53-dependent stress response, and p53 deficiency can promote continued proliferation despite insufficient autophagy and DNA damage (Altman et al. 2011).

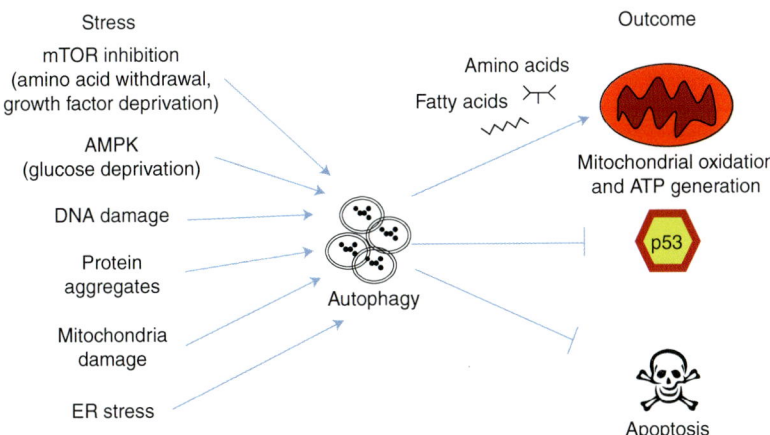

Figure 2. Autophagy responds to a variety of stresses to prevent cell death. Autophagy can be induced by a variety of nutrient stresses and other cellular insults. Induced autophagy can provide nutrients for mitochondrial oxidation, suppress p53 pathway activation, and ultimately delay apoptosis.

As discussed above, growth factor-mediated activation of the PI3K-Akt-mTORC1 pathway suppresses autophagy in healthy cells in nutrient-replete conditions. mTORC1 inhibits autophagy by phosphorylating and inhibiting ULK1 and ULK2, key upstream regulators of autophagy (Fig. 1) (Ganley et al. 2009; Jung et al. 2009). Growth factor or nutrient withdrawal (particularly withdrawal of amino acids) inactivates mTORC1 and activates autophagy (Beugnet et al. 2003; Altman et al. 2009). The program of autophagy is performed by a complex pathway of Atg-family proteins and two ubiquitinlike conjugation systems, reviewed in detail in Das et al. (2012). In addition to amino acid-sensitive regulation of autophagy by mTORC1, decreases in ATP resulting from nutrient deprivation can activate the AMPK pathway to promote autophagy. The AMPK heterotrimer is maximally activated when both AMP (adenyl monophosphate) increases and the AMPKα subunit is phosphorylated by the tumor suppressor LKB1 (Hawley et al. 2003; Woods et al. 2003; Shaw et al. 2004; Shaw 2009). AMPK then initiates a cellular program to conserve and generate additional ATP by initiating a G1 cell-cycle arrest, increasing glucose uptake, glycolysis, fatty acid oxidation, and halting protein synthesis and glycogen synthesis (Imamura et al. 2001; Jones et al. 2005; Shaw 2009). AMPK can phosphorylate and activate TSC2 to inhibit mTORC1 and thus indirectly activate autophagy to degrade damaged organelles and mobilize intracellular nutrients (Hoyer-Hansen and Jaattela 2007a; Canto and Auwerx 2010). In addition, several recent studies in *Caenorhabditis elegans* and mammalian cells have shown that AMPK directly associates and phosphorylates ULK1 in response to glucose deprivation or multinutrient deprivation, and that this interaction is essential for induction of autophagy (Fig. 1) (Egan et al. 2011; Kim et al. 2011; Shang et al. 2011).

Autophagic degradation of mitochondria and other organelles, protein aggregates, cytoplasm, and long-lived proteins can yield significant energy to cells (Kelekar 2005). Autophagosomes fuse with lysosomes to digest the contents and release amino acids and free fatty acids back to the cytosol for mitochondrial oxidation (Fig. 2) (Altman et al. 2011; Guo et al. 2011). This metabolic strategy can be highly efficient to maintain bioenergetics for a potential long period, as complete catabolism of a single molecule of palmitate, a long-chain fatty acid that is often a component of the phospholipid bilayer of membranes, can yield up to 104 ATP, compared to 31 from a single molecule of glucose (Salway 2004). Likewise, metabolism of amino acids can also yield large amounts of ATP.

Although a role for autophagy as a nutrient source has been apparent in a variety of genetic experiments using yeast (Abeliovich and Klionsky 2001), direct biochemical evidence for autophagy in mammalian cell metabolism has emerged only recently. Lum et al. (2005) showed that autophagy eventually becomes critical as a nutrient source for survival after growth factor deprivation of Bax/Bak-deficient cells. Most interestingly, this cell death caused by autophagy inhibition could be rescued by addition of exogenous nutrients, strongly indicating that autophagy was functioning as a nutrient-generating process. Similarly, Boya et al. (2005) showed that autophagy was critical for survival under conditions of nutrient limitation. More recently, we examined the potential of autophagy to support metabolism by mass spectrometry measurement of metabolite levels in autophagy-deficient cells. These data highlighted the ability of autophagy to supply growth factor or nutrient-deprived cells with an intracellular source of long-chain fatty acids (Altman et al. 2009, 2011). Likewise, Singh et al. (2009) showed that autophagy is critical in regulating lipid metabolism. This ability of autophagy to produce nutrients can be critical to allow cell survival in conditions of low exogenous nutrients (Degenhardt et al. 2006; Karantza-Wadsworth et al. 2007; Mathew et al. 2007; Altman et al. 2011).

The ability of autophagy to both control cell stress and produce nutrients has led to the recent observation that some cancers may be "addicted" to autophagy. We found that autophagy may suppress cell stress and a p53-dependent pathway of cell death, and that cells transformed by the BCR-Abl oncogene were dependent on autophagy for survival and leukemogenesis (Altman et al. 2011). Similarly, Guo et al. (2011) recently showed that autophagy is necessary for survival and can support mitochondrial metabolism of Ras-transformed cells under conditions of nutrient starvation. In Ras-transformed breast epithelial cells, Lock et al. showed that autophagy increases glucose flux through glycolysis to support survival of matrix-detached cells (Lock et al. 2011). These data all suggest that autophagy can prevent death in the absence of nutrients in part by supporting metabolism from breakdown of intracellular components and may play a broad role in cellular homeostasis and control of stress.

Despite these effects to relieve stress, autophagy induction can in some cases lead to apoptosis rather than protection from cell death. In *Drosophila*, Scott et al. (2007) showed that enforced expression of Atg1, the homolog of ULK1/2, led to apoptosis. Autophagy contributes to cell death of several different tissue types during *Drosophila* development, including the salivary gland, midgut, and reproductive cells (Berry and Baehrecke 2007; Hou et al. 2008; Denton et al. 2009; Nezis et al. 2010). We observed in several mammalian hematopoietic lines that autophagy induction in the absence of growth factor initially provided nutrients but eventually led to apoptosis dependent on direct induction of the proapoptotic protein Bim (Altman et al. 2009). Similarly, Kiyono et al. (2009) recently observed that autophagy induced in hepatocellular carcinoma cells by TGFβ led to induction of Bim and the proapoptotic protein BMF and eventual apoptosis. Finally, other groups showed that autophagy was necessary for apoptosis of rat neurons treated with a glutamate receptor agonist (Wang et al. 2008a) and for apoptosis downstream from death receptor signaling mouse embryonic fibroblasts (Wang et al. 2008b). However, the molecular mechanisms governing the decision of autophagy to act as a cytoprotective process or to induce apoptosis are poorly understood.

BEYOND AUTOPHAGY: METABOLIC STRESS AND APOPTOSIS

When autophagy is unable to provide sufficient additional nutrients or mitigate cell stress, insufficient metabolism can induce apoptosis. Links between metabolism and apoptosis are now widely appreciated in many species, and glucose metabolism has been shown to play a direct role in regulation apoptosis. In *Xenopus* oocytes, nutrient depletion over time leads to reduced NADPH and activation of caspase-2 to induce apoptosis (Nutt et al. 2005, 2009). Likewise, loss of growth signals and diminished Akt/mTOR signaling in mammalian cells leads to internalization of Glut1 that ultimately leads to apoptosis (Vander Heiden et al. 2001; Wieman et al. 2007; Wofford et al. 2008). If Glut1 is overexpressed, glucose uptake can be maintained even after growth factor deprivation, allowing growth factor-independent glucose metabolism that is sufficient to delay apoptosis (Rathmell et al. 2003; Zhao et al. 2007).

Disruption of metabolism leads to a proapoptotic balance of Bcl-2 family proteins that is essential for apoptosis (Fig. 3). If expression of the proapoptotic Bcl-2 family proteins Puma, Bim, or Noxa is decreased, cells can persist for extended periods even in the absence of glucose (Alves et al. 2006; Coloff et al. 2011). Induction or activation of some proapoptotic BH3-only members of the Bcl-2 family in particular are sensitive to metabolic status. Puma is induced by inhibition of glucose metabolism or disruption in glucose availability in a pathway partially dependent on p53-mediated transcription (Zhao et al. 2008; Coloff et al. 2011). Glucose metabolism also regulates Puma protein stability, as loss of glucose stabilized the Puma protein to enhance apoptosis, whereas addition of exogenous nutrients to support mitochondrial metabolism caused Puma degradation and enhanced cell survival (Coloff et al. 2011). Bim expression is also increased when glucose metabolism is blocked, although this may be owing to the onset of ER stress (Puthalakath et al. 2007) and deficiency of the hexosamine pathway rather than an energetic stress.

In contrast to Puma and Bim, the proapoptotic Bcl-2 family protein Noxa is important to promote apoptosis in metabolic stress, but Noxa activity rather than expression appears regulated by cell metabolism. Eldering and colleagues showed in two separate studies that activated T cells up-regulated Noxa, which bound to the antiapoptotic protein Mcl-1. When glucose became limiting, Mcl-1 levels decreased and Noxa promoted apoptosis (Alves et al. 2006; Wensveen et al. 2011). Lowman et al. shed some light on a possible mechanism for this effect by demonstrating that glucose deprivation led to loss of an inhibitory phosphorylation of Noxa, promoting Noxa activation (Lowman et al. 2010). Ultimately the combination of proapoptotic proteins including Puma, Noxa, and also Bim all contribute to cell death.

AMP and LKB1 can activate AMPK on metabolic stress to regulate Bcl-2 family proteins and apoptosis. AMPK and LKB1 can inhibit apoptosis by promoting up-regulation of Bcl-

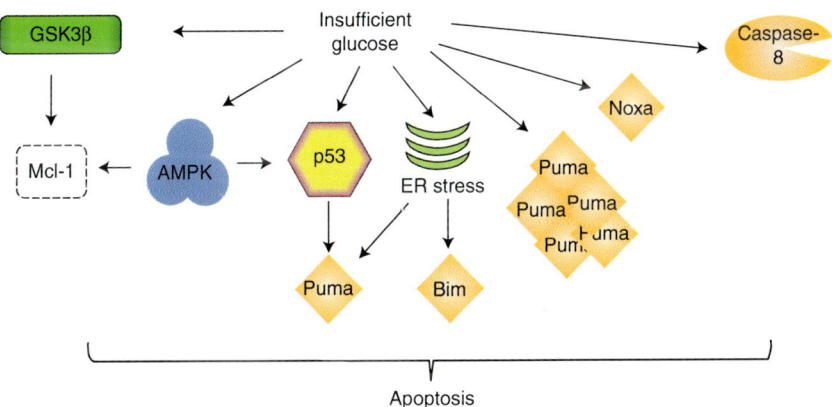

Figure 3. Glucose withdrawal in apoptosis induction. Glucose withdrawal can lead to a shift in the balance of antiapoptotic and proapoptotic proteins to lead to cell death. Activation of GSK3β downstream can lead to loss of Mcl-1. Activation of AMPK can lead to loss of Mcl-1 and activation of p53. Activation of p53 can lead to induction of Puma. ER stress downstream from glucose withdrawal and loss of glycosylation can lead to induction of both Bim and Puma. Glucose withdrawal can also independently lead to Puma accumulation through stabilization of the protein. Finally, glucose withdrawal can lead to activation of Noxa and caspase-8.

xL and suppressing proapoptotic Erk signaling (Cao et al. 2010; Kim et al. 2010). Indeed, LKB1-deficient thymocytes have low Bcl-xL levels and are sensitive to apoptosis (Cao et al. 2010). Pradelli et al. showed that prolonged AMPK activation caused by inhibition of glycolysis suppresses mTOR signaling and leads to decreased translation of Mcl-1, sensitizing cells to apoptosis (Pradelli et al. 2010). Additionally, AMPK can phosphorylate and activate p53 after nutrient stress, with subsequent cell-cycle arrest and apoptosis (Jones et al. 2005; Okoshi et al. 2008).

Glucose deprivation also results in decreased flux through the hexoamine pathway that can lead to diminished protein glycosylation, protein misfolding, and ER stress. Protein glycosylation depends in part on N-acetylglucosamine (GlcNAc) and is important for proper protein folding in the ER (Kaufman et al. 2010). Misfolded proteins invoke the unfolded protein response (UPR) to reduce global protein synthesis and increase production of chaperone proteins to increase ER protein folding (Hoyer-Hansen and Jaattela 2007b). Initially, lack of glycosylation can greatly reduce glycosylation and surface expression of both nutrient transporters and growth factor receptors, which diminishes cell signaling to further exacerbate metabolic stress (Wellen et al. 2010). Ultimately, however, unresolved misfolded proteins lead to apoptotic death through transcription activation of proapoptotic proteins such as Bim and Puma and activation of apoptosis by the UPR-responsive transcription factor Chop/GADD153 (Oyadomari and Mori 2004; Ishihara et al. 2007; Puthalakath et al. 2007).

In some cases loss of glucose metabolism may lead to apoptosis even in cells lacking essential proapoptotic Bcl-2 family proteins. Two studies from Muñoz-Pinedo and colleagues showed that glucose deprivation could lead to apoptosis of Bak and Bax-deficient cells (Munoz-Pinedo et al. 2003; Caro-Maldonado et al. 2010). In this case, caspase-8 was activated to induce apoptosis. Surprisingly, however, caspase-8 activation did not appear to be caused by death receptor signaling and was not Fas-dependent. Rather, caspase-8 appeared to respond to metabolic stress and promote apoptosis through an unknown mechanism. Although poorly understood, this pathway may be critical as a link to promote the cell death of cancer cells that have acquired mutations to resist intrinsic pathways of apoptosis.

HYPOXIA AND CELL DEATH

Oxygen is essential to support mitochondrial metabolism, but ischemia, tissue damage, or poorly vascularized tumor microenvironments can limit oxygen distribution, leading to metabolic stress and potentially apoptosis (Melillo 2007; Rey and Semenza 2010). Under hypoxia, the normally short-lived transcription factors hypoxia-inducible factor $1/2\alpha$ (HIF1α and HIF2α) are stabilized to promote an adaptive response that increases anaerobic metabolic pathways. Regulation of HIF1/2α by oxygen occurs through prolyl hydroxylation in an oxygen-dependent reaction that leads to degradation by the ubiquitin-ligase Von Hippel-Lindau (VHL) in normoxia. In hypoxia, prolyl hydroxylation is reduced and HIF1/2α are stabilized (Rey and Semenza 2010). The HIF transcription factors promote a complex transcriptional program to increase anaerobic metabolic flux and survival by recruiting new vasculature and expressing Glut1 and glycolytic enzymes (Fulda and Debatin 2007). HIF also induces the small BH3-only protein BNIP3, which is not thought to induce apoptosis but rather promotes autophagic targeting of mitochondria, or mitophagy (Tracy et al. 2007; Burton and Gibson 2009). Interestingly, BNIP3 is frequently lost in cancer cells, suggesting that mitophagy may be an impediment to cancer cell response to hypoxia (Lee and Paik 2006).

Hypoxia and HIF1α activation can also inhibit apoptosis on a transcriptional level. HIF1α can up-regulate the antiapoptotic proteins Bcl-2, Bcl-xL, and Survivin and down-regulate the proapoptotic Bid and Bax (Erler et al. 2004; Liu et al. 2008b; Chen et al. 2009a,b). HIF1α activity can also antagonize p53-mediated apoptosis, especially with combined DNA damage (Graeber et al. 1996; Hao et al. 2008; Sendoel et al. 2010). In some settings, however, HIF1α has been found to stabilize p53, leading to subsequent apoptosis (Suzuki et al. 2001; Hansson et al. 2002; Chen et al. 2003; An et al. 2004). The interaction between the HIF1α and p53 may thus be highly context and cell-type dependent. More severe hypoxia or anoxia promotes apoptosis independent of the HIF1α pathway. In these conditions, the electron transport chain collapses, expression of the antiapoptotic protein Mcl-1 is lost, and Bax and Bak are activated to induce apoptosis (Saikumar et al. 1998; Brunelle et al. 2007). Thus, changes in oxygen availability induce an adaptive response that may allow for extended survival, but severe or prolonged oxygen deprivation leads to apoptosis.

p53 IN METABOLIC STRESS

Metabolic stress also leads to p53 activation to induce apoptosis by up-regulating Puma, Noxa, and Bax (Nakano and Vousden 2001; Gottlieb et al. 2002; Rozan and El-Deiry 2007; Coloff et al. 2011). Lee et al. found that glucose deprivation caused the tricarboxylic acid (TCA)-cycle enzyme malate dehydrogenase (MDH) to bind to and activate acetylated p53, leading to cell-cycle arrest and apoptosis (Lee et al. 2009). AMPK activation on glucose withdrawal can also lead to p53 phosphorylation on serine 15 to cause cell-cycle arrest (Jones et al. 2005). This phosphorylation may also play a key role in cellular response to metabolic stress to induce Puma and apoptosis. It is unclear, however, if AMPK directly phosphorylates p53 or if an intermediate kinase is also involved.

Growth factor withdrawal also leads to p53 phosphorylation on serine 15 (18 in mouse) that is metabolically sensitive. We have observed Glut1/Hexokinase overexpressing cells that maintain growth factor-independent glucose metabolism can selectively suppress this phosphorylation and p53 activity after growth factor withdrawal (Mason et al. 2010). Importantly, changes in glucose metabolism also controlled p53 phosphorylation on treatment of BCR-Abl$^+$ leukemic cells with imatinib mesylate, demonstrating that metabolic control of p53 activity may play a general role in the action of kinase inhibition in cancer therapy. This glucose-sensitive regulation of p53 occurred in part through the activation of the lipid-sensitive kinase PKCδ and was independent from canonical DNA damage-induced p53 activation, which led to p53 activation independent of PKCδ and regardless of cellular metabolic state. Romero Rosales et al. (2009) also observed a proapoptotic role for PKCδ downstream from cytokine withdrawal. Importantly, PKCδ can confer sensitivity to apoptosis after chemotherapy or radiation treatment, and the loss of PKCδ has been associated with several kinds of aggressive cancer (Gonelli et al. 2009). Although it does not appear that PKCδ directly phosphorylates p53, this pathway nevertheless provides a metabolically sensitive mechanism for p53 regulation.

Acetylation may also link metabolism to p53 regulation. Acetyl-CoA is required for acetylation and is derived from the metabolism of glucose, the β-oxidation of long-chain fatty acids, or from the conversion of the TCA-cycle intermediate citrate via the ATP citrate lyase (ACL) enzyme (Salway 2004). Thus, levels of acetyl-CoA are sensitive to changes in growth factor input or nutrient availability. Indeed, Wellen et al. (2009) have recently shown that ACL is an important source of acetyl-CoA to support histone acetylation, and loss of growth factor input or glucose availability leads to a global decrease in acetyl-CoA and protein acetylation. p53 acetylation (Gu and Roeder 1997; Sakaguchi et al. 1998; Pearson et al. 2000) is critical for p53 to induce cell-cycle arrest or apoptosis (Lavin and Gueven 2006; Tang et al. 2008) and may be metabolically regulated. Indeed, we have showed that p53 is acetylated downstream from growth factor withdrawal in a metabolically sensitive manner, leading to a transcriptional program distinct from DNA damage induced activation (Mason et al. 2010).

METABOLIC STRESS AND NECROSIS

Ultimately, necrosis may occur when cells do not meet their minimal bioenergetic demands (Jin et al. 2007). In these instances, a collapse of ATP levels may lead to failure of ATP-dependent sodium/potassium exchangers and osmotic stress and cell rupture. Necrosis may occur in metabolically challenging situations such as ischemia/reperfusion, in which tissue is subjected to nutrient deprivation during ischemia followed by an oxidative and ROS burst during reperfusion (Zong and Thompson 2006). Demonstrating the role of ATP in preventing necrosis, Leist et al. (1997, 1999) showed in seminal studies that T cells subjected to various forms of stress died by necrosis rather than apoptosis when ATP was depleted. To respond to metabolic stress, cells activate adaptive pathways to uptake or generate more nutrients, such as autophagy or the HIF pathway. Thus, inhibiting autophagy in the presence of metabolic stress has been shown to lead to metabolic catastrophe and necrosis (Lum et al. 2005; Degenhardt et al. 2006), and a similar effect has been observed in hypoxic cells deprived of proper HIF signaling (Tennant et al. 2009).

Some cell stresses may lead to artificial nutrient depletion that can cause necrosis. In particular, DNA repair performed by poly(ADP-ribose) polymerase (PARP) can also cause necrosis if PARP activity is excessive. PARP requires cytosolic nicotinamide adenine dinucleotide (NAD^+) as a substrate to repair various kinds of DNA damage including single- and double-strand breaks and base excision repair (Sodhi et al. 2010). If DNA damage is extensive, PARP can deplete NAD^+, causing a collapse of glycolysis, loss of ATP production, and subsequent necrosis (Martin et al. 2000; Cipriani et al. 2005). Consistent with this ability of PARP to lead to necrosis, caspase-dependent cleavage of PARP may be a key step to prevent energy depletion and allow apoptosis to occur (Sodhi et al. 2010). Although metabolic collapse and necrosis would seem to be an attractive strategy in which to target cancer cells, excessive necrosis has been shown to lead to inflammation and promotion of advanced tumorigenesis (Balkwill et al. 2005), so treatment strategies that instead cause apoptotic death are often pursued.

CONCLUDING REMARKS

Mammalian cells rely on extracellular nutrient uptake to maintain metabolism and provide bioenergetic precursors for macromolecular synthesis. This uptake is normally under tight control of growth factors, but oncogenes and oncogenic kinases can mimic growth factor signaling to promote cell intrinsic nutrient uptake and metabolism. In particular, activation of Akt can promote glucose, amino acid, and lipid uptake (Edinger and Thompson 2002; Wieman et al. 2007). Nutrient withdrawal, either by direct means, growth factor deprivation, or oncogenic kinase inhibition, leads to metabolic stress and a set of responses that may ultimately result in cell death, and understanding these responses may be critical in efforts to exploit cancer metabolism. It is now clear that autophagy plays a key role to reduce intracellular stress and provide nutrients to replace diminished extracellular nutrient uptake, and inhibition of autophagy can enhance the proapoptotic ability of oncogenic kinase inhibitors (Kamitsuji et al. 2008; Bellodi et al. 2009; Altman et al. 2011; Guo et al. 2011). Continued nutrient deprivation will regulate Bcl-2 family proteins to induce Puma and Bim and activate Noxa while down-regulating Mcl-1 to promote apoptosis. If cells resist apoptosis or if nutrient deprivation is severe, necrosis can occur. Ultimately, understanding links between nutrient stress, autophagy, and cell death through apoptosis or necrosis will be central to inhibiting cancer cell metabolism in novel metabolic cancer therapies.

ACKNOWLEDGMENTS

We thank Drs. Pankuri Goraksha-Hicks, Andrew MacIntyre, and Nancie MacIver of the Rathmell laboratory for support and comments. This work was supported by National Institutes of Health (NIH) R01CA123350 (J.C.R.), the Gabrielle's Angel Foundation (J.C.R.), and the Leukemia and Lymphoma Foundation (J.C.R.).

REFERENCES

*Reference is also in this collection.

Abeliovich H, Klionsky DJ. 2001. Autophagy in yeast: Mechanistic insights and physiological function. *Microbiol Mol Biol Rev* **65:** 463–479.

Altman BJ, Wofford JA, Zhao Y, Coloff JL, Ferguson EC, Wieman HL, Day AE, Ilkayeva O, Rathmell JC. 2009.

Autophagy provides nutrients but can lead to Chop-dependent induction of Bim to sensitize growth factor-deprived cells to apoptosis. *Mol Biol Cell* **20:** 1180–1191.

Altman BJ, Jacobs SR, Mason EF, Michalek RD, Macintyre AN, Coloff JL, Ilkayeva O, Jia W, He YW, Rathmell JC. 2011. Autophagy is essential to suppress cell stress and to allow BCR-Abl-mediated leukemogenesis. *Oncogene* **30:** 1855–1867.

Alves NL, Derks IA, Berk E, Spijker R, van Lier RA, Eldering E. 2006. The Noxa/Mcl-1 axis regulates susceptibility to apoptosis under glucose limitation in dividing T cells. *Immunity* **24:** 703–716.

An J, Muoio DM, Shiota M, Fujimoto Y, Cline GW, Shulman GI, Koves TR, Stevens R, Millington D, Newgard CB. 2004. Hepatic expression of malonyl-CoA decarboxylase reverses muscle, liver and whole-animal insulin resistance. *Nat Med* **10:** 268–274.

Balkwill F, Charles KA, Mantovani A. 2005. Smoldering and polarized inflammation in the initiation and promotion of malignant disease. *Cancer Cell* **7:** 211–217.

Barnes K, McIntosh E, Whetton AD, Daley GQ, Bentley J, Baldwin SA. 2005. Chronic myeloid leukaemia: An investigation into the role of Bcr-Abl-induced abnormalities in glucose transport regulation. *Oncogene* **24:** 3257–3267.

Bellodi C, Lidonnici MR, Hamilton A, Helgason GV, Soliera AR, Ronchetti M, Galavotti S, Young KW, Selmi T, Yacobi R, et al. 2009. Targeting autophagy potentiates tyrosine kinase inhibitor-induced cell death in Philadelphia chromosome-positive cells, including primary CML stem cells. *J Clin Invest* **119:** 1109–1123.

Bernales S, McDonald KL, Walter P. 2006. Autophagy counterbalances endoplasmic reticulum expansion during the unfolded protein response. *PLoS Biol* **4:** e423.

Berry DL, Baehrecke EH. 2007. Growth arrest and autophagy are required for salivary gland cell degradation in *Drosophila*. *Cell* **131:** 1137–1148.

Beugnet A, Tee AR, Taylor PM, Proud CG. 2003. Regulation of targets of mTOR (mammalian target of rapamycin) signalling by intracellular amino acid availability. *Biochem J* **372:** 555–566.

Bjorkoy G, Lamark T, Brech A, Outzen H, Perander M, Overvatn A, Stenmark H, Johansen T. 2005. p62/SQSTM1 forms protein aggregates degraded by autophagy and has a protective effect on huntingtin-induced cell death. *J Cell Biol* **171:** 603–614.

Bodet L, Menoret E, Descamps G, Pellat-Deceunynck C, Bataille R, Le Gouill S, Moreau P, Amiot M, Gomez-Bougie P. 2010. BH3-only protein Bik is involved in both apoptosis induction and sensitivity to oxidative stress in multiple myeloma. *Br J Cancer* **103:** 1808–1814.

Boya P, Gonzalez-Polo RA, Casares N, Perfettini JL, Dessen P, Larochette N, Metivier D, Meley D, Souquere S, Yoshimori T, et al. 2005. Inhibition of macroautophagy triggers apoptosis. *Mol Cell Biol* **25:** 1025–1040.

Brunelle JK, Shroff EH, Perlman H, Strasser A, Moraes CT, Flavell RA, Danial NN, Keith B, Thompson CB, Chandel NS. 2007. Loss of Mcl-1 protein and inhibition of electron transport chain together induce anoxic cell death. *Mol Cell Biol* **27:** 1222–1235.

Burton TR, Gibson SB. 2009. The role of Bcl-2 family member BNIP3 in cell death and disease: NIPping at the heels of cell death. *Cell Death Differ* **16:** 515–523.

Byfield MP, Murray JT, Backer JM. 2005. hVps34 is a nutrient-regulated lipid kinase required for activation of p70 S6 kinase. *J Biol Chem* **280:** 33076–33082.

Canto C, Auwerx J. 2010. AMP-activated protein kinase and its downstream transcriptional pathways. *Cell Mol Life Sci* **67:** 3407–3423.

Cao Y, Li H, Liu H, Zheng C, Ji H, Liu X. 2010. The serine/threonine kinase LKB1 controls thymocyte survival through regulation of AMPK activation and Bcl-XL expression. *Cell Res* **20:** 99–108.

Caro-Maldonado A, Tait SW, Ramirez-Peinado S, Ricci JE, Fabregat I, Green DR, Munoz-Pinedo C. 2010. Glucose deprivation induces an atypical form of apoptosis mediated by caspase-8 in Bax-, Bak-deficient cells. *Cell Death Differ* **17:** 1335–1344.

Chen D, Li M, Luo J, Gu W. 2003. Direct interactions between HIF-1α and Mdm2 modulate p53 function. *J Biol Chem* **278:** 13595–13598.

Chen N, Chen X, Huang R, Zeng H, Gong J, Meng W, Lu Y, Zhao F, Wang L, Zhou Q. 2009a. BCL-xL is a target gene regulated by hypoxia-inducible factor-1α. *J Biol Chem* **284:** 10004–10012.

Chen YQ, Zhao CL, Li W. 2009b. Effect of hypoxia-inducible factor-1α on transcription of survivin in non-small cell lung cancer. *J Exp Clin Cancer Res* **28:** 29.

Chipuk JE, Moldoveanu T, Llambi F, Parsons MJ, Green DR. 2010. The BCL-2 family reunion. *Mol Cell* **37:** 299–310.

Cipriani G, Rapizzi E, Vannacci A, Rizzuto R, Moroni F, Chiarugi A. 2005. Nuclear poly(ADP-ribose) polymerase-1 rapidly triggers mitochondrial dysfunction. *J Biol Chem* **280:** 17227–17234.

Colell A, Ricci JE, Tait S, Milasta S, Maurer U, Bouchier-Hayes L, Fitzgerald P, Guio-Carrion A, Waterhouse NJ, Li CW, et al. 2007. GAPDH and autophagy preserve survival after apoptotic cytochrome c release in the absence of caspase activation. *Cell* **129:** 983–997.

Coloff JL, Mason EF, Altman BJ, Gerriets VA, Liu T, Nichols AN, Zhao Y, Wofford JA, Jacobs SR, Ilkayeva O, et al. 2011. Akt requires glucose metabolism to suppress puma expression and prevent apoptosis of leukemic T cells. *J Biol Chem* **286:** 5921–5933.

Dang CV, Le A, Gao P. 2009. MYC-induced cancer cell energy metabolism and therapeutic opportunities. *Clin Cancer Res* **15:** 6479–6483.

* Das G, Shravage BV, Baehrecke EH. 2012. Regulation and function of autophagy during cell survival and cell death. *Cold Spring Harb Perspect Biol* **4:** a008813.

DeBerardinis RJ, Lum JJ, Thompson CB. 2006. Phosphatidylinositol 3-kinase-dependent modulation of carnitine palmitoyltransferase 1A expression regulates lipid metabolism during hematopoietic cell growth. *J Biol Chem* **281:** 37372–37380.

DeBerardinis RJ, Lum JJ, Hatzivassiliou G, Thompson CB. 2008. The biology of cancer: Metabolic reprogramming fuels cell growth and proliferation. *Cell Metab* **7:** 11–20.

Degenhardt K, Mathew R, Beaudoin B, Bray K, Anderson D, Chen G, Mukherjee C, Shi Y, Gelinas C, Fan Y, et al. 2006. Autophagy promotes tumor cell survival and restricts necrosis, inflammation, and tumorigenesis. *Cancer Cell* **10:** 51–64.

Dennis JW, Nabi IR, Demetriou M. 2009. Metabolism, cell surface organization, and disease. *Cell* **139:** 1229–1241.

Denton D, Shravage B, Simin R, Mills K, Berry DL, Baehrecke EH, Kumar S. 2009. Autophagy, not apoptosis, is essential for midgut cell death in *Drosophila*. *Curr Biol* **19:** 1741–1746.

Duvel K, Yecies JL, Menon S, Raman P, Lipovsky AI, Souza AL, Triantafellow E, Ma Q, Gorski R, Cleaver S, et al. 2010. Activation of a metabolic gene regulatory network downstream of mTOR complex 1. *Mol Cell* **39:** 171–183.

Edinger AL, Thompson CB. 2002. Akt maintains cell size and survival by increasing mTOR-dependent nutrient uptake. *Mol Biol Cell* **13:** 2276–2288.

Egan DF, Shackelford DB, Mihaylova MM, Gelino S, Kohnz RA, Mair W, Vasquez DS, Joshi A, Gwinn DM, Taylor R, et al. 2011. Phosphorylation of ULK1 (hATG1) by AMP-activated protein kinase connects energy sensing to mitophagy. *Science* **331:** 456–461.

Elstrom RL, Bauer DE, Buzzai M, Karnauskas R, Harris MH, Plas DR, Zhuang H, Cinalli RM, Alavi A, Rudin CM, et al. 2004. Akt stimulates aerobic glycolysis in cancer cells. *Cancer Res* **64:** 3892–3899.

Erler JT, Cawthorne CJ, Williams KJ, Koritzinsky M, Wouters BG, Wilson C, Miller C, Demonacos C, Stratford IJ, Dive C. 2004. Hypoxia-mediated down-regulation of Bid and Bax in tumors occurs via hypoxia-inducible factor 1-dependent and -independent mechanisms and contributes to drug resistance. *Mol Cell Biol* **24:** 2875–2889.

Fan Y, Dickman KG, Zong WX. 2010. Akt and c-Myc differentially activate cellular metabolic programs and prime cells to bioenergetic inhibition. *J Biol Chem* **285:** 7324–7333.

Fingar DC, Salama S, Tsou C, Harlow E, Blenis J. 2002. Mammalian cell size is controlled by mTOR and its downstream targets S6K1 and 4EBP1/eIF4E. *Genes Dev* **16:** 1472–1487.

Frech M, Andjelkovic M, Ingley E, Reddy KK, Falck JR, Hemmings BA. 1997. High affinity binding of inositol phosphates and phosphoinositides to the pleckstrin homology domain of RAC/protein kinase B and their influence on kinase activity. *J Biol Chem* **272:** 8474–8481.

Fulda S, Debatin KM. 2007. HIF-1-regulated glucose metabolism: A key to apoptosis resistance? *Cell Cycle* **6:** 790–792.

Galonek HL, Hardwick JM. 2006. Upgrading the BCL-2 network. *Nat Cell Biol* **8:** 1317–1319.

Ganley IG, Lam du H, Wang J, Ding X, Chen S, Jiang X. 2009. ULK1.ATG13.FIP200 complex mediates mTOR signaling and is essential for autophagy. *J Biol Chem* **284:** 12297–12305.

Gao P, Tchernyshyov I, Chang TC, Lee YS, Kita K, Ochi T, Zeller KI, De Marzo AM, Van Eyk JE, Mendell JT, et al. 2009. c-Myc suppression of miR-23a/b enhances mitochondrial glutaminase expression and glutamine metabolism. *Nature* **458:** 762–765.

Gonelli A, Mischiati C, Guerrini R, Voltan R, Salvadori S, Zauli G. 2009. Perspectives of protein kinase C (PKC) inhibitors as anti-cancer agents. *Mini Rev Med Chem* **9:** 498–509.

Gottlieb TM, Leal JF, Seger R, Taya Y, Oren M. 2002. Crosstalk between Akt, p53 and Mdm2: Possible implications for the regulation of apoptosis. *Oncogene* **21:** 1299–1303.

Gottlob K, Majewski N, Kennedy S, Kandel E, Robey RB, Hay N. 2001. Inhibition of early apoptotic events by Akt/PKB is dependent on the first committed step of glycolysis and mitochondrial hexokinase. *Genes Dev* **15:** 1406–1418.

Graeber TG, Osmanian C, Jacks T, Housman DE, Koch CJ, Lowe SW, Giaccia AJ. 1996. Hypoxia-mediated selection of cells with diminished apoptotic potential in solid tumours. *Nature* **379:** 88–91.

Gu W, Roeder RG. 1997. Activation of p53 sequence-specific DNA binding by acetylation of the p53 C-terminal domain. *Cell* **90:** 595–606.

Guo JY, Chen HY, Mathew R, Fan J, Strohecker AM, Karsli-Uzunbas G, Kamphorst JJ, Chen G, Lemmons JM, Karantza V, et al. 2011. Activated Ras requires autophagy to maintain oxidative metabolism and tumorigenesis. *Genes Dev* **25:** 460–470.

Hamanaka RB, Chandel NS. 2010. Mitochondrial reactive oxygen species regulate cellular signaling and dictate biological outcomes. *Trends Biochem Sci* **35:** 505–513.

Hansson LO, Friedler A, Freund S, Rudiger S, Fersht AR. 2002. Two sequence motifs from HIF-1α bind to the DNA-binding site of p53. *Proc Natl Acad Sci* **99:** 10305–10309.

Hao J, Song X, Song B, Liu Y, Wei L, Wang X, Yu J. 2008. Effects of lentivirus-mediated HIF-1α knockdown on hypoxia-related cisplatin resistance and their dependence on p53 status in fibrosarcoma cells. *Cancer Gene Ther* **15:** 449–455.

Hara K, Yonezawa K, Weng QP, Kozlowski MT, Belham C, Avruch J. 1998. Amino acid sufficiency and mTOR regulate p70 S6 kinase and eIF-4E BP1 through a common effector mechanism. *J Biol Chem* **273:** 14484–14494.

Hawley SA, Boudeau J, Reid JL, Mustard KJ, Udd L, Makela TP, Alessi DR, Hardie DG. 2003. Complexes between the LKB1 tumor suppressor, STRAD α/β and MO25 α/β are upstream kinases in the AMP-activated protein kinase cascade. *J Biol* **2:** 28.

Hou YC, Chittaranjan S, Barbosa SG, McCall K, Gorski SM. 2008. Effector caspase Dcp-1 and IAP protein Bruce regulate starvation-induced autophagy during *Drosophila melanogaster* oogenesis. *J Cell Biol* **182:** 1127–1139.

Hoyer-Hansen M, Jaattela M. 2007a. AMP-activated protein kinase: A universal regulator of autophagy? *Autophagy* **3:** 381–383.

Hoyer-Hansen M, Jaattela M. 2007b. Connecting endoplasmic reticulum stress to autophagy by unfolded protein response and calcium. *Cell Death Differ* **14:** 1576–1582.

Imamura K, Ogura T, Kishimoto A, Kaminishi M, Esumi H. 2001. Cell cycle regulation via p53 phosphorylation by a 5′-AMP activated protein kinase activator, 5-aminoimidazole-4-carboxamide-1-β-D-ribofuranoside, in a human hepatocellular carcinoma cell line. *Biochem Biophys Res Commun* **287:** 562–567.

Ishihara T, Hoshino T, Namba T, Tanaka K, Mizushima T. 2007. Involvement of up-regulation of PUMA in nonsteroidal anti-inflammatory drug-induced apoptosis. *Biochem Biophys Res Commun* **356:** 711–717.

Jin S, DiPaola RS, Mathew R, White E. 2007. Metabolic catastrophe as a means to cancer cell death. *J Cell Sci* **120:** 379–383.

Jones RG, Plas DR, Kubek S, Buzzai M, Mu J, Xu Y, Birnbaum MJ, Thompson CB. 2005. AMP-activated protein kinase induces a p53-dependent metabolic checkpoint. *Mol Cell* **18:** 283–293.

Jung CH, Jun CB, Ro SH, Kim YM, Otto NM, Cao J, Kundu M, Kim DH. 2009. ULK-Atg13-FIP200 complexes mediate mTOR signaling to the autophagy machinery. *Mol Biol Cell* **20:** 1992–2003.

Kamitsuji Y, Kuroda J, Kimura S, Toyokuni S, Watanabe K, Ashihara E, Tanaka H, Yui Y, Watanabe M, Matsubara H, et al. 2008. The Bcr-Abl kinase inhibitor INNO-406 induces autophagy and different modes of cell death execution in Bcr-Abl-positive leukemias. *Cell Death Differ* **15:** 1712–1722.

Karantza-Wadsworth V, Patel S, Kravchuk O, Chen G, Mathew R, Jin S, White E. 2007. Autophagy mitigates metabolic stress and genome damage in mammary tumorigenesis. *Genes Dev* **21:** 1621–1635.

Karawajew L, Rhein P, Czerwony G, Ludwig WD. 2005. Stress-induced activation of the p53 tumor suppressor in leukemia cells and normal lymphocytes requires mitochondrial activity and reactive oxygen species. *Blood* **105:** 4767–4775.

Kaufman RJ, Back SH, Song B, Han J, Hassler J. 2010. The unfolded protein response is required to maintain the integrity of the endoplasmic reticulum, prevent oxidative stress and preserve differentiation in β-cells. *Diabetes Obes Metab* **12:** 99–107.

Kelekar A. 2005. Autophagy. *Ann NY Acad Sci* **1066:** 259–271.

Kim E, Goraksha-Hicks P, Li L, Neufeld TP, Guan KL. 2008. Regulation of TORC1 by Rag GTPases in nutrient response. *Nat Cell Biol* **10:** 935–945.

Kim MJ, Park IJ, Yun H, Kang I, Choe W, Kim SS, Ha J. 2010. AMP-activated protein kinase antagonizes pro-apoptotic extracellular signal-regulated kinase activation by inducing dual-specificity protein phosphatases in response to glucose deprivation in HCT116 carcinoma. *J Biol Chem* **285:** 14617–14627.

Kim J, Kundu M, Viollet B, Guan KL. 2011. AMPK and mTOR regulate autophagy through direct phosphorylation of Ulk1. *Nat Cell Biol* **13:** 132–141.

Kiyono K, Suzuki HI, Matsuyama H, Morishita Y, Komuro A, Kano MR, Sugimoto K, Miyazono K. 2009. Autophagy is activated by TGF-β and potentiates TGF-β-mediated growth inhibition in human hepatocellular carcinoma cells. *Cancer Res* **69:** 8844–8852.

Komatsu M, Waguri S, Chiba T, Murata S, Iwata J, Tanida I, Ueno T, Koike M, Uchiyama Y, Kominami E, et al. 2006. Loss of autophagy in the central nervous system causes neurodegeneration in mice. *Nature* **441:** 880–884.

Lavin MF, Gueven N. 2006. The complexity of p53 stabilization and activation. *Cell Death Differ* **13:** 941–950.

Lee H, Paik SG. 2006. Regulation of BNIP3 in normal and cancer cells. *Mol Cells* **21:** 1–6.

Lee SM, Kim JH, Cho EJ, Youn HD. 2009. A nucleocytoplasmic malate dehydrogenase regulates p53 transcriptional activity in response to metabolic stress. *Cell Death Differ* **16:** 738–748.

Leist M, Single B, Castoldi AF, Kuhnle S, Nicotera P. 1997. Intracellular adenosine triphosphate (ATP) concentration: A switch in the decision between apoptosis and necrosis. *J Exp Med* **185:** 1481–1486.

Leist M, Single B, Naumann H, Fava E, Simon B, Kuhnle S, Nicotera P. 1999. Inhibition of mitochondrial ATP generation by nitric oxide switches apoptosis to necrosis. *Exp Cell Res* **249:** 396–403.

Letai A, Bassik MC, Walensky LD, Sorcinelli MD, Weiler S, Korsmeyer SJ. 2002. Distinct BH3 domains either sensitize or activate mitochondrial apoptosis, serving as prototype cancer therapeutics. *Cancer Cell* **2:** 183–192.

Liang XH, Jackson S, Seaman M, Brown K, Kempkes B, Hibshoosh H, Levine B. 1999. Induction of autophagy and inhibition of tumorigenesis by beclin 1. *Nature* **402:** 672–676.

Liu B, Chen Y, St Clair DK. 2008a. ROS and p53: A versatile partnership. *Free Radic Biol Med* **44:** 1529–1535.

Liu L, Ning X, Sun L, Zhang H, Shi Y, Guo C, Han S, Liu J, Sun S, Han Z, et al. 2008b. Hypoxia-inducible factor-1α contributes to hypoxia-induced chemoresistance in gastric cancer. *Cancer Sci* **99:** 121–128.

Lock R, Roy S, Kenific CM, Su JS, Salas E, Ronen SM, Debnath J. 2011. Autophagy facilitates glycolysis during Ras-mediated oncogenic transformation. *Mol Biol Cell* **22:** 165–178.

Lowman XH, McDonnell MA, Kosloske A, Odumade OA, Jenness C, Karim CB, Jemmerson R, Kelekar A. 2010. The proapoptotic function of Noxa in human leukemia cells is regulated by the kinase Cdk5 and by glucose. *Mol Cell* **40:** 823–833.

Lum JJ, Bauer DE, Kong M, Harris MH, Li C, Lindsten T, Thompson CB. 2005. Growth factor regulation of autophagy and cell survival in the absence of apoptosis. *Cell* **120:** 237–248.

Lynch CJ. 2001. Role of leucine in the regulation of mTOR by amino acids: Revelations from structure-activity studies. *J Nutr* **131:** 861S–865S.

Martin DR, Lewington AJ, Hammerman MR, Padanilam BJ. 2000. Inhibition of poly(ADP-ribose) polymerase attenuates ischemic renal injury in rats. *Am J Physiol Regul Integr Comp Physiol* **279:** R1834–R1840.

Mason EF, Zhao Y, Goraksha-Hicks P, Coloff JL, Gannon H, Jones SN, Rathmell JC. 2010. Aerobic glycolysis suppresses p53 activity to provide selective protection from apoptosis upon loss of growth signals or inhibition of BCR-Abl. *Cancer Res* **70:** 8066–8076.

Mathew R, Kongara S, Beaudoin B, Karp CM, Bray K, Degenhardt K, Chen G, Jin S, White E. 2007. Autophagy suppresses tumor progression by limiting chromosomal instability. *Genes Dev* **21:** 1367–1381.

Mathew R, Karp CM, Beaudoin B, Vuong N, Chen G, Chen HY, Bray K, Reddy A, Bhanot G, Gelinas C, et al. 2009. Autophagy suppresses tumorigenesis through elimination of p62. *Cell* **137:** 1062–1075.

Melillo G. 2007. Targeting hypoxia cell signaling for cancer therapy. *Cancer Metastasis Rev* **26:** 341–352.

Michalek RD, Rathmell JC. 2010. The metabolic life and times of a T-cell. *Immunol Rev* **236**: 190–202.

Munoz-Pinedo C, Ruiz-Ruiz C, Ruiz de Almodovar C, Palacios C, Lopez-Rivas A. 2003. Inhibition of glucose metabolism sensitizes tumor cells to death receptor-triggered apoptosis through enhancement of death-inducing signaling complex formation and apical procaspase-8 processing. *J Biol Chem* **278**: 12759–12768.

Nakano K, Vousden KH. 2001. PUMA, a novel proapoptotic gene, is induced by p53. *Mol Cell* **7**: 683–694.

Nezis IP, Shravage BV, Sagona AP, Lamark T, Bjorkoy G, Johansen T, Rusten TE, Brech A, Baehrecke EH, Stenmark H. 2010. Autophagic degradation of dBruce controls DNA fragmentation in nurse cells during late *Drosophila melanogaster* oogenesis. *J Cell Biol* **190**: 523–531.

Niizuma K, Endo H, Chan PH. 2009. Oxidative stress and mitochondrial dysfunction as determinants of ischemic neuronal death and survival. *J Neurochem* **109**: 133–138.

Nobukuni T, Joaquin M, Roccio M, Dann SG, Kim SY, Gulati P, Byfield MP, Backer JM, Natt F, Bos JL, et al. 2005. Amino acids mediate mTOR/raptor signaling through activation of class 3 phosphatidylinositol 3OH-kinase. *Proc Natl Acad Sci* **102**: 14238–14243.

Nutt LK, Margolis SS, Jensen M, Herman CE, Dunphy WG, Rathmell JC, Kornbluth S. 2005. Metabolic regulation of oocyte cell death through the CaMKII-mediated phosphorylation of caspase-2. *Cell* **123**: 89–103.

Nutt LK, Buchakjian MR, Gan E, Darbandi R, Yoon SY, Wu JQ, Miyamoto YJ, Gibbons JA, Andersen JL, Freel CD, et al. 2009. Metabolic control of oocyte apoptosis mediated by 14-3-3ζ-regulated dephosphorylation of caspase-2. *Dev Cell* **16**: 856–866.

Okoshi R, Ozaki T, Yamamoto H, Ando K, Koida N, Ono S, Koda T, Kamijo T, Nakagawara A, Kizaki H. 2008. Activation of AMP-activated protein kinase induces p53-dependent apoptotic cell death in response to energetic stress. *J Biol Chem* **283**: 3979–3987.

Osthus RC, Shim H, Kim S, Li Q, Reddy R, Mukherjee M, Xu Y, Wonsey D, Lee LA, Dang CV. 2000. Deregulation of glucose transporter 1 and glycolytic gene expression by c-Myc. *J Biol Chem* **275**: 21797–21800.

Oyadomari S, Mori M. 2004. Roles of CHOP/GADD153 in endoplasmic reticulum stress. *Cell Death Differ* **11**: 381–389.

Pearson M, Carbone R, Sebastiani C, Cioce M, Fagioli M, Saito S, Higashimoto Y, Appella E, Minucci S, Pandolfi PP, et al. 2000. PML regulates p53 acetylation and premature senescence induced by oncogenic Ras. *Nature* **406**: 207–210.

Plas DR, Talapatra S, Edinger AL, Rathmell JC, Thompson CB. 2001. Akt and Bcl-xL promote growth factor-independent survival through distinct effects on mitochondrial physiology. *J Biol Chem* **276**: 12041–12048.

Pradelli LA, Beneteau M, Chauvin C, Jacquin MA, Marchetti S, Munoz-Pinedo C, Auberger P, Pende M, Ricci JE. 2010. Glycolysis inhibition sensitizes tumor cells to death receptors-induced apoptosis by AMP kinase activation leading to Mcl-1 block in translation. *Oncogene* **29**: 1641–1652.

Pua HH, Guo J, Komatsu M, He YW. 2009. Autophagy is essential for mitochondrial clearance in mature T lymphocytes. *J Immunol* **182**: 4046–4055.

Puthalakath H, O'Reilly LA, Gunn P, Lee L, Kelly PN, Huntington ND, Hughes PD, Michalak EM, McKimm-Breschkin J, Motoyama N, et al. 2007. ER stress triggers apoptosis by activating BH3-only protein Bim. *Cell* **129**: 1337–1349.

Raff MC. 1992. Social controls on cell survival and cell death. *Nature* **356**: 397–400.

Rathmell JC, Vander Heiden MG, Harris MH, Frauwirth KA, Thompson CB. 2000. In the absence of extrinsic signals, nutrient utilization by lymphocytes is insufficient to maintain either cell size or viability. *Mol Cell* **6**: 683–692.

Rathmell JC, Fox CJ, Plas DR, Hammerman PS, Cinalli RM, Thompson CB. 2003. Akt-directed glucose metabolism can prevent Bax conformation change and promote growth factor-independent survival. *Mol Cell Biol* **23**: 7315–7328.

Ren D, Tu HC, Kim H, Wang GX, Bean GR, Takeuchi O, Jeffers JR, Zambetti GP, Hsieh JJ, Cheng EH. 2010. BID, BIM, and PUMA are essential for activation of the BAX- and BAK-dependent cell death program. *Science* **330**: 1390–1393.

Rey S, Semenza GL. 2010. Hypoxia-inducible factor-1-dependent mechanisms of vascularization and vascular remodelling. *Cardiovasc Res* **86**: 236–242.

Robey RB, Hay N. 2006. Mitochondrial hexokinases, novel mediators of the antiapoptotic effects of growth factors and Akt. *Oncogene* **25**: 4683–4696.

Roccio M, Bos JL, Zwartkruis FJ. 2006. Regulation of the small GTPase Rheb by amino acids. *Oncogene* **25**: 657–664.

Romero Rosales K, Peralta ER, Guenther GG, Wong SY, Edinger AL. 2009. Rab7 activation by growth factor withdrawal contributes to the induction of apoptosis. *Mol Biol Cell* **20**: 2831–2840.

Rouschop KM, Ramaekers CH, Schaaf MB, Keulers TG, Savelkouls KG, Lambin P, Koritzinsky M, Wouters BG. 2009. Autophagy is required during cycling hypoxia to lower production of reactive oxygen species. *Radiother Oncol* **92**: 411–416.

Rozan LM, El-Deiry WS. 2007. p53 downstream target genes and tumor suppression: A classical view in evolution. *Cell Death Differ* **14**: 3–9.

Sade H, Sarin A. 2004. Reactive oxygen species regulate quiescent T-cell apoptosis via the BH3-only proapoptotic protein BIM. *Cell Death Differ* **11**: 416–423.

Saikumar P, Dong Z, Patel Y, Hall K, Hopfer U, Weinberg JM, Venkatachalam MA. 1998. Role of hypoxia-induced Bax translocation and cytochrome c release in reoxygenation injury. *Oncogene* **17**: 3401–3415.

Sakaguchi K, Herrera JE, Saito S, Miki T, Bustin M, Vassilev A, Anderson CW, Appella E. 1998. DNA damage activates p53 through a phosphorylation-acetylation cascade. *Genes Dev* **12**: 2831–2841.

Salway JG. 2004. *Metabolism at a glance*. Blackwell, Malden, MA.

Sancak Y, Bar-Peled L, Zoncu R, Markhard AL, Nada S, Sabatini DM. 2010. Ragulator-Rag complex targets mTORC1 to the lysosomal surface and is necessary for its activation by amino acids. *Cell* **141**: 290–303.

Schafer ZT, Kornbluth S. 2006. The apoptosome: Physiological, developmental, and pathological modes of regulation. *Dev Cell* **10:** 549–561.

Scott RC, Juhasz G, Neufeld TP. 2007. Direct induction of autophagy by Atg1 inhibits cell growth and induces apoptotic cell death. *Curr Biol* **17:** 1–11.

Sendoel A, Kohler I, Fellmann C, Lowe SW, Hengartner MO. 2010. HIF-1 antagonizes p53-mediated apoptosis through a secreted neuronal tyrosinase. *Nature* **465:** 577–583.

Shang L, Chen S, Du F, Li S, Zhao L, Wang X. 2011. Nutrient starvation elicits an acute autophagic response mediated by Ulk1 dephosphorylation and its subsequent dissociation from AMPK. *Proc Natl Acad Sci* **108:** 4788–4793.

Shaw RJ. 2009. LKB1 and AMP-activated protein kinase control of mTOR signalling and growth. *Acta Physiol (Oxf)* **196:** 65–80.

Shaw RJ, Kosmatka M, Bardeesy N, Hurley RL, Witters LA, DePinho RA, Cantley LC. 2004. The tumor suppressor LKB1 kinase directly activates AMP-activated kinase and regulates apoptosis in response to energy stress. *Proc Natl Acad Sci* **101:** 3329–3335.

Singh R, Kaushik S, Wang Y, Xiang Y, Novak I, Komatsu M, Tanaka K, Cuervo AM, Czaja MJ. 2009. Autophagy regulates lipid metabolism. *Nature* **458:** 1131–1135.

Sodhi RK, Singh N, Jaggi AS. 2010. Poly(ADP-ribose) polymerase-1 (PARP-1) and its therapeutic implications. *Vascul Pharmacol* **53:** 77–87.

Suzuki H, Tomida A, Tsuruo T. 2001. Dephosphorylated hypoxia-inducible factor 1α as a mediator of p53-dependent apoptosis during hypoxia. *Oncogene* **20:** 5779–5788.

Tal MC, Sasai M, Lee HK, Yordy B, Shadel GS, Iwasaki A. 2009. Absence of autophagy results in reactive oxygen species-dependent amplification of RLR signaling. *Proc Natl Acad Sci* **106:** 2770–2775.

Tandon P, Gallo CA, Khatri S, Barger JF, Yepiskoposyan H, Plas DR. 2011. Requirement for ribosomal protein S6 kinase 1 to mediate glycolysis and apoptosis resistance induced by Pten deficiency. *Proc Natl Acad Sci* **108:** 2361–2365.

Tang Y, Zhao W, Chen Y, Zhao Y, Gu W. 2008. Acetylation is indispensable for p53 activation. *Cell* **133:** 612–626.

Tennant DA, Frezza C, MacKenzie ED, Nguyen QD, Zheng L, Selak MA, Roberts DL, Dive C, Watson DG, Aboagye EO, et al. 2009. Reactivating HIF prolyl hydroxylases under hypoxia results in metabolic catastrophe and cell death. *Oncogene* **28:** 4009–4021.

Tracy K, Dibling BC, Spike BT, Knabb JR, Schumacker P, Macleod KF. 2007. BNIP3 is an RB/E2F target gene required for hypoxia-induced autophagy. *Mol Cell Biol* **27:** 6229–6242.

van Delft MF, Huang DC. 2006. How the Bcl-2 family of proteins interact to regulate apoptosis. *Cell Res* **16:** 203–213.

Vander Heiden MG, Plas DR, Rathmell JC, Fox CJ, Harris MH, Thompson CB. 2001. Growth factors can influence cell growth and survival through effects on glucose metabolism. *Mol Cell Biol* **21:** 5899–5912.

Vander Heiden MG, Cantley LC, Thompson CB. 2009. Understanding the Warburg effect: The metabolic requirements of cell proliferation. *Science* **324:** 1029–1033.

von Harsdorf R, Li PF, Dietz R. 1999. Signaling pathways in reactive oxygen species-induced cardiomyocyte apoptosis. *Circulation* **99:** 2934–2941.

Walensky LD. 2006. BCL-2 in the crosshairs: Tipping the balance of life and death. *Cell Death Differ* **13:** 1339–1350.

Wang Y, Han R, Liang ZQ, Wu JC, Zhang XD, Gu ZL, Qin ZH. 2008a. An autophagic mechanism is involved in apoptotic death of rat striatal neurons induced by the non-N-methyl-D-aspartate receptor agonist kainic acid. *Autophagy* **4:** 214–226.

Wang Y, Singh R, Massey AC, Kane SS, Kaushik S, Grant T, Xiang Y, Cuervo AM, Czaja MJ. 2008b. Loss of macroautophagy promotes or prevents fibroblast apoptosis depending on the death stimulus. *J Biol Chem* **283:** 4766–4777.

Warburg O. 1956. On the origin of cancer cells. *Science* **123:** 309–314.

Wellen KE, Thompson CB. 2010. Cellular metabolic stress: Considering how cells respond to nutrient excess. *Mol Cell* **40:** 323–332.

Wellen KE, Hatzivassiliou G, Sachdeva UM, Bui TV, Cross JR, Thompson CB. 2009. ATP-citrate lyase links cellular metabolism to histone acetylation. *Science* **324:** 1076–1080.

Wellen KE, Lu C, Mancuso A, Lemons JM, Ryczko M, Dennis JW, Rabinowitz JD, Coller HA, Thompson CB. 2010. The hexosamine biosynthetic pathway couples growth factor-induced glutamine uptake to glucose metabolism. *Genes Dev* **24:** 2784–2799.

Wensveen FM, Alves NL, Derks IA, Reedquist KA, Eldering E. 2011. Apoptosis induced by overall metabolic stress converges on the Bcl-2 family proteins Noxa and Mcl-1. *Apoptosis* **16:** 708–721.

Wieman HL, Wofford JA, Rathmell JC. 2007. Cytokine stimulation promotes glucose uptake via phosphatidylinositol-3 kinase/Akt regulation of Glut1 activity and trafficking. *Mol Biol Cell* **18:** 1437–1446.

Wise DR, Thompson CB. 2010. Glutamine addiction: A new therapeutic target in cancer. *Trends Biochem Sci* **35:** 427–433.

Wise DR, DeBerardinis RJ, Mancuso A, Sayed N, Zhang XY, Pfeiffer HK, Nissim I, Daikhin E, Yudkoff M, McMahon SB, et al. 2008. Myc regulates a transcriptional program that stimulates mitochondrial glutaminolysis and leads to glutamine addiction. *Proc Natl Acad Sci* **105:** 18782–18787.

Wofford JA, Wieman HL, Jacobs SR, Zhao Y, Rathmell JC. 2008. IL-7 promotes Glut1 trafficking and glucose uptake via STAT5-mediated activation of Akt to support T-cell survival. *Blood* **111:** 2101–2111.

Woods A, Johnstone SR, Dickerson K, Leiper FC, Fryer LG, Neumann D, Schlattner U, Wallimann T, Carlson M, Carling D. 2003. LKB1 is the upstream kinase in the AMP-activated protein kinase cascade. *Curr Biol* **13:** 2004–2008.

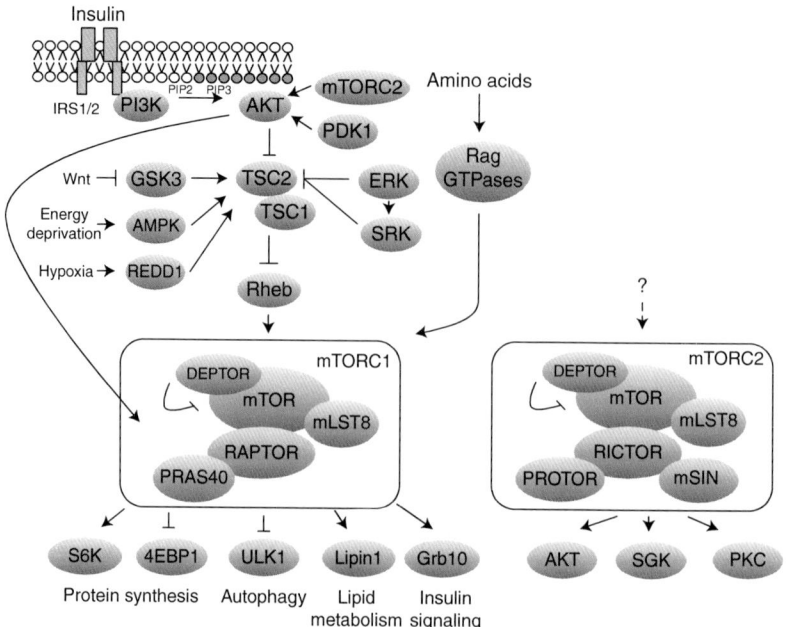

Figure 1. The mTOR pathway. The mTOR kinase exists in two distinct complexes called mTOR complex 1 (mTORC1) and mTORC2. In addition to mTOR, both complexes also contain mLST8 and DEPTOR. RAPTOR and PRAS40 are unique to mTORC1, whereas RICTOR, mSIN1, and PROTOR are specific to mTORC2. mTORC1 is well known to control cell growth. Although the downstream mechanisms by which mTORC1 controls growth are still being elucidated, several direct mTORC1 substrates have now been validated. For example, mTORC1 can regulate growth by directly phosphorylating S6K1 and 4EBP1, two regulators of protein translation; by regulating autophagy through the direct phosphorylation of ULK1; by regulating lipid metabolism at least in part by phosphorylating Lipin1; and by modulating insulin signaling through Grb10. In cells, mTORC1 activity is controlled by nutrient availability, particularly that of amino acids, through a novel pathway requiring the Rag GTPases. In the presence of amino acids, the Rags deliver mTORC1 to the lysosome, which is thought to be the signaling center for mTORC1-driven growth control (not shown but described in text). Growth factor signaling through PI3K–AKT, as well as numerous stresses, such as energy deprivation and hypoxia, can also impinge on mTORC1 activity through the TSC1–TSC2 complex. TSC2 negatively regulates a small GTPase called Rheb that directly activates mTORC1 by an unknown biochemical mechanism. AKT can also activate mTORC1 by directly phosphorylating PRAS40, which relieves an inhibitory function of this subunit. mTORC2 was discovered more recently, and although it is known to be activated in response to growth factors, the mechanisms are unknown. The best described substrates of mTORC2 are AKT and SGK, which require mTORC2-dependent phosphorylation for full biochemical activity. PKC is also regulated by mTORC2.

RICTOR (PROTOR) are specific to mTORC2. Recent work is beginning to reveal insight into the structure of mTORC1 (Yip et al. 2010). A cryo-EM structure of mTORC1 at 26 Å indicates that the complex is a dimer with interlocking mTOR–RAPTOR interactions and where PRAS40 acts as a competitive inhibitor for the binding of mTORC1 substrates to RAPTOR (Wang et al. 2008; Yip et al. 2010). mLST8 associates with the mTOR kinase domain, located in the carboxyl terminus (Kim et al. 2003). In the presence of rapamycin, the mTOR–mLST8 interaction is stable, whereas the drug weakens the mTOR–RAPTOR interaction (Kim and Sabatini 2004; Yip et al. 2010). The structure of mTORC2 remains a mystery.

mTORC1 is the best understood mTOR complex and is well known to control cell autonomous growth by integrating at least four growth regulatory inputs: nutrient availability, growth factor signaling, cellular energy status, and cellular stress levels (for review, see Sengupta et al. 2010; Zoncu et al. 2010). The best described mechanism by which mTORC1 controls growth is by directly phosphorylating two regulators of protein translation, p70-S6 kinase 1 (S6K1) and 4E binding protein 1 (4E-BP1). mTORC1-dependent phosphorylation of S6K at T389 activates its kinase activity toward several substrates involved in mRNA maturation and protein translation. In contrast to S6K1, 4E-BP1 represses translation, and its multisite phosphorylation by mTORC1 decreases its affinity to the translation initiation factor eIF4E, thus activating cap-dependent translation (for review, see Ma and Blenis 2009). Recently, several new substrates for mTORC1 have been identified and characterized. For example, mTORC1 also controls growth by negatively regulating autophagy through the direct phosphorylation of Unc-51-like kinase (ULK1; discussed below) (Chan 2009; Ganley et al. 2009; Hosokawa et al. 2009; Jung et al. 2009). In addition, mTORC1 directly phosphorylates the phosphatidate phosphatase Lipin1 (which regulates lipid metabolism through SREBP1) and the growth factor receptor-bound protein 10 (Grb10; part of a negative feedback loop targeting insulin receptor signaling) (Hsu et al. 2011; Peterson et al. 2011; Yu et al. 2011). In fact, two recent proteomic studies identifying Grb10 as an mTORC1 substrate further suggest that the insulin-stimulated phosphorylation cascade is largely mTOR-dependent (Hsu et al. 2011; Yu et al. 2011). A role for mTORC1 in regulating PGC1α activity and mitochondrial function has also been described (Cunningham et al. 2007; Ramanathan and Schreiber 2009).

How do upstream signals regulate mTORC1? In multicellular organisms, one mechanism by which growth factor signaling regulates mTORC1 is through the PI3K–AKT pathway (for review, see Manning and Cantley 2007). Following phosphoinositide 3-kinase (PI3K) activation, AKT (also known as PKB) is recruited to the membrane, which triggers its phosphorylation and activation. Among its many substrates, AKT directly phosphorylates and inactivates TSC2, which together in a complex with TSC1 negatively regulates mTORC1 activity (Inoki et al. 2002). The TSC1/TSC2 complex inhibits mTORC1 by suppressing the activity of its activator, a small GTPase called Ras homolog enriched in brain (Rheb) (for review, see Manning and Cantley 2003). AKT can also activate mTORC1 by directly phosphorylating PRAS40 and relieving its inhibitory effect on the complex (Sancak et al. 2007; Vander Haar et al. 2007). In addition to being inactivated by PI3K–AKT signaling, the TSC1/TSC2 complex can also be inhibited by other growth-promoting signals including mitogens or cytokines through the action of the mitogen-activated protein kinase (MAPK) ERK1/2 and the p90 ribosomal S6 kinase (RSK) (for review, see Huang and Manning 2008). In contrast, growth inhibitory signals such as energy deprivation and hypoxia can activate TSC1/TSC2 through the AMP-activated protein kinase (AMPK) and regulated in development and DNA damage responses 1 (REDD1), respectively, to suppress mTORC1 activity. In addition, Wnt-regulated glycogen synthase kinase 3β (GSK3β) coordinates with AMPK to inhibit mTOR signaling such that when Wnt signaling is active, GSK3β/AMPK-dependent activation of TSC2 is inhibited to allow mTORC1 activation (Inoki et al. 2006). Thus, the TSC1/TSC2 complex integrates numerous positive and negative growth signals and adjusts mTORC1 activity accordingly.

In contrast to growth factors, nutrient sensing by mTORC1—particularly of amino acids—is an ancient function of the complex conserved from yeast to humans. However, the mechanism by which amino acids regulate mTORC1 is just beginning to be elucidated. Amino acids (in particular, leucine) promote mTORC1 activity independently of TSC2 through a pathway that requires Rheb-dependent activation of the complex (Smith et al. 2005; Sancak and Sabatini 2009). mTORC1 activation by L-leucine is also dependent on glutamine transport into the cell, although glutamine alone has no effect on mTORC1 (DeBerardinis et al. 2007; Nicklin et al. 2009). Although the amino acid-sensing mechanism is just beginning to be revealed, it appears to require a second family of GTPases: the Rag proteins, which bind to mTORC1 and translocate it from an undefined location in the cytoplasm to the lysosome upon amino acid stimulation (Kim et al. 2008; Sancak et al. 2008, 2010). Mammals have four Rag proteins that form heterodimers: RagA or RagB (which are more closely related to each other) with RagC or RagD (which likewise are more closely related). When RagA/B is bound to GTP, RagC/D is bound to GDP and vice versa. In the presence of amino acids, for unknown reasons, RagA/B is GTP bound, and this somehow promotes the interaction and lysosomal recruitment of mTORC1. This interaction may be facilitated by the p62/sequestosome 1, which colocalizes with the Rags at the lysosome and is required for mTORC1 recruitment there (Duran et al. 2011). The Rags are tethered to the lysosome by a complex of proteins also essential for mTORC1 function called the Ragulator (formed by p18, p14, and MP1 proteins) (Sancak et al. 2010). The Ragulator forms the mTORC1 lysosomal docking site. Because a fraction of Rheb also resides at the lysosome, the current model is that the Rags/Ragulator, by an amino-acid-dependent mechanism, bring mTORC1 into close proximity to its activator Rheb-GTP at the lysosomal surface.

Why is the lysosome the site of mTORC1 activation in response to amino acid sufficiency and not the plasma membrane or some other cellular location? Although still a major question, the mystery is beginning to unravel. A complex of lysosomal proteins collectively called the v-ATPase is required for mTORC1 activity (Zoncu et al. 2011). Amino-acid-stimulated recruitment of mTORC1 to the lysosome and its ability to phosphorylate downstream substrates requires the v-ATPase, which interacts directly with the Ragulator. Interestingly, amino acids appear to accumulate inside the lysosomal lumen and signal through the v-ATPase to the Rags and Ragulator to activate mTORC1 by what has been dubbed an "inside-out" mechanism. Exactly how this happens is still under investigation but appears to require ATP hydrolysis and an associated rotation of the v-ATPase stalk. Interestingly, mTORC1 also negatively regulates lysosome biogenesis (Settembre et al. 2012), indicating that mTORC1 might also play an important role in controlling the number of lysosomes in the cell. Importantly, much of the understanding of the nutrient-sensing mechanism has been solved in vitro, and it will be important in future studies to determine the extent to which the amino acid-sensing pathway functions in various tissues and tumor cells.

Compared with mTORC1, mTORC2 was discovered more recently, and we know considerably less about its regulation and function. Growth factors activate mTORC2 at least in part through PI3K signaling, but the mechanism is unknown. Few substrates of mTORC2 have been described (Sparks and Guertin 2010). However, the discovery that mTORC2 directly phosphorylates AKT revealed a key role for mTORC2 in mediating downstream PI3K signaling in response to growth factor activation (Sarbassov et al. 2005). There are three mammalian AKT isoforms (AKT1, AKT2, and AKT3) that have both overlapping and distinct functions. In general, AKT1 is believed to have a more critical role in cell survival (Chen et al. 2001; Cho et al. 2001b), whereas AKT2 regulates glucose homeostasis (Cho et al. 2001a; Garofalo et al. 2003). AKT3 is implicated in brain development (Tschopp et al. 2005). As mentioned above, full AKT activation requires dual phosphorylation by both PDK1 (which phosphorylates AKT1 at T308 in the kinase motif) and by mTORC2 (which phosphorylates AKT1 at S473 in a carboxy-terminal hydrophobic motif). In addition to AKT, another AGC kinase family protein, Serum and glucocorticoid-induced kinase (SGK), is directly phosphorylated by mTORC2 (Garcia-Martinez and Alessi 2008).

Most signals upstream of mTOR appear to target either mTORC1 or mTORC2. However, a mechanism of regulation shared by both complexes may occur through DEPTOR, which is a common subunit of both mTOR complexes (for review, see Zoncu et al. 2010). Although the regulation of DEPTOR is complicated, it appears to be a natural inhibitor of both mTOR complexes because in its absence, S6K, AKT, and SGK activity increases (Peterson et al. 2009). DEPTOR levels are controlled though the ubiquitin-dependent degradation pathway by a mechanism requiring direct phosphorylation of DEPTOR by mTOR in an apparent positive feedback loop (Peterson et al. 2009; Duan et al. 2011; Gao et al. 2011; Zhao et al. 2011). Interestingly, DEPTOR levels are low in many cancers, which may promote mTOR-dependent cell growth, proliferation, and survival (Peterson et al. 2009). However, in a subset of multiple myelomas, DEPTOR is highly expressed. In these cells, DEPTOR overexpression inhibits mTORC1, but this relieves strong negative feedback loops to PI3K that may override its inhibitory effect on mTORC2 and consequently promote AKT-mediated cell survival (discussed below).

TSC2 PROTECTS AGAINST METABOLIC STRESS-INDUCED APOPTOSIS

As mentioned above, the TSC1/TSC2 complex integrates many growth regulatory signals to control mTORC1 activity. Many of these signals convey information regarding the metabolic state of the cell such that when nutrients are limiting, the cell will restrict mTORC1-dependent growth pathways. One of the best examples is that of AMPK-dependent phosphorylation and activation of TSC2, which is required for cell survival under glucose deprivation conditions (Inoki et al. 2003). AMPK is a major sensor of cellular energy status that is activated under conditions of metabolic stress (e.g., glucose deprivation) that decrease ATP production (for review, see Hardie 2007). AMPK is activated by increases in cellular AMP levels to promote catabolic pathways necessary to restore a critical level of ATP required for cell survival. It was initially found that AMPK activation inhibits S6K1 phosphorylation, suggesting a link between energy-sensing pathways and mTORC1 signaling (Kimura et al. 2003). It was subsequently shown that at least one mechanism by which AMPK inhibits mTORC1 is by directly phosphorylating and activating TSC2 (Inoki et al. 2003). Later work found that AMPK also directly phosphorylates RAPTOR, which can also suppress mTORC1 activity (Gwinn et al. 2008). By activating TSC2 and inhibiting mTORC1, AMPK shuts down mTORC1-dependent growth pathways from consuming cellular energy. In the absence of TSC2 function (e.g., in $TSC2^{-/-}$ cells), glucose deprivation results in cell death (Inoki et al. 2003). Consistent with death being driven by mTORC1 pathways, rapamycin treatment prevents the death of $TSC2^{-/-}$ cells when glucose is unavailable in the culture medium (Inoki et al. 2003; Choo et al. 2010).

A number of mechanisms might explain why TSC2-deficient cells are sensitive to apoptosis when deprived of glucose. One possibility is that uncontrolled mTORC1 activity in starving cells continues to drive protein translation through 4E-BP1 and S6K1, and because translation is a major consumer of cellular energy, this exhausts the ATP reserves. However, rapamycin is only a partial mTORC1 inhibitor and relatively inefficient at suppressing translation in mammalian cells, suggesting that other rapamycin-sensitive mTORC1 pathways might also be important (Feldman et al. 2009; Thoreen et al. 2009; Choo et al. 2010).

A second possibility is that mTORC1 activation during nutrient stress promotes p53 synthesis and accumulation (Lee et al. 2007). AMPK has also been shown to stabilize p53 by direct phosphorylation (Jones et al. 2005). Thus, glucose deprivation in normal cells causes AMPK to inhibit mTORC1 and stabilize p53, stalling cell growth and division; but when $TSC2^{-/-}$ cells are glucose deprived, AMPK stabilizes p53 whereas unrestricted mTORC1 signaling drives p53 synthesis. This synergistically results in greatly elevated levels of p53 and subsequently apoptosis. Notably, the connections between p53, mTORC1 activity, and survival are complex (Feng et al. 2005). For instance, p53 also induces the transcription of PTEN, TSC2, and REDD1, which negatively regulate mTORC1 (Stambolic et al. 2001; Ellisen et al. 2002; Feng et al. 2005). In addition, p53 also activates AMPK as well as Sestrin 1 and Sestrin 2, which are also negative regulators of mTORC1 signaling (Feng et al. 2007; Budanov and Karin 2008; Feng 2010).

A third possibility is that in the glucose-deprived state, losing TSC2 function promotes mTORC1-dependent negative feedback loops that suppress PI3K–AKT signaling, squelching critical survival signals. The best described mTORC1 negative feedback loops function through S6K1 and Grb10 (Zoncu et al. 2010; Hsu et al. 2011; Yu et al. 2011). Another mechanism of negative feedback occurs through the unfolded protein response (UPR) (Ozcan et al. 2008). The UPR senses unfolded proteins in the ER lumen and transmits that information to the cell nucleus, where it drives a transcriptional program to reestablish homeostasis (Kozutsumi et al. 1988). In this model, mTOR hyperactivation during glucose deprivation induces ER stress (presumably through increased client load) and therefore, the UPR. It was found that the UPR promotes feedback inhibition of PI3K–AKT signaling, possibly through the JNK kinase (Ozcan et al. 2008). The UPR might also directly activate apoptotic pathways in response to the overwhelming demand on the ER to faithfully regulate protein folding.

Another more recent report finds that the hypersensitivity of $TSC2$-deficient cells to glucose deprivation is not linked to blocking apoptosis, to p53 levels, or to activating autophagy, but rather to rapamycin's ability to decrease metabolic consumption, maintain ATP levels, and suppress AMPK, thus preventing energetic stress (Choo et al. 2010). These investigators also find that $TSC2^{-/-}$ cells deprived of glucose shift to using glutamine as a carbon source and that rapamycin fails to suppress cell death in the absence of glutamine. Therefore, in this model, rapamycin's protective effect is the result of decreasing the bioenergetic demand in order to balance cellular metabolism with the supply of nutrients and to support the shift to a glutamine-based metabolism. This response is at least partially dependent on S6K1, but not on eIF4E, consistent with rapamycin's relative ineffectiveness at blocking 4E-BP1 phosphorylation (Feldman et al. 2009; Thoreen et al. 2009; Choo et al. 2010).

mTORC1 DIRECTLY REGULATES AUTOPHAGY

Although the mTORC1 pathways responsible for triggering apoptosis in cultured $TSC2^{-/-}$ cells deprived of glucose are complex, one clear and conserved connection between mTORC1 and a pathway critical for cell survival upon nutrient deprivation is the discovery that mTORC1 directly regulates autophagy. Normally cells activate autophagy (or macroautophagy) in times of nutrient deprivation to salvage critical nutrients essential for cell survival. By mechanisms still being worked out, autophagy targets proteins and organelles (such as the mitochondria) to the autophagosome, which then delivers the cargo to the lysosome for degradation and recycling of macromolecules (for review, see Chen and Klionsky 2010; Yang and Klionsky 2010; Das et al. 2012). More than 30 different autophagy genes (ATGs) have been identified that regulate autophagy induction, cargo selection, vesicle formation, autophagosome fusion, cargo degradation, and release (for review, see He and Klionsky 2009). In cells, autophagy is a critical survival mechanism under nutrient deprivation conditions, and when inhibited either genetically or pharmacologically, nutrient deprivation can result in apoptosis (Boya et al. 2005). Autophagy is also essential for the survival of newborn mice, which require autophagy to mobilize nutrient stores during a brief starvation period immediately after birth (Kuma et al. 2004). Defective autophagy is also implicated in neuronal degeneration, cancer, and aging-associated pathologies (for review, see Yang and Klionsky 2009).

Although understanding mammalian autophagy regulation is an emerging and intense area of research, it is generally accepted that mTORC1 is a major negative regulator of autophagosome formation. Studies in yeast suggested early on that TORC1 inhibits autophagy. For example, rapamycin treatment activates autophagy in yeast even when they are growing in nutrient-rich conditions (Noda and Ohsumi 1998). Yeast TORC1 directly prevents assembly of an ATG1 kinase-containing complex required for autophagy induction (Kamada et al. 2000). Rapamycin inhibits TORC1's ability to disrupt ATG1 complex assembly, thus activating ATG1 kinase activity. It was thought that mTORC1 also controlled autophagy in mammalian cells, but until recently, the mechanism was vague. In general, rapamycin is less effective at activating autophagy in mammalian cells, although in some cases, rapamycin treatment causes accumulation of insoluble protein aggregates characteristic of those typically associated with failed autophagy (Spilman et al. 2010). In contrast to rapamycin, catalytic (ATP-competitive) inhibitors of mTOR are much more potent activators of autophagy in mammalian cells, indicating the role of mTOR in autophagy regulation is clearly conserved (Thoreen et al. 2009).

A series of reports suggests that the mechanism by which mTORC1 regulates autophagy is at least in part through direct phosphorylation of Unc-51-like kinase (ULK1), the homolog of yeast ATG1 (Chan 2009; Ganley et al. 2009; Hosokawa et al. 2009; Jung et al. 2009). In mammalian cells, ULK1 is the catalytic subunit of a complex containing mAtg13, Focal adhesion kinase-interacting protein of 200 kD (FIP200), and Atg101, all of which are essential for starvation-induced autophagy (Fig. 2). mAtg13 binds ULK1 and mediates the interaction between ULK1 and FIP200, but both FIP200 and mAtg13 appear to regulate ULK1 localization and stability. In nutrient-rich conditions, mTORC1 associates with the ULK1–mAtg13–FIP200 complex through a direct interaction between RAPTOR and ULK1, and this facilitates phosphorylation of both mAtg13 and ULK1 by mTORC1. The function of mTORC1-dependent ULK1 phosphorylation is not entirely clear, but it appears to diminish ULK1 kinase activity, thus reducing autophagic vesicle formation (Kim et al. 2011; Shang et al. 2011).

In the nutrient-deprived state, mTORC1 dissociates from the ULK1 complex, resulting in ULK1 dephosphorylation (Kim et al. 2011; Shang et al. 2011). However, this alone does not result in autophagy activation. For this to occur, ULK1 also requires direct activating phosphorylation by AMPK, emphasizing again the interplay between nutrient and energy-sensing pathways (Egan et al. 2011; Kim and Guan 2011; Kim et al. 2011; Zhao and Klionsky 2011). In starved cells, AMPK tightly binds ULK1, and this interaction is enhanced by rapamycin and disrupted by Rheb overexpression (Behrends et al. 2010; Kim and Guan 2011). In addition, AMPK can also inhibit mTORC1 by phosphorylating and activating TSC2, and directly by phosphorylating the RAPTOR subunit, which inhibits mTORC1 activity (Krause et al. 2002; Inoki et al. 2003; Gwinn et al. 2008). Interestingly, mTORC1 can be reactivated after prolonged starvation by the autolysosomal products generated by autophagy, indicating that a minimal level of mTORC1 activity is required for survival (Liang et al. 2007; Matsui et al. 2007; Herrero-Martin et al. 2009). This latter requirement for mTORC1 might be important for recycling lysosomes (Yu et al. 2010).

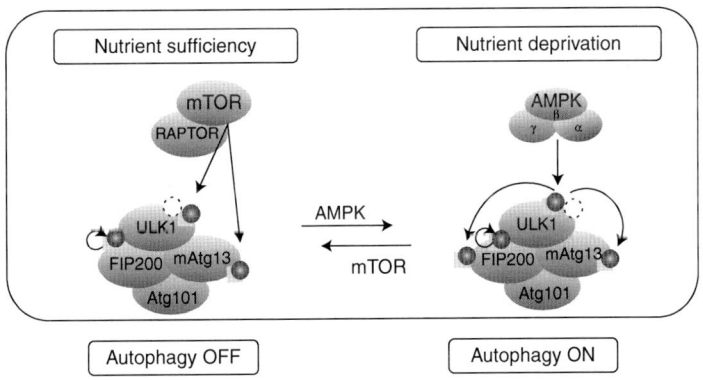

Figure 2. Regulation of the ULK complex by mTORC1 and AMPK. The ULK complex contains ULK1, mAtg13, FIP200, and Atg101. In nutrient-rich conditions, mTORC1 associates with the complex through a direct interaction between RAPTOR and ULK1 and phosphorylates both mAtg13 and ULK1. mTORC1-dependent ULK1 phosphorylation diminishes ULK1 kinase activity, preventing autophagy induction. In the nutrient-deprived state, mTORC1 dissociates from the ULK1 complex, resulting in ULK1 dephosphorylation at the mTORC1-dependent sites and phosphorylation at distinct sites by AMPK. Under these conditions, AMPK also tightly interacts with ULK1 to promote autophagy induction.

The widely held view is that autophagy is downstream from mTORC1 such that when mTORC1 is "OFF" autophagy is "ON." However, the final destination of autophagic cargo including proteins/amino acids is the lysosome, and the amino acid signal that activates mTORC1 was recently shown to emanate from within the lysosome (discussed above). Thus, autophagy should activate mTORC1, which, in fact, has been observed (Liang et al. 2007; Matsui et al. 2007; Herrero-Martin et al. 2009). How can this be reconciled? An alternative view of the relationship between mTORC1 and autophagy might explain this. In this alternative view, autophagy is actually upstream of mTORC1, and AMPK is the main regulator of autophagy that directly activates ULK1 and suppresses mTORC1. For example, nutrient and energy deprivation reduces mTORC1 activity and activates AMPK, and active AMPK promotes autophagy as described above. But without mTORC1 activity, cells would die. Therefore, by delivering amino acids to the lysosome, autophagy actually maintains mTORC1 in a minimally active state. In turn, mTORC1 inhibits autophagy as part of a negative feedback loop to prevent cells from eating themselves to death. Thus, autophagy may actually regulate mTORC1.

How long can cells survive nutrient deprivation? It would seem that prolonged starvation would, in fact, lead cells to consume themselves to death or initiate apoptosis. In fact, under certain conditions, autophagy can kill cells through a process known as autophagic cell death or cell death type II (for review, see Eskelinen 2005). However, feedback mechanisms may exist to prevent cell death caused by prolonged activation of autophagy, and one appears to require mTORC1-dependent phosphorylation of DAP1 (death-associated protein 1) (Koren et al. 2010). Under amino acid starvation (i.e., mTORC1 inhibited), DAP1 is rapidly dephosphorylated, and by an unclear mechanism, this restricts excessive autophagy. Protein phosphatases likely also work together with mTORC1 and AMPK to control autophagy. In fact, mice lacking the protein phosphatase 1 (PP1) regulatory subunit Gadd34 (growth arrest and DNA damaged protein) cannot turn off autophagy because of sustained phosphorylation of the AMPK target site in TSC2 (Uddin et al. 2011). Thus, although the regulation of autophagy is complex, by linking autophagy induction to both mTORC1 inhibition and AMPK activation, cells can tightly regulate cellular energy homeostasis and survive under conditions of metabolic stress.

mTORC2-DEPENDENT CELL SURVIVAL PATHWAYS

In addition to its role in promoting cell growth through the TSC–mTORC1 pathway, AKT has long been thought to promote cell survival directly through several mechanisms including (1) directly phosphorylating and inhibiting pro-apoptotic proteins such as BAD; (2) directly phosphorylating and inhibiting the Forkhead box O (FoxO) transcription factors, which regulate pro-apoptotic genes such as BIM and Fas ligand (also known as CD95L); (3) promoting p53 degradation by activating the murine double minute 2 (MDM2); (4) blocking the glycogen synthase kinase 3 (GSK3)-mediated inhibitory signals to the pro-survival protein Mcl-1; and (5) activating the NF-κB survival pathway via phosphorylating IκB kinase α (IKKα) (for review, see Manning and Cantley 2007). The widespread role of AKT in regulating cell survival pathways predicts that mTORC2 might control some or all of these processes. Although this seems likely, the exact role of mTORC2-mediated AKT-S473 phosphorylation in regulating cell survival is still under investigation.

One unresolved issue is whether mTORC2-dependent AKT phosphorylation is required for all AKT functions, or for only a subset of its targets. Typically, many protein kinases of the protein kinase A/protein kinase G/protein kinase C (AGC) family, such as AKT, serum, and glucocorticoid-induced protein kinase (SGK), and S6K, require prior phosphorylation in their hydrophobic motif (HM), which creates a docking site for PDK1 (Biondi et al. 2002; Frodin et al. 2002). For example, mTORC1-mediated phosphorylation on the HM site of S6K enhances the affinity for PDK1 binding and promotes full S6K activation (for review, see Jacinto and Lorberg 2008). Accordingly, acute in vitro knockdown experiments suggest that mTORC2 is required for both AKT Thr308 and Ser473 phosphorylation (Sarbassov et al. 2005). However, genetic studies indicate that Thr308 is still phosphorylated even in the complete absence of mTORC2 activity (by deleting the *rictor*, *mlst8*, or *sin1* genes), arguing that in other contexts, these events might occur independently (Guertin et al. 2006; Jacinto et al. 2006; Shiota et al. 2006). The reason for this discrepancy remains unclear but may reflect a difference between acute knockdown and chronic knockout experiments, or the existence of a compensatory mechanism. Although phosphorylation at both T308 and S473 is required for maximal AKT activity in vitro, it appears that T308 phosphorylation alone empowers AKT with enough activity to phosphorylate many of its substrates in cultured cells or in tissues (Alessi et al. 1996; Guertin et al. 2006; Jacinto et al. 2006; Yang et al. 2006; Bentzinger et al. 2008; Kumar et al. 2008, 2010; Cybulski et al. 2009; Gu et al. 2011).

To date, only a select few of the predicted AKT substrates have been examined for signaling defects caused by loss of mTORC2 activity. One AKT substrate important in cell survival that appears to require mTORC2 activity is the FoxO1/3a transcription factors (Guertin et al. 2006, 2009; Jacinto et al. 2006; Yang et al. 2006). For example, FoxO1 (T24) and FoxO3a (T32) phosphorylation is decreased upon mTORC2 inactivation in both knockdown and genetic knockout studies. In the absence of AKT-mediated phosphorylation, FoxOs accumulate in the nucleus and activate metabolic and cell survival genes (Biggs et al. 1999; Nakae et al. 1999; Rena et al. 1999; Tang et al. 1999). Interestingly, phosphorylation of other AKT targets such as BAD, TSC2, and GSK3β show little to no effect upon genetic ablation of mTORC2 (Guertin et al. 2006, 2009; Shiota et al. 2006; Yang et al. 2006). Other key survival proteins downstream from AKT such as MDM2, Caspase-9 and IKKα have not yet been investigated. Because FoxOs have critical roles in cell survival, mTORC2 may regulate cell viability through Akt–FoxO pathways.

Because AKT is predicted to activate mTORC1, it is not unreasonable to predict that losing mTORC2-dependent AKT phosphorylation might also decrease mTORC1 activity and induce autophagy. However, despite most models placing mTORC2 upstream of the AKT–TSC2–mTORC1 signaling axis, evidence that this connection is important in vivo is lacking. In fact, it appears that losing mTORC2 activity has minimal effects on mTORC1 signaling in many cell types (Guertin et al. 2006; Jacinto et al. 2006; Shiota et al. 2006; Bentzinger et al. 2008; Kumar et al. 2008, 2010; Cybulski et al. 2009; Gu et al. 2011). Interestingly, mTORC2 may regulate autophagy independently of mTORC1 via the AKT–FoxO3a axis (Mammucari et al. 2007). In fasting skeletal muscle, FoxO3a positively controls transcription of several autophagy-related genes, including LC3 and Bnip3. Tamoxifen-induced activation of recombinant AKT (AKT fused with estrogen receptor) blocks FoxO3a activation and autophagy induction. Conversely, knockdown of RICTOR promotes FoxO3 nuclear retention and autophagosome formation. Another report indicates that insulin signaling also inhibits autophagy in the liver through a FoxO1-mediated mechanism (Liu et al. 2009). Thus, mTORC2 may regulate autophagic survival through AKT-dependent, mTORC1-independent mechanisms.

In vivo studies of conventional *rictor*, *sin1*, and *mlst8* knockout embryos (all of which result in selective mTORC2 ablation) indicate that mTORC2 is essential for progression through mid-embryonic development (Guertin et al. 2006). Although the exact cause of lethality is unknown, increased cell death was not readily apparent in the knockout embryos. To gain further insight into the tissue-specific functions of mTORC2, conditional knockout models of *rictor* have been developed, including skeletal muscle, white adipose tissue, and pancreatic β cells (Bentzinger et al. 2008; Kumar et al. 2008, 2010; Cybulski et al. 2009; Gu et al. 2011). Although some metabolic defects are reported, there is no indication from these studies that mTORC2 loss—at least under otherwise normal physiological conditions—results in increased apoptosis. This may reflect the fact that in all three of these tissues, AKT Thr308 phosphorylation is largely preserved despite decreased mTORC2-dependent Ser473 phosphorylation. Thus, more studies are needed to determine exactly if and how mTORC2 might regulate cell survival.

DOES mTORC2 REGULATE CANCER CELL SURVIVAL?

Although it is not clear exactly which AKT pathways require mTORC2 under normal conditions, several lines of evidence suggest that mTORC2 may be more essential for AKT signaling in cells with oncogenic activation of PI3K activity. For example, in a *PTEN*-deletion-driven mouse model of prostate cancer deleting *RICTOR* blocks tumor formation but has no effect on normal prostate growth or function (Guertin et al. 2009). Interestingly, ablating mTORC2 activity in the *PTEN*-null tumor cells reduces both AKT Ser473 and Thr308 phosphorylation, perhaps suggesting a differential requirement for mTORC2 in prostate cancer cells compared with MEFs. The experiments in the prostate cancer model are reminiscent of genetic studies in *Drosophila*, in which *dRICTOR* is less essential for fly development but is required for phenotypes induced by *PTEN* deletion or PI3K activation (Hietakangas and Cohen 2007). Importantly, it is unclear in these models whether the *rictor*/mTORC2-deficient cells have survival defects.

In vitro studies using several human cancer cell lines further indicate that knocking down RICTOR is toxic to transformed cells with elevated AKT activity. For example, mTORC2 activity is elevated in gliomas and is required for anchorage-independent growth and proliferation in vitro and for tumor growth in a xenograft model (Masri et al. 2007). Moreover, mTORC2 promotes cell cycle progression and anchorage-independent growth of breast (MCF7) and prostate (PC3) cancer cell lines (Hietakangas and Cohen 2008; Guertin et al. 2009). However, regarding a specific role for mTORC2 in cell survival, only one report shows that stable knockdown of RICTOR specifically impairs survival of a cancer cell line in vitro (in this case, colorectal cancer cells) (Gulhati et al. 2009). Cell proliferation is also inhibited in these cells.

One possible explanation for the differential requirement for mTORC2 activity in normal cells versus cancer cells is that under normal conditions, basal AKT activity (Thr308 phosphorylation only) is sufficient for maintaining AKT's essential functions, whereas in *PTEN*-deficient or *PI3KCA* (phosphatidyl inositol 3-kinase catalytic subunit) mutant transformed cells, the demand for AKT signaling is maximal, requiring both T308 and S473 phosphorylation to achieve its full activation potential. Alternatively, the difference may reflect a compensatory signal that reactivates AKT by up-regulated T308 phosphorylation upon prolonged loss of mTORC2 activity, and this pathway might not yet be functional following acute loss of mTORC2 in transient knockdown experiments. These compensatory mechanisms may exist specifically to avoid cell death. Such protective circuits may be cell type specific or only active under specific conditions, and clearly this needs further examination. Nevertheless, studies to date suggest that cancer cells with an abnormally high level of PI3K–AKT activity may have a greater requirement for mTORC2 than otherwise normal cells, and this provides rationale for developing a therapeutic strategy that selectively targets mTORC2. Importantly, however, it remains unclear to what extent mTORC2 activity is required for cancer cell survival in more advanced, therapeutically relevant stages of cancer, and whether a selective mTORC2 inhibitor (if it existed) would have an acceptable therapeutic window.

In addition to AKT, another AGC kinase family protein, SGK, is also directly phosphorylated by mTORC2 (Garcia-Martinez and Alessi 2008). It is reported that SGK can regulate cell viability through activation of MDM2-dependent p53 degradation (Amato et al. 2009). Moreover, like AKT, SGK also exists in three isoforms (SGK1, SGK2, and SGK3) in human and mouse (Brunet et al. 2001). However, in contrast to AKT, even basal SGK activity is controlled by mTORC2 because phosphorylation in its hydrophobic motif by mTORC2 is required for phosphorylation in the kinase domain by PDK1 (Garcia-Martinez and Alessi 2008). mTORC2 regulation of SGK could also have relevance in cancer cell survival as suggested by a recent study that finds SGK3 signaling (but not AKT signaling) downstream from PIK3CA mutations is essential for the survival of certain cancer cells (Vasudevan et al. 2009). Thus, mTORC2 may regulate cell survival through both AKT-dependent and AKT-independent pathways, although definitive mechanisms require further investigation.

DO mTOR INHIBITORS AFFECT CANCER CELL SURVIVAL?

Aberrant PI3K–AKT–mTOR signaling is a common feature of most cancers (Shaw and Cantley 2006; Zoncu et al. 2010; Hanahan and Weinberg 2011). Consequently, there is intense interest in developing mTOR inhibitors as cancer therapeutics. First-generation mTOR inhibitors are based on the chemical structure of rapamycin, but despite the strong rationale for using rapamycin in oncology, this class of drugs has unfortunately had limited success (O'Reilly et al. 2006; Sudarsanam and Johnson 2010; for review, see Guertin and Sabatini 2009; Benjamin et al. 2011). The best albeit modest responses to rapamycin as a therapy have been reported in renal cell carcinoma, mantle cell lymphoma, neuroendocrine tumors of the pancreas, and in treating tuberous sclerosis (caused by mutations in *TSC1* or *TSC2*) (Benjamin et al. 2011). There are several reasons that could explain this including the fact that rapamycin is an allosteric inhibitor of mTOR that binds outside the kinase domain and only partially inhibits mTORC1 activity. For example, rapamycin universally inhibits mTORC1-dependent S6K1 phosphorylation but has only minor and acute inhibitory effects on other mTORC1 substrates including 4E-BP1 (Choo et al. 2008; Feldman et al. 2009; Thoreen et al. 2009). In addition, rapamycin relieves strong negative feedback loops to PI3K that exist downstream from mTORC1 (Fig. 3) (Choo and Blenis 2009). As mentioned above, these feedback loops can function through mTORC1 substrates including S6K1 and Grb10. S6K1 directly phosphorylates insulin receptor substrate 1 (IRS-1), mislocalizing it and targeting it for degradation. The recently discovered mTORC1 substrate Grb10 directly binds to and negatively regulates the insulin and insulin-like growth factor receptors (Hsu et al. 2011; Yu et al. 2011). mTORC1 phosphorylation of Grb10 promotes its stability. Interestingly, Grb10 levels are often decreased in cancer, suggesting that it could have tumor-suppressor functions. In both cases, rapamycin relieves feedback inhibition and promotes PI3K–AKT survival signaling. The clinical relevance of losing feedback inhibition is emphasized in human trials that find rapamycin increases AKT activation in many malignancies (O'Reilly et al. 2006; Tabernero et al. 2008; Sudarsanam and Johnson 2010). Rapamycin can also activate the MAPK pathway, providing another potential avenue to resistance (Carracedo et al. 2008). Thus, as a single agent, rapamycin may actually promote cancer cell survival; however, rapamycin may ultimately prove to be useful in combination with agents such as PI3K, AKT, or MAPK inhibitors.

The discovery of mTORC2, which is generally rapamycin insensitive, and the widespread ineffectiveness of rapamycin as a monotherapy led to development of inhibitors that directly target the mTOR catalytic site (for review, see Guertin and Sabatini 2009). The first to be reported include Torin1, PP242, Ku-0063794, and WYE-354 (Feldman et al. 2009; Garcia-Martinez et al. 2009; Thoreen et al. 2009; Yu et al. 2009). The ATP-competitive inhibitor class more completely inhibits mTORC1 and additionally inhibits mTORC2. Of note, prolonged exposure to rapamycin can inhibit mTORC2 in a subset of cell types, and this might explain some of the clinical successes with the drug (Sarbassov et al. 2006; Gulhati et al. 2009). The mechanism is not entirely understood but may result from rapamycin blocking the assembly of new mTORC2 complexes. The mTOR ATP-competitive inhibitors are just beginning to be tested for clinical efficacy, and it is hoped that they will outperform rapamycin in the clinic. In a few preclinical studies, the mTOR catalytic inhibitors were shown to induce cell death in combination with other inhibitors (Janes et al. 2010; Sini et al. 2010).

Although the preclinical studies with mTOR catalytic site inhibitors are exciting, several

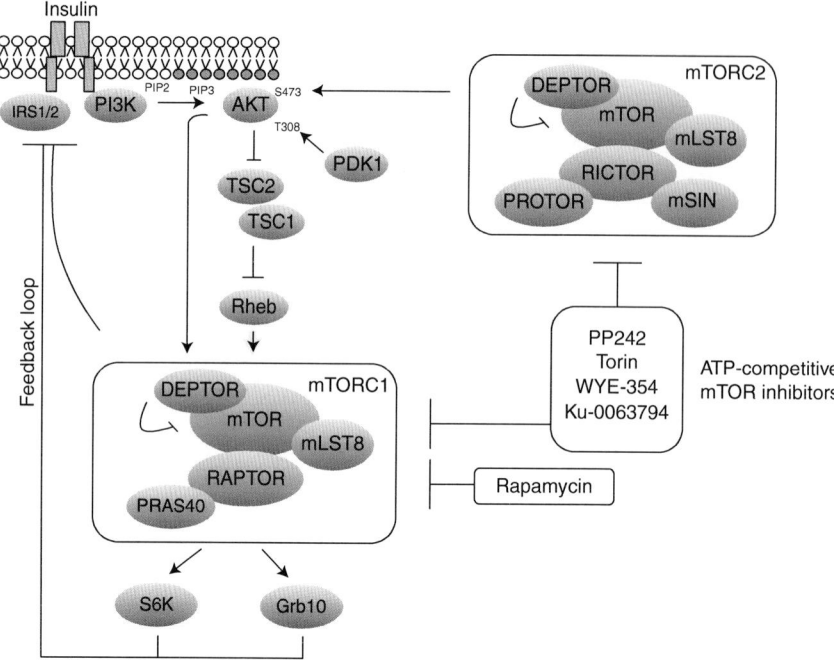

Figure 3. Feedback inhibition of insulin signaling. Several mechanisms of negative feedback inhibition of insulin signaling exist downstream from mTORC1. The best described are mediated by S6K1, which can phosphorylate and inhibit IRS1; and by Grb10, which is stabilized by mTORC1-dependent phosphorylation and suppresses signaling from the insulin and insulin-like growth factor receptors. mTOR inhibitors such as rapamycin and the new generation of ATP-competitive inhibitors can relive these negative feedback loops, resulting in PI3K–AKT activation.

questions regarding their efficacy remain. For example, will feedback activation of PI3K–AKT signaling still promote survival even though mTORC2 is also inhibited? Evidence that this could be problematic comes from studies of the natural mTOR inhibitor DEPTOR (discussed above), which emphasize the fact that losing feedback inhibition by inhibiting mTORC1 can override mTORC2 inhibition with respect to AKT activation (Peterson et al. 2009). Another potential concern is whether mTOR catalytic inhibitors will be well tolerated, although preclinical tests in rodent models are promising (for review, see Benjamin et al. 2011). The mTOR catalytic inhibitors are also more formidable activators of autophagy compared with rapamycin (Chresta et al. 2009; Thoreen et al. 2009). Because autophagy can promote cell survival in nutrient-limiting conditions, increasing autophagic activity could also promote cancer cell survival in the nutrient-deprived tumor microenvironment. In fact, in melanoma cells, inhibiting autophagy in combination with nutrient deprivation induces apoptosis, suggesting that autophagy can protect cancer cells from nutrient-limiting conditions (Sheen et al. 2011).

CONCLUSION

In this article, we review mechanisms by which cells respond to nutrient deprivation that impinge on mTORC1 signaling. We also discuss possible mechanisms by which the less-well-understood mTORC2 might regulate cell survival. Although starvation is clearly detrimental to a cell's ability to maintain long-term homeostasis, nutrient overload also stresses cells by forcing them to elevate their metabolism, increasing damaging reactive oxygen species (ROS) and oxidative stress (for review, see Wellen and Thompson 2010). This is the case in cancer, in which oncogenic pathways drive aberrant nutrient uptake and metabolism; and diabetes, where nutrient overload promotes obesity and insulin resistance. mTOR, by functioning as a point of convergence between a nutrient-sensing pathway and PI3K–AKT signaling (i.e., as part of mTORC1) and as a regulator of AKT itself (i.e., as part of mTORC2), is central to understanding how both normal and cancer cells survive nutrient excess and is a growing area of research. In sum, mTOR integrates growth signals from diverse mechanisms that sense nutrient availability and as part of the response regulates cell survival. Pathways deregulated in many human diseases clearly impinge on mTOR signaling; thus, defining the cell and tissue-specific mechanisms through which mTOR regulates cell survival will be critical to developing therapies to treat cancer and metabolic diseases.

ACKNOWLEDGMENTS

D.A.G. is supported by grants from the NIH (R00 CA129613 and R21 CA161121), the Charles Hood Foundation, the UMass Center for Clinical and Translational Sciences, and the PEW Charitable Trusts.

REFERENCES

*Reference is also in this collection.

Alessi DR, Andjelkovic M, Caudwell B, Cron P, Morrice N, Cohen P, Hemmings BA. 1996. Mechanism of activation of protein kinase B by insulin and IGF-1. *EMBO J* **15:** 6541–6551.

Amato R, D'Antona L, Porciatti G, Agosti V, Menniti M, Rinaldo C, Costa N, Bellacchio E, Mattarocci S, Fuiano G, et al. 2009. Sgk1 activates MDM2 dependent p53 degradation and affects cell proliferation, survival, and differentiation. *J Mol Med (Berl)* **87:** 1221–1239.

Behrends C, Sowa ME, Gygi SP, Harper JW. 2010. Network organization of the human autophagy system. *Nature* **466:** 68–76.

Benjamin D, Colombi M, Moroni C, Hall MN. 2011. Rapamycin passes the torch: A new generation of mTOR inhibitors. *Nat Rev Drug Discov* **10:** 868–880.

Bentzinger CF, Romanino K, Cloetta D, Lin S, Mascarenhas JB, Oliveri F, Xia J, Casanova E, Costa CF, Brink M, et al. 2008. Skeletal muscle-specific ablation of raptor, but not of rictor, causes metabolic changes and results in muscle dystrophy. *Cell Metab* **8:** 411–424.

Biggs WH 3rd, Meisenhelder J, Hunter T, Cavenee WK, Arden KC. 1999. Protein kinase B/Akt-mediated phosphorylation promotes nuclear exclusion of the winged helix transcription factor FKHR1. *Proc Natl Acad Sci* **96:** 7421–7426.

Biondi RM, Komander D, Thomas CC, Lizcano JM, Deak M, Alessi DR, van Aalten DM. 2002. High resolution crystal structure of the human PDK1 catalytic domain defines the regulatory phosphopeptide docking site. *EMBO J* **21:** 4219–4228.

Boya P, Gonzalez-Polo RA, Casares N, Perfettini JL, Dessen P, Larochette N, Metivier D, Meley D, Souquere S, Yoshimori T, et al. 2005. Inhibition of macroautophagy triggers apoptosis. *Mol Cell Biol* **25:** 1025–1040.

Brunet A, Park J, Tran H, Hu LS, Hemmings BA, Greenberg ME. 2001. Protein kinase SGK mediates survival signals by phosphorylating the Forkhead transcription factor FKHRL1 (FOXO3a). *Mol Cell Biol* **21:** 952–965.

Budanov AV, Karin M. 2008. p53 target genes *sestrin1* and *sestrin2* connect genotoxic stress and mTOR signaling. *Cell* **134:** 451–460.

Carracedo A, Ma L, Teruya-Feldstein J, Rojo F, Salmena L, Alimonti A, Egia A, Sasaki AT, Thomas G, Kozma SC, et al. 2008. Inhibition of mTORC1 leads to MAPK pathway activation through a PI3K-dependent feedback loop in human cancer. *J Clin Invest* **118:** 3065–3074.

Chan EY. 2009. mTORC1 phosphorylates the ULK1–mAtg13–FIP200 autophagy regulatory complex. *Sci Signal* **2:** e51.

Chen Y, Klionsky DJ. 2010. The regulation of autophagy—Unanswered questions. *J Cell Sci* **124:** 161–170.

Chen WS, Xu PZ, Gottlob K, Chen ML, Sokol K, Shiyanova T, Roninson I, Weng W, Suzuki R, Tobe K, et al. 2001. Growth retardation and increased apoptosis in mice with homozygous disruption of the *Akt1* gene. *Genes Dev* **15:** 2203–2208.

Cho H, Mu J, Kim JK, Thorvaldsen JL, Chu Q, Crenshaw EB 3rd, Kaestner KH, Bartolomei MS, Shulman GI, Birnbaum MJ. 2001a. Insulin resistance and a diabetes mellitus-like syndrome in mice lacking the protein kinase Akt2 (PKBβ). *Science* **292:** 1728–1731.

Cho H, Thorvaldsen JL, Chu Q, Feng F, Birnbaum MJ. 2001b. Akt1/PKBα is required for normal growth but dispensable for maintenance of glucose homeostasis in mice. *J Biol Chem* **276:** 38349–38352.

Choo AY, Blenis J. 2009. Not all substrates are treated equally: Implications for mTOR, rapamycin-resistance and cancer therapy. *Cell Cycle* **8:** 567–572.

Choo AY, Yoon SO, Kim SG, Roux PP, Blenis J. 2008. Rapamycin differentially inhibits S6Ks and 4E-BP1 to mediate cell-type-specific repression of mRNA translation. *Proc Natl Acad Sci* **105:** 17414–17419.

Choo AY, Kim SG, Vander Heiden MG, Mahoney SJ, Vu H, Yoon SO, Cantley LC, Blenis J. 2010. Glucose addiction of TSC null cells is caused by failed mTORC1-dependent balancing of metabolic demand with supply. *Mol Cell* **38:** 487–499.

Chresta CM, Davies BR, Hickson I, Harding T, Cosulich S, Critchlow SE, Vincent JP, Ellston R, Jones D, Sini P, et al. 2009. AZD8055 is a potent, selective, and orally bioavailable ATP-competitive mammalian target of rapamycin kinase inhibitor with in vitro and in vivo antitumor activity. *Cancer Res* **70:** 288–298.

Cunningham JT, Rodgers JT, Arlow DH, Vazquez F, Mootha VK, Puigserver P. 2007. mTOR controls mitochondrial oxidative function through a YY1–PGC–1α transcriptional complex. *Nature* **450:** 736–740.

Cybulski N, Polak P, Auwerx J, Ruegg MA, Hall MN. 2009. mTOR complex 2 in adipose tissue negatively controls whole-body growth. *Proc Natl Acad Sci* **106:** 9902–9907.

* Das G, Shravage BV, Baehrecke EH. 2012. Regulation and function of autophagy during cell survival and cell death. *Cold Spring Harb Perspect Biol* **4:** a008813.

DeBerardinis RJ, Mancuso A, Daikhin E, Nissim I, Yudkoff M, Wehrli S, Thompson CB. 2007. Beyond aerobic glycolysis: Transformed cells can engage in glutamine metabolism that exceeds the requirement for protein and nucleotide synthesis. *Proc Natl Acad Sci* **104:** 19345–19350.

Duan S, Skaar JR, Kuchay S, Toschi A, Kanarek N, Ben-Neriah Y, Pagano M. 2011. mTOR generates an autoamplification loop by triggering the βTrCP- and CK1α-dependent degradation of DEPTOR. *Mol Cell* **44:** 317–324.

Duran A, Amanchy R, Linares JF, Joshi J, Abu-Baker S, Porollo A, Hansen M, Moscat J, Diaz-Meco MT. 2011. 62 is a key regulator of nutrient sensing in the mTORC1 pathway. *Mol Cell* **44:** 134–146.

Egan DF, Shackelford DB, Mihaylova MM, Gelino S, Kohnz RA, Mair W, Vasquez DS, Joshi A, Gwinn DM, Taylor R, et al. 2011. Phosphorylation of ULK1 (hATG1) by AMP-activated protein kinase connects energy sensing to mitophagy. *Science* **331:** 456–461.

Ellisen LW, Ramsayer KD, Johannessen CM, Yang A, Beppu H, Minda K, Oliner JD, McKeon F, Haber DA. 2002. REDD1, a developmentally regulated transcriptional target of p63 and p53, links p63 to regulation of reactive oxygen species. *Mol Cell* **10:** 995–1005.

Eskelinen EL. 2005. Doctor Jekyll and Mister Hyde: Autophagy can promote both cell survival and cell death. *Cell Death Differ* **12:** 1468–1472.

Feldman ME, Apsel B, Uotila A, Loewith R, Knight ZA, Ruggero D, Shokat KM. 2009. Active-site inhibitors of mTOR target rapamycin-resistant outputs of mTORC1 and mTORC2. *PLoS Biol* **7:** e38.

Feng Z. 2010. 53 regulation of the IGF-1/AKT/mTOR pathways and the endosomal compartment. *Cold Spring Harb Perspect Biol* **2:** a001057.

Feng Z, Zhang H, Levine AJ, Jin S. 2005. The coordinate regulation of the p53 and mTOR pathways in cells. *Proc Natl Acad Sci* **102:** 8204–8209.

Feng Z, Hu W, de Stanchina E, Teresky AK, Jin S, Lowe S, Levine AJ. 2007. The regulation of AMPK β1, TSC2, and PTEN expression by p53: Stress, cell and tissue specificity, and the role of these gene products in modulating the IGF-1–AKT–mTOR pathways. *Cancer Res* **67:** 3043–3053.

Frodin M, Antal TL, Dummler BA, Jensen CJ, Deak M, Gammeltoft S, Biondi RM. 2002. A phosphoserine/threonine-binding pocket in AGC kinases and PDK1 mediates activation by hydrophobic motif phosphorylation. *EMBO J* **21:** 5396–5407.

Ganley IG, Lam du H, Wang J, Ding X, Chen S, Jiang X. 2009. ULK1·ATG13·FIP200 complex mediates mTOR signaling and is essential for autophagy. *J Biol Chem* **284:** 12297–12305.

Gao D, Inuzuka H, Tan MK, Fukushima H, Locasale JW, Liu P, Wan L, Zhai B, Chin YR, Shaik S, et al. 2011. mTOR drives its own activation via SCF(βTrCP)-dependent degradation of the mTOR inhibitor DEPTOR. *Mol Cell* **44:** 290–303.

Garcia-Martinez JM, Alessi DR. 2008. mTOR complex-2 (mTORC2) controls hydrophobic motif phosphorylation and activation of serum and glucocorticoid induced protein kinase-1 (SGK1). *Biochem J* **416:** 375–385.

Garcia-Martinez JM, Moran J, Clarke RG, Gray A, Cosulich SC, Chresta CM, Alessi DR. 2009. Ku-0063794 is a specific inhibitor of the mammalian target of rapamycin (mTOR). *Biochem J* **421:** 29–42.

Garofalo RS, Orena SJ, Rafidi K, Torchia AJ, Stock JL, Hildebrandt AL, Coskran T, Black SC, Brees DJ, Wicks JR, et al. 2003. Severe diabetes, age-dependent loss of adipose tissue, and mild growth deficiency in mice lacking Akt2/PKBβ. *J Clin Invest* **112:** 197–208.

Gu Y, Lindner J, Kumar A, Yuan W, Magnuson MA. 2011. Rictor/mTORC2 is essential for maintaining a balance between β-cell proliferation and cell size. *Diabetes* **60:** 827–837.

Guertin DA, Sabatini DM. 2009. The pharmacology of mTOR inhibition. *Sci Signal* **2:** e24.

Guertin DA, Stevens DM, Thoreen CC, Burds AA, Kalaany NY, Moffat J, Brown M, Fitzgerald KJ, Sabatini DM. 2006. Ablation in mice of the mTORC components *raptor*, *rictor*, or *mLST8* reveals that mTORC2 is required for signaling to Akt-FOXO and PKCα, but not S6K1. *Dev Cell* **11:** 859–871.

Guertin DA, Stevens DM, Saitoh M, Kinkel S, Crosby K, Sheen JH, Mullholland DJ, Magnuson MA, Wu H, Sabatini DM. 2009. mTOR complex 2 is required for the development of prostate cancer induced by Pten loss in mice. *Cancer Cell* **15:** 148–159.

Gulhati P, Cai Q, Li J, Liu J, Rychahou PG, Qiu S, Lee EY, Silva SR, Bowen KA, Gao T, et al. 2009. Targeted inhibition of Mammalian target of rapamycin signaling inhibits tumorigenesis of colorectal cancer. *Clin Cancer Res* **15:** 7207–7216.

Gwinn DM, Shackelford DB, Egan DF, Mihaylova MM, Mery A, Vasquez DS, Turk BE, Shaw RJ. 2008. AMPK phosphorylation of raptor mediates a metabolic checkpoint. *Mol Cell* **30:** 214–226.

Hanahan D, Weinberg RA. 2011. Hallmarks of cancer: The next generation. *Cell* **144:** 646–674.

Hardie DG. 2007. AMP-activated/SNF1 protein kinases: Conserved guardians of cellular energy. *Nat Rev Mol Cell Biol* **8:** 774–785.

He C, Klionsky DJ. 2009. Regulation mechanisms and signaling pathways of autophagy. *Annu Rev Genet* **43:** 67–93.

Herrero-Martin G, Hoyer-Hansen M, Garcia-Garcia C, Fumarola C, Farkas T, Lopez-Rivas A, Jaattela M. 2009.

TAK1 activates AMPK-dependent cytoprotective autophagy in TRAIL-treated epithelial cells. *EMBO J* **28:** 677–685.

Hietakangas V. 2008. TOR complex 2 is needed for cell cycle progression and anchorage-independent growth of MCF7 and PC3 tumor cells. *BMC Cancer* **8:** 282.

Hietakangas V, Cohen SM. 2007. Re-evaluating AKT regulation: Role of TOR complex 2 in tissue growth. *Genes Dev* **21:** 632–637.

Hosokawa N, Hara T, Kaizuka T, Kishi C, Takamura A, Miura Y, Iemura S, Natsume T, Takehana K, Yamada N, et al. 2009. Nutrient-dependent mTORC1 association with the ULK1–Atg13–FIP200 complex required for autophagy. *Mol Biol Cell* **20:** 1981–1991.

Hsu PP, Kang SA, Rameseder J, Zhang Y, Ottina KA, Lim D, Peterson TR, Choi Y, Gray NS, Yaffe MB, et al. 2011. The mTOR-regulated phosphoproteome reveals a mechanism of mTORC1-mediated inhibition of growth factor signaling. *Science* **332:** 1317–1322.

Huang J, Manning BD. 2008. The TSC1–TSC2 complex: A molecular switchboard controlling cell growth. *Biochem J* **412:** 179–190.

Inoki K, Li Y, Zhu T, Wu J, Guan KL. 2002. TSC2 is phosphorylated and inhibited by Akt and suppresses mTOR signalling. *Nat Cell Biol* **12:** 12.

Inoki K, Zhu T, Guan KL. 2003. TSC2 mediates cellular energy response to control cell growth and survival. *Cell* **115:** 577–590.

Inoki K, Ouyang H, Zhu T, Lindvall C, Wang Y, Zhang X, Yang Q, Bennett C, Harada Y, Stankunas K, et al. 2006. TSC2 integrates Wnt and energy signals via a coordinated phosphorylation by AMPK and GSK3 to regulate cell growth. *Cell* **126:** 955–968.

Jacinto E, Lorberg A. 2008. TOR regulation of AGC kinases in yeast and mammals. *Biochem J* **410:** 19–37.

Jacinto E, Facchinetti V, Liu D, Soto N, Wei S, Jung SY, Huang Q, Qin J, Su B. 2006. SIN1/MIP1 maintains rictor–mTOR complex integrity and regulates Akt phosphorylation and substrate specificity. *Cell* **127:** 125–137.

Janes MR, Limon JJ, So L, Chen J, Lim RJ, Chavez MA, Vu C, Lilly MB, Mallya S, Ong ST, et al. 2010. Effective and selective targeting of leukemia cells using a TORC1/2 kinase inhibitor. *Nat Med* **16:** 205–213.

Jones RG, Plas DR, Kubek S, Buzzai M, Mu J, Xu Y, Birnbaum MJ, Thompson CB. 2005. AMP-activated protein kinase induces a p53-dependent metabolic checkpoint. *Mol Cell* **18:** 283–293.

Jung CH, Jun CB, Ro SH, Kim YM, Otto NM, Cao J, Kundu M, Kim DH. 2009. ULK–Atg13–FIP200 complexes mediate mTOR signaling to the autophagy machinery. *Mol Biol Cell* **20:** 1992–2003.

Kamada Y, Funakoshi T, Shintani T, Nagano K, Ohsumi M, Ohsumi Y. 2000. Tor-mediated induction of autophagy via an Apg1 protein kinase complex. *J Cell Biol* **150:** 1507–1513.

Kim J, Guan KL. 2011. Regulation of the autophagy initiating kinase ULK1 by nutrients: Roles of mTORC1 and AMPK. *Cell Cycle* **10:** 1337–1338.

Kim DH, Sabatini DM. 2004. Raptor and mTOR: Subunits of a nutrient-sensitive complex. *Curr Top Microbiol Immunol* **279:** 259–270.

Kim DH, Sarbassov dos D, Ali SM, Latek RR, Guntur KV, Erdjument-Bromage H, Tempst P, Sabatini DM. 2003. GβL, a positive regulator of the rapamycin-sensitive pathway required for the nutrient-sensitive interaction between raptor and mTOR. *Mol Cell* **11:** 895–904.

Kim E, Goraksha-Hicks P, Li L, Neufeld TP, Guan KL. 2008. Regulation of TORC1 by Rag GTPases in nutrient response. *Nat Cell Biol* **10:** 935–945.

Kim J, Kundu M, Viollet B, Guan KL. 2011. AMPK and mTOR regulate autophagy through direct phosphorylation of Ulk1. *Nat Cell Biol* **13:** 132–141.

Kimura N, Tokunaga C, Dalal S, Richardson C, Yoshino K, Hara K, Kemp BE, Witters LA, Mimura O, Yonezawa K. 2003. A possible linkage between AMP-activated protein kinase (AMPK) and mammalian target of rapamycin (mTOR) signalling pathway. *Genes Cells* **8:** 65–79.

Koren I, Reem E, Kimchi A. 2010. DAP1, a novel substrate of mTOR, negatively regulates autophagy. *Curr Biol* **20:** 1093–1098.

Kozutsumi Y, Segal M, Normington K, Gething MJ, Sambrook J. 1988. The presence of malfolded proteins in the endoplasmic reticulum signals the induction of glucose-regulated proteins. *Nature* **332:** 462–464.

Krause U, Bertrand L, Maisin L, Rosa M, Hue L. 2002. Signalling pathways and combinatory effects of insulin and amino acids in isolated rat hepatocytes. *Eur J Biochem* **269:** 3742–3750.

Kuma A, Hatano M, Matsui M, Yamamoto A, Nakaya H, Yoshimori T, Ohsumi Y, Tokuhisa T, Mizushima N. 2004. The role of autophagy during the early neonatal starvation period. *Nature* **432:** 1032–1036.

Kumar A, Harris TE, Keller SR, Choi KM, Magnuson MA, Lawrence JC Jr. 2008. Muscle-specific deletion of rictor impairs insulin-stimulated glucose transport and enhances basal glycogen synthase activity. *Mol Cell Biol* **28:** 61–70.

Kumar A, Lawrence JC Jr, Jung DY, Ko HJ, Keller SR, Kim JK, Magnuson MA, Harris TE. 2010. Fat cell-specific ablation of *rictor* in mice impairs insulin-regulated fat cell and whole-body glucose and lipid metabolism. *Diabetes* **59:** 1397–1406.

Lee CH, Inoki K, Karbowniczek M, Petroulakis E, Sonenberg N, Henske EP, Guan KL. 2007. Constitutive mTOR activation in TSC mutants sensitizes cells to energy starvation and genomic damage via p53. *EMBO J* **26:** 4812–4823.

Liang J, Shao SH, Xu ZX, Hennessy B, Ding Z, Larrea M, Kondo S, Dumont DJ, Gutterman JU, Walker CL, et al. 2007. The energy sensing LKB1–AMPK pathway regulates p27^{kip1} phosphorylation mediating the decision to enter autophagy or apoptosis. *Nat Cell Biol* **9:** 218–224.

Liu HY, Han J, Cao SY, Hong T, Zhuo D, Shi J, Liu Z, Cao W. 2009. Hepatic autophagy is suppressed in the presence of insulin resistance and hyperinsulinemia: Inhibition of FoxO1-dependent expression of key autophagy genes by insulin. *J Biol Chem* **284:** 31484–31492.

Ma XM, Blenis J. 2009. Molecular mechanisms of mTOR-mediated translational control. *Nat Rev Mol Cell Biol* **10:** 307–318.

Mammucari C, Milan G, Romanello V, Masiero E, Rudolf R, Del Piccolo P, Burden SJ, Di Lisi R, Sandri C, Zhao J, et al.

2007. FoxO3 controls autophagy in skeletal muscle in vivo. *Cell Metab* **6:** 458–471.

Manning BD, Cantley LC. 2003. Rheb fills a GAP between TSC and TOR. *Trends Biochem Sci* **28:** 573–576.

Manning BD, Cantley LC. 2007. AKT/PKB signaling: Navigating downstream. *Cell* **129:** 1261–1274.

Masri J, Bernath A, Martin J, Jo OD, Vartanian R, Funk A, Gera J. 2007. mTORC2 activity is elevated in gliomas and promotes growth and cell motility via overexpression of rictor. *Cancer Res* **67:** 11712–11720.

Matsui Y, Takagi H, Qu X, Abdellatif M, Sakoda H, Asano T, Levine B, Sadoshima J. 2007. Distinct roles of autophagy in the heart during ischemia and reperfusion: Roles of AMP-activated protein kinase and Beclin 1 in mediating autophagy. *Circ Res* **100:** 914–922.

Nakae J, Park BC, Accili D. 1999. Insulin stimulates phosphorylation of the Forkhead transcription factor FKHR on serine 253 through a Wortmannin-sensitive pathway. *J Biol Chem* **274:** 15982–15985.

Nicklin P, Bergman P, Zhang B, Triantafellow E, Wang H, Nyfeler B, Yang H, Hild M, Kung C, Wilson C, et al. 2009. Bidirectional transport of amino acids regulates mTOR and autophagy. *Cell* **136:** 521–534.

Noda T, Ohsumi Y. 1998. Tor, a phosphatidylinositol kinase homologue, controls autophagy in yeast. *J Biol Chem* **273:** 3963–3966.

O'Reilly KE, Rojo F, She QB, Solit D, Mills GB, Smith D, Lane H, Hofmann F, Hicklin DJ, Ludwig DL, et al. 2006. mTOR inhibition induces upstream receptor tyrosine kinase signaling and activates Akt. *Cancer Res* **66:** 1500–1508.

Ozcan U, Ozcan L, Yilmaz E, Duvel K, Sahin M, Manning BD, Hotamisligil GS. 2008. Loss of the tuberous sclerosis complex tumor suppressors triggers the unfolded protein response to regulate insulin signaling and apoptosis. *Mol Cell* **29:** 541–551.

Peterson TR, Laplante M, Thoreen CC, Sancak Y, Kang SA, Kuehl WM, Gray NS, Sabatini DM. 2009. DEPTOR is an mTOR inhibitor frequently overexpressed in multiple myeloma cells and required for their survival. *Cell* **137:** 873–886.

Peterson TR, Sengupta SS, Harris TE, Carmack AE, Kang SA, Balderas E, Guertin DA, Madden KL, Carpenter AE, Finck BN, et al. 2011. mTOR complex 1 regulates Lipin 1 localization to control the SREBP pathway. *Cell* **146:** 408–420.

Ramanathan A, Schreiber SL. 2009. Direct control of mitochondrial function by mTOR. *Proc Natl Acad Sci* **106:** 22229–22232.

Rena G, Guo S, Cichy SC, Unterman TG, Cohen P. 1999. Phosphorylation of the transcription factor Forkhead family member FKHR by protein kinase B. *J Biol Chem* **274:** 17179–17183.

Sancak Y, Sabatini DM. 2009. Rag proteins regulate amino-acid-induced mTORC1 signalling. *Biochem Soc Trans* **37:** 289–290.

Sancak Y, Thoreen CC, Peterson TR, Lindquist RA, Kang SA, Spooner E, Carr SA, Sabatini DM. 2007. PRAS40 is an insulin-regulated inhibitor of the mTORC1 protein kinase. *Mol Cell* **25:** 903–915.

Sancak Y, Peterson TR, Shaul YD, Lindquist RA, Thoreen CC, Bar-Peled L, Sabatini DM. 2008. The Rag GTPases bind raptor and mediate amino acid signaling to mTORC1. *Science* **320:** 1496–1501.

Sancak Y, Bar-Peled L, Zoncu R, Markhard AL, Nada S, Sabatini DM. 2010. Ragulator–Rag complex targets mTORC1 to the lysosomal surface and is necessary for its activation by amino acids. *Cell* **141:** 290–303.

Sarbassov DD, Guertin DA, Ali SM, Sabatini DM. 2005. Phosphorylation and regulation of Akt/PKB by the rictor-mTOR complex. *Science* **307:** 1098–1101.

Sarbassov DD, Ali SM, Sengupta S, Sheen JH, Hsu PP, Bagley AF, Markhard AL, Sabatini DM. 2006. Prolonged rapamycin treatment inhibits mTORC2 assembly and Akt/PKB. *Mol Cell* **22:** 159–168.

Sengupta S, Peterson TR, Sabatini DM. 2010. Regulation of the mTOR complex 1 pathway by nutrients, growth factors, and stress. *Mol Cell* **40:** 310–322.

Settembre C, Zoncu R, Medina DL, Vetrini F, Erdin S, Erdin S, Huynh T, Ferron M, Karsenty G, Vellard MC, et al. 2012. A lysosome-to-nucleus signalling mechanism senses and regulates the lysosome via mTOR and TFEB. *EMBO J* **31:** 1095–1108.

Shang L, Chen S, Du F, Li S, Zhao L, Wang X. 2011. Nutrient starvation elicits an acute autophagic response mediated by Ulk1 dephosphorylation and its subsequent dissociation from AMPK. *Proc Natl Acad Sci* **108:** 4788–4793.

Shaw RJ, Cantley LC. 2006. Ras, PI(3)K and mTOR signalling controls tumour cell growth. *Nature* **441:** 424–430.

Sheen JH, Zoncu R, Kim D, Sabatini DM. 2011. Defective regulation of autophagy upon leucine deprivation reveals a targetable liability of human melanoma cells in vitro and in vivo. *Cancer Cell* **19:** 613–628.

Shiota C, Woo JT, Lindner J, Shelton KD, Magnuson MA. 2006. Multiallelic disruption of the *rictor* gene in mice reveals that mTOR complex 2 is essential for fetal growth and viability. *Dev Cell* **114:** 583–589.

Sini P, James D, Chresta C, Guichard S. 2010. Simultaneous inhibition of mTORC1 and mTORC2 by mTOR kinase inhibitor AZD8055 induces autophagy and cell death in cancer cells. *Autophagy* **6:** 553–554.

Smith EM, Finn SG, Tee AR, Browne GJ, Proud CG. 2005. The tuberous sclerosis protein TSC2 is not required for the regulation of the mammalian target of rapamycin by amino acids and certain cellular stresses. *J Biol Chem* **280:** 18717–18727.

Sparks CA, Guertin DA. 2010. Targeting mTOR: Prospects for mTOR complex 2 inhibitors in cancer therapy. *Oncogene* **29:** 3733–3744.

Spilman P, Podlutskaya N, Hart MJ, Debnath J, Gorostiza O, Bredesen D, Richardson A, Strong R, Galvan V. 2010. Inhibition of mTOR by rapamycin abolishes cognitive deficits and reduces amyloid-β levels in a mouse model of Alzheimer's disease. *PLoS ONE* **5:** e9979.

Stambolic V, MacPherson D, Sas D, Lin Y, Snow B, Jang Y, Benchimol S, Mak TW. 2001. Regulation of PTEN transcription by p53. *Mol Cell* **8:** 317–325.

Sudarsanam S, Johnson DE. 2010. Functional consequences of mTOR inhibition. *Curr Opin Drug Discov Devel* **13:** 31–40.

Tabernero J, Rojo F, Calvo E, Burris H, Judson I, Hazell K, Martinelli E, Ramon y Cajal S, Jones S, Vidal L, et al. 2008. Dose- and schedule-dependent inhibition of the mammalian target of rapamycin pathway with everolimus: A phase I tumor pharmacodynamic study in patients with advanced solid tumors. *J Clin Oncol* **26:** 1603–1610.

Tang ED, Nunez G, Barr FG, Guan KL. 1999. Negative regulation of the Forkhead transcription factor FKHR by Akt. *J Biol Chem* **274:** 16741–16746.

Thoreen CC, Kang SA, Chang JW, Liu Q, Zhang J, Gao Y, Reichling LJ, Sim T, Sabatini DM, Gray NS. 2009. An ATP-competitive mammalian target of rapamycin inhibitor reveals rapamycin-resistant functions of mTORC1. *J Biol Chem* **284:** 8023–8032.

Tschopp O, Yang ZZ, Brodbeck D, Dummler BA, Hemmings-Mieszczak M, Watanabe T, Michaelis T, Frahm J, Hemmings BA. 2005. Essential role of protein kinase Bγ (PKBγ/Akt3) in postnatal brain development but not in glucose homeostasis. *Development* **132:** 2943–2954.

Uddin MN, Ito S, Nishio N, Suganya T, Isobe K. 2011. Gadd34 induces autophagy through the suppression of the mTOR pathway during starvation. *Biochem Biophys Res Commun* **407:** 692–698.

Vander Haar E, Lee SI, Bandhakavi S, Griffin TJ, Kim DH. 2007. Insulin signalling to mTOR mediated by the Akt/PKB substrate PRAS40. *Nat Cell Biol* **9:** 316–323.

Vasudevan KM, Barbie DA, Davies MA, Rabinovsky R, McNear CJ, Kim JJ, Hennessy BT, Tseng H, Pochanard P, Kim SY, et al. 2009. AKT-independent signaling downstream of oncogenic PIK3CA mutations in human cancer. *Cancer Cell* **16:** 21–32.

Wang L, Harris TE, Lawrence JC Jr. 2008. Regulation of proline-rich Akt substrate of 40 kDa (PRAS40) function by mammalian target of rapamycin complex 1 (mTORC1)-mediated phosphorylation. *J Biol Chem* **283:** 15619–15627.

Wellen KE, Thompson CB. 2010. Cellular metabolic stress: Considering how cells respond to nutrient excess. *Mol Cell* **40:** 323–332.

Yang Z, Klionsky DJ. 2009. Mammalian autophagy: Core molecular machinery and signaling regulation. *Curr Opin Cell Biol* **22:** 124–131.

Yang Z, Klionsky DJ. 2010. Eaten alive: A history of macroautophagy. *Nat Cell Biol* **12:** 814–822.

Yang Q, Inoki K, Ikenoue T, Guan KL. 2006. Identification of Sin1 as an essential TORC2 component required for complex formation and kinase activity. *Genes Dev* **20:** 2820–2832.

Yip CK, Murata K, Walz T, Sabatini DM, Kang SA. 2010. Structure of the human mTOR complex I and its implications for rapamycin inhibition. *Mol Cell* **38:** 768–774.

Yu K, Toral-Barza L, Shi C, Zhang WG, Lucas J, Shor B, Kim J, Verheijen J, Curran K, Malwitz DJ, et al. 2009. Biochemical, cellular, and in vivo activity of novel ATP-competitive and selective inhibitors of the mammalian target of rapamycin. *Cancer Res* **69:** 6232–6240.

Yu L, McPhee CK, Zheng L, Mardones GA, Rong Y, Peng J, Mi N, Zhao Y, Liu Z, Wan F, et al. 2010. Termination of autophagy and reformation of lysosomes regulated by mTOR. *Nature* **465:** 942–946.

Yu Y, Yoon SO, Poulogiannis G, Yang Q, Ma XM, Villen J, Kubica N, Hoffman GR, Cantley LC, Gygi SP, et al. 2011. Phosphoproteomic analysis identifies Grb10 as an mTORC1 substrate that negatively regulates insulin signaling. *Science* **332:** 1322–1326.

Zhao M, Klionsky DJ. 2011. AMPK-dependent phosphorylation of ULK1 induces autophagy. *Cell Metab* **13:** 119–120.

Zhao Y, Xiong X, Sun Y. 2011. DEPTOR, an mTOR inhibitor, is a physiological substrate of SCF(βTrCP) E3 ubiquitin ligase and regulates survival and autophagy. *Mol Cell* **44:** 304–316.

Zoncu R, Efeyan A, Sabatini DM. 2010. mTOR: From growth signal integration to cancer, diabetes and ageing. *Nat Rev Mol Cell Biol* **12:** 21–35.

Zoncu R, Bar-Peled L, Efeyan A, Wang S, Sancak Y, Sabatini DM. 2011. mTORC1 senses lysosomal amino acids through an inside-out mechanism that requires the vacuolar H^+-ATPase. *Science* **334:** 678–683.

Oncogenes in Cell Survival and Cell Death

Jake Shortt[1] and Ricky W. Johnstone[1,2]

[1]Cancer Therapeutics Program, Gene Regulation Laboratory, Peter MacCallum Cancer Centre, East Melbourne 3002, Victoria, Australia

[2]Department of Pathology, University of Melbourne, Parkville 3052, Victoria, Australia

Correspondence: ricky.johnstone@petermac.org

The transforming effects of proto-oncogenes such as *MYC* that mediate unrestrained cell proliferation are countered by "intrinsic tumor suppressor mechanisms" that most often trigger apoptosis. Therefore, cooperating genetic or epigenetic effects to suppress apoptosis (e.g., overexpression of *BCL2*) are required to enable the dual transforming processes of unbridled cell proliferation and robust suppression of apoptosis. Certain oncogenes such as *BCR-ABL* are capable of concomitantly mediating the inhibition of apoptosis and driving cell proliferation and therefore are less reliant on cooperating lesions for transformation. Accordingly, direct targeting of BCR-ABL through agents such as imatinib have profound antitumor effects. Other oncoproteins such as MYC rely on the anti-apoptotic effects of cooperating oncoproteins such as BCL2 to facilitate tumorigenesis. In these circumstances, where the primary oncogenic driver (e.g., *MYC*) cannot yet be therapeutically targeted, inhibition of the activity of the cooperating antiapoptotic protein (e.g., BCL2) can be exploited for therapeutic benefit.

Proto-oncogenes perform physiological functions that are necessary for normal cellular homeostasis. In particular, proto-oncogenes govern the processes of growth, proliferation, and survival that a cancer cell can exploit to gain competitive advantages over its non-neoplastic counterparts (Hanahan and Weinberg 2011). During malignant transformation, an emerging clone must circumvent the antineoplastic countermeasures that usually regulate the activity of proto-oncogenes (Hanahan and Weinberg 2011). Tumor cells can then use the beneficial properties of an oncogene (e.g., enhanced proliferation) without the negative effects of "fail-safe" countermeasures that have been otherwise overridden (Lowe et al. 2004). Indeed "intrinsic tumor suppression" activities such as cellular senescence or apoptosis are triggered when a cell is driven to uncontrolled proliferation through the inappropriate activity of an oncogene (Lowe et al. 2004). One of the best examples of such a fail-safe mechanism is detailed later in this review through the focus on *c-MYC* that concomitantly induces cell proliferation and apoptosis. In this instance, cooperating lesions that inhibit apoptosis (e.g., loss of function of p53 or overexpression of BCL2) are required to facilitate *MYC*-driven tumorigenesis (Fig. 1). In contrast, an oncogene such as *BCL2* may have the capacity to inhibit apoptosis but

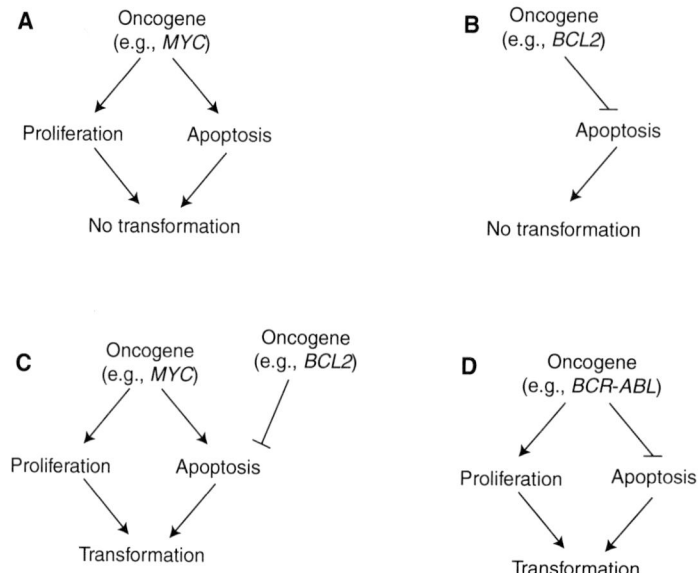

Figure 1. Cellular transformation through concomitant stimulation of cell proliferation and inhibition of apoptosis. (*A*) Oncogenes such as *MYC* induce cellular proliferation; however, intrinsic "fail-safe" apoptotic mechanisms such as those induced by the ARF/MDM2/P53 pathway counteract the mitotic stimulus mediated by MYC and suppress transformation. (*B*) Oncogenes such as *BCL2* are potent inhibitors of apoptosis but poor inducers of cell proliferation and are, therefore, insufficient to drive tumorigenesis as a single oncogenic event. (*C*) The cooperative activity of oncogenes such as *MYC* and *BCL2* suppress apoptosis and drive proliferation resulting in cellular transformation. (*D*) Certain oncogenes such as *BCR-ABL* can activate signaling pathways that simultaneously induce cell proliferation and suppress apoptosis, thereby leading to transformation.

does not contain strong intrinsic mitogenic activity and therefore cooperating oncogenes that drive cell proliferation (e.g., *RAS*, *MYC*) are required for cellular transformation (Strasser et al. 1990; Lee et al. 2007) (Fig. 1). Finally, there are oncoproteins that engage multiple signal transduction pathways (e.g., *BCR-ABL*) that can concomitantly activate both cell proliferation and cell survival (reviewed by Druker 2008). Such oncoproteins may therefore be less reliant on secondary genetic or epigenetic hits to initiate tumorigenesis (Fig. 1).

Oncogenes are identifiable by their capacity to transform a cell in the context of deregulated expression or function. Mechanisms of oncogene activation are diverse and include upregulated expression of a normal gene product, expression of mutant protein with enhanced stability or altered functionality or altered recruitment or subcellular localization of a normal gene product through interaction with an aberrantly expressed or mutant binding partner. Perhaps the simplest illustrations of deregulated oncogene expression in malignant transformation can be found in B-cell lymphomas. Here, as a lymphocyte undergoes genomic V(D)J rearrangement to generate antibody specificities, oncogenes may be switched on by "accidental" rearrangement alongside powerful immunoglobulin gene enhancer elements (Kuppers and Dalla-Favera 2001). The resulting "switch-translocations" define lymphomas at a chromosomal level, and strongly implicate the rearranged oncogenes as initiating lesions in lymphomagenesis (Table 1). Classical examples include immunoglobulin heavy chain (*IGH*)-*cMYC* translocations in Burkitt lymphoma and *IGH-BCL2* translocations in follicular lymphoma. *MYC* and *BCL2* are considered prototypic oncogenes and their diverse mechanisms of transformation will be highlighted in this review. Neoplastic transformation is a multistep

Table 1. Examples of oncogenes revealed by IGH-switch translocations

Oncogene	Disease	Gene product (protein) function	Effects
BCL2	Follicular lymphoma	Prevents mitochondrial outer membrane permeabilization	Antiapoptotic
BCL6	Diffuse large B-cell lymphoma	Transcription factor	Proliferative, differentiation block
BCL10	Extranodal marginal zone lymphoma	Caspase-recruiting domain containing protein	Activation of prosurvival signaling (e.g., NF-κB), proapoptotic
CCND1	Mantle cell lymphoma	Activates cyclin-dependent kinases	Proliferative
CDK6	Splenic marginal zone lymphoma	Cyclin-dependent kinase	Proliferative
FGFR3	Plasma cell myeloma	Receptor tyrosine kinase	Proliferative, activation of prosurvival signaling (e.g., PI3K, MAPK)
cMAF	Plasma cell myeloma	Transcription factor	Proliferative, altered interactions with tumor microenvironment
MALT1	Extranodal marginal zone lymphoma	Protease/paracaspase	Activation of prosurvival signaling (e.g., NF-κB)
cMYC	Burkitt lymphoma	Transcription factor	Proliferative, proapoptotic, prosenescent

process involving cooperating activation of oncogenes and/or silencing of tumor suppressors in order for a clone to achieve full malignant potential (Hanahan and Weinberg 2000). Accordingly, deregulation of a single oncogene is usually insufficient to transform a cell.

INITIATING LESIONS AND THE CONCEPT OF "ONCOGENE ADDICTION"

The genetic complexity of neoplastic transformation varies from tumor to tumor. Whereas lymphomas arising from IGH switch translocations may be relatively monomorphic in their genetic make-up, carcinomas often have complex aneuploid DNA harboring multiple aberrancies (Kinzler and Vogelstein 1996, 1998). As such, the initial oncogenic event that spawned a genetically complex cancer can be difficult to determine. However, certain malignancies clearly evolve in a stepwise fashion that builds on an initiating oncogenic "hit." As this stepwise progression rests on the beneficial effects of a primary oncogene, the developing cancer becomes increasingly dependent on that oncogene for viability. At this point, the cancer is considered "oncogene-addicted" because of its heightened reliance on the offending gene products (Weinstein 2002). Selective targeting of an oncogene of addiction is a major goal in cancer therapeutics, in which oncogene dependency represents a potential vulnerability in cancer cell survival (Weinstein 2002).

MECHANISMS OF COOPERATIVE ONCOGENIC TRANSFORMATION

Oncogenes show diversity in their proneoplastic effects and in the regulatory fail-safes that protect against transformation. Some oncogenes such as BCR-ABL possess a near-complete repertoire of proneoplastic properties, requiring less assistance from cooperative mutations (reviewed by Perrotti et al. 2010). Other oncogenes have a narrow spectrum of biological activity (e.g., BCL2) or simultaneously trigger cell proliferation and cell death/senescence (e.g., MYC) and therefore cannot transform a cell without cooperating lesions that blunt or inhibit the intrinsic tumor suppressor mechanisms. In this way, the disruption of a specific prosurvival effector can combine with broader oncogenic hits to potently accelerate the development of cancer. For example, deregulation of MYC drives cell growth and proliferation, but its net effect is countered by increased rates of apoptosis (Evan et al. 1992). If deregulated, *Myc* activity is then combined with *Bcl2* overexpression, the apoptotic fail-safe is removed in a synergistic route to transformation (Strasser et al. 1990) (Fig. 1). This paradigm is illustrated by the ability of *MYC* to cooperate with a range of antiapoptotic proteins in experimental models of lymphomagenesis (Beverly and Varmus 2009; Whitecross et al. 2009; Campbell et al. 2010). Moreover in the clinic, "double-hit" lymphomas containing *MYC* and *BCL2* translocations are recognized as a distinct entity with poorer prognosis (Aukema et al. 2011).

MYC—AN EXAMPLE OF A TWO-EDGED SWORD IN ONCOGENESIS

Deregulation of *MYC* gene expression is one of the most frequently encountered events in human cancer, implicated in over half of all malignancies (reviewed by Vita and Henriksson 2006). Signal transduction pathways activated in response to nutrients, growth factors, and mitogenic stimuli relay signals from phosphorylation cascades directly to MYC. For example, ERK, the canonical downstream effector of the RAS/RAF/MEK/ERK pathway stabilizes MYC by phosphorylation of serine 62 (Sears 2004; Gustafson and Weiss 2010). Conversely, glycogen synthase kinase (GSK)-3β mediated phosphorylation of Myc at serine 58 has a destabilizing effect (Sears 2004). GSK-3β is antagonized by AKT, a nodal effector in the phosphotidylinositol-3-kinase (PI3K) pathway. Therefore, RAS/RAF and PI3K/AKT signaling cooperate to promote MYC protein stability in response to optimal growth conditions.

The *MYC* gene promoter receives input from many transcription factors and is normally switched on in response to mitogenic stimuli. In particular *MYC* is transactivated by E2F transcription factors released from RB in response to phosphorylation by cyclin/cyclin-dependent kinase (CDK) complexes (Oswald et al. 1994). MYC gene-targets are implicated in most aspects of cellular physiology, however, those involved in growth and proliferation feature prominently. The proliferative effects of MYC are evident at multiple points in cell-cycle regulation. For example, MYC activates transcription of D-type cyclins and CDKs to promote cell cycle check point progression (Hermeking et al. 2000; Bouchard et al. 2001). At the same time, MYC represses the expression of CIP/KIP family cyclin/CDK inhibitors (Seoane et al. 2002). To prepare the cell for division, MYC augments the processes of ribosomal biogenesis (Gomez-Roman et al. 2006) and 5′ cap-dependent translation (Lin et al. 2009), allowing a cell to manufacture protein and accumulate biomass. MYC also modulates specific metabolic pathways integral to cell growth, such as glucose and iron homeostasis (Adhikary and Eilers 2005).

Unbridled MYC activity is a potent oncogenic stimulus, requiring stringent regulation at every possible level from transcriptional control to protein stability. MYC also possesses inherent proapoptotic and prosenescent properties that must be overcome before neoplastic transformation can occur. As discussed in detail below, MYC activates p53-driven apoptosis programs through the ARF tumor suppressor protein (Lowe et al. 2004). These dual proneoplastic and cancer-protective properties make MYC a "two-edged sword" in oncogenesis.

CANCER-PROTECTIVE PROPERTIES OF MYC

MYC-Induced Apoptosis

MYC's inherent proapoptotic properties safeguard against malignant transformation. To a large extent, MYC's apoptotic activity is underpinned by its capacity to initiate a DNA-damage response (DDR) and activate the MDM2-ARF-p53 tumor suppressor pathway (Lowe et al. 2004; Meyer and Penn 2008). Proposed mechanisms for DDR activation include generation of genotoxic reactive oxygen species and formation of aberrant DNA replication intermediates (Vafa et al. 2002). Upregulation of ARF in response to MYC stabilizes p53 expression by antagonism of MDM2. Augmented p53 activity counteracts the proliferative effects of MYC and activates p53-dependent apoptotic effectors such as PUMA and Noxa. MYC also up-regulates the p53-independent expression of proapoptotic BIM (Hemann et al. 2004). The importance of MYC's interaction with the DDR and apoptosis is highlighted in experimental models of MYC-transformation and human Burkitt lymphoma. Genetic interrogation of these cancers reveals the majority contain perturbations in the MDM2-ARF-p53 pathway as a means to bypass MYC-induced apoptosis (Eischen et al. 1999). Accordingly, genetic crosses of *Myc* transgenic mice with *p53-*, *p19 Arf-*, or *Atm*-deficient mice results in accelerated tumorigenesis by inhibition of *Myc*-induced apoptosis (Schmitt et al. 1999; Reimann et al. 2007). Moreover, knockdown or knockout of *Puma* (Hemann et al. 2004) and to a lesser extent *Noxa* (Michalak et al. 2009) also accelerates *Myc*-induced lymphomagenesis. Conversely, genetic manipulation to overexpress Bcl2 family proteins (Schmitt et al. 2002; Beverly and Varmus 2009; Whitecross et al. 2009; Campbell et al. 2010) or delete *Bim* (Egle et al. 2004) in preneoplastic *Myc*-overexpressing stem cells greatly enhances the oncogenic activity of *Myc* and concomitantly relieves the pressure to mutate *p53* or p19 *Arf* in *Myc*-driven tumorigenesis. These data provide compelling evidence that suppression of *Myc*-induced apoptosis in the presence of a strong pro-proliferation signal by *Myc* cooperate to drive tumorigenesis.

ONCOGENIC SIGNAL TRANSDUCTION PATHWAYS AND REGULATION OF APOPTOSIS

Diverse signal transduction pathways can be activated by upregulated expression of oncoproteins or gain-of-function mutations leading to abnormal protein activity. For example, gain-of-function receptor mutations can initiate ligand-independent receptor tyrosine kinase (RTK) or G-protein coupled receptor (GPCR) activity. Similarly, activating mutations of *RAS* and *PI3K* can amplify the responses to upstream receptor ligation or cause ligand-independent pathway activation (Vivanco and Sawyers 2002; Engelman 2009). Moreover, chromosomal translocations can give rise to aberrant fusion proteins with oncogenic enzymatic activity (e.g., *BCR-ABL*). A common feature of these oncogenic signaling molecules is their ability to simultaneously stimulate cellular proliferation and suppress apoptosis. Although reductionist scientific approaches describe canonical linear signaling cascades, in reality complex networks of communicating proteins exist within a cell. Oncogenic activation of one prosurvival "pathway" can lead to compensatory downregulation of another. For this reason, a combination of upstream genetic lesions (e.g., mutation of *PI3K* and *RAS*) can cooperate by circumventing regulatory modulation by negative feedback loops between networks. It is now recognized that virtually all cancers show either activation of an oncogenic kinase and/or silencing of a tumor-suppressor phosphatase. Below we will provide examples of oncoproteins that either alone or in combination activate a diverse array of signaling pathways to provide the dual transforming effect of uncontrolled cell proliferation and inhibition of apoptosis.

BCR-ABL AND THE CHRONIC MYELOPROLIFERATIVE NEOPLASMS

The initiating lesion of CML arises by reciprocal translocation of *c-ABL1* on chromosome 9 with the *BCR* gene on chromosome 22 to generate the BCR-ABL fusion protein with greatly enhanced cABL-1 kinase activity. Acquisition of this *BCR-ABL* fusion is sufficient to transform hematopoietic stem cells (Kharas et al. 2008; Perrotti et al. 2010). The effects of constitutive BCR-ABL activity are pleiotropic and include enhanced cellular proliferation, evasion of apoptosis, and increased capacity for cell self-renewal (Druker 2008). Enhanced proliferation and survival are variably attributable to concomitant and coordinated activation of PI3K, JAK/STAT, and RAS/RAF/MEK/ERK pathways (Druker 2008). At the same time, BCR-ABL sets the scene for acquisition of new oncogenic lesions by promoting genomic instability and downregulating tumor suppressors (Nowicki et al. 2004; Koptyra et al. 2008). Clonal evolution to blast crisis is most commonly accompanied by mutations of the MDM2-ARF-p53 axis and transcription factors including *RUNX1* and *IKZF1* (Perrotti et al. 2010). Although these secondary lesions cluster within the terminal phenotypes of myeloid and lymphoid blast crises, they do not follow a consistent pattern between patients. Therefore, it appears deregulated BCR-ABL expression provides a "fertile soil" for further oncogenic devolution, and these secondary events occur stochastically and additively to convey a more aggressive phenotype.

GROWTH-FACTOR RECEPTOR MUTATIONS IN EPITHELIAL CANCERS

Deregulated expression or activity of members of the epidermal growth factor receptor (EGFR) family underlies development of a range of epithelial tumors across numerous tissues (Wheeler et al. 2010). These cell surface proteins function as RTKs, and EGFR (ERBB1/HER1) and HER2/NEU (ERBB2) are the most extensively studied members of the family. EGFR activation results in an explosion of intracellular signaling including triggering of the RAS/RAF/MEK/ERK and PI3K/AKT/MTOR pathways, as well as activation of Src tyrosine kinases, PLCγ, and STATs 3 and 5 (Marmor et al. 2004). The net result of simultaneous activation of these pathways is enhanced cellular proliferation, suppression of apoptosis, and cellular transformation that is conceptually and somewhat mechanistically similar to that described above for *BCR-ABL* (Fig. 1).

PI3K AND AKT

PI3K is a lipid kinase that integrates afferent signals from GPCRs, RTKs, RAS, and other drivers of oncogenic signal transduction to activate downstream targets implicated in growth, proliferation, and evasion of apoptosis (reviewed by Engelman et al. 2006). As such, PI3K plays an important role in facilitating GPCR, RTK, and RAS-initiated oncogenesis. Indeed, pharmacological or genetic silencing of PI3K severely impairs the ability of *BCR-ABL* to transform hematopoietic stem cells (Kharas et al. 2008) and induces apoptosis in epithelial tumors "addicted" to EGFR and HER2 (Faber et al. 2009). However, PI3K is much more than a passive yet important conduit in oncogenesis, as aberrant PI3K activation by mutation or gene amplification is observed in a range of solid-organ malignancies.

Within the PI3K family, class IA PI3K is the most closely associated with cancer. Class IA PI3K consists of a catalytic (p110) and regulatory (p85) subunit. There are four p110 isoforms (α, β, γ, δ) encoded by the *PIK3CA*, *PIK3CB*, *PIK3CG*, and *PIK3CD* gene loci, respectively. The *PIK3R1* gene encodes the p85 regulatory subunit that has a suppressive effect on p110 catalytic activity that is relieved by an interaction with phosphorylated Tyr residues on RTKs. *PIK3CA* and *PIK3R1* have been identified as oncogenes and somatic-activating mutations of *PIK3CA* occur in 30% of epithelial cancers and precursor lesions (reviewed by Engelman 2009). *PIK3CA* mutations increase lipid kinase activity in the absence of receptor activation, promoting cell transformation in vivo (Engelman 2009). The oncogenic activity of PI3K has largely been attributed to its ability to activate mammalian target of rapamycin complex (MTORC)-1 and AKT (Engelman 2009). MTORC-1 is considered a master controller of cell growth, as it modulates the processes of ribosomal biogenesis and 5′ cap-dependent translation (Schmelzle and Hall 2000).

AKT itself is an oncogene, first identified as the cellular homolog of a virally encoded protein (vAKT) implicated in the etiology of rodent T-cell lymphoma (Staal et al. 1977). AKT has both proliferative and antiapoptotic effects whereas negatively regulating tumor suppressors including p53, GSK-3β, and the forkhead-box family of transcription factors (FOXOs) (Liu et al. 2009). The proliferative effects of AKT are conveyed by antagonism of p53-mediated growth arrest and inhibition of CIP/KIP family cyclin/CDK inhibitors. AKT directly inhibits apoptosis by promoting the cytoplasmic sequestration of BAD (Datta et al. 2000) and down-regulating the transcription of BIM (which is normally transactivated by FOXOs). AKT also stabilizes the levels of endogenous IAP proteins (Dan et al. 2004), whereas AKT-mediated phosphorylation of caspases can increase resistance to cleavage (Cardone et al. 1998).

The functional relationship between MYC, AKT, and MTOR provides a further example of the duality of MYC and the potential for MYC to cooperate with other oncogenes (Fig. 2). MYC transcriptionally represses TSC-2 expression (Ravitz et al. 2007), the regulatory complex that bridges AKT and MTORC-1 in canonical PI3K signaling (Astrinidis and Henske 2005). TSC-2 downregulates MTORC-1 but is required for full activation of the second rapamycin insensitive MTOR complex (MTORC-2) (Huang et al. 2008). As MTORC-2 cooperates with PI3K to activate AKT, the net effect of MYC expression is to promote MTORC-1 while suppressing MTORC-2/AKT. This "binary switch" protects against the oncogenic effects of MTORC-1 activation by dampening upstream AKT phosphorylation. Interestingly, chronic pharmacological MTORC-1 antagonism can prevent the onset of lymphoma in experimental *Myc*-driven lymphoma. Conversely genetic activation of AKT by myristoylation (similar to vAKT) dramatically accelerates *Myc*-driven transformation, putatively by counteracting the normally suppressive effects of MYC on MTORC-2/AKT (Wendel et al. 2004).

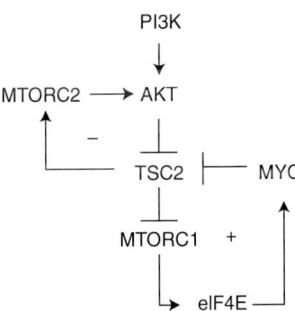

Figure 2. The interplay between MYC AND PI3K/MTOR. The *MYC*-regulated transcriptional network interacts with canonical PI3K/AKT/MTOR signaling at multiple levels with opposing "counter-regulatory effects." In one example, *MYC* represses the activity of TSC to activate MTORC-1 and enhance the cell's capacity to translate *MYC*-mRNA in what is postulated to represent an oncogenic feed-forward loop (+). The concurrent regulatory "fail-safe" is that TSC is required for full activation of the AKT hydrophobic motif-kinase, MTORC-2. In this way, MYC activates MTORC-1 while dampening MTORC-2/AKT activity (−). Consequently, oncogenic activation of AKT by MTORC-2-independent mechanisms dramatically accelerates *MYC*-driven oncogenesis.

ONCOGENIC RAS

The *RAS* proto-oncogenes (*HRAS*, *NRAS*, *KRAS*) encode small monomeric GTPases. Wild-type RAS proteins are conformationally activated when bound to GTP and become inactive by the process of hydrolyzing GTP to GDP. Oncogenic RAS is resistant to GAPs, locking the protein into its GTP-bound active conformation. Activating mutations in *RAS* genes occur in >30% of all tumors (Fernandez-Medarde and Santos 2011). GTP-bound RAS binds and activates a range of effector proteins, most notably PI3K (p110) and RAF (Castellano and Downward 2010). Mutant *RAS* and *PIK3CA* have the potential to cooperate in PI3K pathway activation and colorectal tumors with genetic lesions in both the RAS/RAF/MEK/ERK and PI3K pathways have been reported (Yuan and Cantley 2008). In contrast, mutations of *RAS* and *RAF* are thought to be mutually exclusive in human cancer, suggesting functional redundancy (Sensi et al. 2006). RAF is a Ser/Thr kinase that initiates a phosphorylation cascade including MEK and MAPK. Oncogenic *RAF* mutations are implicated in the pathogenesis of melanoma and colorectal cancer (Davies et al. 2002). Engagement of the PI3K/AKT is a major mechanism of RAS-mediated inhibition of apoptosis (Castellano and Downward 2010). In addition, various downstream effectors including TBK1, AKT, and Rac can facilitate RAS-induced activation of NF-κB signaling to suppress apoptosis that may be p53-dependent and -independent depending on the cellular context (Mayo et al. 1997; Cox and Der 2003; Barbie et al. 2009; Meylan et al. 2009). Through multiple molecular mechanisms, RAS concomitantly enhances cellular proliferation and inhibits apoptosis (Castellano and Downward 2010). However, cooperating genetic lesions are required for RAS-induced transformation, indicating the presence of intrinsic fail-safe mechanisms to suppress tumorigenesis. In the case of RAS, induction of senescence may play a protective role against its combined oncogenic proliferative and antiapoptotic effects in vivo. Whereas ectopic expression of RAS in normal cells promotes a senescent phenotype, this is dependent to a varying extent of intact p53/DDR detection mechanisms (Serrano et al. 1997). Hence, loss of DDR-related tumor suppressors is one cooperative mechanism by which RAS-induced senescence may be overcome in transformation.

INDUCTION OF APOPTOSIS THROUGH TARGETING ONCOGENES

If suppression of apoptosis is key for oncoproteins to mediate their tumorigenic effects then one would predict that targeted inhibition of the transforming oncoprotein, or of the cooperating protein that suppresses oncogene-induced apoptosis, would result in a robust proapoptotic response. This is indeed the case when the Abl-kinase inhibitor, imatinib, is used against *BCR-ABL*-addicted cells (O'Dwyer and Druker 2000), when the EGFR inhibitor, gefitinib, is used to treat tumors addicted to activated EGFR (Takeuchi and Ito 2010), and when the PI3K/AKT pathway is targeted using MTOR inhibitors in the context of constitutive AKT activation (Wendel et al. 2004). At present, direct pharmacological inhibitors of RAS and MYC have not yet been developed. However, transgenic mouse models of inducible *Myc*- or *Ras*-induced oncogenesis, showed dramatic apoptosis and subsequent tumor regression when these proteins were "turned off" (Chin et al. 1999; Felsher and Bishop 1999; Jain et al. 2002) indicating that this would be an effective cancer therapy strategy should these proteins or key downstream effectors be targeted. Moreover, we and others have shown that *Myc*-driven tumors that coexpress BCL2 or BCLxl resulting in accelerated tumorigenesis are highly sensitive to apoptosis ABT-737, the small molecule inhibitor of BCL2 and BCLxl, resulting in an effective therapeutic response (Mason et al. 2008; Whitecross et al. 2009). This provides proof-of-principle that targeting a second, cooperating apoptotic lesion may be effective against tumors that are addicted to a different "driver" oncogene.

CONCLUDING REMARKS

It is well recognized that two important cooperating hallmarks of tumorigenesis are unrestrained cell proliferation and resistance to apoptosis. Whereas certain proto-oncogenes such as *BCR-ABL* may be capable of concomitantly inducing both of these important effects, others such as *MYC* must circumvent intrinsic failsafe mechanisms that have evolved to counter unrestrained cell proliferation. Accordingly, genetic lesions that cooperate with *MYC* to facilitate cellular transformation are most often those that inactivate regulatory proapoptotic pathways such as the ARF/MDM2/p53 axis. Understanding the molecular interplay between genes and signaling pathways that cooperate to drive cell proliferation and inhibit apoptosis have provided important mechanistic insight into the complex process of cellular transformation. Importantly however, such scientific detail has provided the impetus to develop novel therapeutic agents that target the "apoptotic arm" of a driver or cooperating oncogene. Accordingly, successful treatment of an "oncogene-addicted" tumor may not necessarily require direct inhibition of the proto-oncogene itself, but may be achieved through inactivation of the cooperating antiapoptotic countermeasure that is by definition essential for survival of the neoplastic cell.

ACKNOWLEDGMENTS

R.W.J. is a principal research fellow of the National Health and Medical Research Council of Australia (NHMRC) and is supported by NHMRC Program and Project Grants, the Susan G. Komen Breast Cancer Foundation, the Prostate Cancer Foundation of Australia, Cancer Council Victoria, the Victorian Cancer Agency, the Leukemia Foundation of Australia, Victorian Breast Cancer Research Consortium, and the Australian Rotary Health Foundation. J.S. is supported by the Leukaemia Foundation of Australia and the Cooperative Research Centre for Biomedical Imaging Development.

REFERENCES

Adhikary S, Eilers M. 2005. Transcriptional regulation and transformation by Myc proteins. *Nat Rev Mol Cell Biol* **6:** 635–645.

Astrinidis A, Henske EP. 2005. Tuberous sclerosis complex: Linking growth and energy signaling pathways with human disease. *Oncogene* **24:** 7475–7481.

Aukema SM, Siebert R, Schuuring E, van Imhoff GW, Kluin-Nelemans HC, Boerma EJ, Kluin PM. 2011. Double-hit B-cell lymphomas. *Blood* **117:** 2319–2331.

Barbie DA, Tamayo P, Boehm JS, Kim SY, Moody SE, Dunn IF, Schinzel AC, Sandy P, Meylan E, Scholl C, et al. 2009. Systematic RNA interference reveals that oncogenic KRAS-driven cancers require TBK1. *Nature* **462:** 108–112.

Beverly LJ, Varmus HE. 2009. MYC-induced myeloid leukemogenesis is accelerated by all six members of the antiapoptotic BCL family. *Oncogene* **28:** 1274–1279.

Bouchard C, Dittrich O, Kiermaier A, Dohmann K, Menkel A, Eilers M, Luscher B. 2001. Regulation of cyclin D2 gene expression by the Myc/Max/Mad network: Myc-dependent TRRAP recruitment and histone acetylation at the cyclin D2 promoter. *Genes Dev* **15:** 2042–2047.

Campbell KJ, Bath ML, Turner ML, Vandenberg CJ, Bouillet P, Metcalf D, Scott CL, Cory S. 2010. Elevated Mcl-1 perturbs lymphopoiesis, promotes transformation of hematopoietic stem/progenitor cells, and enhances drug resistance. *Blood* **116:** 3197–3207.

Cardone MH, Roy N, Stennicke HR, Salvesen GS, Franke TF, Stanbridge E, Frisch S, Reed JC. Regulation of cell death protease caspase-9 by phosphorylation. *Science* **282:** 1318–1321.

Castellano E, Downward J. 2010. RAS interaction with PI3K: More than just another effector pathway. *Genes Cancer* **2:** 261–274.

Chin L, Tam A, Pomerantz J, Wong M, Holash J, Bardeesy N, Shen Q, O'Hagan R, Pantginis J, Zhou H, et al. 1999. Essential role for oncogenic Ras in tumour maintenance. *Nature* **400:** 468–472.

Cox AD, Der CJ. 2003. The dark side of Ras: Regulation of apoptosis. *Oncogene* **22:** 8999–9006.

Dan HC, Sun M, Kaneko S, Feldman RI, Nicosia SV, Wang HG, Tsang BK, Cheng JQ. 2004. Akt phosphorylation and stabilization of X-linked inhibitor of apoptosis protein (XIAP). *J Biol Chem* **279:** 5405–5412.

Datta SR, Katsov A, Hu L, Petros A, Fesik SW, Yaffe MB, Greenberg ME. 2000. 14-3-3 proteins and survival kinases cooperate to inactivate BAD by BH3 domain phosphorylation. *Mol Cell* **6:** 41–51.

Davies H, Bignell GR, Cox C, Stephens P, Edkins S, Clegg S, Teague J, Woffendin H, Garnett MJ, Bottomley W, et al. 2002. Mutations of the BRAF gene in human cancer. *Nature* **417:** 949–954.

Druker BJ. 2008. Translation of the Philadelphia chromosome into therapy for CML. *Blood* **112:** 4808–4817.

Egle A, Harris AW, Bouillet P, Cory S. 2004. Bim is a suppressor of Myc-induced mouse B cell leukemia. *Proc Natl Acad Sci* **101:** 6164–6169.

Eischen CM, Weber JD, Roussel MF, Sherr CJ, Cleveland JL. 1999. Disruption of the ARF-Mdm2-p53 tumor suppressor pathway in Myc-induced lymphomagenesis. *Genes Dev* **13:** 2658–2669.

Engelman JA. 2009. Targeting PI3K signalling in cancer: Opportunities, challenges and limitations. *Nat Rev Cancer* **9:** 550–562.

Engelman JA, Luo J, Cantley LC. 2006. The evolution of phosphatidylinositol 3-kinases as regulators of growth and metabolism. *Nat Rev Genet* **7:** 606–619.

Evan GI, Wyllie AH, Gilbert CS, Littlewood TD, Land H, Brooks M, Waters CM, Penn LZ, Hancock DC. 1992. Induction of apoptosis in fibroblasts by c-myc protein. *Cell* **69:** 119–128.

Faber AC, Li D, Song Y, Liang MC, Yeap BY, Bronson RT, Lifshits E, Chen Z, Maira SM, Garcia-Echeverria C, et al. 2009. Differential induction of apoptosis in HER2 and EGFR addicted cancers following PI3K inhibition. *Proc Natl Acad Sci* **106:** 19503–19508.

Felsher DW, Bishop JM. 1999. Reversible tumorigenesis by MYC in hematopoietic lineages. *Mol Cell* **4** 199–207.

Fernandez-Medarde A, Santos E. 2011. Ras in cancer and developmental diseases. *Genes Cancer* **2:** 344–358.

Gomez-Roman N, Felton-Edkins ZA, Kenneth NS, Goodfellow SJ, Athineos D, Zhang J, Ramsbottom BA, Innes F, Kantidakis T, Kerr ER, et al. 2006. Activation by c-Myc of transcription by RNA polymerases I, II and III. *Biochem Soc Symp* 141–154.

Gustafson WC, Weiss WA. 2010. Myc proteins as therapeutic targets. *Oncogene* **29:** 1249–1259.

Hanahan D, Weinberg RA. 2000. The hallmarks of cancer. *Cell* **100:** 57–70.

Hanahan D, Weinberg RA. 2011. Hallmarks of cancer: The next generation. *Cell* **144:** 646–674.

Hemann MT, Zilfou JT, Zhao Z, Burgess DJ, Hannon GJ, Lowe SW. 2004. Suppression of tumorigenesis by the p53 target PUMA. *Proc Natl Acad Sci* **101:** 9333–9338.

Hermeking H, Rago C, Schuhmacher M, Li Q, Barrett JF, Obaya AJ, O'Connell BC, Mateyak MK, Tam W, Kohlhuber F, et al. 2000. Identification of CDK4 as a target of c-MYC. *Proc Natl Acad Sci* **97:** 2229–2234.

Huang J, Dibble CC, Matsuzaki M, Manning BD. 2008. The TSC1-TSC2 complex is required for proper activation of mTOR complex 2. *Mol Cell Biol* **28:** 4104–4115.

Jain M, Arvanitis C, Chu K, Dewey W, Leonhardt E, Trinh M, Sundberg CD, Bishop JM, Felsher DW. 2002. Sustained loss of a neoplastic phenotype by brief inactivation of MYC. *Science* **297:** 102–104.

Kharas MG, Janes MR, Scarfone VM, Lilly MB, Knight ZA, Shokat KM, Fruman DA. 2008. Ablation of PI3K blocks BCR-ABL leukemogenesis in mice, and a dual PI3K/mTOR inhibitor prevents expansion of human BCR-ABL[+] leukemia cells. *J Clin Invest* **118:** 3038–3050.

Kinzler KW, Vogelstein B. 1996. Lessons from hereditary colorectal cancer. *Cell* **87:** 159–170.

Kinzler KW, Vogelstein B. 1998. Landscaping the cancer terrain. *Science* **280:** 1036–1037.

Koptyra M, Cramer K, Slupianek A, Richardson C, Skorski T. 2008. BCR/ABL promotes accumulation of chromosomal aberrations induced by oxidative and genotoxic stress. *Leukemia* **22:** 1969–1972.

Kuppers R, Dalla-Favera R. 2001. Mechanisms of chromosomal translocations in B cell lymphomas. *Oncogene* **20:** 5580–5594.

Lee S, Chari NS, Kim HW, Wang X, Roop DR, Cho SH, DiGiovanni J, McDonnell TJ. 2007. Cooperation of Ha-ras and Bcl-2 during multistep skin carcinogenesis. *Mol Carcinog* **46:** 949–957.

Lin CJ, Malina A, Pelletier J. 2009. c-Myc and eIF4F constitute a feedforward loop that regulates cell growth: Implications for anticancer therapy. *Cancer Res* **69:** 7491–7494.

Liu P, Cheng H, Roberts TM, Zhao JJ. 2009. Targeting the phosphoinositide 3-kinase pathway in cancer. *Nat Rev Drug Discov* **8:** 627–644.

Lowe SW, Cepero E, Evan G. 2004. Intrinsic tumour suppression. *Nature* **432:** 307–315.

Marmor MD, Skaria KB, Yarden Y. 2004. Signal transduction and oncogenesis by ErbB/HER receptors. *Int J Radiat Oncol Biol Phys* **58:** 903–913.

Mason KD, Vandenberg CJ, Scott CL, Wei AH, Cory S, Huang DC, Roberts AW. 2008. In vivo efficacy of the Bcl-2 antagonist ABT-737 against aggressive Myc-driven lymphomas. *Proc Natl Acad Sci* **105:** 17961–17966.

Mayo MW, Wang CY, Cogswell PC, Rogers-Graham KS, Lowe SW, Der CJ, Baldwin AS Jr. 1997. Requirement of NF-κB activation to suppress p53-independent apoptosis induced by oncogenic Ras. *Science* **278:** 1812–1815.

Meyer N, Penn LZ. 2008. Reflecting on 25 years with MYC. *Nat Rev Cancer* **8:** 976–990.

Meylan E, Dooley AL, Feldser DM, Shen L, Turk E, Ouyang C, Jacks T. 2009. Requirement for NF-κB signalling in a mouse model of lung adenocarcinoma. *Nature* **462:** 104–107.

Michalak EM, Jansen ES, Happo L, Cragg MS, Tai L, Smyth GK, Strasser A, Adams JM, Scott CL. 2009. Puma and to a lesser extent Noxa are suppressors of Myc-induced lymphomagenesis. *Cell Death Differ* **16:** 684–696.

Nowicki MO, Falinski R, Koptyra M, Slupianek A, Stoklosa T, Gloc E, Nieborowska-Skorska M, Blasiak J, Skorski T. 2004. BCR/ABL oncogenic kinase promotes unfaithful repair of the reactive oxygen species-dependent DNA double-strand breaks. *Blood* **104:** 3746–3753.

O'Dwyer ME, Druker BJ. 2000. STI571: An inhibitor of the BCR-ABL tyrosine kinase for the treatment of chronic myelogenous leukaemia. *Lancet Oncol* **1:** 207–211.

Oswald F, Lovec H, Moroy T, Lipp M. 1994. E2F-dependent regulation of human MYC: Trans-activation by cyclins D1 and A overrides tumour suppressor protein functions. *Oncogene* **9:** 2029–2036.

Perrotti D, Jamieson C, Goldman J, Skorski T. 2010. Chronic myeloid leukemia: Mechanisms of blastic transformation. *J Clin Invest* **120:** 2254–2264.

Ravitz MJ, Chen L, Lynch M, Schmidt EV. 2007. c-myc repression of TSC2 contributes to control of translation initiation and Myc-induced transformation. *Cancer Res* **67:** 11209–11217.

Reimann M, Loddenkemper C, Rudolph C, Schildhauer I, Teichmann B, Stein H, Schlegelberger B, Dorken B, Schmitt CA. 2007. The Myc-evoked DNA damage response accounts for treatment resistance in primary lymphomas in vivo. *Blood* **110:** 2996–3004.

Schmelzle T, Hall MN. 2000. TOR, a central controller of cell growth. *Cell* **103:** 253–262.

Schmitt CA, McCurrach ME, de Stanchina E, Wallace-Brodeur RR, Lowe SW. 1999. INK4a/ARF mutations accelerate lymphomagenesis and promote chemoresistance by disabling p53. *Genes Dev* **13:** 2670–2677.

Schmitt CA, Fridman JS, Yang M, Baranov E, Hoffman RM, Lowe SW. 2002. Dissecting p53 tumor suppressor functions in vivo. *Cancer Cell* **1:** 289–298.

Sears RC. 2004. The life cycle of C-myc: From synthesis to degradation. *Cell Cycle* **3:** 1133–1137.

Sensi M, Nicolini G, Petti C, Bersani I, Lozupone F, Molla A, Vegetti C, Nonaka D, Mortarini R, Parmiani G, et al. 2006. Mutually exclusive NRASQ61R and BRAFV600E mutations at the single-cell level in the same human melanoma. *Oncogene* **25:** 3357–3364.

Seoane J, Le HV, Massague J. 2002. Myc suppression of the p21(Cip1) Cdk inhibitor influences the outcome of the p53 response to DNA damage. *Nature* **419:** 729–734.

Serrano M, Lin AW, McCurrach ME, Beach D, Lowe SW. 1997. Oncogenic ras provokes premature cell senescence associated with accumulation of p53 and p16INK4a. *Cell* **88:** 593–602.

Staal SP, Hartley JW, Rowe WP. 1977. Isolation of transforming murine leukemia viruses from mice with a high incidence of spontaneous lymphoma. *Proc Natl Acad Sci* **74:** 3065–3067.

Strasser A, Harris AW, Bath ML, Cory S. 1990. Novel primitive lymphoid tumours induced in transgenic mice by cooperation between myc and bcl-2. *Nature* **348:** 331–333.

Takeuchi K, Ito F. 2010. EGF receptor in relation to tumor development: Molecular basis of responsiveness of cancer cells to EGFR-targeting tyrosine kinase inhibitors. *FEBS J* **277:** 316–326.

Vafa O, Wade M, Kern S, Beeche M, Pandita TK, Hampton GM, Wahl GM. 2002. c-Myc can include DNA damage, increase reactive oxygen species, and mitigate p53 function: A mechanism for oncogene-induced genetic instability. *Mol Cell* **9:** 1031–1044.

Vita M, Henriksson M. 2006. The Myc oncoprotein as a therapeutic target for human cancer. *Semin Cancer Biol* **16:** 318–330.

Vivanco I, Sawyers CL. 2002. The phosphatidylinositol 3-kinase AKT pathway in human cancer. *Nat Rev Cancer* **2:** 489–501.

Weinstein IB. 2002. Cancer. Addiction to oncogenes—The Achilles heal of cancer. *Science* **297:** 63–64.

Wendel HG, De Stanchina E, Fridman JS, Malina A, Ray S, Kogan S, Cordon-Cardo C, Pelletier J, Lowe SW. 2004. Survival signalling by Akt and eIF4E in oncogenesis and cancer therapy. *Nature* **428:** 332–337.

Wheeler DL, Dunn EF, Harari PM. 2010. Understanding resistance to EGFR inhibitors-impact on future treatment strategies. *Nat Rev Clin Oncol* **7:** 493–507.

Whitecross KF, Alsop AE, Cluse LA, Wiegmans A, Banks KM, Coomans C, Peart MJ, Newbold A, Lindemann RK, Johnstone RW. 2009. Defining the target specificity of ABT-737 and synergistic antitumor activities in combination with histone deacetylase inhibitors. *Blood* **113:** 1982–1991.

Yuan TL, Cantley LC. 2008. PI3K pathway alterations in cancer: Variations on a theme. *Oncogene* **27:** 5497–5510.

The Role of the Apoptotic Machinery in Tumor Suppression

Alex R.D. Delbridge[1,2], Liz J. Valente[1,2], and Andreas Strasser[1,2]

[1]The Walter and Eliza Hall Institute, Parkville 3050, Melbourne, Australia
[2]Department of Medical Biology, The University of Melbourne, Parkville 3050, Melbourne, Australia
Correspondence: strasser@wehi.edu.au

Multicellular organisms have evolved processes to prevent abnormal proliferation or inappropriate tissue infiltration of cells, and these tumor suppressive mechanisms serve to prevent tissue hyperplasia, tumor development, and metastatic spread of tumors. These include potentially reversible processes such as cell cycle arrest and cellular senescence, as well as apoptotic cell death, which in contrast eliminates dangerous cells that may initiate tumor development. Tumor suppressive processes are organized as complex, extensive signaling networks, controlled by central "nodes." These "nodes" are prominent tumor suppressors, such as P53 or PTEN, whose loss is responsible for the development of the majority of human cancers. In this review we discuss the processes by which some of these prominent tumor suppressors trigger apoptotic cell death and how this process protects us from cancer development.

A malignant tumor is characterized by the ability to expand in an uncontrolled manner, destroy normal tissue architecture, and ultimately undergo metastatic spread (Hanahan and Weinberg 2000). Although the number of mutations required for neoplastic transformation may vary, all tumors are reliant on two critical mechanisms for their development; the activation of oncogenes that promote proliferation and survival of cancer cells, as well as the inactivation of tumor suppressor genes that normally repress development and growth of tumors (Hanahan and Weinberg 2000).

Oncogenes can be activated via multiple mechanisms, including chromosomal translocations, deletions or insertions, as well as point mutations. One such example is the translocation between chromosomes 9 and 22 that is present in most cases of chronic myeloid leukemia. The juxtaposition of the *BCR* and *c-ABL* genes results in the production of an abnormal BCR-ABL fusion protein with constitutive kinase activity (Deininger et al. 2005). However, in other cancer-causing chromosomal translocations, such as the t[8;14] translocation in Burkitt's lymphoma, the coding sequence of the oncogene, *c-MYC*, is unchanged; rather its activation results from deregulated expression in B lymphoid cells as a consequence of its proximity to the *IGH* gene enhancer (Cory et al. 1987). Tumorigenesis promoted by deregulated kinase activity frequently results from the acquisition of point mutations. In this context, a single amino acid substitution can dramatically enhance kinase activity by preventing binding of negative regulators or "locking" the catalytic domain in the active conformation. This is exemplified by the *BRAF(V600E)* mutation frequently observed in melanoma or colon carcinoma (Poulikakos and Rosen 2011) and the activating mutations in *EGF-R* observed in lung adenocarcinoma (Sharma et al. 2007).

Analogous to the activation of oncogenes, tumor suppressor genes can be inactivated through multiple mechanisms, including large-scale chromosomal alterations or point mutations. However, in most cases both alleles of the gene must be compromised to abolish gene function, unless the mutated protein can act in a dominant-negative fashion to block the activity of its wild-type counterpart.

Multicellular organisms have evolved a plethora of mechanisms to restrain the growth or even eliminate aberrant cells—these processes can all function as tumor suppressors. Notably, of the attributes that cells must acquire to become cancerous ("hallmarks of cancer") discussed by Hanahan and Weinberg (2000), several relate to escape from regulatory processes that would normally suppress tumor growth. They include cell cycle arrest, cellular senescence, and cell death; of these only cell death is irreversible, all others can (at least potentially) be reversed. In this review, we describe the mechanisms by which tumor suppressors that are disabled in a broad range and large fraction of cancers trigger cell death, and how components of the apoptotic machinery can themselves act as tumor suppressors.

APOPTOSIS AS A MEDIATOR OF TUMOR SUPPRESSION

Apoptosis, also known as programmed cell death, is a highly regulated program of ordered cellular destruction that facilitates the removal of damaged or superfluous cells. This process is critical for many physiological processes, including embryonic development and tissue homeostasis in adulthood (Strasser et al. 2000; Hotchkiss et al. 2009). In vertebrates, apoptosis can be initiated by two distinct, albeit ultimately converging, signaling pathways (Strasser et al. 1995), termed "BCL-2-regulated" ("intrinsic," "mitochondrial," "stress-induced") (Chipuk and Green 2008) and "death receptor-" (Strasser et al. 2009) induced apoptosis (Fig. 1). In both pathways, cell demolition is mediated by aspartate-specific cysteine proteases (caspases) that proteolyze hundreds of cellular proteins (Timmer and Salvesen 2007). Cell surface "death receptors" (members of the TNF-R family with an intracellular "death domain," e.g., FAS, TNF-R1) can trigger apoptosis by direct activation of caspases, through adaptor protein (FADD, TRADD)-mediated activation (via conformational change) of initiator caspase-8, which then proteolytically activates effector caspases (caspase-3, -6, and -7) (Strasser et al. 2009) (Fig. 1). Mutations in FAS or its ligand, FASL, perturb peripheral lymphoid homeostasis, ultimately leading to severe lymphadenopathy, a systemic autoimmune disease and a predisposition to hematopoietic malignancy in both mice (Watanabe-Fukunaga et al. 1992; O'Reilly et al. 2009) and humans (Rieux-Laucat et al. 1995; Drappa et al. 1996).

The "BCL-2-regulated" apoptotic pathway can be activated by developmental cues and a broad range of cytotoxic insults, including cytokine deprivation or DNA damage. This pathway is regulated by the BCL-2 protein family and relies (at least in part) on the initiator caspase-9 (and its adaptor APAF-1) to activate the effector caspases (Marsden et al. 2002; Shi 2002; Riedl and Shi 2004) (Fig. 1). The BCL-2 protein family is composed of one antiapoptotic and two proapoptotic subgroups that regulate commitment to cell death through complex protein–protein interactions (Youle and Strasser 2008). The antiapoptotic members (BCL-2, BCL-X_L, BCL-W, MCL-1, and A1) share four BH (BCL-2 Homology) domains and are essential for cell survival, functioning in a cell type-specific manner, albeit with significant functional overlap (Youle and Strasser 2008). Proapoptotic BAX and BAK share remarkable structural similarity with their prosurvival relatives, but they have largely overlapping functions during the execution of apoptosis and are

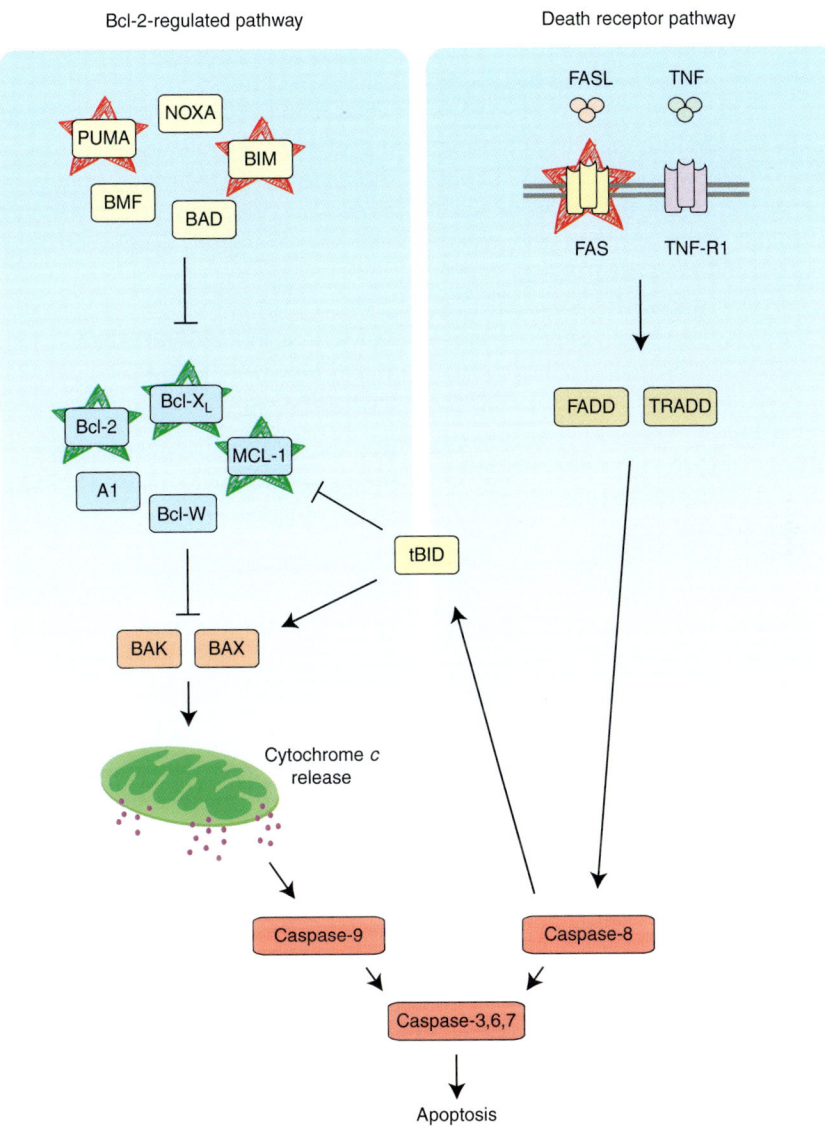

Figure 1. Apoptosis can be initiated by activation of two distinct, albeit ultimately converging, pathways, the "Bcl-2-regulated pathway" (*left*; also known as the "stress" or "mitochondrial" pathway) and the "death receptor pathway" (*right*). As the name suggests, protein–protein interactions between members of the Bcl-2 family govern activation of the Bcl-2-regulated pathway, whereas binding of their cognate ligands activates the death receptors (e.g., FAS, TNF-R1). Cellular demolition is performed by the effector caspases that act downstream of both pathways. Proteins that have been identified as suppressors or oncogenes are indicated in red or green, respectively.

essential for mitochondrial outer membrane permeabilization (MOMP) with consequent release of apoptogenic molecules (e.g., cytochrome *c*, DIABLO/Smac) and activation of caspase-9 (Lindsten et al. 2000; Chipuk and Green 2008). The proapoptotic BH3-only proteins (BIM, PUMA, BID, BAD, BMF, HRK, BIK, and NOXA) share with each other and the BCL-2 family overall only the BH3 domain and are required for initiation of apoptosis

(Huang and Strasser 2000). These proteins are activated transcriptionally and/or post-translationally and exert stimulus-specific as well as cell type-specific actions. For example, BIM is critical for cytokine deprivation-induced apoptosis (Bouillet et al. 1999), whereas PUMA and, to a lesser extent, NOXA trigger the apoptosis activated by the tumor suppressor P53 (Jeffers et al. 2003; Villunger et al. 2003; Michalak et al. 2008). BH3-only proteins are thought to activate BAX/BAK either through direct binding and/or indirectly by binding to their repressors, the prosurvival BCL-2 proteins (Chipuk and Green 2008; Merino et al. 2009).

Deregulated expression of prosurvival BCL-2 proteins (e.g., because of the t[14;18], *BCL2*; *IGH* chromosomal translocation in human follicular center lymphoma) promotes tumorigenesis by sustaining the viability of cells undergoing neoplastic transformation, thereby facilitating the acquisition of additional oncogenic mutations (Vaux et al. 1988; Strasser et al. 1990, 1993). Similarly, loss of proapoptotic BCL-2 family members in isolation is not potently transforming, but it is consistent with the notion that cancer cells must acquire the ability to evade apoptosis; such defects promote tumorigenesis when they occur in concert with additional oncogenic mutations. For example, homozygous deletion of the *BIM* gene was found in ~20% of human mantle cell lymphoma cases and, accordingly, loss of *Bim* accelerated Eμ-*MYC*-driven lymphomagenesis in mice (Egle et al. 2004). Moreover, ~40% of human Burkitt's lymphomas fail to express PUMA (Garrison et al. 2008), and loss of PUMA can also (like loss of BIM) accelerate lymphoma development in Eμ-*MYC* transgenic mice (Garrison et al. 2008; Michalak et al. 2009). In addition, loss of the BH3-only proteins NOXA (Michalak et al. 2010) and BMF (Labi et al. 2008) as well as loss of BAX (Eischen et al. 2001) were found to promote development of lymphoma in different experimental mouse models, but so far deregulation of these genes has not been detected in human cancers. As mentioned above, defects in apoptosis by themselves are not potently transforming. Therefore, mutations of oncogenes or tumor suppressor genes that encode master regulators of multiple pathways (including apoptosis) are frequently observed in human cancers because one (or two) oncogenic lesions will simultaneously activate several tumorigenic processes. In the following sections, we discuss the mechanisms by which tumor suppressors trigger apoptosis and their importance for prevention of neoplastic disease.

TUMOR SUPPRESSORS AND THEIR MECHANISMS FOR INDUCING APOPTOSIS

P53, Orchestrator of the Cellular Response to DNA Damage, Hypoxia, and Oncogenic Stress

The tumor suppressor *P53*, a transcriptional regulator, is mutated in ~50% of human cancers and in those cancers that lack *P53* mutations, P53 signaling is often defective because of the acquisition of some other mutation(s) (Vousden and Lane 2007). P53 can be activated by a broad range of cytotoxic stress signals, including DNA damage, hypoxia, and activation of certain oncogenes (e.g., *c-MYC*) (Vousden and Lane 2007). In the majority of unstressed cells, *P53* mRNA is expressed constitutively, but P53 protein levels are low. This is mainly attributable to the action of MDM2, an E3 ubiquitin ligase, which targets P53 for K48 linkage-mediated ubiquitination and proteasomal degradation (Haupt et al. 1997; Kubbutat et al. 1997). Stress-induced P53 activation involves stabilization of the P53 protein. This is mediated primarily through inhibition of MDM2 by the tumor suppressor ARF (Vousden and Lane 2007). However, in addition, multiple posttranslational modifications, including phosphorylation, acetylation, and neddylation, also affect P53 stability, its binding to target genes, and/or its transcriptional activity (Vousden and Lane 2007). Upon activation, P53 binds as a homotetramer to specific sequences within the regulatory regions of a broad range of target genes (Riley et al. 2008) and thereby triggers a multitude of effector pathways, including cell cycle arrest (Livingstone et al. 1992), cellular senescence (Metz et al. 1995), coordination of DNA repair, and apoptosis (Vousden and Lane 2007) (Fig. 2). In addition, P53 triggers a process that regulates its own activity that operates mainly through direct transcriptional activation of MDM2 (Barak et al. 1993; Wu et al. 1993). The critical importance of appropriate P53/MDM2 feedback regulation was revealed by the discovery that loss of MDM2 causes embryonic lethality in mice and that this can be prevented by concomitant loss of P53 (Jones et al. 1995; Montes de Oca Luna et al. 1995).

It remains unclear why a specific effector pathway will dominate in a particular cell in response to P53 activation. For example, why does low-dose γ-irradiation elicit cell G_1/S cycle arrest and DNA repair in fibroblasts but apoptosis in thymocytes? It is possible that parallel signaling pathways, active in some cells but not others, can modulate the overall outcome of P53 activation. In addition, the various post-translational modifications may affect the preference of P53 for different target genes, thereby determining which effector pathway will predominate. To induce apoptosis, P53 directly transcriptionally up-regulates the expression of several proapoptotic BCL-2 family members, the BH3-only proteins PUMA (Nakano and Vousden 2001; Yu et al. 2001) and NOXA (Oda et al. 2000) as well as the multi-BH domain proapoptotic protein BAX (Miyashita and Reed 1995). Experiments with gene-targeted mice have shown that P53-induced apoptosis in a broad range of cell types is mediated predominantly by PUMA and to lesser extent by NOXA (Jeffers et al. 2003; Villunger et al. 2003; Erlacher et al. 2005; Naik et al. 2007; Michalak et al. 2008). BAX is expressed in P53-deficient cells, and many P53-independent apoptotic stimuli such as glucocorticoids rely on BAX (plus BAK) for cell killing (Lindsten et al. 2000). Thus, the P53-mediated induction of BAX probably serves to increase the efficiency of apoptosis signaling but does not determine whether a cell will live or die. As mentioned, P53 activates diverse effector pathways and it is not yet clear which one(s) is/are critical for tumor suppression. Loss of P53's apoptotic pathway, through loss of PUMA (Garrison et al. 2008; Michalak et al. 2009) and/or NOXA (Michalak et al. 2010), can accelerate lymphoma development elicited by MYC overexpression or low-dose γ-irradiation, respectively. However, in contrast to P53-deficient mice (Donehower et al. 1992), animals lacking both PUMA and NOXA are not tumor prone, although their cells display comparable resistance to P53-dependent apoptotic stimuli (e.g., γ-irradiation) as those lacking P53 itself (Michalak et al. 2008). Thus, apoptosis induction does not account for all of the tumor suppressive action of P53. Loss

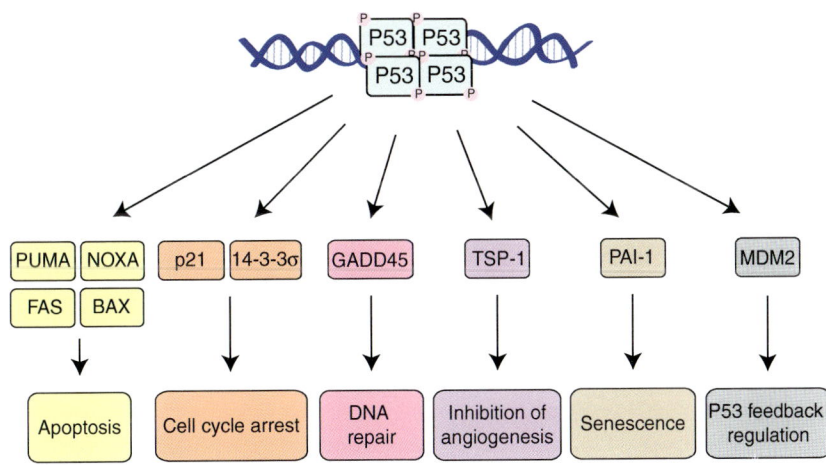

Figure 2. The tumor suppressor P53 acts as a transcriptional regulator. It has the capacity to activate diverse cellular processes. Stimulus and cell type-specific effects determine which particular effector pathway(s) will dominate.

of P21, the major effector of P53-mediated G_1/S cell cycle arrest, also does not constitute a potently transforming oncogenic lesion (Deng et al. 1995). It therefore appears that it is the composite loss of several effector pathways (coordination of DNA repair, cell cycle arrest, apoptosis, and/or senescence) and not the loss of a single process that accounts for the high incidence and rapid onset of tumors when P53 is mutated in mice (Donehower et al. 1992) and in humans (Li-Fraumeni syndrome [Srivastava et al. 1990]). Recent evidence has emerged that suggests that signaling resulting from acute P53 activation is distinct from its ability to effect tumor suppression (Brady et al. 2011). Furthermore, there is now evidence that even combined defects in P53-mediated induction of apoptosis, cell cycle arrest, and cell senescence are all dispensable for P53-mediated tumor suppression (Li et al. 2012). This may indicate that regulation of cellular metabolism and perhaps coordination of DNA repair may be critical for the tumor suppressive action of P53.

Retinoblastoma Protein, a Negative Regulator of Cell Cycle Entry

The retinoblastoma protein (RB) is the central element in a tumor suppression network that is interconnected with that of P53 through the regulation of cell cycle arrest and also through activation by oncogenic stress (Lee et al. 1987). *Rb* was the first tumor suppressor identified in human cancer following the realization that inheritance of one mutated copy of *Rb* followed by mutation or loss of the wild-type allele resulted in the development of familial retinoblastoma in early childhood (Lee et al. 1987). Accordingly, heterozygous loss of RB in mice results in the development of pituitary tumors (that have lost the wt *Rb* allele), but curiously not retinoblastoma (Jacks et al. 1992). As for P53, the critical importance of the RB protein in the regulation of cell proliferation came from the discovery that DNA tumor viruses, such as the human papillomaviruses (HPV), promote host cell survival and proliferation and thereby viral replication by encoding specific inhibitors of these two tumor suppressors (Levine 2009). At that time, it was thought that most cancers were of viral origin; although this has not proven to be the case, the studies that have followed have illuminated RB (and P53) as critical components of the tumor suppression network.

RB controls cell proliferation through its regulation of the E2F family of transcription factors. In its active form, RB binds and sequesters the E2F proteins, which play critical roles in S phase entry. Upon phosphorylation, RB releases the E2F proteins allowing them to regulate transcription of their target genes. In addition to its role in cell cycle regulation, RB is also critical for the differentiation programs of certain tissues; however, the relevance, if any, of these activities for tumor suppression is currently unclear. This division of labor is achieved through the presence of distinct subsets within the E2F family. E2F1-3 promote cell cycle progression and are regulated by RB (Lees et al. 1993), whereas E2F4-8 act as transcriptional repressors and promote cell cycle exit and differentiation. In addition to its role in controlling P53 protein levels, MDM2 can also regulate the process by which E2F1 and its cofactor, DP1, acts to promote cell division (Martin et al. 1995; Xiao et al. 1995). The ability of MDM2 to potentiate E2F1 activity relies on inhibition of SKP2-mediated degradation of E2F1 (Zhang et al. 2005). The critical role that the activator E2Fs play in the control of proliferation is underscored by the finding that the loss of their negative regulator, RB, and its homologs abrogates the G_1/S checkpoint and thereby promotes cellular immortalization (Dannenberg et al. 2000; Sage et al. 2000).

In addition to their role in promoting proliferation, paradoxically, the activator E2Fs can also promote apoptosis, at least in certain settings. This is particularly clear in the context of DNA damage when E2F1 is stabilized post-translationally through ATM-, ATR-, CHK1-, and CHK2-mediated phosphorylation, as well as by acetylation (Lin et al. 2001; Pediconi et al. 2003; Stevens et al. 2003; Urist et al. 2004). These modifications appear to drive E2F1 to up-regulate proapoptotic genes (Hershko and Ginsberg 2004), such as the BH3-only gene *Bim* (O'Connor et al. 1998). This duality of function is thought to represent a fail-safe mechanism whereby cells that have sustained DNA damage are predisposed to undergo apoptosis following E2F1 activation rather than DNA synthesis and proliferation. Indeed, E2F-driven proliferation can impose an "oncogene activation"-like stress on cells stimulating P53 in an ARF-dependent manner (Bates et al. 1998). The relevance of this signaling network for tumorigenesis is exemplified by studies of an E2F1-driven mouse model of skin cancer. Overexpression of E2F1 resulted in hyperplasia in the epidermis but this was held in check by a concomitant increase in apoptosis (Pierce et al. 1998a). Loss of P53 prevented this apoptosis and caused progression to skin carcinoma (Pierce et al. 1998b).

The realization that loss of RB, through consequent activation of the E2Fs, can have proapoptotic effects may at least in part explain why loss of RB activity commonly occurs late in tumor development and is associated with progression rather than initiation (Polager and Ginsberg 2009). Thus, apoptotic signaling may first need to be compromised (e.g., by loss of P53) so that the oncogenic stress imposed by RB inactivation does not eliminate the cells undergoing transformation.

Phosphatase with Tensin Homology, the Critical Safety Catch of the PI3K/AKT Signaling Pathway

PTEN (phosphatase with tensin homology), a lipid phosphatase, is the key negative regulator of the phosphatidylinositol 3-kinase (PI3K) signaling pathway that promotes cell survival and proliferation and is frequently deregulated in various human cancers (Stambolic et al. 1998) (Fig. 3). Whereas the PI3K pathway is extensive and can direct diverse cellular processes, PTEN inhibits this pathway by breaking down PI3K's active second messenger molecule, PIP_3. It is also noteworthy that PTEN is a direct P53 target and thus represents a constituent of the wider P53 tumor suppression network (Stambolic et al. 2001). PI3K can be activated by receptor tyrosine kinases (RTKs) as well as RAS. PI3K in turn activates AKT (also known as PKB), which regulates a broad range of cellular processes,

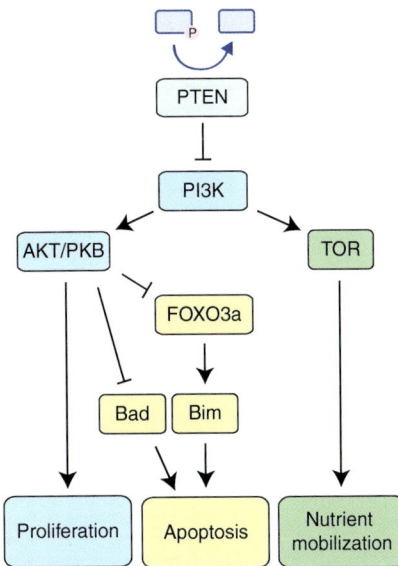

Figure 3. PTEN acts as the endogenous antagonist of the PI3K-regulated growth pathway. PTEN breaks down the key signaling intermediate PIP_3, thereby inhibiting downstream signaling mediated by AKT and TOR. Thus, PTEN negates PI3K's ability to drive proliferation, facilitate nutrient mobilization, and inhibit apoptosis.

including proliferation, nutrient mobilization, and cell survival.

AKT was reported to directly inhibit apoptosis by phosphorylating and thereby causing sequestration of the BH3-only protein BAD (del Peso et al. 1997). However, given that loss of BAD (Ranger et al. 2003) or even combined loss of BAD and BIM (Kelly et al. 2010) had only minimal effects on cell survival indicates that this process is not critical for sustaining survival of cells undergoing transformation. AKT has also been shown to promote cell survival by phosphorylating FOXO3a, thereby preventing this transcription factor from activating its target genes, such as the proapoptotic BH3-only gene *Bim* (O'Connor et al. 1998; Dijkers et al. 2000). Because *BIM* deletions are found in human mantle cell lymphomas (Tagawa et al. 2005) and BIM loss promotes lymphomagenesis in mice (Egle et al. 2004), this pathway is more likely to be critical for tumor suppression.

AKT also activates TOR signaling by inhibiting its negative regulators TSC1 and TSC2, thereby promoting RHEB activity (Potter et al. 2002). RHEB allows TOR to act as a nutrient sensor and liberate additional glucose and amino acids to promote cell growth and proliferation (Garami et al. 2003). This metabolic process, inhibited by PTEN, is likely to intersect with the control of cell survival.

The inappropriate activation of TOR signaling has particular relevance to patients with Tuberous Sclerosis Complex disease, which is caused by inherited germline mutations in either *TSC1* or *TSC2* that act in an autosomal dominant manner. This disease is characterized by skin, brain, kidney, and heart abnormalities, with brain tumors accounting for most of the morbidity and mortality. Inherited mutations in PTEN are the cause of the PTEN Hamartoma Tumor Syndrome. These patients are predisposed to the development of cancers of the breast, thyroid, and endometrium, of which breast tumors are the most common with a lifetime risk of 25%–50%. Accordingly, in mice loss of one *Pten* allele resulted in hyperplasia in multiple organs, such as the skin and prostate, which progressed to colon adenocarcinoma, gonadostromal tumors, teratomas, thyroid carcinoma, and lymphoma (Di Cristofano et al. 1998; Suzuki et al. 1998).

In conclusion, acting as a safety catch at the apex of the PI3K/AKT signaling pathway, PTEN plays a critical role in tumor suppression by preventing inappropriate activation of cellular metabolism (via effects on TOR), proliferation and survival (via effects on proapoptotic BH3-only proteins).

CYLD, a Negative Regulator of the NF-κB Signaling Pathway

CYLD is a member of the USP subfamily of deubiquitinases (DUBs), first identified as the gene mutated in the inherited condition familial cylindromatosis (Bignell et al. 2000) that is characterized by the development of benign skin tumors. In these patients, one mutated copy of *CYLD* is inherited and the wild-type allele is commonly lost during neoplastic progression.

Although ubiquitination was first identified as a process for targeting protein substrates for proteasome-mediated degradation, conjugation of ubiquitin moieties to substrates by linkages other than K48 (particularly K63 linkages) can activate signal transduction pathways (Wertz and Dixit 2010). Like many other cell signaling processes, ubiquitination is a reversible process, and whereas ubiquitin chains are assembled on target proteins by E3 ligases, they can also be hydrolyzed by DUBs. The existence of ~100 DUBs in humans presumably allows highly specific regulation of a broad range of signaling pathways.

CYLD acts as a specific negative regulator of NF-κB signaling through its interaction with NEMO, the regulatory subunit of the IKK proteins, and TRAF2, an adaptor for several TNF-R family members (Brummelkamp et al. 2003; Kovalenko et al. 2003; Trompouki et al. 2003) (Fig. 4). CYLD antagonizes the conjugation of K63 ubiquitin chains on TRAF2 and thereby prevents IKK activation, which is required for nuclear import and functional activation of NF-κB complexes (e.g., REL/P50). Transcriptional

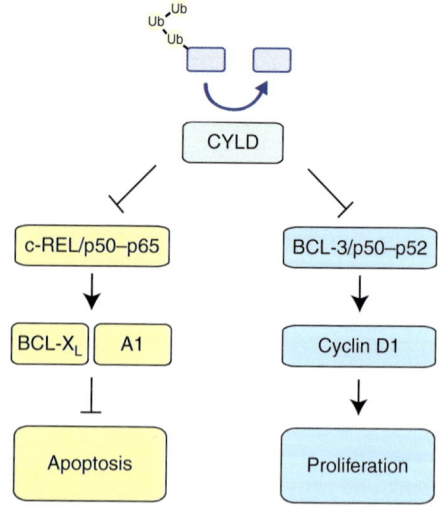

Figure 4. CYLD acts proximal to the membrane as a negative regulator of TNF-induced NF-κB signaling. The deubiquitinase activity of CYLD allows it to hydrolyze K63-linked ubiquitin chains, which are critical for the activity of multiple components within the NF-κB signaling network, and thereby acts as a suppressor by inhibiting proliferation and promoting apoptosis.

targets of NF-κB can modulate the duration and intensity of signaling forming either feed-forward or negative feedback loops. Whereas IκBα induction constitutes a negative feedback loop, BCL3 acts to potentiate signaling through the noncanonical NF-κB pathway. This process is of particular relevance to tumorigenesis, because cyclin D1, which requires BCL3 and NF-κB for induction and is repressed by CYLD, has been shown to promote tumor cell proliferation (Massoumi et al. 2006).

In addition to driving cellular proliferation, NF-κB also promotes cell survival, at least in part via transcriptional induction of antiapoptotic Bcl-2 family members (BCL-2, BCL-X_L, and A1) (Grumont et al. 1999; Lee et al. 1999; Zong et al. 1999; Chen et al. 2000; Grossmann et al. 2000) as well as several IAP (inhibitor of apoptosis) proteins (Chu et al. 1997; Stehlik et al. 1998; Wang et al. 1998). These NF-κB targets are known to be essential for the survival of various cell types in response to diverse stress stimuli. Collectively, these data indicate that CYLD suppresses tumorigenesis by reducing cell cycling (e.g., via repression of CYCLIN D1 expression) and by reducing expression of antiapoptotic regulators. Indeed, CYLD is frequently deleted in cases of multiple myeloma showing hyperactive NF-κB and abnormalities in cell cycling and apoptosis induction (Annunziata et al. 2007).

PTPN12, a Safety Catch for Several Oncogenic Receptor Kinases

PTPN12 has recently been described as a tumor suppressor in HER2, estrogen receptor and progesterone receptor negative (so-called "triple negative") breast cancer (TNBC) (Sun et al. 2011). PTPN12 expression was found to be low or undetectable in primary breast cancer samples, and the locus from which it is expressed is more frequently lost in TNBC compared to other breast cancer subtypes (Sun et al. 2011). PTPN12 encodes a receptor tyrosine phosphatase that can dephosphorylate, and thereby inactivate, several kinase-containing surface receptors (e.g., ER, HER2, EGF-R) that promote breast cancer development when abnormally activated or overexpressed. Knock-down of the PTPN12 phosphatase by RNAi enhanced anchorage-independent growth of breast epithelial cells in culture and the formation of abnormal acini in 3D culture conditions (Sun et al. 2011). Interestingly, the enhanced proliferation and acini formation of MCF10A breast epithelial derived cells resulting from reduced PTPN12 function was not accompanied by increased apoptosis (Sun et al. 2011). Mutants of PTPN12 deficient in phosphatase activity could not inhibit cell growth and acini formation, indicating that PTPN12's catalytic activity is required for tumor suppression. Consistent with the hypothesis that PTPN12 acts as a safety catch for HER2 and EGFR signaling for cell growth, pharmacological or shRNA-mediated inhibition of these receptors reduced the efficiency of transformation achieved through PTPN12 knockdown (Sun et al. 2011).

Loss of PTPN12 is thought to promote breast cancer development by allowing unrestrained signaling from the HER2, EGF-R, and PDGF-R receptors (Sun et al. 2011). These receptor tyrosine kinases all stimulate pathways for cell proliferation and, through activation of AKT as well as ERK, they probably also promote cell survival by repressing expression and/or function of the BH3-only proteins BIM and BAD (Fig. 5). Interestingly, it has been shown that repression of BIM (and to a lesser extent BAD) is critical for the sustained survival of certain other cancers that are driven by other oncogenic kinases—for example, BCR-ABL in CML (Kuroda et al. 2006), mutant EGF-R in lung cancer (Costa et al. 2007; Cragg et al. 2007), mutant B-RAF in melanoma, and colon carcinoma (Cragg et al. 2008).

These data show that oncogenic hyperactivation of several receptor tyrosine kinases that promote abnormal cell proliferation and survival can be achieved more readily through loss of a single common "safety catch" than through coordinate activation of multiple cell surface receptors.

CLOSING REMARKS

Multicellular organisms require mechanisms to ensure appropriate cellularity of their tissues

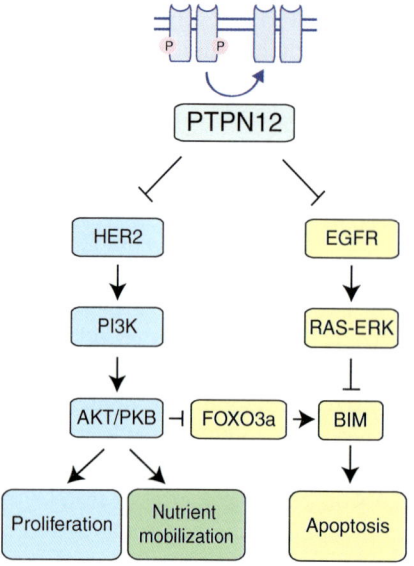

Figure 5. PTPN12 is a recently described tumor suppressor with phosphatase activity. It negatively regulates signaling through multiple oncogenic receptor kinases, such as EGFR and HER2. By limiting the activity of these receptors, PTPN12 blocks their ability to promote proliferation, nutrient mobilization, and inhibit apoptosis.

and, consequently, these processes impose barriers against tumor development. Some tumor suppressive mechanisms, such as cell cycle arrest or cellular senescence, facilitate repair and recovery of damaged cells. However, because of their reversible nature, these mechanisms can be subverted allowing outgrowth of clonal offspring. In contrast, apoptotic cell death eliminates cells, and although this unequivocally prevents clonal outgrowth of mutated cells, it comes at a cost to the organism. Some cells once lost cannot be readily replaced, and if the threshold for apoptosis initiation is set too low, cells may be eliminated unnecessarily and tissue degeneration may ensue. So, in tissues that are critical for function of the organism but have only low regenerative capacity (e.g., brain), processes such as cell cycle arrest, senescence, and differentiation are more appropriate mechanisms for tumor suppression than apoptosis. Conversely, in tissues with considerable regenerative capacity, such as the immune system, apoptosis would be the preferred mechanism for tumor suppression.

Although apoptosis is clearly a potent mechanism for tumor suppression and must be evaded to facilitate neoplastic transformation and sustained tumor expansion, other mechanisms are also critical to protect us from cancer. Thus, if escape from cell cycle arrest, senescence, apoptosis, and the mobilization of nutrients are all required for tumorigenesis to proceed, then mutational activation or inactivation of oncogenes or tumor suppressor genes, respectively, that regulate a number of these processes will enable cells to circumvent homeostatic controls in a more efficient manner than if each of these processes must be deregulated independently through multiple oncogenic events. The prevalence of mutations in certain tumor suppressors, particularly P53 and PTEN, in diverse human cancers identifies them as "Achilles heels" in the cell's tumor suppression network, in which loss or inactivation of these targets can confer on cells undergoing neoplastic transformation multiple advantages rather than enhanced survival or increased cycling alone. Deciphering the complex networks for tumor suppression is not only critical for understanding the development and growth of cancer, but also has ramifications for cancer therapy because many tumor suppressors, and their processes for apoptosis induction in particular, also regulate the responses of tumor cells to anticancer therapy (Johnstone et al. 2002).

ACKNOWLEDGMENTS

The authors thank all present and past members of the apoptosis research programs at WEHI, particularly Drs. J. Adams, S. Cory, D. Vaux, D. Huang, P. Colman, P. Bouillet, A. Harris, R. Kluck, J. Silke, and C. Scott, for their outstanding contributions to cell death research and for stimulating discussions. Research in the authors' laboratories is supported by fellowships and grants from the Australian NHMRC (257502, 461299), Cancer Council of Victoria, NIH (CA 043540), Leukemia and Lymphoma Society (LLS SCOR 7413), and the JDRF/NHMRC (466658).

REFERENCES

Annunziata CM, Davis RE, Demchenko Y, Bellamy W, Gabrea A, Zhan F, Lenz G, Hanamura I, Wright G, Xiao W, et al. 2007. Frequent engagement of the classical and alternative NF-κB pathways by diverse genetic abnormalities in multiple myeloma. *Cancer Cell* **12**: 115–130.

Barak Y, Juven T, Haffner R, Oren M. 1993. *mdm2* expression is induced by wild type p53 activity. *EMBO J* **12**: 461–468.

Bates S, Phillips AC, Clark PA, Stott F, Peters G, Ludwig RL, Vousden KH. 1998. p14ARF links the tumour suppressors RB and p53. *Nature* **395**.

Bignell GR, Warren W, Seal S, Takahashi M, Rapley E, Barfoot R, Green H, Brown C, Biggs PJ, Lakhani SR, et al. 2000. Identification of the familial cylindromatosis tumour-suppressor gene. *Nat Genet* **25**: 160–165.

Bouillet P, Metcalf D, Huang DCS, Tarlinton DM, Kay TWH, Köntgen F, Adams JM, Strasser A. 1999. Pro-apoptotic Bcl-2 relative Bim required for certain apoptotic responses, leukocyte homeostasis, and to preclude autoimmunity. *Science* **286**: 1735–1738.

Brady CA, Jiang D, Mello SS, Johnson TM, Jarvis LA, Kozak MM, Broz DK, Basak S, Park EJ, McLaughlin ME, et al. 2011. Distinct p53 transcriptional programs dictate acute DNA-damage responses and tumour suppression. *Cell* **145**: 571–583.

Brummelkamp TR, Nijman SM, Dirac AM, Bernards R. 2003. Loss of the cylindromatosis tumour suppressor inhibits apoptosis by activating NF-κB. *Nature* **424**: 797–801.

Chen C, Edelstein LC, Gelinas C. 2000. The Rel/NF-κB family directly activates expression of the apoptosis inhibitor Bcl-x_L. *Mol Cell Biol* **20**: 2687–2695.

Chipuk JE, Green DR. 2008. How do BCL-2 proteins induce mitochondrial outer membrane permeabilization? *Trends Cell Biol* **18**: 157–164.

Chu Z-L, McKinsey TA, Liu L, Gentry JJ, Malim MH, Ballard DW. 1997. Suppression of tumour necrosis factor-induced cell death by inhibitor of apoptosis c-IAP2 is under NF-κB control. *Proc Natl Acad Sci* **94**: 10057–10062.

Cory S, Harris AW, Langdon WY, Alexander WS, Corcoran LM, Palmiter RD, Pinkert CA, Brinster RL, Adams JM. 1987. The *myc* oncogene and lymphoid neoplasia: From translocations to transgenic mice. *Modern Trends Human Leuk* **VII**: 248–251.

Costa DB, Halmos B, Kumar A, Schumer ST, Huberman MS, Boggon TJ, Tenen DG, Kobayashi S. 2007. BIM mediates EGFR tyrosine kinase inhibitor-induced apoptosis in lung cancers with oncogenic EGFR mutations. *PLoS Med* **4**: e315.

Cragg MS, Kuroda J, Puthalakath H, Huang DCS, Strasser A. 2007. Gefitinib-induced killing of NSCLC cell lines expressing mutant *EGFR* requires Bim and can be enhanced by BH3 mimetics. *PLoS Med* **4**: 1681–1689.

Cragg MS, Jansen ES, Cook M, Harris C, Strasser A, Scott CL. 2008. Treatment of B-RAF mutant human tumour cells with a MEK inhibitor requires Bim and is enhanced by a BH3 mimetic. *J Clin Invest* **118**: 3651–3659.

Dannenberg JH, van Rossum A, Schuijff L, te Riele H. 2000. Ablation of the retinoblastoma gene family deregulates G_1 control causing immortalization and increased cell turnover under growth-restricting conditions. *Genes Dev* **14**: 3051–3064.

Deininger M, Buchdunger E, Druker BJ. 2005. The development of imatinib as a therapeutic agent for chronic myeloid leukemia. *Blood* **105**: 2640–2653.

del Peso L, González-Garcia M, Page C, Herrera R, Nuñez G. 1997. Interleukin-3–induced phosphorylation of BAD through the protein kinase Akt. *Science* **278**: 687–689.

Deng C, Zhang P, Harper JW, Elledge SJ, Leder P. 1995. Mice lacking p21CIP1/WAF1 undergo normal development, but are defective in G1 checkpoint control. *Cell* **82**: 675–684.

Di Cristofano A, Pesce B, Cordon-Cardo C, Pandolfi PP. 1998. Pten is essential for embryonic development and tumour suppression. *Nat Genet* **19**: 348–355.

Dijkers PF, Medema RH, Lammers JJ, Koenderman L, Coffer PJ. 2000. Expression of the pro-apoptotic Bcl-2 family member Bim is regulated by the forkhead transcription factor FKHR-L1. *Curr Biol* **10**: 1201–1204.

Donehower LA, Harvey M, Slagle BL, McArthur MJ, Montgomery CAJ, Butel JS, Bradley A. 1992. Mice deficient for p53 are developmentally normal but are susceptible to spontaneous tumours. *Nature* **356**: 215–221.

Drappa J, Vaishnaw AK, Sullivan KE, Chu JL, Elkon KB. 1996. *Fas* gene mutations in the Canale–Smith syndrome, an inherited lymphoproliferative disorder associated with autoimmunity. *N Engl J Med* **335**: 1643–1649.

Egle A, Harris AW, Bouillet P, Cory S. 2004. Bim is a suppressor of Myc-induced mouse B cell leukemia. *Proc Natl Acad Sci* **101**: 6164–6169.

Eischen CM, Roussel MF, Korsmeyer SJ, Cleveland JL. 2001. Bax loss impairs Myc-induced apoptosis and circumvents the selection of p53 mutations during Myc-mediated lymphomagenesis. *Mol Cell Biol* **21**: 7653–7662.

Erlacher M, Michalak EM, Kelly PN, Labi V, Niederegger H, Coultas L, Adams JM, Strasser A, Villunger A. 2005. BH3-only proteins Puma and Bim are rate-limiting for γ-radiation and glucocorticoid-induced apoptosis of lymphoid cells in vivo. *Blood* **106**: 4131–4138.

Garami A, Zwartkruis FJ, Nobukuni T, Joaquin M, Roccio M, Stocker H, Kozma SC, Hafen E, Bos JL, Thomas G. 2003. Insulin activation of Rheb, a mediator of mTOR/S6K/4E-BP signaling, is inhibited by TSC1 and 2. *Mol Cell* **11**: 1457–1466.

Garrison SP, Jeffers JR, Yang C, Nilsson JA, Hall MA, Rehg JE, Yue W, Yu J, Zhang L, Onciu M, et al. 2008. Selection against PUMA gene expression in Myc-driven B-cell lymphomagenesis. *Mol Cell Biol* **28**: 5391–5402.

Grossmann M, O'Reilly LA, Gugasyan R, Strasser A, Adams JM, Gerondakis S. 2000. The anti-apoptotic activities of rel and RelA required during B-cell maturation involve the regulation of Bcl-2 expression. *EMBO J* **19**: 6351–6360.

Grumont RJ, Rourke IJ, Gerondakis S. 1999. Rel-dependent induction of *A1* transcription is required to protect B cells from antigen receptor ligation-induced apoptosis. *Gen Dev* **13:** 400–411.

Hanahan D, Weinberg RA. 2000. The hallmarks of cancer. *Cell* **100:** 57–70.

Haupt Y, Maya R, Kazaz A, Oren M. 1997. Mdm2 promotes the rapid degradation of p53. *Nature* **387:** 296–299.

Hershko T, Ginsberg D. 2004. Up-regulation of Bcl-2 homology 3 (BH3)-only proteins by E2F1 mediates apoptosis. *J Biol Chem* **279:** 8627–8634.

Hotchkiss RS, Strasser A, McDunn JE, Swanson PE. 2009. Cell death. *N Engl J Med* **361:** 1570–1583.

Huang DCS, Strasser A. 2000. BH3-only proteins—Essential initiators of apoptotic cell death. *Cell* **103:** 839–842.

Jacks T, Fazeli A, Schmitt EM, Bronson RT, Goodell MA, Weinberg RA. 1992. Effects of an *Rb* mutation in the mouse. *Nature* **359:** 295–300.

Jeffers JR, Parganas E, Lee Y, Yang C, Wang J, Brennan J, MacLean KH, Han J, Chittenden T, Ihle JN, et al. 2003. Puma is an essential mediator of p53–dependent and -independent apoptotic pathways. *Cancer Cell* **4:** 321–328.

Johnstone RW, Ruefli AA, Lowe SW. 2002. Apoptosis: A link between cancer genetics and chemotherapy. *Cell* **108:** 153–164.

Jones SN, Roe AE, Donehower LA, Bradley A. 1995. Rescue of embryonic lethality in Mdm2-deficient mice by absence of p53. *Nature* **378:** 206–208.

Kelly PN, White MJ, Goschnick MW, Fairfax KA, Tarlinton DM, Kinkel SA, Bouillet P, Adams JM, Kile BT, Strasser A. 2010. Individual and overlapping roles of BH3-only proteins Bim and Bad in apoptosis of lymphocytes and platelets and in suppression of thymic lymphoma development. *Cell Death Differ* **17:** 1655–1664.

Kovalenko A, Chable-Bessia C, Cantarella G, Israel A, Wallach D, Courtois G. 2003. The tumour suppressor CYLD negatively regulates NF-κB signalling by deubiquitination. *Nature* **424:** 801–805.

Kubbutat MH, Jones SN, Vousden KH. 1997. Regulation of p53 stability by Mdm2. *Nature* **387:** 299–303.

Kuroda J, Puthalakath H, Cragg MS, Kelly PN, Bouillet P, Huang DC, Kimura S, Ottmann OG, Druker BJ, Villunger A, et al. 2006. Bim and Bad mediate imatinib-induced killing of Bcr/Abl$^+$ leukemic cells, and resistance due to their loss is overcome by a BH3 mimetic. *Proc Natl Acad Sci* **103:** 14907–14912.

Labi V, Erlacher M, Kiessling S, Manzl C, Frenzel A, O'Reilly L, Strasser A, Villunger A. 2008. Loss of the BH3-only protein Bmf impairs B cell homeostasis and accelerates γ irradiation-induced thymic lymphoma development. *J Exp Med* **205:** 641–655.

Lee W-H, Bookstein R, Hong F, Young L-J, Shew J-Y, Lee EY-HP. 1987. Human retinoblastoma susceptibility gene: cloning, identification and sequence. *Science* **235:** 1394–1399.

Lee HH, Dadgostar H, Cheng Q, Shu J, Cheng G. 1999. NF-κB-mediated up-regulation of Bcl-x and Bfl-1/A1 is required for CD40 survival signaling in B lymphocytes. *Proc Natl Acad Sci* **96:** 9136–9141.

Lees JA, Saito M, Vidal M, Valentine M, Look T, Harlow E, Dyson N, Helin K. 1993. The retinoblastoma protein binds to a family of E2F transcription factors. *Mol Cell Biol* **13:** 7813–7825.

Levine AJ. 2009. The common mechanisms of transformation by the small DNA tumour viruses: The inactivation of tumour suppressor gene products: p53. *Virology* **384:** 285–293.

Li T, Kon N, Jiang L, Tan M, Ludwig T, Zhao Y, Baer R, Gu W. 2012. Tumour suppression in the absence of p53-mediated cell-cycle arrest, apoptosis, and senescence. *Cell* **149:** 1269–1283.

Lin WC, Lin FT, Nevins JR. 2001. Selective induction of E2F1 in response to DNA damage, mediated by ATM-dependent phosphorylation. *Genes Dev* **15:** 1833–1844.

Lindsten T, Ross AJ, King A, Zong W, Rathmell JC, Shiels HA, Ulrich E, Waymire KG, Mahar P, Frauwirth K, et al. 2000. The combined functions of proapoptotic Bcl-2 family members Bak and Bax are essential for normal development of multiple tissues. *Mol Cell* **6:** 1389–1399.

Livingstone LR, White A, Sprouse J, Livanos E, Jacks T, Tlsty TD. 1992. Altered cell cycle arrest and gene amplification potential accompany loss of wild-type p53. *Cell* **70:** 923–935.

Marsden V, O'Connor L, O'Reilly LA, Silke J, Metcalf D, Ekert P, Huang DCS, Cecconi F, Kuida K, Tomaselli KJ, et al. 2002. Apoptosis initiated by Bcl-2-regulated caspase activation independently of the cytochrome c/Apaf–1/caspase–9 apoptosome. *Nature* **419:** 634–637.

Martin K, Trouche D, Hagemeier C, Sorensen TS, La Thangue NB, Kouzarides T. 1995. Stimulation of E2F1/DP1 transcriptional activity by MDM2 oncoprotein. *Nature* **375:** 691–694.

Massoumi R, Chmielarska K, Hennecke K, Pfeifer A, Fassler R. 2006. Cyld inhibits tumour cell proliferation by blocking Bcl-3-dependent NF-κB signaling. *Cell* **125:** 665–677.

Merino D, Giam M, Hughes PD, Siggs OM, Heger K, O'Reilly LA, Adams JM, Strasser A, Lee EF, Fairlie WD, et al. 2009. The role of BH3-only protein Bim extends beyond inhibiting Bcl-2-like prosurvival proteins. *J Cell Biol* **186:** 355–362.

Metz T, Harris AW, Adams JM. 1995. Absence of p53 allows direct immortalization of hematopoietic cells by the *myc* and *raf* oncogenes. *Cell* **82:** 29–36.

Michalak EM, Villunger A, Adams JM, Strasser A. 2008. In several cell types the tumour suppressor p53 induces apoptosis largely via Puma but Noxa can contribute. *Cell Death Differ* **15:** 1019–1029.

Michalak EM, Jansen ES, Happo L, Cragg MS, Tai L, Smyth GK, Strasser A, Adams JM, Scott CL. 2009. Puma and to a lesser extent Noxa are suppressors of Myc-induced lymphomagenesis. *Cell Death Differ* **16:** 684–696.

Michalak EM, Vandenberg CJ, Delbridge ARD, Wu L, Scott CL, Adams JM, Strasser A. 2010. Apoptosis-promoted tumorigenesis: γ-Irradiation-induced thymic lymphomagenesis requires Puma-driven leukocyte death. *Genes Dev* **24:** 1608–1613.

Miyashita T, Reed JC. 1995. Tumour suppressor p53 is a direct transcriptional activator of the human *bax* gene. *Cell* **80:** 293–299.

Montes de Oca Luna R, Wagner DS, Lozano G. 1995. Rescue of early embryonic lethality in mdm2-deficient mice by deletion of p53. *Nature* **378:** 203–206.

Naik E, Michalak EM, Villunger A, Adams JM, Strasser A. 2007. UV-radiation triggers apoptosis of fibroblasts and skin keratinocytes mainly via the BH3-only protein Noxa. *J Cell Biol* **176:** 415–424.

Nakano K, Vousden KH. 2001. *PUMA*, a novel proapoptotic gene, is induced by p53. *Mol Cell* **7:** 683–694.

O'Connor L, Strasser A, O'Reilly LA, Hausmann G, Adams JM, Cory S, Huang DCS. 1998. Bim: A novel member of the Bcl-2 family that promotes apoptosis. *EMBO J* **17:** 384–395.

Oda E, Ohki R, Murasawa H, Nemoto J, Shibue T, Yamashita T, Tokino T, Taniguchi T, Tanaka N. 2000. Noxa, a BH3-only member of the bcl-2 family and candidate mediator of p53–induced apoptosis. *Science* **288:** 1053–1058.

O'Reilly LA, Tai L, Lee L, Kruse EA, Grabow S, Fairlie WD, Haynes NM, Tarlinton DM, Zhang JG, Belz GT, et al. 2009. Membrane-bound Fas ligand only is essential for Fas-induced apoptosis. *Nature* **461:** 659–663.

Pediconi N, Ianari A, Costanzo A, Belloni L, Gallo R, Cimino L, Porcellini A, Screpanti I, Balsano C, Alesse E, et al. 2003. Differential regulation of E2F1 apoptotic target genes in response to DNA damage. *Nat Cell Biol* **5:** 552–558.

Pierce AM, Fisher SM, Conti CJ, Johnson DG. 1998a. Deregulated expression of E2F1 induces hyperplasia and cooperates with ras in skin tumour development. *Oncogene* **16:** 1267–1276.

Pierce AM, Gimenez-Conti IB, Schneider-Broussard R, Martinez LA, Conti CJ, Johnson DG. 1998b. Increased E2F1 activity induces skin tumours in mice heterozygous and nullizygous for p53. *Proc Natl Acad Sci* **95:** 8858–8863.

Polager S, Ginsberg D. 2009. p53 and e2f: Partners in life and death. *Nat Rev Cancer* **9:** 738–748.

Potter CJ, Pedraza LG, Xu T. 2002. Akt regulates growth by directly phosphorylating Tsc2. *Nat Cell Biol* **4:** 658–665.

Poulikakos PI, Rosen N. 2011. Mutant BRAF melanomas—Dependence and resistance. *Cancer Cell* **19:** 11–15.

Ranger AM, Zha J, Harada H, Datta SR, Danial NN, Gilmore AP, Kutok JL, Le Beau MM, Greenberg ME, Korsmeyer SJ. 2003. Bad-deficient mice develop diffuse large B cell lymphoma. *Proc Natl Acad Sci* **100:** 9324–9329.

Riedl SJ, Shi Y. 2004. Molecular mechanisms of caspase regulation during apoptosis. *Nat Rev Mol Cell Biol* **5:** 897–907.

Rieux-Laucat F, Le Deist F, Hivroz C, Roberts IAG, Debatin KM, Fischer A, de Villartay JP. 1995. Mutations in Fas associated with human lymphoproliferative syndrome and autoimmunity. *Science* **268:** 1347–1349.

Riley T, Sontag E, Chen P, Levine A. 2008. Transcriptional control of human p53-regulated genes. *Nat Rev Mol Cell Biol* **9:** 402–412.

Sage J, Mulligan GJ, Attardi LD, Miller A, Chen S, Williams B, Theodorou E, Jacks T. 2000. Targeted disruption of the three *Rb*-related genes leads to loss of G_1 control and immortalization. *Genes Dev* **14:** 3037–3050.

Sharma SV, Bell DW, Settleman J, Haber DA. 2007. Epidermal growth factor receptor mutations in lung cancer. *Nat Rev Cancer* **7:** 169–181.

Shi Y. 2002. Mechanisms of caspase activation and inhibition during apoptosis. *Mol Cell* **9:** 459–470.

Srivastava S, Zou ZQ, Pirollo K, Plattner W, Chang EH. 1990. Germ-line transmission of a mutated p53 gene in a cancer-prone family with Li-Fraumeni syndrome. *Nature* **348:** 747–749.

Stambolic V, Suzuki A, de la Pompa JL, Brothers GM, Mirtsos C, Sasaki T, Ruland J, Penninger JM, Siderovski DP, Mak TW. 1998. Negative regulation of PKB/Akt-dependent cell survival by the tumour suppressor PTEN. *Cell* **95:** 29–39.

Stambolic V, MacPherson D, Sas D, Lin Y, Snow B, Jang Y, Benchimol S, Mak TW. 2001. Regulation of PTEN transcription by p53. *Mol Cell* **8:** 317–325.

Stehlik C, de Martin R, Kumabashiri I, Schmid JA, Binder BR, Lipp J. 1998. Nuclear factor NF-κB-regulated X-chromosome-linked iap gene expression protects endothelial cells from tumour necrosis factor alpha-induced apoptosis. *J Exp Med* **188:** 211–216.

Stevens C, Smith L, La Thangue NB. 2003. Chk2 activates E2F-1 in response to DNA damage. *Nat Cell Biol* **5:** 401–409.

Strasser A, Harris AW, Bath ML, Cory S. 1990. Novel primitive lymphoid tumours induced in transgenic mice by cooperation between *myc* and *bcl-2*. *Nature* **348:** 331–333.

Strasser A, Harris AW, Cory S. 1993. Eμ-*bcl-2* transgene facilitates spontaneous transformation of early pre-B and immunoglobulin-secreting cells but not T cells. *Oncogene* **8:** 1–9.

Strasser A, Harris AW, Huang DCS, Krammer PH, Cory S. 1995. Bcl-2 and Fas/APO-1 regulate distinct pathways to lymphocyte apoptosis. *EMBO J* **14:** 6136–6147.

Strasser A, O'Connor L, Dixit VM. 2000. Apoptosis signaling. *Ann Rev Biochem* **69:** 217–245.

Strasser A, Jost PJ, Nagata S. 2009. The many roles of FAS receptor signaling in the immune system. *Immunity* **30:** 180–192.

Sun T, Aceto N, Meerbrey KL, Kessler JD, Zhou C, Migliaccio I, Nguyen DX, Pavlova NN, Botero M, Huang J, et al. 2011. activation of multiple proto-oncogenic tyrosine kinases in breast cancer via loss of the PTPN12 phosphatase. *Cell* **144:** 703–718.

Suzuki A, de la Pompa JL, Stambolic V, Elia AJ, Sasaki T, del Barco Barrantes I, Ho A, Wakeham A, Itie A, Khoo W, et al. 1998. High cancer susceptibility and embryonic lethality associated with mutation of the *PTEN* tumour suppressor gene in mice. *Curr Biol* **8:** 1169–1178.

Tagawa H, Karnan S, Suzuki R, Matsuo K, Zhang X, Ota A, Morishima Y, Nakamura S, Seto M. 2005. Genome-wide array-based CGH for mantle cell lymphoma: Identification of homozygous deletions of the proapoptotic gene BIM. *Oncogene* **24:** 1348–1358.

Timmer JC, Salvesen GS. 2007. Caspase substrates. *Cell Death Differ* **14:** 66–72.

Trompouki E, Hatzivassiliou E, Tsichritzis T, Farmer H, Ashworth A, Mosialos G. 2003. CYLD is a deubiquitinating enzyme that negatively regulates NF-κB activation by TNFR family members. *Nature* **424:** 793–796.

Urist M, Tanaka T, Poyurovsky MV, Prives C. 2004. p73 induction after DNA damage is regulated by checkpoint kinases Chk1 and Chk2. *Genes Dev* **18:** 3041–3054.

Vaux DL, Cory S, Adams JM. 1988. Bcl-2 gene promotes haemopoietic cell survival and cooperates with c-*myc* to immortalize pre-B cells. *Nature* **335:** 440–442.

Villunger A, Michalak EM, Coultas L, Mullauer F, Bock G, Ausserlechner MJ, Adams JM, Strasser A. 2003. p53- and drug-induced apoptotic responses mediated by BH3-only proteins Puma and Noxa. *Science* **302:** 1036–1038.

Vousden KH, Lane DP. 2007. p53 in health and disease. *Nat Rev Mol Cell Biol* **8:** 275–283.

Wang C-Y, Mayo MW, Korneluk RG, Goeddel DV, Baldwin AS Jr. 1998. NF-κB antiapoptosis: Induction of TRAF1 and TRAF2 and c-IAP1 and c-IAP2 to suppress caspase-8 activation. *Science* **281:** 1680–1683.

Watanabe-Fukunaga R, Brannan CI, Copeland NG, Jenkins NA, Nagata S. 1992. Lymphoproliferation disorder in mice explained by defects in Fas antigen that mediates apoptosis. *Nature* **356:** 314–317.

Wertz IE, Dixit VM. 2010. Regulation of death receptor signaling by the ubiquitin system. *Cell Death Differ* **17:** 14–24.

Wu X, Bayle JH, Olson D, Levine AJ. 1993. The p53–mdm-2 autoregulatory feedback loop. *Genes Dev* **7:** 1126–1132.

Xiao ZX, Chen J, Levine AJ, Modjtahedi N, Xing J, Sellers WR, Livingston DM. 1995. Interaction between the retinoblastoma protein and the oncoprotein MDM2. *Nature* **375:** 694–698.

Youle RJ, Strasser A. 2008. The BCL-2 protein family: Opposing activities that mediate cell death. *Nat Rev Mol Cell Biol* **9:** 47–59.

Yu J, Zhang L, Hwang PM, Kinzler KW, Vogelstein B. 2001. PUMA induces the rapid apoptosis of colorectal cancer cells. *Mol Cell* **7:** 673–682.

Zhang Z, Wang H, Li M, Rayburn ER, Agrawal S, Zhang R. 2005. Stabilization of E2F1 protein by MDM2 through the E2F1 ubiquitination pathway. *Oncogene* **24:** 7238–7247.

Zong WX, Edelstein LC, Chen C, Bash J, Gelinas C. 1999. The prosurvival Bcl-2 homolog Bfl-1/A1 is a direct transcriptional target of NF-κB that blocks TNFα-induced apoptosis. *Genes Dev* **13:** 382–387.

The Role of Apoptosis-Induced Proliferation for Regeneration and Cancer

Hyung Don Ryoo[1] and Andreas Bergmann[2]

[1]Department of Cell Biology, New York University School of Medicine, New York, New York 10016
[2]Department of Cancer Biology, University of Massachusetts Medical School, Worcester, Massachusetts 01605
Correspondence: abergman@mdanderson.org

Genes dedicated to killing cells must have evolved because of their positive effects on organismal survival. Positive functions of apoptotic genes have been well established in a large number of biological contexts, including their role in eliminating damaged and potentially cancerous cells. More recently, evidence has suggested that proapoptotic proteins—mostly caspases—can induce proliferation of neighboring surviving cells to replace dying cells. This process, which we will refer to as "apoptosis-induced proliferation," may be critical for stem cell activity and tissue regeneration. Depending on the caspases involved, at least two distinct types of apoptosis-induced proliferation can be distinguished. One of these types have been studied using a model in which cells have initiated cell death, but are prevented from executing it because of effector caspase inhibition, thereby generating "undead" cells that emit persistent mitogen signaling and overgrowth. Such conditions are likely to contribute to certain forms of cancer. In this review, we summarize the current knowledge of apoptosis-induced proliferation and discuss its relevance for tissue regeneration and cancer.

Why is cell death beneficial in certain contexts, but not others? Obviously, the answer to this question is dependent on the developmental status of the tissue and may have to do with the tissue's ability to replace dead cells. In general, cells in proliferating tissues are readily replaced after tissue damage, whereas cells in differentiated tissue are much harder to renew, although examples have been reported (see below). The reason for this difference is obvious. Because cells in differentiated tissues are largely postmitotic, only a few cells will be able to reenter the cell cycle up on loss of vital cells. However, the ability to replace dying cells in proliferating tissues is so robust that cell death phenotypes are often overlooked. For example, in classical studies in *Drosophila*, it was observed that ionizing radiation induces massive cell death during larval stages, but that the emerging flies appear largely normal—because of compensatory proliferation (CP) (Haynie and Bryant 1977; Jaklevic and Su 2004; Perez-Garijo et al. 2004). Also, experimentally induced homozygous patches of cell lethal mutations eliminate a significant fraction of cells, often without any obvious effect on tissue size or morphology—because of CP (Stowers and Schwarz 1999; Newsome et al. 2000). In another dramatic example, it was found that massive cell death caused by Ricin expression in one lineage-restricted compartment of the *Drosophila* wing imaginal discs triggered cells in the neighboring compartment to undergo enhanced cell proliferation, suggesting that a diffusible mitogenic signal is being emitted from the dying compartment (Milan et al. 1997). The ability to replace lost cells through CP appears to be conserved in mammals as well. When mice are subject to ionizing irradiation, actively proliferating cells such as those of the hematopoietic lineage or of the digestive system undergo massive cell death (Down et al. 1991). Although the loss of large numbers of cells compromise the health of these animals in the short term, those cells are quickly replenished to restore tissue function. In fact, if the ensuing cell death is blocked, for example through the disruption of p53, the animals suffer from deleterious consequences like cancer (Vousden 2000). This indicates that cell death serves as a beneficial quality control mechanism.

More recent work has shown that—at least in some cases—CP requires cell death-inducing genes, most notably caspases (Fan and Bergmann 2008a; Galliot and Chera 2010). Therefore, we refer to the induction of CP by apoptotic caspases specifically as "apoptosis-induced proliferation" and such examples will be extensively reviewed in this article.

Apoptosis-induced proliferation is beneficial for the organism, as it allows tissues to easily eliminate damaged or potentially dangerous cells and replace them with the progeny of healthy neighbors. When such compensatory mechanisms are no longer available, apoptosis may lose its beneficial effect to the body. In this review, we discuss recent progress in our understanding of how apoptosis can trigger proliferation. Special emphasis will be put on the role of apoptotic caspases in this process.

CASPASES ARE CRITICAL EXECUTIONERS OF APOPTOSIS

Across phyla, caspases are known as critical executioners of apoptosis. These proteins belong to a family of cysteine proteases that proteolytically cleave substrates after aspartic acid residues. Although their activity can be deadly, most cells containing these proteins do not die because caspases are synthesized as inactive zymogens. These proteases gain catalytic activity mostly on activation by upstream apoptotic signals (Thornberry and Lazebnik 1998). Caspases can be divided into initiator and effector caspases (Fig. 1). Initiator caspases, such as caspase-8, caspase-9, and *Drosophila* Dronc are characterized by long prodomains, which carry protein interaction motifs for upstream apoptotic signaling (Kumar 2007). Activated initiator caspases cleave and activate effector caspases such as caspase-3, caspase-7, and *Drosophila* DrICE and Dcp-1 (Fig. 1). Effector caspases are thought to cleave a large number of downstream substrates that collectively bring about the morphological features of apoptotic cells (Kerr et al. 1972; Dix et al. 2008; Mahrus et al. 2008). Elimination of caspases abolishes most apoptosis in

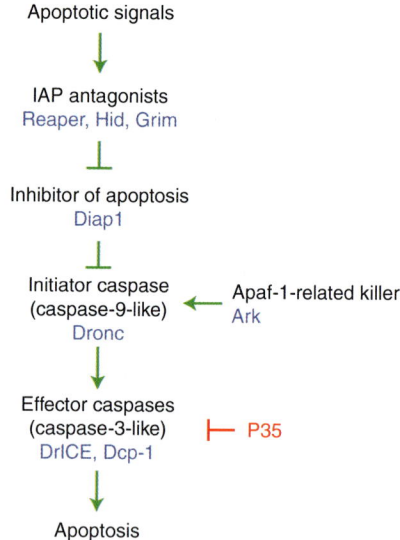

Figure 1. The apoptosis pathway in *Drosophila*. In living cells, the caspase-9-like initiator caspase Dronc is inhibited by *Drosophila* inhibitor of apoptosis protein 1 (Diap1). Apoptosis is induced by activation of the IAP-antagonists Reaper, Hid, and Grim that induce ubiquitylation and degradation of Diap1. Dronc is released from Diap1 inhibition and forms the apoptosome holoenzyme together with the adaptor protein Ark. Dronc is activated in this complex, thereby proteolytically activating effector caspases DrICE and Dcp-1 causing apoptosis. The Baculovirus caspase inhibitor P35 acts specifically on DrICE and Dcp-1.

model organisms such as *Caenorhabditis elegans* and *Drosophila* (Yuan et al. 1993; Chew et al. 2004; Daish et al. 2004; Xu et al. 2005, 2006). Likewise, knockout of caspases in mice significantly reduces the amount of cell death in that organism (Colussi and Kumar 1999). Given the importance of caspases in apoptosis, blocking their function is a good way to test whether cell death has a causal effect in triggering CP. A number of studies in various contexts have shown that, indeed, this is the case.

THE ROLE OF CASPASES FOR COMPENSATORY PROLIFERATION IN *DROSOPHILA*

It is clear that, in many tissues, injury-provoked cell death is accompanied by CP. Conceptually, this can happen through at least two ways. One possibility is that injury-provoked signaling may induce apoptosis-independent pathways that contribute to CP. Alternatively, proliferation is directly "instructed" by the dying cells. There is a considerable amount of data supporting the latter concept.

Genetic studies in *Drosophila* have identified two distinct pathways for apoptosis-induced proliferation, which require different sets of caspases. The first pathway requires the initiator caspase Dronc, but is independent of effector caspases. We will henceforth refer to this as the effector caspase-independent pathway of apoptosis-induced proliferation. This pathway and the independence of effector caspases was discovered in studies of "undead" cells (Huh et al. 2004; Perez-Garijo et al. 2004; Ryoo et al. 2004), in which cell death is initiated, but its execution is blocked by expression of the caspase inhibitor P35 (Clem et al. 1991; Xue and Horvitz 1995). In *Drosophila*, P35 specifically inhibits the effector caspases DrICE and Dcp-1, but not the initiator caspase Dronc (Fig. 1) (Hawkins et al. 2000; Meier et al. 2000). Thus, undead cells activate Dronc, but cannot execute apoptosis because of effector caspase inhibition by P35. Such undead cells caused hyperplastic growth (Fig. 2), most likely because of the persistence of mitogenic signals that would have otherwise been short-lived, if cells were allowed to undergo a natural course of cell death. Because P35 does not inhibit Dronc, this experimental setup revealed a nonapoptotic role of Dronc in inducing proliferation of neighboring cells (Fig. 3A). Such role of Dronc in growth

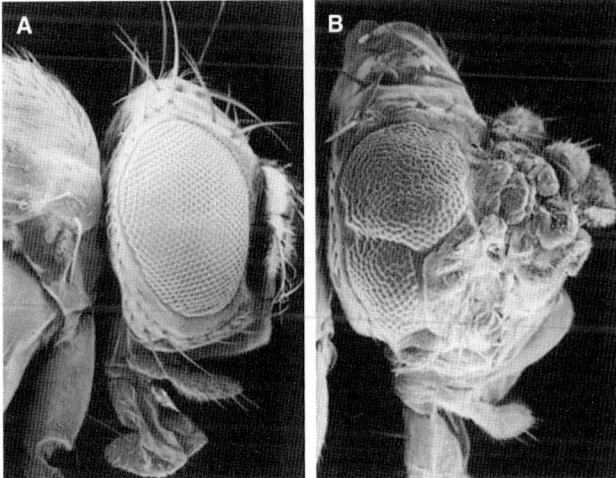

Figure 2. Hyperplastic overgrowth by undead cells. Scanning EM photographs showing eyes and heads from a normal fly (A) and a fly containing a large number of undead cells in the head (B). The head capsule in B is strongly overgrown. The eye in B is reduced in size because of increased *wingless* signaling that inhibits eye formation. Genotype in B: *ey-Flp*; *tub > GFP > Gal4*; *UAS-hid UAS-p35* (> marks FLP recognition target sites).

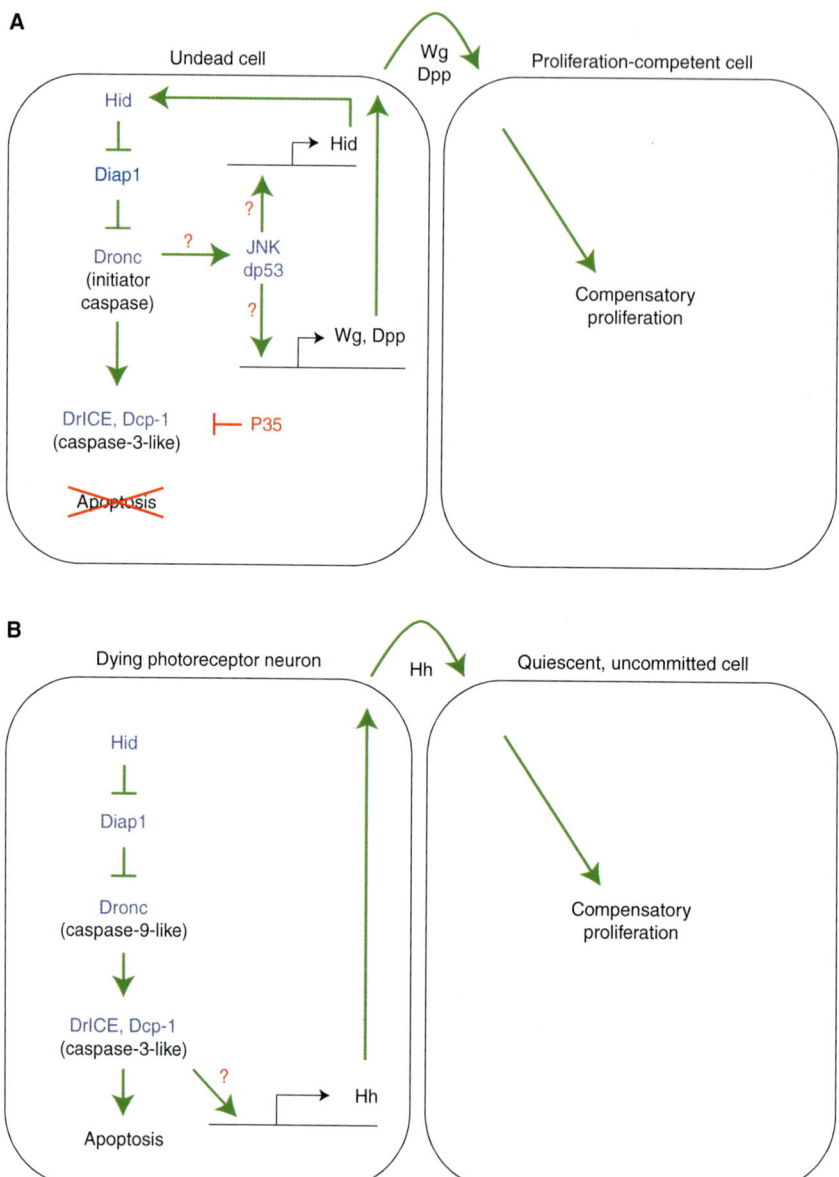

Figure 3. Distinct types of apoptosis-induced proliferation in *Drosophila*. (*A*) The effector caspase-independent pathway. Induction of apoptosis in the presence of P35 produces undead cells because of effector caspase inhibition. The initiator caspase Dronc remains active in undead cells and stimulates the Jun amino-terminal kinase (JNK) pathway. JNK together with *Drosophila* p53 (dp53) induces expression of the IAP antagonist Hid, establishing a positive feedback loop of apoptosis amplification. JNK and dp53 also stimulate the expression of the mitogens Wg and Dpp that promote neighboring cells to undergo compensatory proliferation. This mechanism leads to hyperplastic overgrowth shown in Figure 2. Question marks denote points of uncertainty. (*B*) The effector caspase-dependent pathway. Expression of the IAP antagonist Hid in the postmitotic, differentiating field of the eye retina induces apoptosis. However, dying photoreceptor neurons are able to induce expression of the mitogen Hedgehog (Hh), which stimulates neighboring cells that are postmitotic, but have not initiated a differentiation program (uncommitted) to reenter the cell cycle and undergo compensatory proliferation. This pathway is mediated by the effector caspases DrICE and Dcp-1. Expression of P35 does block this type of apoptosis-induced proliferation. Question marks denote open questions.

promotion was directly confirmed by *dronc* mutants or dominant negative Dronc constructs (Huh et al. 2004; Kondo et al. 2006).

The second pathway of apoptosis-induced proliferation is used in a different developmental context and is dependent on effector caspases (hence, we refer to it as effector caspase-dependent apoptosis-induced proliferation pathway) (Fig. 3B). In contrast to the effector caspase-independent pathway, expression of P35 does inhibit apoptosis-induced proliferation in this context, as do double mutants of the effector caspases *drICE* and *dcp-1*, again illustrating a direct requirement of caspases for apoptosis-induced proliferation (Fan and Bergmann 2008a,b). Therefore, blocking effector caspases has different effects on these two forms of apoptosis-induced proliferation. In the first case, it causes overgrowth (Fig. 2) with potential relevance for cancer; in the second one, it blocks apoptosis-induced proliferation, which has severe consequences for regenerative growth.

Why are there at least two distinct mechanisms of apoptosis-induced proliferation? The difference may lie in the developmental state of the apoptotic tissue. The effector caspase-independent pathway occurs in the proliferating regions with largely undifferentiated cells of wing and eye imaginal discs, whereas the effector caspase-dependent pathway is employed in postmitotic eye tissue containing many differentiating cells (Fan and Bergmann 2008b). Nevertheless, in the latter, although postmitotic, only cells that have not initiated a differentiation process can be induced to reenter the cell cycle. Differentiating cells such as photoreceptor neurons have lost the ability to reenter the cell cycle. In fact, differentiating photoreceptor neurons are the cells that emit mitogenic signals for apoptosis-induced proliferation of unspecified, yet postmitotic cells (Fan and Bergmann 2008b).

THE ROLE OF CASPASES FOR REGENERATIVE GROWTH IN VARIOUS MODEL ORGANISMS

In addition to the pioneering studies in *Drosophila*, the requirement of caspases for proliferation in regenerative growth has been shown in several regeneration models. The freshwater polyp Hydra is a classical model famously studied by eighteenth century zoologist Abraham Trembley, who showed that this animal has a remarkable regenerative capacity. When Hydra polyps were cut into half, he found that the two parts grow back to perfect polyps: the foot fragment grows a head, and the head fragment a foot (Trembley 1744). Only head regeneration requires new proliferation. Correlating with this, only the head-regenerating tip induces apoptosis at the amputation site in an Erk/MAPK-dependent manner (Chera et al. 2009, 2011; Galliot and Chera 2010). Importantly, if cell death was blocked with an effector caspase inhibitor, cell proliferation and head regeneration was impaired, demonstrating an essential requirement of caspases for regenerative growth in this organism (Fig. 4). Moreover, these investigators took the analysis a step further and caused ectopic apoptosis in the foot-regenerating tip. Strikingly, this treatment not only induced cell proliferation, but changed the regeneration program from foot to head, resulting—remarkably—in an ectopic head (Chera et al. 2009; Galliot and Chera 2010). These experiments underscore the requirement of apoptotic caspases for regenerative growth.

Similar observations have been made in vertebrates as well. The *Xenopus* tadpole is able to regenerate its tail after amputation. Twelve hours after amputation, apoptotic cells are detected at the site of injury. The regeneration of the lost tissue is dependent on the presence of these apoptotic cells, as effector caspase inhibitors block the ensuing cell proliferation and overall tissue regeneration (Fig. 4) (Tseng et al. 2007).

Mouse embryonic fibroblasts (MEFs) also have similar properties. A recent study has found that unstressed stem cell populations are stimulated to divide when mixed with MEFs that have been exposed to a lethal dose of irradiation. As in other organisms, this growth-promoting property of irradiated MEFs can be attributed to effector caspases, as eliminating caspase-3 and caspase-7 function blocks stem cell growth (Li et al. 2010). Furthermore, in intact animals, wound healing and liver regeneration after partial hepatectomy are dependent on caspase-3

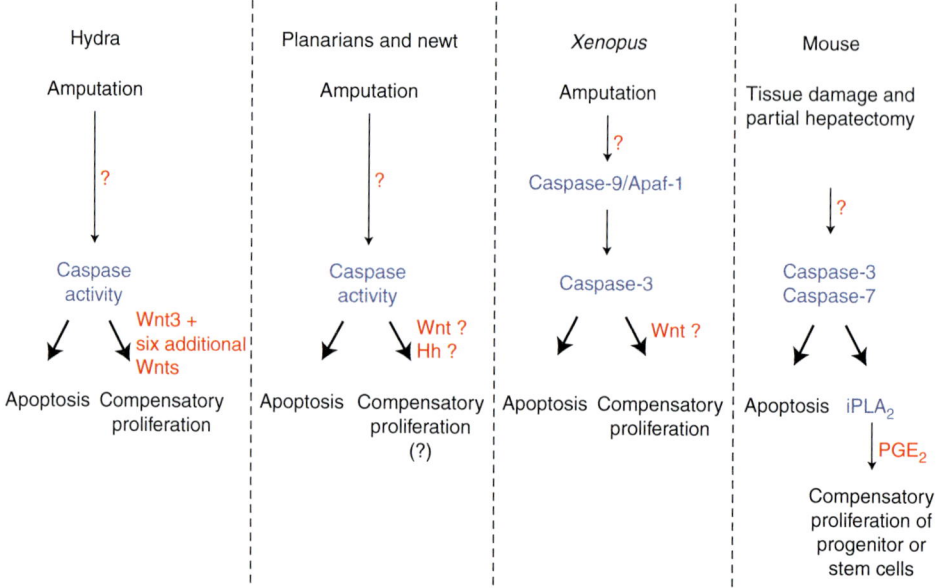

Figure 4. Effector caspases stimulate regeneration in several model organisms. Schematic summary of the experiments in several model organisms establishing a role of apoptotic caspases for regenerative growth. In most of these cases, this type of apoptosis-induced proliferation is effector caspase-dependent. Inhibition of effector caspases blocks regeneration. For details, see text. Question marks denote points of uncertainty.

and caspase-7, because targeted inactivation of these caspases inhibits these regenerative processes (Fig. 4) (Li et al. 2010).

Therefore, similar to the observations made in *Drosophila*, these experiments suggest that apoptotic cells play an active role, possibly instructing the neighboring cells to undergo cell proliferation for regenerative processes and establish a causal role of apoptotic caspases in this process. In each of these cases, effector caspases were either inhibited or targeted by gene inactivation, resulting in loss of regeneration. This is similar to the second pathway of apoptosis-induced proliferation identified in *Drosophila*.

IS THERE ENOUGH TIME FOR APOPTOTIC CELLS TO EXPRESS MITOGENS?

Caspases are executioners of apoptosis that can dismantle a cell within hours. Is there really enough time for caspases to induce the transcription of mitogens, that would in turn, become translated in the cytoplasm, and then travel through the secretory pathway? A careful analysis of the literature suggests that, in fact, the kinetics of cell death do not appear so rapid that gene expression would be prohibited.

Caspases are well known for their ability to rapidly execute cell death. This is certainly the case when researchers inflict their cells with strong death stimulus, leading to full-blown caspase activation in a short period of time. However, it is also clear that many cells have subthreshold caspase activity that does not kill cells. This is well documented when fluorescent caspase sensors are expressed in living cells. Clearly, caspase activity is detected in cells that remain healthy (Kanuka et al. 2005). In addition, there is evidence of Dronc activity in living cells of *Drosophila*. This initiator caspase can be activated by its adaptor protein Ark (ortholog of Apaf-1 in mammals) (Fig. 1), which forms the multimeric apoptosome complex that serves as a platform to activate the initiator caspase Dronc (Yu et al. 2006). However, overexpression of Ark does not trigger apoptosis in vivo. Instead, it mildly activates Dronc activity that leads to Dronc protein instability (Shapiro et al. 2008). These experiments show that caspases can be active to perform certain functions without inducing apoptosis. Low levels of caspase activity in fact plays various nonapoptotic roles, including sensory neuron development (Geisbrecht and Montell 2004; Kanuka et al. 2005; Kuranaga et al. 2006; Oshima et al. 2006; Ouyang et al. 2011), cell migration (Geisbrecht and Montell 2004), cytoskeletal rearrangements (Kuranaga et al. 2006), spermatid differentiation (De Maria et al. 1999; Arama et al. 2003; Huh et al. 2004; Arama et al. 2007; Bader et al. 2011), and cell fate decision (De Maria et al. 1999; and reviewed by Miura 2012). Together, these observations indicate that caspase activity can be maintained below the threshold of apoptosis execution, and conceptually, this may allow cells to signal for proliferation prior to the onset of the rapid cell death program.

Even when cells increase caspase activity to trigger apoptosis, there may be a significant lag period. In *Drosophila*, cell death is initiated by IAP-antagonists, Reaper, Hid, and Grim (Fig. 1). These proteins directly bind and antagonize Diap1, which in turn directly inhibits the caspases Dronc and DrICE. Although the initial reports indicated that Reaper, Hid, and Grim are sufficient to initiate apoptosis (Grether et al. 1995; Chen et al. 1996; White et al. 1996), more recent evidence indicates that they require downstream signaling. These include the observation that Reaper activates JNK signaling, and without this downstream signaling event, Reaper remains as a poor inducer of apoptosis (Igaki et al. 2002; Shlevkov and Morata 2012). Independent observations report that overexpression of one IAP-antagonist leads to the transcriptional activation of another (Sandu et al. 2010), implying that these proteins induce a transcriptional response to further enhance caspase activity. In fact, overexpressed IAP-antagonists are poor killer proteins on their own, without the transcriptional induction of other endogenous IAP-antagonists (Sandu et al. 2010).

These observations paint a picture of the events that precede apoptosis and apoptosis-induced proliferation. The literature suggests that apoptosis is initiated by the expression of a small amount of apoptosis-initiating proteins. This could be the IAP-antagonists in *Drosophila*, and BH3-only domain proteins or death ligands in mammals. The initial expression of such proapoptotic protein is not enough for a full-blown caspase activation for apoptosis, but sufficient to induce a signaling pathway that can orchestrate cell death and proliferation. In the case of *Drosophila*, this pathway is mediated by JNK and p53, which in turn has two different sets of transcriptional targets. One set are the proapoptotic genes for apoptosis, and the other set are mitogens that instruct proliferation (Fig. 3A). In such a scenario, there is plenty of time for apoptotic cells to induce proliferation. These considerations raise the next question: How do caspases induce proliferation?

SIGNALING PATHWAYS WITHIN DYING CELLS: THE ROLE OF JNK SIGNALING

First clues about the signaling pathways involved in apoptosis-induced proliferation came from genetic studies in *Drosophila*. Key for this analysis was to keep dying cells in an undead condition by P35 expression. In these cells, signaling pathways and mitogen expression persists in making their identification possible. This treatment helped gain mechanistic understanding of the effector caspase-independent pathway of apoptosis-induced proliferation, in which Dronc initiates the proliferation response (Fig. 3A) (Huh et al. 2004; Perez-Garijo et al. 2004; Ryoo et al. 2004).

How does Dronc induce proliferation? The principle mediator of apoptosis-induced proliferation is JNK signaling (Ryoo et al. 2004). In fact, activation of JNK signaling independently of apoptosis is sufficient for induction of proliferation (Fig. 3A) (Perez-Garijo et al. 2005, 2009; Smith-Bolton et al. 2009; Bergantinos et al. 2010; Warner et al. 2010; Suissa et al. 2011). However, JNK activity is also tightly associated with stress and apoptosis. As such, JNK signaling acts upstream of the apoptotic pathway and can induce the expression of proapoptotic genes (Moreno et al. 2002). Therefore, JNK signaling acts both upstream and downstream from the apoptotic pathway, forming a feed-forward loop for apoptosis amplification under stress conditions (Fig. 3A) (Wells et al. 2006; Shlevkov and Morata 2012). Supporting this notion, JNK markers are autonomously induced in undead cells, and tissue growth can be blocked when a JNK inhibitor is overexpressed (Ryoo et al. 2004). Because of this strong proapoptotic effect of JNK signaling, its role in apoptosis-induced proliferation can only be uncovered if cells are rendered undead by effector caspase inhibition such as P35. Nevertheless, JNK induced in undead cells requires Dronc, as this pathway is blocked in *dronc* mutant cells (Kondo et al. 2006; Wells et al. 2006; Shlevkov and Morata 2012). How Dronc activates JNK for apoptosis-induced proliferation is currently unknown, but evidence for nonapoptotic substrates of Dronc has been provided (Fan and Bergmann 2010).

In addition to JNK, *Drosophila* p53 (Dp53) was found to be required for the Dronc/JNK amplification loop and for growth of undead tissue (Fig. 3A) (Wells et al. 2006; Wells and Johnston 2012). Similarly, a planarian p53 ortholog has been implicated in stem cell renewal and proliferation (Pearson and Sanchez Alvarado 2010). The function of Dp53 for apoptosis-induced proliferation is independent of the DNA-damage sensing pathway as the DNA-damage checkpoint genes *atm* and *chk2* are not required for proliferation (Wells et al. 2006). Both JNK and Dp53 constitute a positive feedback loop that leads to activation of *hid* and *reaper* expression (Fig. 3A) (Brodsky et al. 2000; Wells et al. 2006; Fan et al. 2010; Shlevkov and Morata 2012; Wells and Johnston 2012). How p53 affects JNK signaling remains unclear, but a recent study provides a possible mechanistic basis. In both humans and *Drosophila*, p53 physically binds to phosphorylated JNK, thereby protecting them from the inactivating phosphatases (Gowda et al. 2012). Based on these observations, it is most likely that the p53/JNK complex serves as a signaling mediator that connects Dronc activation in the cytoplasm to gene expression in the nucleus.

The intracellular signaling events in the second effector caspase-dependent pathway of apoptosis-induced proliferation are unknown, except that it requires the effector caspases DrICE and Dcp-1 (Fig. 3B) (Fan and Bergmann 2008b). JNK and Dp53 are not involved in this pathway, which poses an interesting problem, because Dronc is also necessary for this type of proliferation. Why Dronc does not activate JNK signaling in this context is unknown.

MITOGEN SIGNALING FOR COMPENSATORY PROLIFERATION IN *DROSOPHILA*

How are apoptotic cells actively signaling to the surviving neighbors to undergo compensatory cell proliferation? For such signaling to occur, the apoptotic cell must express a secretory mitogen, or at least a membrane anchored ligand that can instruct the neighboring cells to undergo extra proliferation. Now, there are a number of well-documented studies indicating that this is indeed the case.

Again, early evidence that apoptotic cells induce mitogens came from studies in *Drosophila*. The proliferation pathway in P35-expressing undead cells induces the expression of the Wnt family member *wingless* and the TGF-β/BMP member *dpp* (Fig. 3A) (Huh et al. 2004; Perez-Garijo et al. 2004; Ryoo et al. 2004), which encode secretory proteins with strong mitogenic properties. At least in the wing imaginal discs, *wingless* and *dpp* were attributable to the overgrowth phenotype as elimination of these genes blocked abnormal growth (Perez-Garijo et al. 2005, 2009). These results can be most simply interpreted by the fact that P35 is not able to block Dronc, which is responsible for nonautonomous proliferation (Fig. 3A). Although most of the experiments were performed in a rather artificial setup in which cell death was inhibited by P35, mitogen expression was confirmed in settings where apoptotic cells were allowed to undergo its natural death course as well (Perez-Garijo et al. 2004; Ryoo et al. 2004; Smith-Bolton et al. 2009). In addition, such expression of mitogens can explain how apoptosis can fuel proliferation.

Although developmentally older tissues, especially after completion of the cell proliferation phase, are less able to compensate for cell loss (Smith-Bolton et al. 2009), the second form of apoptosis-induced proliferation in *Drosophila*, effector caspase-dependent proliferation, was observed in a largely postmitotic tissue. This tissue is the differentiating retina in the *Drosophila* eye during late larval stages. At that stage, a large number of postmitotic cells has formed, which will be instructed to differentiate first into photoreceptor neurons followed by cone, pigment, and bristle cells in a very specific temporal order (Voas and Rebay 2004). When apoptosis is induced in this tissue, proliferation is induced as well. However, this proliferation is restricted to cells that have not initiated differentiation yet (Fan and Bergmann 2008b) suggesting that these cells—although postmitotic under normal conditions—are still competent to reenter the cell cycle. Differentiating cells such as photoreceptor neurons do not reenter the cell cycle under apoptotic conditions and may have lost mitotic potential.

However, dying photoreceptor neurons produce and secrete Hedgehog (Hh), which is the mitogen for apoptosis-induced proliferation in this system (Fig. 3B) (Fan and Bergmann 2008b). Interestingly, the postmitotic cells do not respond to Hh signaling immediately. It takes them about 6–12 h before they reenter the cell cycle. The reason for this delay is unknown, but it is very similar to the time it takes quiescent cells in mammals to become mitotically active again (Coller 2007). Thus, genetic studies in *Drosophila* may uncover mechanisms of cell cycle reentry, which may be very relevant for mammalian cells and tumor initiation.

The genes required in the surviving, proliferating cells are less well characterized. In an approach to identify such genes, Hariharan and colleagues conducted an unbiased genetic screen (Gerhold et al. 2011). They analyzed the ability of tissues to recover size after massive apoptosis was triggered by a temperature-sensitive cell lethal mutation under nonpermissive conditions. They reported that mutations in the genes *Ribonucleoside diphosphate reductase large subunit* (*RnrL*), the GDP-mannose 4, 6 dehydratase *Gmd*, and the putative transcription factor *bunched* (*bun*) specifically affect CP without affecting growth under normal conditions (Gerhold et al. 2011). As yet, how these genes affect CP remains to be elucidated. It is also unknown whether apoptotic cells trigger the surviving cells to proliferate, i.e., whether this is a form of apoptosis-induced proliferation.

MITOGEN SIGNALING FOR APOPTOSIS-INDUCED PROLIFERATION IN OTHER REGENERATION MODELS

Since the original reports from *Drosophila*, a number of studies have established that the ability of apoptotic cells to induce mitogen expression is conserved across phyla and involves signals of the Wnt, TGF-β, and Hh families. In the Hydra regeneration model, it was shown that apoptotic cells express Wnt3, a homolog of the *Drosophila wingless* gene, and blocking Wnt3 induction indeed impairs head regeneration (Chera et al. 2009; Galliot and Chera 2010). Interestingly, Wnt3 initiates a Wnt cascade that involves six additional Wnt signaling factors for head regeneration (Fig. 4) (Lengfeld et al. 2009). In other models of regeneration, such as planarians and newt, Wnt, TGF-β, and Hh signaling have been implicated in regeneration responses (Fig. 4) (Reddien et al. 2007; Gurley et al. 2008; Petersen and Reddien 2008, 2009; Rink et al. 2009). In these animals, induction of massive apoptosis has been observed at the amputation site (Hwang et al. 2004; Vlaskalin et al. 2004; Pellettieri and Sanchez Alvarado 2007), but it is unknown at present whether apoptotic cells are the source of the regenerating mitogens.

In vertebrates, Wnt signaling is also critical for regeneration, but its function is more complex and can have even opposing functions. For example, fin regeneration in zebrafish is enhanced by Wnt8 expression, but suppressed by Wnt5b (Stoick-Cooper et al. 2007). In addition to Wnt signaling, prostaglandin E2 (PGE2) has also been implicated in apoptosis-induced proliferation (Fig. 4). In zebrafish, the hematopoietic lineage readily undergoes cell death after irradiation. However, when these fish are treated with a stabilized derivative of PGE2, hematopoietic stem cells and bone marrow recovery is enhanced (North et al. 2007). The role of PGE2 in apoptosis-induced proliferation was further validated in a mouse model (Li et al. 2010). The investigators of that study used a cell culture system in which various stem cells were incubated with stressed or nonstressed MEFs. They found that the stem cells grew faster only when the coincubated MEFs were previously exposed to a lethal dose of ionizing irradiation. Furthermore, the growth boosting activity from apoptotic MEFs was dependent on caspase-3 and caspase-7. Under these conditions, caspase-3 and caspase-7 activate calcium-independent phospholipase A2 (iPLA2), an enzyme in the pathway of PGE2 synthesis (Fig. 4). Consistently, knockdown of iPLA2 impairs the ability of apoptotic cells to stimulate apoptosis-induced proliferation (Li et al. 2010).

These reports establish the importance of Wnt and PGE2 in apoptosis-induced proliferation. In fact, Wnt and PGE2 signaling are intimately linked. Hints to a relationship were revealed in the 1990s through the analysis of APC^{min} mice, in which unrestrained Wnt signaling underlies the development of innumerable colonic polyps and ultimately colon cancer. Such formation of polyps is inhibited when prostaglandin biosynthesis is impaired (Giardiello et al. 1993). This relationship is also conserved in hematopoietic stem cells (HSC). These cells are particularly well known for their ability to repopulate after irradiation-induced apoptosis. In the regenerating HSCs, Wnt-induced β-catenin signaling and PGE2 are required. In fact, PGE2 treatment activates Wnt-responsive reporters within the regenerating HSCs. Espistatic experiments show that PGE2 regulates Wnt signaling by controlling the cAMP/PKA pathway (Goessling et al. 2009).

Although these studies have established the importance of mitogenic signaling that originates from apoptotic cells, many details remain to be worked out. For example, there remains some disagreement about the role of *wingless* in apoptosis-induced proliferation in *Drosophila*. In one particular study, where regeneration capacity after massive cell death was analyzed, it was shown that RNAi-mediated knockdown of *wingless* blocks cell proliferation associated with apoptosis (Smith-Bolton et al. 2009). In a different study, however, the tissue's ability to recover its size after X-ray irradiation-induced cell death was reportedly not affected by the loss of *wingless* (Perez-Garijo et al. 2009). Whether the role of *wingless* is dependent on the nature of the apoptotic stress, or the experimental setup remains to be resolved.

There is a possibility that other signaling proteins function redundantly with *wingless* (and *dpp*) in apoptosis-induced proliferation. Supporting this thought, a recent report has implicated a role of the Hippo/Yorkie pathway in apoptosis-induced proliferation. This is a growth regulatory pathway that is regulated by cell adhesion proteins and apico-basal polarity determinants (Bennett and Harvey 2006; Cho et al. 2006; Silva et al. 2006; Willecke et al. 2006; Chen et al. 2010; Grzeschik et al. 2010; Ling et al. 2010). Downstream from these factors are kinase complexes, represented by the Hippo kinase, and the transcription factor Yorkie. When Hippo signaling is active, Yorkie is phosphorylated and remains in the cytoplasm. This transcription factor translocates to the nucleus when the Hippo pathway is inactive, thereby inducing growth and antiapoptotic genes. Independent studies on *Drosophila* tissue regeneration have found that Yorkie translocates to the nucleus in the regenerating cells (Cai et al. 2010; Sun and Irvine 2011). Such regulation of Yorkie activity in living cells depends on JNK activity in dying cells (Sun and Irvine 2011). Thus, it is conceivable that JNK signaling in apoptotic cells induces a Hippo pathway ligand, in addition to other known mitogens.

THE ROLE OF APOPTOSIS-INDUCED COMPENSATORY PROLIFERATION IN CANCER

Evasion from apoptosis is one of the hallmarks of cancer (Hanahan and Weinberg 2000). Initially, the concept of "evasion from apoptosis" implied that tumor cells autonomously escape apoptotic signals either by the immune system or by therapeutic regimens (Hanahan and Weinberg 2000). However, the discovery of apoptosis-induced proliferation adds another layer to this concept, the nonautonomous induction of proliferation, which may have the opposite cancer-promoting activity. As outlined above, genetic studies in *Drosophila* have revealed two distinct types of apoptosis-induced proliferation. Of particular interest for cancer development is effector caspase-independent apoptosis-induced proliferation. In this case, undead cells generated by P35-mediated inhibition or genetic ablation of effector caspases promote overgrowth, which may contribute to tumor development. Many tumor cells have acquired at least some resistance to apoptosis and may resemble undead cells. They initiate the cell death program, but cannot or only slowly execute it, because an essential component of the cell death pathway is mutated or otherwise inhibited. For example, the breast cancer-derived MCF-7 cell line harbors an inactivating point mutation in the gene encoding the effector caspase caspase-3 (Kurokawa et al. 1999). Furthermore, loss of caspase-3 mRNA and protein has been reported in samples from breast, ovarian, and cervical cancers, as well as neuroblastoma (Devarajan et al. 2002; Iolascon et al. 2003). Large-scale analysis of caspase genes in many different tumors reveal several alterations of caspase-3 and other caspases (reviewed by Ghavami et al. 2009). In addition, caspase inhibitors such as IAPs and c-FLIP are frequently overexpressed in cancer cells (Tamm et al. 2000; LaCasse et al. 2008; Yang 2008). Thus, in these cases, loss or inhibition of caspase-3 may not only confer resistance to apoptosis-inducing chemotherapy, but these undead tumor cells may emit excessive amounts of mitogens and thus may cause even more severe tumor growth.

It is unknown whether caspase-9, the mammalian ortholog of the initiator caspase Dronc, has a similar nonapoptotic function for cell proliferation. However, the CD95/Fas/Apo-1 extrinsic cell death pathway may have such a function. This pathway has been extensively studied for its apoptotic role. CD95 activation recruits FADD and caspase-8 into the death-inducing signaling complex (DISC) that triggers apoptosis by activation of caspase-3 (Fig. 5A) (Nagata 1999; Krammer 2000). However, CD95 can also promote proliferation and tumor growth in a nonapoptotic function (Fig. 5A) (Owen-Schaub et al. 1993; Desbarats et al. 1999; Peter et al. 2007; Chen et al. 2010). Mouse models show that liver and ovarian cancers depend on CD95 function (Chen et al. 2010). CD95 has also been reported to be required for liver regeneration after partial hepatectomy (Desbarats and Newell 2000). In this case, liver regeneration is impaired in *lpr* mice, which have reduced cell surface CD95. Interestingly, *lpr-cg* mutant mice that carry a point mutation in the death domain of CD95 rendering it unable to interact with FADD and induce apoptosis but is expressed at normal cell surface levels (Kimura and Matsuzawa 1994), have an even increased rate of regeneration (Desbarats and Newell 2000). This observation suggests that the apoptosis- and proliferation-promoting functions of CD95 can be uncoupled (Fig. 5A). It also suggests that inhibition of the apoptosis-promoting function of CD95 may augment the tumor-promoting function of CD95 as has been implied in ovarian cancer (Baldwin et al. 1999). For example, it is conceivable to speculate that inhibition of caspase-8 by c-FLIP unleashes the tumor-promoting function of CD95 (Safa and Pollok 2011; Jing et al. 2012). Interestingly, the proliferation- and tumor-promoting function of CD95 is mediated by JNK, Fos, and Egr1 (Fig. 5A) (Chen et al. 2010). The mechanism of JNK activation in this system is unknown, and may not involve caspases, but it illustrates the same principle as originally observed for apoptosis-induced proliferation in *Drosophila* (Huh et al. 2004; Ryoo et al. 2004).

Finally, not only undead tumor cells, but also "genuine" apoptotic tumor cells can promote apoptosis-induced proliferation. Recently, it was shown that tumor cells that were induced to die by radiotherapy stimulate tumor regrowth (Huang et al. 2011). In this case, dying tumor cells were not kept in an undead condition, but were able to secrete signaling molecules for tumor regrowth. In fact, caspase-3 was required for tumor regrowth as caspase-3 deficiency rendered the tumors more sensitive to radiotherapy. This proliferation- and tumor-promoting activity of caspase-3 is mediated through cleavage and activation of cytosolic calcium-independent phospholipase A2 (iPLA2) that ultimately produces prostaglandin E2 (Fig. 4) (Li et al. 2010; Huang et al. 2011).

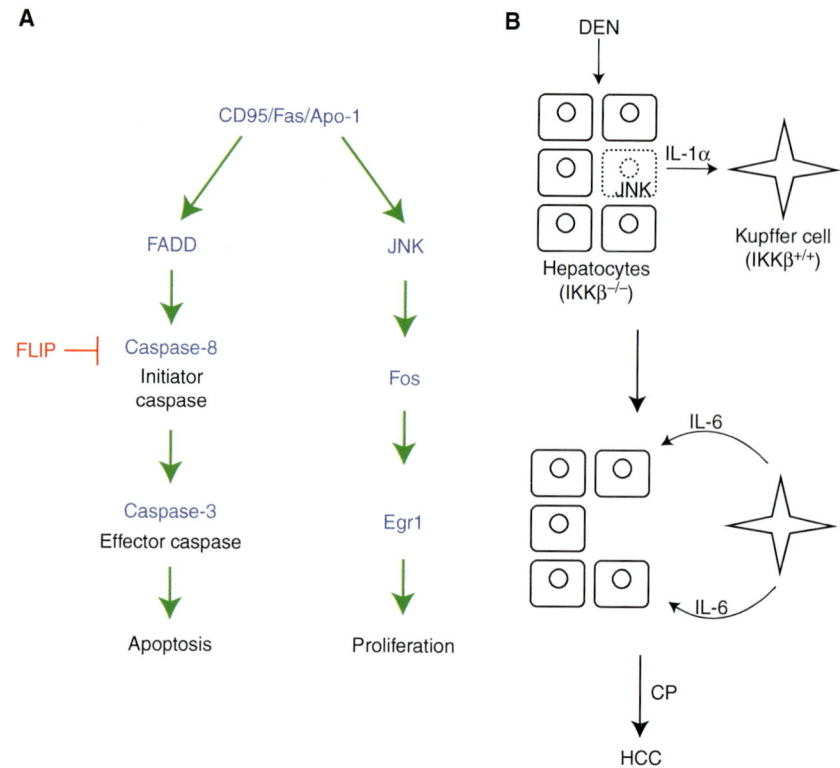

Figure 5. Potential models of apoptosis-induced proliferation in mammals. (A) The CD95/Fas/Apo1 pathway in apoptosis and proliferation. The CD95/Fas/Apo1 receptor induces apoptosis through the very well-studied FADD/caspase-8 pathway. FLIP is a caspase-8 analog and acts as a dominant negative for caspase-8 activation. The proliferation component of CD95/Fas/Apo1 is less characterized, but appears to engage the JNK pathway. (B) Role of compensatory proliferation for development of hepatocarcinoma (HCC). Loss of IKKβ or IKKγ renders hepatocytes extremely sensitive to carcinogens such as diethylnitrosoamine (DEN), causing JNK-induced cell death (stippled cell). However, JNK also induces expression of interleukin-1α (IL-1α). IL-1α induces the production of IL-6 in Kupffer cells. IL-6 feeds back onto surviving hepatocytes and induces proliferation and subsequently HCC. Interestingly, because IL-1α induces NF-κB activation, Kupffer cells need to be wild-type for IKKβ and IKKγ for apoptosis-induced proliferation and HCC to occur. (Panel based on data modified from He and Karin 2011.)

Therefore, radio- and chemotherapy may be ineffective or even counter-productive because apoptotic tumor cells can induce proliferation of surviving tumor cells. Moreover, the same investigators examined untreated tumor samples from patients with head and neck as well as advanced stage breast cancer for evidence of activated caspase-3. Surprisingly, samples with high levels of activated caspase-3 correlated with higher recurrence and shorter survival times of the patients (Huang et al. 2011), consistent with the concept of apoptosis-induced proliferation.

It is unknown how caspase-3 is activated in these tumors.

Apoptosis-induced proliferation may also be the underlying cause of cell and tissue type-specific differences for tumor development. For example, NF-κB activation was found to be a tumor promoter in many tumors including colitis-associated cancer (CAC) and melanoma in mice (Greten et al. 2004; Yang et al. 2010). In contrast, loss of NF-κB activity augments squamous cell carcinoma and carcinogen-induced hepatocellular carcinoma (HCC) (Dajee et al.

2003; Maeda et al. 2005; Sakurai et al. 2008). At least for HCC, this difference may be explained by apoptosis-induced proliferation. The activity of NF-κB is controlled by the IKK complex, which targets IκB for phosphorylation and degradation (Karin et al. 2006). IKKβ and IKKγ, the catalytic and regulatory subunits of IKK, are required for the survival of hepatocytes (Li et al. 1999). Treatment of IKKβ$^{-/-}$ and IKKγ$^{-/-}$ hepatocytes with carcinogens such as diethylnitrosamine (DEN) causes accumulation of radical oxygen species (ROS) resulting in hepatocyte cell death, which triggers apoptosis-induced proliferation and HCC in the following manner (Fig. 5B) (Maeda et al. 2005; Sakurai et al. 2008). Dying hepatocytes increase JNK activity and release interleukin-1α (IL-1α) (Sakurai et al. 2008). IL-1α stimulates Kupffer cells, which in turn secrete IL-6 (Maeda et al. 2005). Finally, IL-6 promotes proliferation of surviving hepatocytes causing HCC (Fig. 5B). Thus, Kupffer cells represent an intermediary step for apoptosis-induced proliferation. The common theme here is that dying cells cause apoptosis-induced proliferation and cancer through activation of JNK signaling, similar to *Drosophila*. However, the secreted mitogens and cytokines are different and may vary in a cell type-specific manner. Interestingly, IL-1α stimulates IKK/NF-κB activity in Kupffer cells. Thus, if Kupffer cells are also IKKβ$^{-/-}$ and IKKγ$^{-/-}$, apoptosis-induced proliferation and HCC are blocked (Maeda et al. 2005), providing another example of cell type-specific differences regarding tumorigenesis.

These examples illustrate that apoptosis-induced proliferation may critically contribute to tumor development and tumor growth.

CONCLUDING REMARKS

The traditional view that apoptotic cells only display signals for engulfment at the cell surface, but otherwise shut down metabolically is an oversimplification. These cells are very capable of communicating with neighboring cells and prepare them for their departure. In this review, we focused on the ability of apoptotic cells to induce proliferation of surviving cells. This is directly dependent on apoptotic caspases and may have significant implications for wound healing and regeneration after tissue loss. Obviously, apoptosis-induced proliferation may be critical for tumor initiation and tumor development. However, apoptotic cells may also influence neighboring surviving cells in other ways. For example, while apoptotic cells die, they can increase the apoptotic resistance of neighboring cells in some cases (Herz et al. 2006). The mechanism of this process is much less understood, but will be of significance for the understanding of tumor development. It is also conceivable that apoptotic cells change the shape and size of surviving cells, induce cellular migration, or may even change their differentiation status. We expect that progress in these areas will have important clinical implications, especially for regenerative medicine and cancer.

ACKNOWLEDGMENTS

We thank our colleagues, especially Hermann Steller and Yun Fan, for stimulating discussions and advice. This work is supported by grants from the National Institutes of Health (NIH) GM068016 to A.B. and GM079425 and EY020866 to H.D.R.

REFERENCES

*Reference is also in this collection.

Arama E, Agapite J, Steller H. 2003. Caspase activity and a specific cytochrome C are required for sperm differentiation in *Drosophila Dev Cell* **4:** 687–697.

Arama E, Bader M, Rieckhof GE, Steller H. 2007. A ubiquitin ligase complex regulates caspase activation during sperm differentiation in *Drosophila*. *PLoS Biol* **5:** e251.

Bader M, Benjamin S, Wapinski OL, Smith DM, Goldberg AL, Steller H. 2011. A conserved F box regulatory complex controls proteasome activity in *Drosophila*. *Cell* **145:** 371–382.

Baldwin RL, Tran H, Karlan BY. 1999. Primary ovarian cancer cultures are resistant to Fas-mediated apoptosis. *Gynecol Oncol* **74:** 265–271.

Bennett FC, Harvey KF. 2006. Fat cadherin modulates organ size in *Drosophila* via the Salvador/Warts/Hippo signaling pathway. *Curr Biol* **16:** 2101–2110.

Bergantinos C, Corominas M, Serras F. 2010. Cell death-induced regeneration in wing imaginal discs requires JNK signalling. *Development* **137:** 1169–1179.

Brodsky MH, Nordstrom W, Tsang G, Kwan E, Rubin GM, Abrams JM. 2000. *Drosophila* p53 binds a damage response element at the reaper locus. *Cell* **101**: 103–113.

Cai J, Zhang N, Zheng Y, de Wilde RF, Maitra A, Pan D. 2010. The Hippo signaling pathway restricts the oncogenic potential of an intestinal regeneration program. *Genes Dev* **24**: 2383–2388.

Chen P, Nordstrom W, Gish B, Abrams JM. 1996. grim, a novel cell death gene in *Drosophila*. *Genes Dev* **10**: 1773–1782.

Chen L, Park SM, Tumanov AV, Hau A, Sawada K, Feig C, Turner JR, Fu YX, Romero IL, Lengyel E, et al. 2010. CD95 promotes tumour growth. *Nature* **465**: 492–496.

Chera S, Ghila L, Dobretz K, Wenger Y, Bauer C, Buzgariu W, Martinou JC, Galliot B. 2009. Apoptotic cells provide an unexpected source of Wnt3 signaling to drive hydra head regeneration. *Dev Cell* **17**: 279–289.

Chera S, Ghila L, Wenger Y, Galliot B. 2011. Injury-induced activation of the MAPK/CREB pathway triggers apoptosis-induced compensatory proliferation in hydra head regeneration. *Dev Growth Differ* **53**: 186–201.

Chew SK, Akdemir F, Chen P, Lu WJ, Mills K, Daish T, Kumar S, Rodriguez A, Abrams JM. 2004. The apical caspase dronc governs programmed and unprogrammed cell death in *Drosophila*. *Dev Cell* **7**: 897–907.

Cho E, Feng Y, Rauskolb C, Maitra S, Fehon R, Irvine KD. 2006. Delineation of a Fat tumor suppressor pathway. *Nat Genet* **38**: 1142–1150.

Clem RJ, Fechheimer M, Miller LK. 1991. Prevention of apoptosis by a baculovirus gene during infection of insect cells. *Science* **254**: 1388–1390.

Coller HA. 2007. What's taking so long? S-phase entry from quiescence versus proliferation. *Nat Rev Mol Cell Biol* **8**: 667–670.

Colussi PA, Kumar S. 1999. Targeted disruption of caspase genes in mice: What they tell us about the functions of individual caspases in apoptosis. *Immunol Cell Biol* **77**: 58–63.

Daish TJ, Mills K, Kumar S. 2004. *Drosophila* caspase DRONC is required for specific developmental cell death pathways and stress-induced apoptosis. *Dev Cell* **7**: 909–915.

Dajee M, Lazarov M, Zhang JY, Cai T, Green CL, Russell AJ, Marinkovich MP, Tao S, Lin Q, Kubo Y, et al. 2003. NF-κB blockade and oncogenic Ras trigger invasive human epidermal neoplasia. *Nature* **421**: 639–643.

De Maria R, Zeuner A, Eramo A, Domenichelli C, Bonci D, Grignani F, Srinivasula SM, Alnemri ES, Testa U, Peschle C. 1999. Negative regulation of erythropoiesis by caspase-mediated cleavage of GATA-1. *Nature* **401**: 489–493.

Desbarats J, Newell MK. 2000. Fas engagement accelerates liver regeneration after partial hepatectomy. *Nat Med* **6**: 920–923.

Desbarats J, Wade T, Wade WF, Newell MK. 1999. Dichotomy between naive and memory $CD4^+$ T cell responses to Fas engagement. *Proc Natl Acad Sci* **96**: 8104–8109.

Devarajan E, Sahin AA, Chen JS, Krishnamurthy RR, Aggarwal N, Brun AM, Sapino A, Zhang F, Sharma D, Yang XH, et al. 2002. Down-regulation of caspase 3 in breast cancer: A possible mechanism for chemoresistance. *Oncogene* **21**: 8843–8851.

Dix MM, Simon GM, Cravatt BF. 2008. Global mapping of the topography and magnitude of proteolytic events in apoptosis. *Cell* **134**: 679–691.

Down JD, Tarbell NJ, Thames HD, Mauch PM. 1991. Syngeneic and allogeneic bone marrow engraftment after total body irradiation: Dependence on dose, dose rate, and fractionation. *Blood* **77**: 661–669.

Fan Y, Bergmann A. 2008a. Apoptosis-induced compensatory proliferation. The Cell is dead. Long live the Cell! *Trends Cell Biol* **18**: 467–473.

Fan Y, Bergmann A. 2008b. Distinct mechanisms of apoptosis-induced compensatory proliferation in proliferating and differentiating tissues in the *Drosophila* eye. *Dev Cell* **14**: 399–410.

Fan Y, Bergmann A. 2010. The cleaved-Caspase-3 antibody is a marker of Caspase-9-like DRONC activity in *Drosophila*. *Cell Death Differ* **17**: 534–539.

Fan Y, Lee TV, Xu D, Chen Z, Lamblin AF, Steller H, Bergmann A. 2010. Dual roles of *Drosophila* p53 in cell death and cell differentiation. *Cell Death Differ* **17**: 912–921.

Galliot B, Chera S. 2010. The Hydra model: Disclosing an apoptosis-driven generator of Wnt-based regeneration. *Trends Cell Biol* **20**: 514–523.

Geisbrecht ER, Montell DJ. 2004. A role for *Drosophila* IAP1-mediated caspase inhibition in Rac-dependent cell migration. *Cell* **118**: 111–125.

Gerhold AR, Richter DJ, Yu AS, Hariharan IK. 2011. Identification and characterization of genes required for compensatory growth in *Drosophila*. *Genetics* **189**: 1309–1326.

Ghavami S, Hashemi M, Ande SR, Yeganeh B, Xiao W, Eshraghi M, Bus CJ, Kadkhoda K, Wiechec E, Halayko AJ, et al. 2009. Apoptosis and cancer: Mutations within caspase genes. *J Med Genet* **46**: 497–510.

Giardiello FM, Hamilton SR, Krush AJ, Piantadosi S, Hylind LM, Celano P, Booker SV, Robinson CR, Offerhaus GJ. 1993. Treatment of colonic and rectal adenomas with sulindac in familial adenomatous polyposis. *N Engl J Med* **328**: 1313–1316.

Goessling W, North TE, Loewer S, Lord AM, Lee S, Stoick-Cooper CL, Weidinger G, Puder M, Daley GQ, Moon RT, et al. 2009. Genetic interaction of PGE2 and Wnt signaling regulates developmental specification of stem cells and regeneration. *Cell* **136**: 1136–1147.

Gowda PS, Zhou F, Chadwell LV, McEwen DG. 2012. p53 binding prevents phosphatase-mediated inactivation of diphosphorylated JUN N-terminal kinase. *J Biol Chem* **287**: 17554–17567.

Greten FR, Eckmann L, Greten TF, Park JM, Li ZW, Egan LJ, Kagnoff MF, Karin M. 2004. IKKβ links inflammation and tumorigenesis in a mouse model of colitis-associated cancer. *Cell* **118**: 285–296.

Grether ME, Abrams JM, Agapite J, White K, Steller H. 1995. The head involution defective gene of *Drosophila* melanogaster functions in programmed cell death. *Genes Dev* **9**: 1694–1708.

Grzeschik NA, Parsons LM, Allott ML, Harvey KF, Richardson HE. 2010. Lgl, aPKC, and Crumbs regulate the Salvador/Warts/Hippo pathway through two distinct mechanisms. *Curr Biol* **20**: 573–581.

Gurley KA, Rink JC, Sanchez Alvarado A. 2008. β-catenin defines head versus tail identity during planarian regeneration and homeostasis. *Science* **319:** 323–327.

Hanahan D, Weinberg RA. 2000. The hallmarks of cancer. *Cell* **100:** 57–70.

Hawkins CJ, Yoo SJ, Peterson EP, Wang SL, Vernooy SY, Hay BA. 2000. The *Drosophila* caspase DRONC cleaves following glutamate or aspartate and is regulated by DIAP1, HID, and GRIM. *J Biol Chem* **275:** 27084–27093.

Haynie JL, Bryant PJ. 1977. The effects of X-rays on the proliferation dynamics of cells in the imaginal wing disc of *Drosophila* melanogaster. *Roux's Arch Dev Biol* **183:** 85–100.

He G, Karin M. 2011. NF-κB and STAT3—key players in liver inflammation and cancer. *Cell Res* **21:** 159–168.

Herz HM, Chen Z, Scherr H, Lackey M, Bolduc C, Bergmann A. 2006. vps25 mosaics display non-autonomous cell survival and overgrowth, and autonomous apoptosis. *Development* **133:** 1871–1880.

Huang Q, Li F, Liu X, Li W, Shi W, Liu FF, O'Sullivan B, He Z, Peng Y, Tan AC, et al. 2011. Caspase 3-mediated stimulation of tumor cell repopulation during cancer radiotherapy. *Nat Med* **17:** 860–866.

Huh JR, Guo M, Hay BA. 2004. Compensatory proliferation induced by cell death in the *Drosophila* wing disc requires activity of the apical cell death caspase Dronc in a nonapoptotic role. *Curr Biol* **14:** 1262–1266.

Hwang JS, Kobayashi C, Agata K, Ikeo K, Gojobori T. 2004. Detection of apoptosis during planarian regeneration by the expression of apoptosis-related genes and TUNEL assay. *Gene* **333:** 15–25.

Igaki T, Kanda H, Yamamoto-Goto Y, Kanuka H, Kuranaga E, Aigaki T, Miura M. 2002. Eiger, a TNF superfamily ligand that triggers the *Drosophila* JNK pathway. *EMBO J* **21:** 3009–3018.

Iolascon A, Borriello A, Giordani L, Cucciolla V, Moretti A, Monno F, Criniti V, Marzullo A, Criscuolo M, Ragione FD. 2003. Caspase 3 and 8 deficiency in human neuroblastoma. *Cancer Genet Cytogenet* **146:** 41–47.

Jaklevic BR, Su TT. 2004. Relative contribution of DNA repair, cell cycle checkpoints, and cell death to survival after DNA damage in *Drosophila* larvae. *Curr Biol* **14:** 23–32.

Jing G, Yuan K, Liang Q, Sun Y, Mao X, McDonald JM, Chen Y. 2012. Reduced CaM/FLIP binding by a single point mutation in c-FLIP(L) modulates Fas-mediated apoptosis and decreases tumorigenesis. *Lab Invest* **92:** 82–90.

Kanuka H, Kuranaga E, Takemoto K, Hiratou T, Okano H, Miura M. 2005. *Drosophila* caspase transduces Shaggy/GSK-3β kinase activity in neural precursor development. *EMBO J* **24:** 3793–3806.

Karin M, Lawrence T, Nizet V. 2006. Innate immunity gone awry: Linking microbial infections to chronic inflammation and cancer. *Cell* **124:** 823–835.

Kerr JF, Wyllie AH, Currie AR. 1972. Apoptosis: A basic biological phenomenon with wide-ranging implications in tissue kinetics. *Brit J Cancer* **26:** 239–257.

Kimura M, Matsuzawa A. 1994. Autoimmunity in mice bearing lprcg: A novel mutant gene. *Int Rev Immunol* **11:** 193–210.

Kondo S, Senoo-Matsuda N, Hiromi Y, Miura M. 2006. DRONC coordinates cell death and compensatory proliferation. *Mol Cell Biol* **26:** 7258–7268.

Krammer PH. 2000. CD95's deadly mission in the immune system. *Nature* **407:** 789–795.

Kumar S. 2007. Caspase function in programmed cell death. *Cell Death Differ* **14:** 32–43.

Kuranaga E, Kanuka H, Tonoki A, Takemoto K, Tomioka T, Kobayashi M, Hayashi S, Miura M. 2006. *Drosophila* IKK-related kinase regulates nonapoptotic function of caspases via degradation of IAPs. *Cell* **126:** 583–596.

Kurokawa H, Nishio K, Fukumoto H, Tomonari A, Suzuki T, Saijo N. 1999. Alteration of caspase-3 (CPP32/Yama/apopain) in wild-type MCF-7, breast cancer cells. *Oncol Rep* **6:** 33–37.

LaCasse EC, Mahoney DJ, Cheung HH, Plenchette S, Baird S, Korneluk RG. 2008. IAP-targeted therapies for cancer. *Oncogene* **27:** 6252–6275.

Lengfeld T, Watanabe H, Simakov O, Lindgens D, Gee L, Law L, Schmidt HA, Ozbek S, Bode H, Holstein TW. 2009. Multiple Wnts are involved in Hydra organizer formation and regeneration. *Dev Biol* **330:** 186–199.

Li Q, Van Antwerp D, Mercurio F, Lee KF, Verma IM. 1999. Severe liver degeneration in mice lacking the IκB kinase 2 gene. *Science* **284:** 321–325.

Li F, Huang Q, Chen J, Peng Y, Roop DR, Bedford JS, Li CY. 2010. Apoptotic cells activate the "phoenix rising" pathway to promote wound healing and tissue regeneration. *Sci Signal* **3:** ra13.

Ling C, Zheng Y, Yin F, Yu J, Huang J, Hong Y, Wu S, Pan D. 2010. The apical transmembrane protein Crumbs functions as a tumor suppressor that regulates Hippo signaling by binding to Expanded. *Proc Natl Acad Sci* **107:** 10532–10537.

Maeda S, Kamata H, Luo JL, Leffert H, Karin M. 2005. IKKβ couples hepatocyte death to cytokine-driven compensatory proliferation that promotes chemical hepatocarcinogenesis. *Cell* **121:** 977–990.

Mahrus S, Trinidad JC, Barkan DT, Sali A, Burlingame AL, Wells JA. 2008. Global sequencing of proteolytic cleavage sites in apoptosis by specific labeling of protein N termini. *Cell* **134:** 866–876.

Meier P, Silke J, Leevers SJ, Evan GI. 2000. The *Drosophila* caspase DRONC is regulated by DIAP1. *EMBO J* **19:** 598–611.

Milan M, Campuzano S, Garcia-Bellido A. 1997. Developmental parameters of cell death in the wing disc of *Drosophila*. *Proc Natl Acad Sci* **94:** 5691–5696.

*Miura M. 2012. Apoptotic and non-apoptotic caspase functions in animal development. *Cold Spring Harb Perspect Biol* **4:** a008664.

Moreno E, Yan M, Basler K. 2002. Evolution of TNF signaling mechanisms: JNK-dependent apoptosis triggered by Eiger, the *Drosophila* homolog of the TNF superfamily. *Curr Biol* **12:** 1263–1268.

Nagata S. 1999. Fas ligand-induced apoptosis. *Annu Rev Genet* **33:** 29–55.

Newsome TP, Asling B, Dickson BJ. 2000. Analysis of *Drosophila* photoreceptor axon guidance in eye-specific mosaics. *Development* **127:** 851–860.

North TE, Goessling W, Walkley CR, Lengerke C, Kopani KR, Lord AM, Weber GJ, Bowman TV, Jang IH, Grosser T, et al. 2007. Prostaglandin E2 regulates vertebrate haematopoietic stem cell homeostasis. *Nature* **447:** 1007–1011.

Oshima K, Takeda M, Kuranaga E, Ueda R, Aigaki T, Miura M, Hayashi S. 2006. IKK epsilon regulates F actin assembly and interacts with *Drosophila* IAP1 in cellular morphogenesis. *Curr Biol* **16:** 1531–1537.

Ouyang Y, Petritsch C, Wen H, Jan L, Jan YN, Lu B. 2011. Dronc caspase exerts a non-apoptotic function to restrain phospho-Numb-induced ectopic neuroblast formation in *Drosophila*. *Development* **138:** 2185–2196.

Owen-Schaub LB, Meterissian S, Ford RJ. 1993. Fas/APO-1 expression and function on malignant cells of hematologic and nonhematologic origin. *J Immunother Emphasis Tumor Immunol* **14:** 234–241.

Pearson BJ, Sanchez Alvarado A. 2010. A planarian p53 homolog regulates proliferation and self-renewal in adult stem cell lineages. *Development* **137:** 213–221.

Pellettieri J, Sanchez Alvarado A. 2007. Cell turnover and adult tissue homeostasis: From humans to planarians. *Annu Rev Genet* **41:** 83–105.

Perez-Garijo A, Martin FA, Morata G. 2004. Caspase inhibition during apoptosis causes abnormal signalling and developmental aberrations in *Drosophila*. *Development* **131:** 5591–5598.

Perez-Garijo A, Martin FA, Struhl G, Morata G. 2005. Dpp signaling and the induction of neoplastic tumors by caspase-inhibited apoptotic cells in *Drosophila*. *Proc Natl Acad Sci* **102:** 17664–17669.

Perez-Garijo A, Shlevkov E, Morata G. 2009. The role of Dpp and Wg in compensatory proliferation and in the formation of hyperplastic overgrowths caused by apoptotic cells in the *Drosophila* wing disc. *Development* **136:** 1169–1177.

Peter ME, Budd RC, Desbarats J, Hedrick SM, Hueber AO, Newell MK, Owen LB, Pope RM, Tschopp J, Wajant H, et al. 2007. The CD95 receptor: Apoptosis revisited. *Cell* **129:** 447–450.

Petersen CP, Reddien PW. 2008. Smed-βcatenin-1 is required for anteroposterior blastema polarity in planarian regeneration. *Science* **319:** 327–330.

Petersen CP, Reddien PW. 2009. A wound-induced Wnt expression program controls planarian regeneration polarity. *Proc Natl Acad Sci* **106:** 17061–17066.

Reddien PW, Bermange AL, Kicza AM, Sanchez Alvarado A. 2007. BMP signaling regulates the dorsal planarian midline and is needed for asymmetric regeneration. *Development* **134:** 4043–4051.

Rink JC, Gurley KA, Elliott SA, Sanchez Alvarado A. 2009. Planarian Hh signaling regulates regeneration polarity and links Hh pathway evolution to cilia. *Science* **326:** 1406–1410.

Ryoo HD, Gorenc T, Steller H. 2004. Apoptotic cells can induce compensatory cell proliferation through the JNK and the Wingless signaling pathways. *Dev Cell* **7:** 491–501.

Safa AR, Pollok KE. 2011. Targeting the anti-apoptotic protein c-FLIP for cancer therapy. *Cancers* **3:** 1639–1671.

Sakurai T, He G, Matsuzawa A, Yu GY, Maeda S, Hardiman G, Karin M. 2008. Hepatocyte necrosis induced by oxidative stress and IL-1 α release mediate carcinogen-induced compensatory proliferation and liver tumorigenesis. *Cancer Cell* **14:** 156–165.

Sandu C, Ryoo HD, Steller H. 2010. *Drosophila* IAP antagonists form multimeric complexes to promote cell death. *J Cell Biol* **190:** 1039–1052.

Shapiro PJ, Hsu HH, Jung H, Robbins ES, Ryoo HD. 2008. Regulation of the *Drosophila* apoptosome through feedback inhibition. *Nat Cell Biol* **10:** 1440–1446.

Shlevkov E, Morata G. 2012. A dp53/JNK-dependant feedback amplification loop is essential for the apoptotic response to stress in *Drosophila*. *Cell Death Differ* **19:** 451–460.

Silva E, Tsatskis Y, Gardano L, Tapon N, McNeill H. 2006. The tumor-suppressor gene fat controls tissue growth upstream of expanded in the hippo signaling pathway. *Curr Biol* **16:** 2081–2089.

Smith-Bolton RK, Worley MI, Kanda H, Hariharan IK. 2009. Regenerative growth in *Drosophila* imaginal discs is regulated by Wingless and Myc. *Dev Cell* **16:** 797–809.

Stoick-Cooper CL, Weidinger G, Riehle KJ, Hubbert C, Major MB, Fausto N, Moon RT. 2007. Distinct Wnt signaling pathways have opposing roles in appendage regeneration. *Development* **134:** 479–489.

Stowers RS, Schwarz TL. 1999. A genetic method for generating *Drosophila* eyes composed exclusively of mitotic clones of a single genotype. *Genetics* **152:** 1631–1639.

Suissa Y, Ziv O, Dinur T, Arama E, Gerlitz O. 2011. The NAB-Brk signal bifurcates at JNK to independently induce apoptosis and compensatory proliferation. *J Biol Chem* **286:** 15556–15564.

Sun G, Irvine KD. 2011. Regulation of Hippo signaling by Jun kinase signaling during compensatory cell proliferation and regeneration, and in neoplastic tumors. *Dev Biol* **350:** 139–151.

Tamm I, Kornblau SM, Segall H, Krajewski S, Welsh K, Kitada S, Scudiero DA, Tudor G, Qui YH, Monks A, et al. 2000. Expression and prognostic significance of IAP-family genes in human cancers and myeloid leukemias. *Clin Cancer Res* **6:** 1796–1803.

Thornberry NA, Lazebnik Y. 1998. Caspases: Enemies within. *Science* **281:** 1312–1316.

Trembley A. 1744. *Meoires pour sevir al'hisoire d'un genre de polyps d'eao douce*. JuH Verbeek, Leiden.

Tseng AS, Adams DS, Qiu D, Koustubhan P, Levin M. 2007. Apoptosis is required during early stages of tail regeneration in Xenopus laevis. *Dev Biol* **301:** 62–69.

Vlaskalin T, Wong CJ, Tsilfidis C. 2004. Growth and apoptosis during larval forelimb development and adult forelimb regeneration in the newt (*Notophthalmus viridescens*). *Dev Genes Evol* **214:** 423–431.

Voas MG, Rebay I. 2004. Signal integration during development: Insights from the *Drosophila* eye. *Dev Dyn* **229:** 162–175.

Vousden KH. 2000. p53: Death star. *Cell* **103:** 691–694.

Warner SJ, Yashiro H, Longmore GD. 2010. The Cdc42/Par6/aPKC polarity complex regulates apoptosis-induced compensatory proliferation in epithelia. *Curr Biol* **20:** 677–686.

Wells BS, Johnston LA. 2012. Maintenance of imaginal disc plasticity and regenerative potential in *Drosophila* by 53. *Dev Biol* **361**: 263–276.

Wells BS, Yoshida E, Johnston LA. 2006. Compensatory proliferation in *Drosophila* imaginal discs requires Dronc-dependent p53 activity. *Curr Biol* **16**: 1606–1615.

White K, Tahaoglu E, Steller H. 1996. Cell killing by the *Drosophila* gene reaper. *Science* **271**: 805–807.

Willecke M, Hamaratoglu F, Kango-Singh M, Udan R, Chen CL, Tao C, Zhang X, Halder G. 2006. The fat cadherin acts through the hippo tumor-suppressor pathway to regulate tissue size. *Curr Biol* **16**: 2090–2100.

Xu D, Li Y, Arcaro M, Lackey M, Bergmann A. 2005. The CARD-carrying caspase Dronc is essential for most, but not all, developmental cell death in *Drosophila*. *Development* **132**: 2125–2134.

Xu D, Wang Y, Willecke R, Chen Z, Ding T, Bergmann A. 2006. The effector caspases drICE and dcp-1 have partially overlapping functions in the apoptotic pathway in *Drosophila*. *Cell Death Differ* **13**: 1697–1706.

Xue D, Horvitz HR. 1995. Inhibition of the *Caenorhabditis elegans* cell-death protease CED-3 by a CED-3 cleavage site in baculovirus p35 protein. *Nature* **377**: 248–251.

Yang JK. 2008. FLIP as an anti-cancer therapeutic target. *Yonsei Med J* **49**: 19–27.

Yang J, Splittgerber R, Yull FE, Kantrow S, Ayers GD, Karin M, Richmond A. 2010. Conditional ablation of Ikkb inhibits melanoma tumor development in mice. *J Clin Invest* **120**: 2563–2574.

Yu X, Wang L, Acehan D, Wang X, Akey CW. 2006. Three-dimensional structure of a double apoptosome formed by the *Drosophila* Apaf-1 related killer. *J Mol Biol* **355**: 577–589.

Yuan J, Shaham S, Ledoux S, Ellis HM, Horvitz HR. 1993. The *C. elegans* cell death gene ced-3 encodes a protein similar to mammalian interleukin-1 β-converting enzyme. *Cell* **75**: 641–652.

Fueling the Flames: Mammalian Programmed Necrosis in Inflammatory Diseases

Francis Ka-Ming Chan

Department of Pathology, Immunology and Virology Program, University of Massachusetts Medical School, Worcester, Massachusetts 01655

Correspondence: francis.chan@umassmed.edu

Programmed necrosis or necroptosis is an inflammatory form of cell death driven by TNF-like death cytokines, toll-like receptors, and antigen receptors. Unlike necrosis induced by physical trauma, a dedicated pathway is involved in programmed necrosis. In particular, a kinase complex composed of the receptor interacting protein kinase 1 (RIPK1) and RIPK3 is a central step in necrotic cell death. Assembly and activation of this RIPK1–RIPK3 "necrosome" is critically controlled by protein ubiquitination, phosphorylation, and caspase-mediated cleavage events. The molecular signals cumulate in formation of intracellular vacuoles, organelle swelling, internal membrane leakage, and eventually plasma membrane rupture. These morphological changes can result in spillage of intracellular adjuvants to promote inflammation and further exacerbate tissue injury. Because of the inflammatory nature of necrosis, it is an attractive pathway for therapeutic intervention in acute inflammatory diseases.

Necrosis and tissue inflammation are two tightly linked phenomena. Cell injury induced by excessive trauma such as heat shock and osmotic shock can result in cell death with "necrotic morphology." These relatively nonspecific means to trigger necrosis have contributed to the notion that necrosis is caused by excessive insults and does not involve elaborate intracellular signaling pathways. In contrast to the notion that necrosis is associated with harmful pathologies, recent work indicates that necrosis can have beneficial roles in certain biological responses. Proteomic approaches and RNA interference screens have identified several crucial regulators of necrosis induced by TNF-like death cytokines. Because a dedicated molecular circuitry is involved, the terms "programmed necrosis" and "necroptosis" have been used to distinguish these types of necrotic cell death from necrosis induced by physical trauma or insults. Here I will discuss the molecular pathway that regulates programmed necrosis/necroptosis. For the sake of simplicity, we will use the term necrosis to refer to programmed necrosis induced by defined death cytokines.

MORPHOLOGICAL FEATURES OF NECROSIS

Similar to apoptosis, necrosis is best defined by its characteristic morphologies. Necrotic cells are typically marked by organelle and cell swelling, and plasma membrane leakage. These features resemble "oncosis" (Majno and Joris

1995), a term that was used to describe cell death with organelle and cell swelling more than one hundred years ago by the German pathologist F. von Recklinghausen (von Recklinghausen 1910). By electron microscopy, numerous small vacuoles can be seen in early necrotic cells (Fig. 1A). The small vacuoles appear to swell or coalesce with each other to form larger vacuoles (Fig. 1B). The identity of these vacuoles is currently unknown, but may be the result of swelling of ER membranes or lysosomes (Vanden Berghe et al. 2010). In addition to the vacuoles, mitochondrial distension can be readily detected (Fig. 1A). Whether these intracellular changes are the cause or consequence of necrosis signaling awaits further experimental evidence. It is noteworthy that the larger vacuoles often show some lesions on their membranes (Fig. 1B). These intracellular membrane lesions can be detected in cells with intact plasma membranes. Thus, internal membrane damage precedes plasma membrane leakage in necrosis. Based on these morphological changes, I can consider necrosis as cell death through "internal bleeding."

CROSS TALK BETWEEN NECROSIS AND OTHER CELL DEATH PATHWAYS

In the laboratory, caspase inhibitors are often used to optimally induce necrosis. Because caspases are key initiators and effectors for apoptosis, the requirement for caspase inhibition for necrosis suggests that molecular cross talk between the two cell death pathways exists. In addition to priming cells to necrosis, caspase inhibition can activate autophagy in some situations (Yu et al. 2004). Although autophagy is predominantly a survival response to stress signals, it can cause cell death under certain conditions, such as that during development of the salivary gland in Drosophila (Berry and Baehrecke 2007). One of the functions of autophagy is to remove damaged proteins and organelles, such as that found in necrotic cells. Leakage of damaged proteins from necrotic cells can elicit a proinflammatory response in the tissue milieu, resulting in further damage from infiltrating immune effector cells. As such, when death receptors are turned on in the presence caspase inhibition, autophagy may represent a

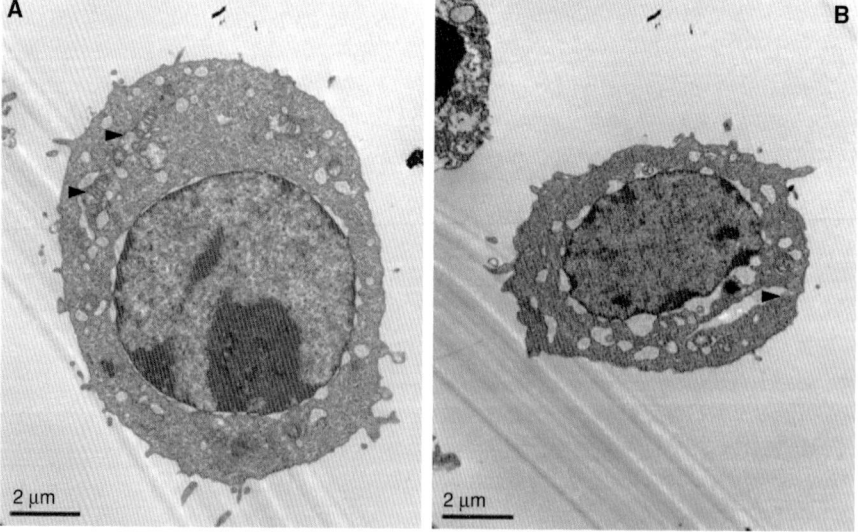

Figure 1. Programmed necrosis is marked by extensive organelle and cell swelling. (*A*) Formation of intracellular vacuoles and swelling of mitochondria in FADD$^{-/-}$ T cells stimulated with plate-bound anti-CD3 (T-cell receptor) antibody. The arrows show the swelling of mitochondria. (*B*) Membrane rupture of large intracellular vacuoles in anti-CD3 activated FADD$^{-/-}$ T cells. The arrow highlights the region of membrane leakage. Note that the plasma membrane is still intact.

last resort for cells to avoid the damaging effects of necrosis. In this scenario, apoptosis, autophagy, and necrosis may represent a continuum of cell death modules with increasing propensity to drives inflammation. It implies that the pathways that regulate these different cell death programs are intimately linked. This progression of cell death modules is an important mechanism for multicellular organisms to control and limit the deleterious consequences of cell death.

We now know that many cell surface receptors can induce cellular necrosis. These include cytokines in the TNF superfamily, toll-like receptors (He et al. 2011; McComb et al. 2012), T-cell receptor (Ch'en et al. 2008; Cho et al. 2011; Zhang et al. 2011), DNA alkylating agents (Tu et al. 2009), and certain cytotoxic drugs (He et al. 2009; Zhang et al. 2009). These upstream triggers use a core pathway to induce necrosis that involves the receptor interacting protein kinase 3 (RIPK3). Depending on the upstream activator, RIPK3 can engage different binding partners to induce necrosis. Because the pathway mediated by TNF–TNF receptor ligation is the best characterized, I will use it to illustrate certain core principles.

NECROSIS IS TIGHTLY REGULATED BY RIPK1 UBIQUITINATION

TNF is a pleiotropic cytokine that triggers diverse biological responses including de novo gene expression and cell death by apoptosis or necrosis. As its name implies (tumor necrosis factor), TNF was made famous because it could induce necrosis in solid tumors (Carswell et al. 1975). The strong proinflammatory effects of TNF, however, preclude its use in the clinic. The receptor interacting protein kinase 1 (RIPK1) is a key regulator of the switch between NF-κB activation and cell death induction by TNF. RIPK1 is a serine/threonine kinase that contains a death domain at the carboxyl terminus. It was originally identified in a yeast two-hybrid screen as an interacting partner to the death domain of Fas/CD95/APO-1 and thus was thought to be a crucial adaptor that mediates Fas-induced apoptosis (Stanger et al. 1995). However, it was not until recently that RIPK1 was shown to be part of the signaling complex that mediates Fas-induced apoptosis (Geserick et al. 2009), especially that through membrane-anchored FasL (Morgan et al. 2009). In contrast to the early connotation that RIPK1 is a death inducer, Goeddel and Ting showed that RIPK1 is recruited to the TNF receptor 1 (TNF-R1) and has an important role in activation of the prosurvival transcription factor NF-κB (Hsu et al. 1996; Ting et al. 1996). Accordingly, RIPK1$^{-/-}$ cells are defective for TNF-induced NF-κB activation (Kelliher et al. 1998), although this early finding has recently been challenged (Wong et al. 2010). Nonetheless, it is clear that RIPK1, especially when it is polyubiquitinated, has a prosurvival function.

RIPK1 ubiquitination occurs when it is recruited to the membrane-associated TNF-R1 (p55/p60) complex, termed "Complex I" by Micheau and Tschopp (Micheau and Tschopp 2003). The E3 ligases, cIAP-1 and cIAP-2, are responsible for RIPK1 polyubiquitination. Several reports indicate that K63 ubiquitination at Lys 377 of RIPK1 is essential for recruitment of NEMO and activation of the IKK complex (Ea et al. 2006; Li et al. 2006). However, recent studies using proteomic approaches (Mollah et al. 2007), linkage-specific ubiquitin antibodies (Newton et al. 2008), and ubiquitin replacement strategy (Xu et al. 2009) clearly show that ubiquitination at sites other than K377 as well as other types of ubiquitin linkage can also occur. Polyubiquitinated RIPK1 can promote cell survival through NF-κB activation by recruiting downstream factors such as NEMO and the IKK complex. In addition, it can serve as a gatekeeper to prevent assembly of the death-inducing cytoplasmic "Complex II" as the membrane bound Complex I become internalized (O'Donnell et al. 2007). This explains why Smac mimetics or IAP antagonists, which target cIAP-1, cIAP-2, and XIAP for autoubiquitination and degradation to promote assembly of the ripoptosome (Petersen et al. 2007; Varfolomeev et al. 2007; Vince et al. 2007; Bertrand et al. 2008), can further enhance TNF-induced necrosis. It also is consistent with the fact that siRNA knock-down of the RIPK1 deubiquitinase cylindromatosis (CYLD) can protect cells against necrosis (O'Donnell et al. 2007; Hitomi et al. 2008; Vanlangenakker et al. 2011). Although it has yet to be proven experimentally, it is tempting to speculate that polyubiquitin chains on RIPK1 may sterically prevent binding of downstream necrosis mediators.

CASPASE-8 INHIBITS NECROSIS BY TARGETING ESSENTIAL NECROSOME COMPONENTS

Besides necrosis, blocking RIPK1 polyubiquitination has a similar enhancing effect to apoptosis. This indicates that additional signals are required to discriminate between the two death inducing responses. From the early days, it is clear that tetra-peptide caspase inhibitors or viral caspase inhibitors can greatly enhance cellular necrosis (Vercammen et al. 1998). The molecular basis for this requirement lies in part to the fact that necrosis mediators such as RIPK1, RIPK3, and CYLD are substrates of caspase-8 (Lin et al. 1999; Chan et al. 2003; Feng et al. 2007; O'Donnell et al. 2011). In the case of RIPK1, cleavage of RIPK1 at D324 separates the amino terminal kinase domain from the carboxyl terminus that contains the death domain and the RIP homotypic interaction motif (RHIM) (see below). Therefore, caspase-8 cleavage will prevent RIPK1 from phosphorylating its downstream substrate RIPK3 (Fig. 2).

RIPK3: THE MASTERMIND OF NECROSIS

In addition to promoting necrosis, the kinase function of RIPK1 is also required for ripoptosome assembly, caspase-8 activation and apoptosis induced by TNF and Smac mimetic (Wang et al. 2008; Feoktistova et al. 2011; Tenev et al. 2011). Because RIPK3 is not activated unless caspase-8 is inhibited, distinct RIPK1 substrates are involved in TNF and Smac mimetic induced apoptosis. RIPK3 was identified in several RNA interference screens to be a critical partner of RIPK1 in necrosis (Cho et al. 2009; He et al. 2009). Both RIPK1 and RIPK3 are phosphorylated during necrosis, perhaps through trans-phosphorylation. They formed a tight and stable complex termed the "necrosome" (Vandenabeele et al. 2010). As discussed in the previous section, formation of the necrosome requires caspase inhibition (Cho et al. 2009; He et al. 2009; Zhang et al. 2009), because RIPK1 and RIPK3 are both substrates of caspase-8 (Lin et al. 1999; Feng et al. 2007).

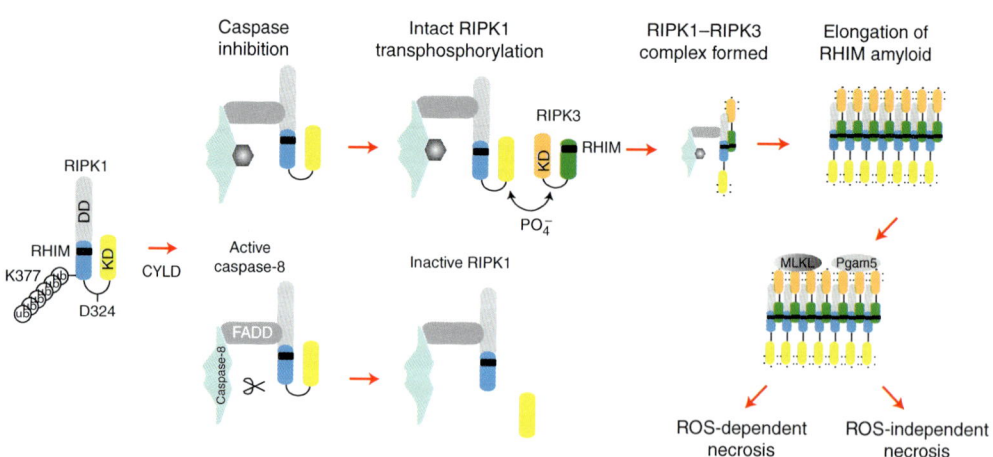

Figure 2. Schematic diagram highlighting the major regulatory steps in the necrosis pathway. De-ubiquitination of RIPK1 by CYLD facilitates assembly of the FADD-caspase-8-RIPK1 ripoptosome. Inhibition of caspase-8 allows RIPK1–RIPK3 necrosome assembly. The negative charge that results from transphosphorylation of RIPK1 and RIPK3 in the kinase domains relieves the inhibition on the RHIM, thereby promoting the amyloidal scaffold to form. This then allows recruitment of downstream RIPK3 substrates including MLKL and Pgam5.

Assembly of the RIPK1–RIPK3 necrosome also requires an intact RHIM. The RHIM is an emerging protein–protein interaction domain found in several other adaptors with functions in innate immune and cell death signaling (Fig. 3). Unlike many other protein–protein interaction motifs, the RHIM homology is ill-defined with only a core of mostly hydrophobic residues that are highly conserved among different RHIMs (Fig. 3). In the case of RIPK1 and RIPK3, the core RHIM tetra-peptide (I/V)Q(I/V)G and its flanking residues are predicted β-sheet structures. Interestingly, recombinant RIPK1 and RIPK3 fragments containing the RHIM assemble in a 1:1 ratio to form large filamentous fibrils. Biophysical studies show that the RHIM core and its flanking sequences behave as amyloid-like fibrils (Li et al. 2012). Mutagenesis of the RHIM core sequences show that this amyloidal assembly is crucial for activation of RIPK1 and RIPK3 kinase activity, necrosome clusters formation, and necrosis induction (Li et al. 2012) (Fig. 2). Intriguingly, truncated RIPK1 and RIPK3 lacking the kinase domains, but not full-length RIP kinases, spontaneously assemble into amyloid-like fibrils. The kinase domains may "mask" the RHIM to prevent amyloid complex formation before activation. Upon phosphorylation of residues in the kinase domains, the RHIM may become unmasked because of negative charge repulsion to allow the RHIM amyloid structures to form (Fig. 2). Because mutations in the RHIM also affect kinase activity, kinase activation and amyloid assembly may be mutually reinforcing reactions. Consistent with this model, both RIPK1 and RIPK3 are phosphorylated at multiple sites in the kinase domains (Degterev et al. 2008; He et al. 2009; Sun et al. 2012).

As I alluded to earlier, RIPK3 can partner with other RHIM-containing adaptors to trigger necrosis. For example, TLR3 and TLR4 induced macrophage necrosis requires the action of RIPK3 and TRIF (He et al. 2011). The viral necrosis inhibitor M45 prevents RIPK3-DAI (DNA activator of interferon) induced necrosis during MCMV infection to inhibit premature host cell death before completion of viral replication (Upton et al. 2008, 2010). Formation of these noncanonical necrosomes also requires RHIM–RHIM interactions. At present, it is unclear if similar amyloidal structures are involved in assembly of these noncanonical necrosomes. However, it is noteworthy that RHIM–RHIM interaction between RIPK1, RIPK3, and TRIF is also required for TLR3 and TLR4 induced NF-κB activation (Meylan et al. 2004; Kaiser et al. 2008; Rebsamen et al. 2009). This raises the interesting and important question of whether the amyloidal assembly of RHIM-containing complexes is strictly involved in necrosis signaling. Because transmission of the necrotic signal requires recruitment of downstream RIPK3 substrates to the necrosome, it seems reasonable to speculate that RHIM-containing amyloid complexes can signal for necrosis and NF-κB in a context dependent manner. This is unlike in neurodegenerative diseases in which amyloid complexes appear to be directly responsible for cell injury.

TRACKING THE ELUSIVE KILLER—DOWNSTREAM EVENTS IN NECROSIS

The mixed lineage kinase domain-like (MLKL) was identified as a kinase substrate of RIPK3 using biochemical purification and RNA interference screen (Sun et al. 2012; Zhao et al. 2012).

Figure 3. RHIM-containing proteins are critical regulators of cell death and innate immune signaling. (*Left*) Schematic diagram of protein adaptors with RHIM. KD, kinase domain; DD, death domain; Z, Z-DNA-binding domain; TIR, Toll/IL-1R domain. The black boxes denote the RHIM. (*Middle*) Sequence alignment of core RHIM sequences from different adaptors. Predicted β sheet is underlined using the Robson-Garnier method.

MLKL has a kinase domain in the carboxyl terminus and an amino terminal coil–coil domain. RIPK3 binds to and phosphorylate the kinase domain of MLKL. Although the name implies that MLKL is a pseudokinase, Liu and colleagues show that mutations of the conserved active site abolished the necrosis inducing function of MLKL (Zhao et al. 2012). Therefore, MLKL may function as the third kinase in the cascade. The significance of MLKL is further revealed through identification of a necrosis inhibitor termed "necrosulfonamide" (NSA) through a small molecule library screen. NSA inhibits necrosis by covalently modifying cysteine 86 that is uniquely present in human MLKL, but not in mouse MLKL (Sun et al. 2012). Thus, unlike necrostatin-1 and its derivatives (Degterev et al. 2005), which target and inhibit RIPK1 kinase activity in both human and mouse cells (Degterev et al. 2008), NSA is a human cell specific inhibitor of necrosis.

Unlike necrostatin-1, NSA did not interfere with RIPK1–RIPK3 necrosome formation. Instead, NSA blocks recruitment of phosphoglycerate mutase family member 5 short (Pgam5s) to the necrosome. Similar to MLKL, both isoforms of Pgam5 Pgam5s and Pgam5L, were identified through biochemical purification as RIPK3 substrates and binding partners (Wang et al. 2012). Pgam5 functions as a phosphatase, not a mutase, as its name might imply (Takeda et al. 2009). Overexpression of Pgam5 causes mitochondrial fragmentation (Lo and Hannink 2008), a phenomenon observed in some necrotic cells. Because inhibition with NSA prevented the recruitment of Pgam5s to RIPK3, MLKL appears to function as a key adaptor that links the RIPK1–RIPK3 necrosome to downstream effectors. Pgam5s has a strong propensity to localize to the mitochondria, suggesting that Pgam5 might tether the RIPK1–RIPK3 necrosome to induce mitochondria fragmentation. Pgam5 dephosphorylates and activates the mitochondria fission factor Drp-1 (Wang et al. 2012), raising the interesting possibility that the necrosome can engage the mitochondria fission machinery to execute necrosis.

The discovery of Pgam5 suggests that similar to apoptosis, the mitochondria may be involved in necrosis signaling. One widely popular view is that reactive oxygen species (ROS) produced by damaged mitochondria may be critical mediators for the execution of necrosis. However, in many cells such as lymphocytes, ROS scavengers do not protect against TNF-induced necrosis (T McQuade and FK-M Chan, unpubl.). This suggests that Pgam5 and other factors that can potentially promote mitochondrial ROS generation may not be universally required for necrosis in all cell types.

Although ROS is not required for necrosis in lymphocytes, ROS scavengers are effective inhibitors of necrosis in many cell types. There are two major sources of cellular ROS: those generated at the mitochondria as a result of oxidative phosphorylation, and those that are generated by membrane associated NADPH oxidases (NOX). Evidence that support a role for both sources of ROS in necrosis can be found (Goossens et al. 1999). Some reports suggest that direct activation of plasma membrane associated NADPH oxidases is a major source of ROS in necrotic cells (Kim et al. 2007; Yazdanpanah et al. 2009). In contrast, ROS production was abrogated in cells lacking essential necrosome components such as RIPK3 and MLKL (Cho et al. 2009; Sun et al. 2012). Because the necrosome is formed in the cytosol subsequent to dissolution of the membrane signaling complex, it argues against a role for the plasma membrane associated NOX in ROS generation. Paradoxically, necrosis induced by hydrogen peroxide, which generates ROS once it enters the cell, is impaired in the absence of Pgam5 (Wang et al. 2012). These results suggest that besides being a product of necrosome activation, ROS may function in a feed-forward manner to promote necrosome function. It is clear that more work is needed to determine the precise source and role of ROS in necrosis.

NECROSIS IN PATHOGEN-INDUCED INFLAMMATION

It is well known to pathologists that necrosis is often associated with inflammation. As discussed in previous sections, protein adaptors that contain RHIM domains share functions in cell death and/or innate immune signaling. This molecular signature suggests that the RIP kinases and necrosis have broad roles in innate immunity and inflammation. The first example that highlights this paradigm comes from study of host defense against vaccinia virus infection. Vaccinia virus, like other poxviruses, encodes many immune evasion genes. One of these gene products, B13R or Spi2, is a serpin that inhibits caspase-1 and caspase-8, and is similar functionally to CrmA from cowpox virus (Zhou et al. 1997). Despite the inhibition of caspase-8, vaccinia virus infected cells are still sensitive to the cytotoxic effect of TNF (Li and Beg 2000). TNF-induced cell death of vaccinia virus infected cells show morphology that resembles necrosis and requires intact RIPK1 and RIPK3 functions (Chan et al. 2003; Cho et al. 2009). Consistent with these in vitro infection results, RIPK3$^{-/-}$ mice show reduced necrosis and inflammation, greatly increased viral replication, and succumb to virus 4–5 d postinfection (Cho et al. 2009). Thus, in poxvirus infection, host cell necrosis during the innate phase of the immune reaction limits the viral factory. This is crucial for host survival before virus-specific T and B lymphocytes are mobilized in high enough number to control the virus.

The sensitization of vaccinia virus infected cells to TNF-induced necrosis is reminiscent of the effect of peptide-based caspase inhibitors. This suggests that other viruses that encode caspase inhibitors may similarly prime cells to necrosis. Murine cytomegalovirus (MCMV) encodes three different types of viral cell death inhibitors, vICA (inhibitor of caspase-8-induced apoptosis), vMIA (mitochondria inhibitor of apoptosis) and vIRA (inhibitor of RIP activation) (reviewed in Mocarski et al. 2012). Productive infection and replication of viral progenies require the action of all three inhibitors. vIRA or M45 encodes a protein with homology to ribonucleotide reductase (Brune et al. 2001). However, it has no enzymatic activity (Lembo et al. 2004). Rather, the RHIM domain within M45 is crucial for binding to cellular factors that also contain RHIM domains, including RIPK1, RIPK3, and DAI (Kaiser et al. 2008; Upton et al. 2008, 2012; Rebsamen et al. 2009). Recombinant virus that encodes a defective vIRA with tetra-alanine substitutions of the RHIM fails to establish productive infection in cells and in mice because of premature cell death by necrosis. Significantly, productive infection is restored when the RHIM mutant MCMV infects RIPK3$^{-/-}$ mice (Upton et al. 2010). The necrotic cell death induced on mutant MCMV infection is not driven by TNF or RIPK1. Instead, RIPK3 pairs with another RHIM-containing adaptor DAI to induce necrosis in the absence of a functional M45/vIRA (Upton et al. 2012). Similar to poxviruses, necrosis is a defense mechanism of the host against MCMV. However, in contrast to vaccinia virus, MCMV has developed an effective strategy to circumvent host cell necrosis.

The MCMV results tell us that active suppression of host cell necrosis is an important survival mechanism of the virus. This suggests that other viruses may also encode necrosis inhibitors. Viral FLIPs are orthologs of cellular caspase-8/10 that contain tandem death effector domains, but lack the enzymatic domains. They were identified first as caspase and apoptosis inhibitors (Bertin et al. 1997; Hu et al. 1997; Thome et al. 1997). A subset of vFLIPs, namely MC159 from Molluscum contagiosum virus and E8 from Equine herpesvirus, were found to also inhibit TNF-induced programmed necrosis (Chan et al. 2003). In contrast to M45, which inhibits necrosis through RHIM-mediated interaction with RIPK3 (Upton et al. 2010), the molecular basis by which vFLIPs inhibit necrosis is not fully understood. Nevertheless, it appears that viral inhibition of necrosis may be a common immune evasion strategy used by many viruses.

NECROSIS IN STERILE INFLAMMATION

Besides pathogen-induced injury and inflammation, necrosis also plays key roles in sterile injury. It is no surprise that many necrosis-induced sterile injury models involve experimental inhibition of caspases or its upstream adaptor FADD. For instance, retinal detachment induced photoreceptor necrosis in the presence of caspase inhibitor is blocked in RIPK3$^{-/-}$ cells

(Trichonas et al. 2010). A large number of studies have been performed using caspase-8$^{-/-}$ or FADD$^{-/-}$ mice. Germline inactivation of these genes results in embryonic lethality that can be rescued by deletion of RIPK1 or RIPK3 (Kaiser et al. 2011; Oberst et al. 2011; Zhang et al. 2011; Dillon et al. 2012). Deletion of these apoptosis mediators in lymphocytes causes defective antigen driven clonal expansion owing to massive necrosis. Again, the defect can be fixed by deletion of RIPK1 or RIPK3 (Ch'en et al. 2011; Lu et al. 2011; Zhang et al. 2011). Interestingly, mice deficient in FADD/caspase-8 and RIPK3 developed a lymphoproliferative disease resembling that of lpr mice or the human Autoimmune Lymphoproliferative Syndromes (ALPS), which are caused by mutations in Fas/CD95/APO-1 (Su and Lenardo 2008). Hence, these genetic experiments show that Fas-induced caspase-dependent apoptosis and RIPK1/RIPK3-dependent necrosis coordinately regulate lymphocyte homeostasis to prevent autoimmune lymphoproliferation. Keratinocyte- or intestinal epithelium-specific deletion of results in severe spontaneous inflammation in the respective tissues that can be corrected by deletion of RIPK3 (Kovalenko et al. 2009; Bonnet et al. 2011; Gunther et al. 2011; Welz et al. 2011). These results reveal a surprising and unexpected function for FADD and caspase-8: that they are prosurvival factors that suppress the deleterious effects of necrosis, possibly by cleavage and inactivation of RIPK1 and RIPK3.

Sterile necrotic injury can also occur in animals with normal expression of FADD and caspase-8. Repeated injections of cerulein cause acute pancreatitis marked by loss of acinar cells in the pancreas. RIPK3$^{-/-}$ mice are protected from cerulein-induced pancreatitis and acinar cell necrosis (He et al. 2009; Zhang et al. 2009). Myocardial infarction, ischemia induced brain injury, and renal ischemia/reperfusion injury all show hallmarks of necrosis. Administration of the RIPK1 inhibitor necrostatin-1 significantly ameliorates the tissue damage in these injury models (Degterev et al. 2005; Lim et al. 2007; Smith et al. 2007; Northington et al. 2011; Linkermann et al. 2012). Although TNF and other inflammatory cytokines are often elevated in these cases, it is not clear if they are directly responsible for the necrotic cell death. It is plausible that necrosis can ensue without engaging cell surface death receptors in some of these situations. This will be analogous to the "intrinsic" apoptosis induced in response to genotoxic stress.

THERAPEUTIC OPPORTUNITIES

As discussed in the previous section, necrostatin-1 and its derivatives have shown promise in treating acute injury in several models. This raises the possibility that interfering with necrosis can be an effective strategy to treat certain acute pathologies. Crucially, RIPK1$^{-/-}$ mice suffer from postnatal lethality (Kelliher et al. 1998). In addition, RIPK1 has important functions in innate immune signaling (Meylan et al. 2004; Vivarelli et al. 2004) and DNA damage response (Hur et al. 2003; Janssens et al. 2005; Biton and Ashkenazi 2011; Yang et al. 2011). This suggests that long-term use of RIPK1 inhibitors such as necrostatin-1 may be problematic. However, kinases are generally good drug targets and RIPK3-specific inhibitors are attractive therapeutic agents. Several issues need to be taken into account when necrosis is considered as potential therapeutic targets. First, there appears to be a "yin–yang" relationship between apoptosis and necrosis. As we have discussed already, blocking apoptosis often sensitizes cells to necrosis. Inhibiting necrosis appears to have fewer deleterious effects. For example, RIPK3$^{-/-}$ mice are developmentally normal (Newton et al. 2004). However, inhibiting necrosis can exacerbate apoptosis, as recently shown in cFLIP$^{-/-}$RIPK3$^{-/-}$ mice (Dillon et al. 2012). Secondly, RIPK3 has recently been shown to participate in inflammasome activation and inflammatory cytokine production (Vince et al. 2012). Finally, an earlier report shows that RIPK3 was highly induced during cutaneous wound repair (Adams et al. 2007). Thus, RIPK3 inhibitors may impair immune functions and wound healing. Other therapeutic opportunities include targeting downstream effectors such as MLKL. In this light, it will be of great interest to determine the utility of NSA in treating acute and chronic diseases.

CONCLUDING REMARKS

We have witnessed a renaissance in necrosis research in the last decade. The discovery of key upstream regulators and demonstration of their functional relevance in development and disease pathologies means that necrosis is no longer an afterthought for biologists. Much is still to be performed to fully understand the biochemical pathway of necrosis. As more investigators join the race to decipher the pathway, one can anticipate with confidence that there will be more exciting discoveries and surprises along the way.

ACKNOWLEDGMENTS

I thank former and present members of the Chan laboratory and many colleagues for discussion. This work is supported by grants from the NIH (AI083497 and AI088502).

REFERENCES

Adams S, Pankow S, Werner S, Munz B. 2007. Regulation of NF-κB activity and keratinocyte differentiation by the RIP4 protein: Implications for cutaneous wound repair. *J Invest Dermatol* **127**: 538–544.

Berry DL, Baehrecke EH. 2007. Growth arrest and autophagy are required for salivary gland cell degradation in *Drosophila*. *Cell* **131**: 1137–1148.

Bertin J, Armstrong RC, Ottilie S, Martin DA, Wang Y, Banks S, Wang GH, Senkevich TG, Alnemri ES, Moss B, et al. 1997. Death effector domain-containing herpesvirus and poxvirus proteins inhibit both Fas- and TNFR1-induced apoptosis. *Proc Natl Acad Sci* **94**: 1172–1176.

Bertrand MJ, Milutinovic S, Dickson KM, Ho WC, Boudreault A, Durkin J, Gillard JW, Jaquith JB, Morris SJ, Barker PA. 2008. cIAP1 and cIAP2 facilitate cancer cell survival by functioning as E3 ligases that promote RIP1 ubiquitination. *Mol Cell* **30**: 689–700.

Biton S, Ashkenazi A. 2011. NEMO and RIP1 control cell fate in response to extensive DNA damage via TNF-α feedforward signaling. *Cell* **145**: 92–103.

Bonnet MC, Preukschat D, Welz PS, van Loo G, Ermolaeva MA, Bloch W, Haase I, Pasparakis M. 2011. The adaptor protein FADD protects epidermal keratinocytes from necroptosis in vivo and prevents skin inflammation. *Immunity* **35**: 572–582.

Brune W, Menard C, Heesemann J, Koszinowski UH. 2001. A ribonucleotide reductase homolog of cytomegalovirus and endothelial cell tropism. *Science* **291**: 303–305.

Carswell EA, Old LJ, Kassel RL, Green S, Fiore N, Williamson B. 1975. An endotoxin-induced serum factor that causes necrosis of tumors. *Proc Natl Acad Sci* **72**: 3666–3670.

Chan FK, Shisler J, Bixby JG, Felices M, Zheng L, Appel M, Orenstein J, Moss B, Lenardo MJ. 2003. A role for tumor necrosis factor receptor-2 and receptor-interacting protein in programmed necrosis and antiviral responses. *J Biol Chem* **278**: 51613–51621.

Ch'en IL, Beisner DR, Degterev A, Lynch C, Yuan J, Hoffmann A, Hedrick SM. 2008. Antigen-mediated T cell expansion regulated by parallel pathways of death. *Proc Natl Acad Sci* **105**: 17463–17468.

Ch'en IL, Tsau JS, Molkentin JD, Komatsu M, Hedrick SM. 2011. Mechanisms of necroptosis in T cells. *J Exp Med* **208**: 633–641.

Cho YS, Challa S, Moquin D, Genga R, Ray TD, Guildford M, Chan FK. 2009. Phosphorylation-driven assembly of the RIP1-RIP3 complex regulates programmed necrosis and virus-induced inflammation. *Cell* **137**: 1112–1123.

Cho Y, McQuade T, Zhang HB, Zhang JK, Chan FKM. 2011. RIP1-dependent and independent effects of necrostatin-1 in necrosis and T cell activation. *Plos ONE* **6**.

Degterev A, Huang Z, Boyce M, Li Y, Jagtap P, Mizushima N, Cuny GD, Mitchison TJ, Moskowitz MA, Yuan J. 2005. Chemical inhibitor of nonapoptotic cell death with therapeutic potential for ischemic brain injury. *Nat Chem Biol* **1**: 112–119.

Degterev A, Hitomi J, Germscheid M, Ch'en IL, Korkina O, Teng X, Abbott D, Cuny GD, Yuan C, Wagner G, et al. 2008. Identification of RIP1 kinase as a specific cellular target of necrostatins. *Nat Chem Biol* **4**: 313–321.

Dillon CP, Oberst A, Weinlich R, Janke LJ, Kang TB, Ben-Moshe T, Mak TW, Wallach D, Green DR. 2012. Survival function of the FADD-CASPASE-8-cFLIPL complex. *Cell Rep* **1**: 401–407.

Ea CK, Deng L, Xia ZP, Pineda G, Chen ZJ. 2006. Activation of IKK by TNFα requires site-specific ubiquitination of RIP1 and polyubiquitin binding by NEMO. *Mol Cell* **22**: 245–257.

Feng S, Yang Y, Mei Y, Ma L, Zhu DE, Hoti N, Castanares M, Wu M. 2007. Cleavage of RIP3 inactivates its caspase-independent apoptosis pathway by removal of kinase domain. *Cell Signal* **19**: 2056–2067.

Feoktistova M, Geserick P, Kellert B, Dimitrova DP, Langlais C, Hupe M, Cain K, MacFarlane M, Hacker G, Leverkus M. 2011. cIAPs block Ripoptosome formation, a RIP1/caspase-8 containing intracellular cell death complex differentially regulated by cFLIP isoforms. *Mol Cell* **43**: 449–463.

Geserick P, Hupe M, Moulin M, Wong WW, Feoktistova M, Kellert B, Gollnick H, Silke J, Leverkus M. 2009. Cellular IAPs inhibit a cryptic CD95-induced cell death by limiting RIP1 kinase recruitment. *J Cell Biol* **187**: 1037–1054.

Goossens V, De Vos K, Vercammen D, Steemans M, Vancompernolle K, Fiers W, Vandenabeele P, Grooten J. 1999. Redox regulation of TNF signaling. *Biofactors* **10**: 145–156.

Gunther C, Martini E, Wittkopf N, Amann K, Weigmann B, Neumann H, Waldner MJ, Hedrick SM, Tenzer S, Neurath MF, et al. 2011. Caspase-8 regulates TNF-α-induced epithelial necroptosis and terminal ileitis. *Nature* **477**: 335–339.

He S, Wang L, Miao L, Du F, Zhao L, Wang X. 2009. Receptor interacting protein kinase-3 determines cellular necrotic response to TNF-α. *Cell* **137:** 1100–1111.

He S, Liang Y, Shao F, Wang X. 2011. Toll-like receptors activate programmed necrosis in macrophages through a receptor-interacting kinase-3-mediated pathway. *Proc Natl Acad Sci* **108:** 20054–20059.

Hitomi J, Christofferson DE, Ng A, Yao J, Degterev A, Xavier RJ, Yuan J. 2008. Identification of a molecular signaling network that regulates a cellular necrotic cell death pathway. *Cell* **135:** 1311–1323.

Hsu H, Huang J, Shu HB, Baichwal V, Goeddel DV. 1996. TNF-dependent recruitment of the protein kinase RIP to the TNF receptor-1 signaling complex. *Immunity* **4:** 387–396.

Hu S, Vincenz C, Buller M, Dixit VM. 1997. A novel family of viral death effector domain-containing molecules that inhibit both CD-95- and tumor necrosis factor receptor-1-induced apoptosis. *J Biol Chem* **272:** 9621–9624.

Hur GM, Lewis J, Yang Q, Lin Y, Nakano H, Nedospasov S, Liu ZG. 2003. The death domain kinase RIP has an essential role in DNA damage-induced NF-κB activation. *Genes Dev* **17:** 873–882.

Janssens S, Tinel A, Lippens S, Tschopp J. 2005. PIDD mediates NF-κB activation in response to DNA damage. *Cell* **123:** 1079–1092.

Kaiser WJ, Upton JW, Mocarski ES. 2008. Receptor-interacting protein homotypic interaction motif-dependent control of NF-κB activation via the DNA-dependent activator of IFN regulatory factors. *J Immunol* **181:** 6427–6434.

Kaiser WJ, Upton JW, Long AB, Livingston-Rosanoff D, Daley-Bauer LP, Hakem R, Caspary T, Mocarski ES. 2011. RIP3 mediates the embryonic lethality of caspase-8-deficient mice. *Nature* **471:** 368–372.

Kelliher MA, Grimm S, Ishida Y, Kuo F, Stanger BZ, Leder P. 1998. The death domain kinase RIP mediates the TNF-induced NF-κB signal. *Immunity* **8:** 297–303.

Kim YS, Morgan MJ, Choksi S, Liu ZG. 2007. TNF-induced activation of the Nox1 NADPH oxidase and its role in the induction of necrotic cell death. *Mol Cell* **26:** 675–687.

Kovalenko A, Kim JC, Kang TB, Rajput A, Bogdanov K, Dittrich-Breiholz O, Kracht M, Brenner O, Wallach D. 2009. Caspase-8 deficiency in epidermal keratinocytes triggers an inflammatory skin disease. *J Exp Med* **206:** 2161–2177.

Lembo D, Donalisio M, Hofer A, Cornaglia M, Brune W, Koszinowski U, Thelander L, Landolfo S. 2004. The ribonucleotide reductase R1 homolog of murine cytomegalovirus is not a functional enzyme subunit but is required for pathogenesis. *J Virol* **78:** 4278–4288.

Li M, Beg AA. 2000. Induction of necrotic-like cell death by tumor necrosis factor α and caspase inhibitors: Novel mechanism for killing virus-infected cells. *J Virol* **74:** 7470–7477.

Li H, Kobayashi M, Blonska M, You Y, Lin X. 2006. Ubiquitination of RIP is required for tumor necrosis factor α-induced NF-κB activation. *J Biol Chem* **281:** 13636–13643.

Li J, McQuade T, Siemer AB, Napetschnig J, Moriwaki K, Hsiao YS, Damko E, Moquin D, Walz T, McDermott A, et al. 2012. The RIP1/RIP3 necrosome forms a functional amyloid signaling complex required for programmed necrosis. *Cell* **150:** 339–350.

Lim SY, Davidson SM, Mocanu MM, Yellon DM, Smith CC. 2007. The cardioprotective effect of necrostatin requires the cyclophilin-D component of the mitochondrial permeability transition pore. *Cardiovasc Drugs Ther* **21:** 467–469.

Lin Y, Devin A, Rodriguez Y, Liu ZG. 1999. Cleavage of the death domain kinase RIP by caspase-8 prompts TNF-induced apoptosis. *Genes Dev* **13:** 2514–2526.

Linkermann A, Brasen JH, Himmerkus N, Liu S, Huber TB, Kunzendorf U, Krautwald S. 2012. Rip1 (receptor-interacting protein kinase 1) mediates necroptosis and contributes to renal ischemia/reperfusion injury. *Kidney Int* **81:** 751–761.

Lo SC, Hannink M. 2008. PGAM5 tethers a ternary complex containing Keap1 and Nrf2 to mitochondria. *Exp Cell Res* **314:** 1789–1803.

Lu JV, Weist BM, van Raam BJ, Marro BS, Nguyen LV, Srinivas P, Bell BD, Luhrs KA, Lane TE, Salvesen GS, et al. 2011. Complementary roles of Fas-associated death domain (FADD) and receptor interacting protein kinase-3 (RIPK3) in T-cell homeostasis and antiviral immunity. *Proc Natl Acad Sci* **108:** 15312–15317.

Majno G, Joris I. 1995. Apoptosis, oncosis, and necrosis. An overview of cell death. *Am J Pathol* **146:** 3–15.

McComb S, Cheung HH, Korneluk RG, Wang S, Krishnan L, Sad S. 2012. cIAP1 and cIAP2 limit macrophage necroptosis by inhibiting Rip1 and Rip3 activation. *Cell Death Differ* doi: 10.1038/cdd.2012.59.

Meylan E, Burns K, Hofmann K, Blancheteau V, Martinon F, Kelliher M, Tschopp J. 2004. RIP1 is an essential mediator of Toll-like receptor 3-induced NF-κB activation. *Nat Immunol* **5:** 503–507.

Micheau O, Tschopp J. 2003. Induction of TNF receptor I-mediated apoptosis via two sequential signaling complexes. *Cell* **114:** 181–190.

Mocarski ES, Upton JW, Kaiser WJ. 2012. Viral infection and the evolution of caspase 8-regulated apoptotic and necrotic death pathways. *Nat Rev Immunol* **12:** 79–88.

Mollah S, Arnott D, Phung Q, Wertz I, Dixit V, Kayagaki N, Lill J. 2007. Targeted mass spectrometric strategy for global mapping of ubiquitination on proteins. *Mol Cell Proteomics* **6:** 58–58.

Morgan MJ, Kim YS, Liu ZG. 2009. Membrane-bound Fas ligand requires RIP1 for efficient activation of caspase-8 within the death-inducing signaling complex. *J Immunol* **183:** 3278–3284.

Newton K, Sun X, Dixit VM. 2004. Kinase RIP3 is dispensable for normal NF-κBs, signaling by the B-cell and T-cell receptors, tumor necrosis factor receptor 1, and Toll-like receptors 2 and 4. *Mol Cell Biol* **24:** 1464–1469.

Newton K, Matsumoto ML, Wertz IE, Kirkpatrick DS, Lill JR, Tan J, Dugger D, Gordon N, Sidhu SS, Fellouse FA, et al. 2008. Ubiquitin chain editing revealed by polyubiquitin linkage-specific antibodies. *Cell* **134:** 668–678.

Northington FJ, Chavez-Valdez R, Graham EM, Razdan S, Gauda EB, Martin LJ. 2011. Necrostatin decreases oxidative damage, inflammation, and injury after neonatal HI. *J Cereb Blood Flow Metab* **31:** 178–189.

Oberst A, Dillon CP, Weinlich R, McCormick LL, Fitzgerald P, Pop C, Hakem R, Salvesen GS, Green DR. 2011. Catalytic activity of the caspase-8-FLIP(L) complex inhibits RIPK3-dependent necrosis. *Nature* **471:** 363–367.

O'Donnell MA, Legarda-Addison D, Skountzos P, Yeh WC, Ting AT. 2007. Ubiquitination of RIP1 regulates an NF-κB-independent cell-death switch in TNF signaling. *Curr Biol* **17:** 418–424.

O'Donnell MA, Perez-Jimenez E, Oberst A, Ng A, Massoumi R, Xavier R, Green DR, Ting AT. 2011. Caspase 8 inhibits programmed necrosis by processing CYLD. *Nat Cell Biol* **13:** 1437–1442.

Petersen SL, Wang L, Yalcin-Chin A, Li L, Peyton M, Minna J, Harran P, Wang X. 2007. Autocrine TNFα signaling renders human cancer cells susceptible to Smac-mimetic-induced apoptosis. *Cancer Cell* **12:** 445–456.

Rebsamen M, Heinz LX, Meylan E, Michallet MC, Schroder K, Hofmann K, Vazquez J, Benedict CA, Tschopp J. 2009. DAI/ZBP1 recruits RIP1 and RIP3 through RIP homotypic interaction motifs to activate NF-κB. *EMBO Rep* **10:** 916–922.

Smith CC, Davidson SM, Lim SY, Simpkin JC, Hothersall JS, Yellon DM. 2007. Necrostatin: A potentially novel cardioprotective agent? *Cardiovasc Drugs Ther* **21:** 227–233.

Stanger BZ, Leder P, Lee TH, Kim E, Seed B. 1995. RIP: A novel protein containing a death domain that interacts with Fas/APO-1 (CD95) in yeast and causes cell death. *Cell* **81:** 513–523.

Su HC, Lenardo MJ. 2008. Genetic defects of apoptosis and primary immunodeficiency. *Immunol Allergy Clin North Am* **28:** 329–351, ix.

Sun L, Wang H, Wang Z, He S, Chen S, Liao D, Wang L, Yan J, Liu W, Lei X, et al. 2012. Mixed lineage kinase domain-like protein mediates necrosis signaling downstream of RIP3 kinase. *Cell* **148:** 213–227.

Takeda K, Komuro Y, Hayakawa T, Oguchi H, Ishida Y, Murakami S, Noguchi T, Kinoshita H, Sekine Y, Iemura S, et al. 2009. Mitochondrial phosphoglycerate mutase 5 uses alternate catalytic activity as a protein serine/threonine phosphatase to activate ASK1. *Proc Natl Acad Sci* **106:** 12301–12305.

Tenev T, Bianchi K, Darding M, Broemer M, Langlais C, Wallberg F, Zachariou A, Lopez J, MacFarlane M, Cain K, et al. 2011. The Ripoptosome, a signaling platform that assembles in response to genotoxic stress and loss of IAPs. *Mol Cell* **43:** 432–448.

Thome M, Schneider P, Hofmann K, Fickenscher H, Meinl E, Neipel F, Mattmann C, Burns K, Bodmer JL, Schroter M, et al. 1997. Viral FLICE-inhibitory proteins (FLIPs) prevent apoptosis induced by death receptors. *Nature* **386:** 517–521.

Ting AT, Pimentel-Muinos FX, Seed B. 1996. RIP mediates tumor necrosis factor receptor 1 activation of NF-κB but not Fas/APO-1-initiated apoptosis. *EMBO J* **15:** 6189–6196.

Trichonas G, Murakami Y, Thanos A, Morizane Y, Kayama M, Debouck CM, Hisatomi T, Miller JW, Vavvas DG. 2010. Receptor interacting protein kinases mediate retinal detachment-induced photoreceptor necrosis and compensate for inhibition of apoptosis. *Proc Natl Acad Sci* **107:** 21695–21700.

Tu HC, Ren D, Wang GX, Chen DY, Westergard TD, Kim H, Sasagawa S, Hsieh JJ, Cheng EH. 2009. The p53–cathepsin axis cooperates with ROS to activate programmed necrotic death upon DNA damage. *Proc Natl Acad Sci* **106:** 1093–1098.

Upton JW, Kaiser WJ, Mocarski ES. 2008. Cytomegalovirus M45 cell death suppression requires receptor-interacting protein (RIP) homotypic interaction motif (RHIM)-dependent interaction with RIP1. *J Biol Chem* **283:** 16966–16970.

Upton JW, Kaiser WJ, Mocarski ES. 2010. Virus inhibition of RIP3-dependent necrosis. *Cell Host Microbe* **7:** 302–313.

Upton JW, Kaiser WJ, Mocarski ES. 2012. DAI/ZBP1/DLM-1 complexes with RIP3 to mediate virus-induced programmed necrosis that is targeted by murine cytomegalovirus vIRA. *Cell Host Microbe* **11:** 290–297.

Vandenabeele P, Galluzzi L, Vanden Berghe T, Kroemer G. 2010. Molecular mechanisms of necroptosis: an ordered cellular explosion. *Nat Rev Mol Cell Biol* **11:** 700–714.

Vanden Berghe T, Vanlangenakker N, Parthoens E, Deckers W, Devos M, Festjens N, Guerin CJ, Brunk UT, Declercq W, Vandenabeele P. 2010. Necroptosis, necrosis and secondary necrosis converge on similar cellular disintegration features. *Cell Death Differ* **17:** 922–930.

Vanlangenakker N, Vanden Berghe T, Bogaert P, Laukens B, Zobel K, Deshayes K, Vucic D, Fulda S, Vandenabeele P, Bertrand MJ. 2011. cIAP1 and TAK1 protect cells from TNF-induced necrosis by preventing RIP1/RIP3-dependent reactive oxygen species production. *Cell Death Differ* **18:** 656–665.

Varfolomeev E, Blankenship JW, Wayson SM, Fedorova AV, Kayagaki N, Garg P, Zobel K, Dynek JN, Elliott LO, Wallweber HJ, et al. 2007. IAP antagonists induce autoubiquitination of c-IAPs, NF-κB activation, and TNFα-dependent apoptosis. *Cell* **131:** 669–681.

Vercammen D, Beyaert R, Denecker G, Goossens V, Van Loo G, Declercq W, Grooten J, Fiers W, Vandenabeele P. 1998. Inhibition of caspases increases the sensitivity of L929 cells to necrosis mediated by tumor necrosis factor. *J Exp Med* **187:** 1477–1485.

Vince JE, Wong WW, Khan N, Feltham R, Chau D, Ahmed AU, Benetatos CA, Chunduru SK, Condon SM, McKinlay M, et al. 2007. IAP antagonists target cIAP1 to induce TNFα-dependent apoptosis. *Cell* **131:** 682–693.

Vince JE, Wong WW, Gentle I, Lawlor KE, Allam R, O'Reilly L, Mason K, Gross O, Ma S, Guarda G, et al. 2012. Inhibitor of apoptosis proteins limit RIP3 kinase-dependent interleukin-1 activation. *Immunity* **36:** 215–227.

Vivarelli MS, McDonald D, Miller M, Cusson N, Kelliher M, Geha RS. 2004. RIP links TLR4 to Akt and is essential for cell survival in response to LPS stimulation. *J Exp Med* **200:** 399–404.

von Recklinghausen F. 1910. *Untersuchungen über rachitis und osteomalazie*. Verlag Gustav Fischer Jena, Stuttgart.

Wang L, Du F, Wang X. 2008. TNF-α induces two distinct caspase-8 activation pathways. *Cell* **133:** 693–703.

Wang Z, Jiang H, Chen S, Du F, Wang X. 2012. The mitochondrial phosphatase PGAM5 functions at the convergence point of multiple necrotic death pathways. *Cell* **148**: 228–243.

Welz PS, Wullaert A, Vlantis K, Kondylis V, Fernandez-Majada V, Ermolaeva M, Kirsch P, Sterner-Kock A, van Loo G, Pasparakis M. 2011. FADD prevents RIP3-mediated epithelial cell necrosis and chronic intestinal inflammation. *Nature* **477**: 330–334.

Wong WW, Gentle IE, Nachbur U, Anderton H, Vaux DL, Silke J. 2010. RIPK1 is not essential for TNFR1-induced activation of NF-κB. *Cell Death Differ* **17**: 482–487.

Xu M, Skaug B, Zeng W, Chen ZJ. 2009. A ubiquitin replacement strategy in human cells reveals distinct mechanisms of IKK activation by TNFα and IL-1β. *Mol Cell* **36**: 302–314.

Yang Y, Xia F, Hermance N, Mabb A, Simonson S, Morrissey S, Gandhi P, Munson M, Miyamoto S, Kelliher MA. 2011. A cytosolic ATM/NEMO/RIP1 complex recruits TAK1 to mediate the NF-κB and p38 mitogen-activated protein kinase (MAPK)/MAPK-activated protein 2 responses to DNA damage. *Mol Cell Biol* **31**: 2774–2786.

Yazdanpanah B, Wiegmann K, Tchikov V, Krut O, Pongratz C, Schramm M, Kleinridders A, Wunderlich T, Kashkar H, Utermohlen O, et al. 2009. Riboflavin kinase couples TNF receptor 1 to NADPH oxidase. *Nature* **460**: 1159–1163.

Yu L, Alva A, Su H, Dutt P, Freundt E, Welsh S, Baehrecke EH, Lenardo MJ. 2004. Regulation of an ATG7-beclin 1 program of autophagic cell death by caspase-8. *Science* **304**: 1500–1502.

Zhang DW, Shao J, Lin J, Zhang N, Lu BJ, Lin SC, Dong MQ, Han J. 2009. RIP3, an energy metabolism regulator that switches TNF-induced cell death from apoptosis to necrosis. *Science* **325**: 332–336.

Zhang H, Zhou X, McQuade T, Li J, Chan FK, Zhang J. 2011. Functional complementation between FADD and RIP1 in embryos and lymphocytes. *Nature* **471**: 373–376.

Zhao J, Jitkaew S, Cai Z, Choksi S, Li Q, Luo J, Liu ZG. 2012. Mixed lineage kinase domain-like is a key receptor interacting protein 3 downstream component of TNF-induced necrosis. *Proc Natl Acad Sci* **109**: 5322–5327.

Zhou Q, Snipas S, Orth K, Muzio M, Dixit VM, Salvesen GS. 1997. Target protease specificity of the viral serpin CrmA. Analysis of five caspases. *J Biol Chem* **272**: 7797–7800.

Regulation and Function of Autophagy during Cell Survival and Cell Death

Gautam Das[1], Bhupendra V. Shravage[1], and Eric H. Baehrecke

Department of Cancer Biology, University of Massachusetts Medical School, Worcester, Massachusetts 01605

Correspondence: eric.baehrecke@umassmed.edu

Autophagy is an important catabolic process that delivers cytoplasmic material to the lysosome for degradation. Autophagy promotes cell survival by elimination of damaged organelles and proteins aggregates, as well as by facilitating bioenergetic homeostasis. Although autophagy has been considered a cell survival mechanism, recent studies have shown that autophagy can promote cell death. The core mechanisms that control autophagy are conserved between yeast and humans, but animals also possess genes that regulate autophagy that are not present in yeast. These regulatory differences may be explained by the need to control autophagy in a cell context-specific manner in multicellular animals, such as during cell survival and cell death. Autophagy was thought to be a bulk cytoplasmic degradation mechanism, but recent studies have shown that specific cargo is recruited for degradation. This suggests the possibility that either cell survival or death may be regulated by selective autophagic clearance of cytoplasmic material. Here we summarize the mechanisms that regulate autophagy and how they may contribute to cell survival and death.

Autophagy (self-eating) is an evolutionarily conserved catabolic process that is used to deliver cytoplasmic materials, including organelles and proteins, to the lysosome for degradation. Three types of autophagy have been described, including macroautophagy, microautophagy, and chaperone-mediated autophagy (Mizushima and Komatsu 2011). Although macroautophagy involves the fusion of the double membrane autophagosome and lysosomes, microautophagy is poorly understood and thought to involve direct uptake of material by the lysosome via a process that appears similar to pinocytosis. By contrast, chaperone-mediated autophagy is a biochemical mechanism to import proteins into the lysosome; it depends on a signature sequence and interaction with protein chaperones. Here we will focus on macroautophagy (hereafter called autophagy) because of our knowledge of this process in cell survival and cell death.

Autophagy was likely first observed when electron microscopy was used to observe "dense bodies" containing mitochondria in mouse kidneys (Clark 1957). Five years later, it was reported that rat hepatocytes exposed to glucagon possessed membrane-bound vesicles that were rich in mitochondria and endoplasmic reticulum

---

[1]G.D. and B.V.S. contributed equally to this manuscript.

Copyright © 2013 Cold Spring Harbor Laboratory Press; all rights reserved
Cite this article as *Cold Spring Harb Perspect Biol* doi: 10.1101/cshperspect.a008813

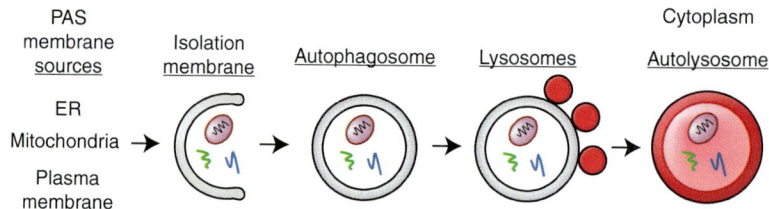

Figure 1. Macroautophagy (autophagy) delivers cytoplasmic cargo to lysosomes for degradation and involves membrane formation and fusion. The isolation membrane is initiated from a membrane source known as the phagophore assembly site (PAS). The isolation membrane surrounds cargo, including organelles and proteins, to form a double membrane autophagosome. Autophagosomes fuse with lysosomes to form autolysosomes in which the cargo is degraded by lysosomal hydrolases.

(Ashford and Porter 1962). Almost simultaneously, it was shown that these membrane-bound vesicles contained lysosomal hydrolases (Novikoff and Essner 1962). In 1965 de Duve coined the term "autophagy" (Klionsky 2008).

The delivery of cytoplasmic material to the lysosome by autophagy involves membrane formation and fusion events (Fig. 1). First an isolation membrane, also known as a phagophore, must be initiated from a membrane source known as the phagophore assembly site (PAS). de Duve suggested that the smooth endoplasmic reticulum could be the source of autophagosome membrane (de Duve and Wattiaux 1966), and subsequent studies have supported this possibility (Dunn 1990; Axe et al. 2008). Although controversial, mitochondria and plasma membrane could also supply membranes for the formation of the autophagosomes under different conditions (Hailey et al. 2010; Ravikumar et al. 2010). The elongating isolation membrane surrounds cargo that is ultimately enclosed in the double membrane autophagosome. Once the autophagosome is formed, it fuses with lysosomes (known as the vacuole in yeasts and plants) to form autolysosomes in which the cargo is degraded by lysosomal hydrolases. At this stage lysosomes must reform so that subsequent autophagy may occur (Yu et al. 2010).

AUTOPHAGY GENES

Autophagy is best characterized in the yeast *Saccharomyces cerevisiae*, in which genetic screens resulted in the identification of genes that are required for autophagy. Screens for yeast mutants with defects in either autophagic structures, degradation of cytoplasmic proteins, or possessing altered cytoplasm to vacuole targeting resulted in the identification of *Apg*, *Aut*, and *Cvt* mutants (Tsukada and Ohsumi 1993; Thumm et al. 1994; Harding et al. 1995). The recognition that some of these mutations were in common genes ultimately resulted in the renaming of these autophagy regulators as *Atg* genes (Harding et al. 1996; Klionsky et al. 2003). Over 30 autophagy genes have been identified in yeast, and many of these genes are conserved in animals (Weidberg et al. 2010).

Autophagy is regulated by Atg1 and its interacting proteins, Vps34 and its interacting proteins, and two ubiquitin-like conjugation systems (Fig. 2). Atg1 (Ulk1 and 2 in mammals) is a serine–threonine protein kinase, and its kinase activity is required for autophagy (Matsuura et al. 1997; Kamada et al. 2000). Atg13 is the regulatory sub-unit of the Atg1 kinase complex that also includes FIP200 and Atg101 in animals (Weidberg et al. 2010). Atg1, Atg13 and FIP200 (Atg17) are present in yeast. However, the other Atg1 complex components, including Atg11, Atg20, Atg24, Atg29, and Atg31, do not appear to be encoded by animal genomes. Atg1 is necessary for the induction of autophagy in different cell types, and expression of Atg1 is also sufficient for the induction of autophagy in *Drosophila* (Scott et al. 2004, 2007; Berry and Baehrecke 2007; Chan et al. 2009).

The Vps34 regulatory complex is comprised of the lipid kinase Vps34 (also known as class III phosphatidylinositol 3 (PI3) kinase), Atg6 (known as Beclin1 in mammals), and the protein

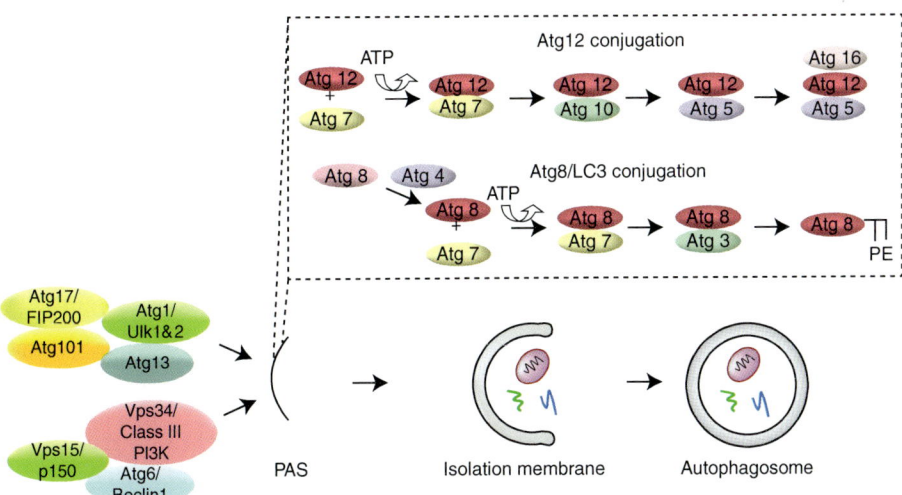

Figure 2. Core pathways that regulate autophagy. Atg1 and its interacting proteins, Vps34 and its interacting proteins, and two ubiquitin-like conjugation systems are required for the elongation of the isolation membrane and formation of an autophagosome.

kinase Vps15 (p150 in mammals) (Simonsen and Tooze 2009). This core complex regulates the formation of PI3 phosphate (PI3P) lipids, and is required for multiple intracellular vesicle trafficking pathways, including endocytosis and autophagy. The Vps34 complex has different proteins associated with it that are thought to be specific to the vesicle process that is regulated, and in the context of autophagy these include Atg14 and Vps38 (UVRAG in mammals). The Vps34 complex components have been localized to the PAS, and are required for the formation of autophagosomes (Juhász et al. 2008). Although some studies have suggested that mTOR is in a common regulatory pathway with Vps34 (Byfield et al. 2005; Nobukuni et al. 2005), others have suggested that these complexes function in parallel genetic pathways (Juhász et al. 2008; Jaber et al. 2012).

Two ubiquitin-like conjugation pathways are required for autophagy, and involve the ubiquitin-like proteins Atg8 (LC3 in mammals) and Atg12 (reviewed in Ohsumi 2001). The carboxy-terminal glycine of Atg8 is covalently bound to phosphatidylethanolamine (PE) following processing by the cysteine protease Atg4 (Ichimura et al. 2000; Kabeya et al. 2000; Kirisako et al. 2000), whereas Atg12 is ultimately associated with Atg5 and Atg16 (Kuma et al. 2002). Both Atg8 and Atg12 conjugation systems use a common E1-like activating enzyme Atg7 (Tanida et al. 1999). Although the Atg8 conjugation system uses Atg3 as an E2-like conjugating enzyme, the Atg12 conjugation system uses Atg10 and associates with Atg5 and Atg16. Atg8-PE is associated with both the isolation membrane and autophagosome, whereas the Atg12, Atg5, and Atg16 complex is only associated with the isolation membrane and disassociates on formation of the autophagosome.

Although autophagy was long considered a bulk degradation process with limited specificity, recent studies have clearly shown that specific cargoes are recruited to autophagosomes for destruction (reviewed in Johansen and Lamark 2011). Several factors have been identified that are required for selection of proteins as cargo for autophagosomes, including p62/SQSTM1/Ref(2)P, Nbr1, and Alfy (Bjørkøy et al. 2005; Kirkin et al. 2009; Filimonenko et al. 2010). Autophagosomes can also consume large cargoes, including peroxisomes (pexophagy) (Manjithaya et al. 2010), mitochondria (mitophagy) (Elmore et al. 2001), ribosomes (ribophagy) (Kraft et al. 2008), and lipid droplets (lipophagy) (Singh et al. 2009). Although the elimination of these organelles may all influence cell survival and death, the removal of mitochondria is particularly interesting in this context given the role of this organelle in bioenergetics and the regulation of cell death. In yeast, Atg32 targets mitochondria to autophagosomes (Kanki et al. 2009; Okamoto et al. 2009), but this protein does not appear to be present in animals. In animals, Parkin and Nix mediate the selective recruitment of mitochondria to autophagosomes (Narendra et al. 2008; Novak et al. 2010). Recruitment of Parkin to damaged mitochondria requires PINK1 (Narendra et al. 2010), and the association of mutations in these genes with Parkinson disease families raises interesting possibilities about the role of autophagy in neurodegeneration.

The formation of autophagosomes is succeeded by docking and fusion with the lysosomes to form the autolysosome, and this process uses the Rab-SNARE system and other molecules that regulate membrane fusion (Nair et al. 2011). Unlike yeast, in which a single vacuole (lysosome) fuses with all autophagosomes, multiple lysosomes fuse with each autophagosome in animals (Yu et al. 2010). Subsequently, lysosomal hydrolases degrade the cargo, and the resulting macromolecules are released into the cytosol for further recycling. Therefore, the rate of autophagy (also known as autophagic flux) depends on both the number of autophagosomes that are formed and the degradative capacity of lysosomes and turnover of autophagic cargo within the cell. This is an important consideration during experimentation when increased numbers of autophagosomes may not necessarily indicate greater autophagic flux, as it may also reflect decreased degradation capacity. Once autolysosmes form, lysosome number is restored, and this process depends on mTOR function (Yu et al. 2010).

AUTOPHAGY GENES THAT ARE SPECIFIC TO MULTICELLULAR ANIMALS

Our knowledge of the core molecular mechanisms controlling autophagy is based on studies in yeast. However, several recent studies indicate that the regulation of autophagy may differ in multicellular animals. As mentioned above, the components and regulation of the Atg1 complex differs between yeast and animals (reviewed in Weidberg et al. 2010). In addition, novel regulators of the Vps34 complex are restricted to higher animals, with the best example being AMBRA1 that is present in mammals but absent in invertebrates (Fimia et al. 2007). It remains to be determined if some of the elegant emerging mechanisms for the regulation of autophagy in mammalian cells, including roles for Bcl-2 (Pattingre et al. 2005), lipid phosphatases (Vergne et al. 2009), and other factors, are conserved in diverse taxa.

The most comprehensive genetic screen for genes that are required for autophagy in animals was conducted by Zhang and colleagues (Tian et al. 2010). They screened for mutations that inhibited clearance of PGL granules in nematode embryos. In addition to the identification of many known core autophagy genes, they identified four ectopic PGL granule (*epg*) genes named epg-2, -3, -4, and -5 that are specific to multicellular animals. Although *epg-3, -4*, and *-5* are required for starvation-induced autophagy, *epg-2* mediates the recognition of cargo (e.g., aggregates of P granule proteins) for delivery to autophagosomes. *epg-2* encodes a protein that appears to be specific to nematodes. By contrast, *epg-3* encodes a protein that is conserved in *Arabidopsis*, *Drosophila*, and mammals, but no similar protein is present in *S. cerevisiae*. Like *epg-3*, *epg-4* encodes a protein that is conserved in plants and animals, but no similar protein is present in yeast. *epg-5* encodes a protein that is conserved in *Drosophila* and is known as VMP1 in mammals (Dusetti et al. 2002), but no similar proteins are present in either *Dictyostelium*, *Arabidopsis*, or *S. cerevisiae*. It is interesting to note that human homologs of the genes identified in this study have been implicated in cancer and other diseases (Gu et al. 2000; Dusetti et al. 2002; Sjöblom et al. 2006).

It is logical that multicellular animals may need specialized mechanisms for the regulation of autophagy in different situations. Although autophagy is a conserved catabolic process, this process may be adapted for use in specific cell contexts, such as cell survival and cell death. Although differences in autophagy may be specified at the level of recruitment of specific cargoes to autophagosomes, it is also possible that different types of autophagy are regulated by distinct activation and repression mechanisms. For example, the conserved immuno receptor Draper is required for autophagy in dying salivary glands in *Drosophila*, but not for autophagy in the fatbody where this process promotes nutrient utilization and cell survival (McPhee et al. 2010). A lack of experimental animal models to study autophagy in specific cell contexts is a limitation facing this research field.

AUTOPHAGY IN CELL SURVIVAL AND NUTRIENT UTILIZATION

Autophagy is involved in maintaining cellular homeostasis. Therefore, it is important to understand the regulation of basal autophagy under normal nutrient conditions. A recent genome-wide screen identified many genes that either suppress or enhance basal autophagy, including a mTOR-independent mechanism for the regulation of autophagy (Lipinski et al. 2010). Another high-throughput study identified numerous proteins that interact with the proteins known to regulate autophagy under basal conditions, thus providing a comprehensive parts list that will enable the dissection of the molecular mechanisms underlying basal autophagy (Behrends et al. 2010).

Studies in yeast pioneered our understanding of the genes that control autophagy, and much of this work has focused on stress-induced autophagy under nutrient-limiting conditions in which catabolism promotes cell survival. Like yeast, autophagy is induced by nutrient limitation in animals, and this influences the bioenergetics of the cell and possibly the organism (Lum et al. 2005). The importance of autophagy during animal starvation is exemplified by the important study showing that mice deficient for Atg5 appear almost normal at birth but die within 1 day of birth (Kuma et al. 2004).

AMPK (SNF1 in yeast) and mTOR are conserved kinases that sense energy and nutrient stress. Both of these kinases influence the activity of Atg1 (ULKs in mammals) to regulate autophagy (Fig. 3) (Samari and Seglen 1998; Kamada et al. 2000; Wang et al. 2001; Scott et al. 2004; Meley et al. 2006). mTOR is inhibited

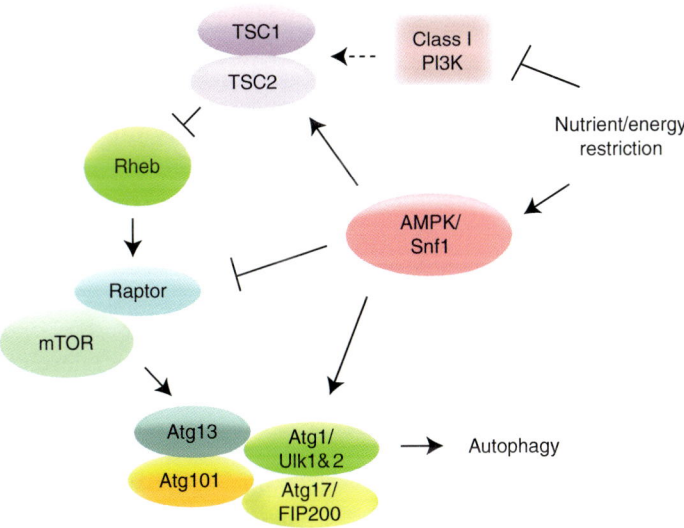

Figure 3. AMPK and mTOR are conserved kinases that sense energy and nutrient stress and influence the activity of Atg1 to regulate autophagy. mTOR influences the activity of the Atg1 complex and autophagy. AMPK regulates autophagy by inhibition of mTOR by phosphorylation of TSC2 and Raptor. AMPK can also influence autophagy by phosphorylation of Atg1/Ulk1.

upon withdrawal of growth factors, such as insulin or insulin-like growth factors, by a cascade of phosphorylation reactions involving Class I PI3K, Akt, TSC1/TSC2, and Rheb (Wullschleger et al. 2006). In mammalian cells, mTOR can also be regulated by a novel mechanism involving localization to the lysosome (Sancak et al. 2010). AMPK regulates autophagy by inhibition of mTOR by phosphorylation of TSC2 and Raptor (Inoki et al. 2003; Gwinn et al. 2008). Furthermore, three recent papers show that AMPK-dependent phosphorylation of Ulk1 can regulate autophagy in nutrient-limiting conditions (Lee et al. 2010; Egan et al. 2011; Kim et al. 2011), although the details of these studies vary. A direct interaction between AMPK and Ulk1 has also been shown (Behrends et al. 2010). The molecular players of starvation-induced autophagy have been studied in considerable detail when compared to autophagy that is induced by several other stresses, such as endoplasmic reticulum stress and hypoxia. Another important challenge that eukaryotic cells face is to combat microorganisms and hostile environments, and autophagy plays a major role in cellular defense and survival under these conditions (Deretic 2011).

In addition to sensing stress, multicellular organisms appear to use developmental signals to regulate autophagy. It is possible that these signals, including hormones, are activated as part of a systemic stress response, but it is also possible that signals during development induce autophagy to regulate cell remodeling. In the context of some animals, this remodeling may be the most efficient method to recycle material for development while maintaining organism fitness and survival. Given the important role of autophagy in stress responses and maintenance of cellular homeostasis, more work is needed to understand if autophagy that is induced by developmental signals is part of a stress program.

AUTOPHAGY IN CELL DEATH

Schweichel and Merker identified three types of cell death based on the role and location of lysosomes inside the cell (Schweichel and Merker 1973). Type II, later called autophagic cell death, is distinguished from type I (apoptotic) cell death by the presence of abundant autophagic structures in the dying cell, a lack of phagocyte recruitment, and, in some instances, by caspase-independence (Schweichel and Merker 1973; Clarke 1990; Baehrecke 2005). The functional contribution of autophagy to cell death has been a subject of great controversy. The reason for controversy appears to be related to the historical focus on autophagy as a cell survival process that is described above. In addition, until relatively recently limited empirical studies had been done to test whether autophagy genes actually facilitate cell death.

Multiple experimental systems have contributed to our recent understanding of autophagy and cell death. *Dictyostelium discoideum*, for example, lacks apoptosis machinery that could participate in nonapoptotic cell death making this a simpler system for the interpretation of the role of autophagy in cell death. *Dictyostelium* exists as a unicellular organism when it is grown on rich media. Upon starvation, however, thousands of cells aggregate to form a multicellular fruiting body in which stalks support balls of spores. These stalk cells undergo developmental cell death via autophagy, as mutations in *Atg* genes prevent the death of stalk cells (Otto et al. 2003; Kosta et al. 2004). One limitation of this system is that *Dictyostelium* lacks apoptosis machinery, and an understanding of the relationship between autophagy and cell death in a system with intact apoptosis machinery is important to our understanding of how to modulate autophagy for therapeutic purposes in humans.

The contribution of autophagy to cell death has been studied most in *Drosophila* in which apoptosis machinery is involved in the death of multiple cell types (Ryoo and Baehrecke 2010). In *Drosophila*, an increase in a steroid hormone triggers the destruction of obsolete tissues at the end of larval development (Jiang et al. 1997). Dying larval midgut and salivary gland cells display markers of apoptosis, such as DNA fragmentation, acridine orange staining, and elevated levels of proapoptotic gene RNAs (Jiang et al. 1997; Lee and Baehrecke 2001; Lee et al. 2002,

2003). These cells also possess large numbers of autophagosomes and elevated levels of *Atg* RNAs (Lee and Baehrecke 2001; Lee et al. 2002, 2003; Li and White 2003; Denton et al. 2009). Surprisingly, midgut degradation is neither disrupted by expression of the pan-caspase inhibitor p35 nor by mutation of multiple caspases, indicating that apoptosis is dispensable for developmental midgut degradation (Denton et al. 2009). Interestingly, midgut destruction is blocked in animals with impaired *Atg1*, *Atg2*, or *Atg18* function, directly implicating autophagy as a crucial process in steroid-induced degradation of midgut cells (Denton et al. 2009). Caspase deficiency does not enhance the *Atg* mutant midgut phenotypes, indicating that autophagic cell death in the midgut is caspase-independent (Denton et al. 2009).

In contrast to the *Drosophila* midgut, destruction of larval salivary glands requires both caspases and autophagy (Berry and Baehrecke 2007). Mutations in either *Atg8* or *Atg18* in addition to decreased function of a number of other *Atg* genes, all lead to the incomplete degradation of larval salivary glands. Similarly, *Atg* genes are required for cell death in the *Drosophila* amnioserosa and ovarian tissue (Hou et al. 2008; Mohseni et al. 2009; Nezis et al. 2009, 2010). It is important to note that the roles and relationship of autophagy and caspases in dying salivary gland, amnioserosa, and ovarian cells in flies is cell context-specific (discussed below). In addition, although larval salivary gland cell death requires both caspases and autophagy for completion of cell clearance, Atg1-induced autophagy in salivary glands is sufficient to induce premature cell death in a caspase-independent manner (Berry and Baehrecke 2007). *Atg1* overexpression is also sufficient to cause cell death in the fat body and imaginal discs, but this death depends on caspase activity (Scott et al. 2007).

Studies in the nematode *C. elegans* also indicate that autophagy contributes to cell death (Kang et al. 2007). *gbp-2* mutants show hyperactive muscarinic acetylcholine signaling in the pharyngeal muscle (You et al. 2006), are sensitive to starvation, and induce excess autophagy and cell death. This phenotype can be partially suppressed by either *beclin-1* or *Atg-7* RNAi indicating that autophagy contributes to cell death.

Autophagy is also observed in dying cells throughout mammalian development, including the regression of the corpus luteum, the involution of mammary and prostate gland and the regression of Mullerian duct structures during male genital development (reviewed in Clarke 1990). Studies of derived mammalian cell lines have shown that *Atg* genes are required for cell death that occurs in the absence of caspase activity (Shimizu et al. 2004; Yu et al. 2004), but no studies to date have shown that autophagy is required for the death of mammalian cells in vivo. However, studies of *beclin1* mutant murine ES cells that form embroid bodies indicates that autophagy is required for lipid signaling that is required for clearance of dying cells (Qu et al. 2007).

AUTOPHAGY, CASPASES, AND CONTEXT SPECIFICITY FOR CELL DEATH

Autophagy promotes cell survival by catabolism of intracellular resources to maintain bioenergetics under nutrient limiting conditions. Furthermore, the elimination of damaged organelles and toxic protein aggregates by autophagy promotes cell survival. Therefore, loss of *Atg* gene function can promote cell death by apoptosis (Boya et al. 2005). In addition, autophagy can promote cell death, but this appears to occur in a cell type and context specific fashion (McPhee and Baehrecke 2009).

The cell context-specific function of autophagy in cell death has been best described in *Drosophila* in which autophagic cell death is known to occur in multiple cell types (Fig. 4). During larval salivary gland degradation, autophagy and caspases cooperate to efficiently clear dying cells. Therefore, impaired function of either autophagy or caspases results in partially degraded salivary gland cells, whereas decreased function of both of these processes results in intact salivary glands. These data indicate that autophagy and caspases function in parallel genetic pathways to degrade salivary glands. By contrast, caspases act upstream of autophagy to direct both starvation-induced

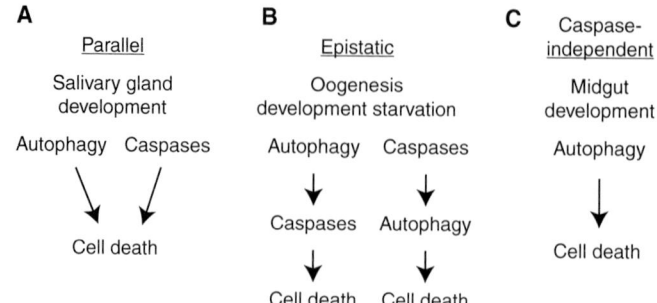

Figure 4. The relationship between autophagy and caspases is cell context specific during cell death. (A) During cell death of *Drosophila* larval salivary glands, autophagy and caspases function in parallel genetic pathways. (B) Autophagy degrades the inhibitor of apoptosis (IAP) protein Bruce enabling caspase activation during fly cell death in oogenesis. By contrast, starvation-induced autophagy leads to degeneration of egg chambers during oogenesis, and the caspase DCP-1 and IAP protein Bruce are required for autophagy to occur in this context. These studies indicate that autophagy and caspases function in an epistatic regulatory hierarchy. (C) Autophagy is essential for cell death during fly midgut cell death, whereas caspases do not appear to play a significant role in the elimination of this tissue.

ovarian cell death (Hou et al. 2008) and degradation of amnioserosa embryonic membrane (Mohseni et al. 2009). In addition, autophagy selectively degrades the caspase inhibitor dBruce to activate caspases and execute cell death in the *Drosophila* ovary (Nezis et al. 2009, 2010). As discussed above, autophagy plays a more prominent role in the death of fly midgut cells (Denton et al. 2009). Combined, these data indicate that multiple possible relationships exist between autophagy and caspases in dying fly cells, and it is important to determine if this is true in other organisms, including humans in which manipulation of autophagy could have therapeutic benefits.

Given the paucity of physiological in vivo models for autophagy and cell death in mammals, it is useful to consider what is known about this relationship in different types of cell lines that may reflect context-specificity. In mammalian cells, most reports of the involvement of autophagy in the execution of death are in cells that possess altered apoptotic pathways (Levine and Yuan 2005; Levine and Kroemer 2009). Treatment of MCF-7 mammary cancer cells that lack caspase-3 with 4-hydroxytamoxifen triggers cell death with autophagy, suggesting the possibility that autophagy can compensate for defects in apoptosis (Bursch et al. 1996).

Lenardo and colleagues described a requirement for autophagy genes during cell death, and reported that U937 monocyte and L929 fibrosarcoma cells use *Beclin1* and *Atg7* for non-apoptotic cell death induced by caspase-8 inhibition (Yu et al. 2004). In another study, Tsujimoto and colleagues showed that $Bax^{-/-}$ and $Bak^{-/-}$ double knockout mouse embryonic fibroblasts undergo cell death accompanied by large scale autophagy, and this death was inhibited by knockdown of either *Atg5* or *Beclin1* (Shimizu et al. 2004). Ryan and colleagues showed that p53-induced cell death is mediated by a stress-induced regulator of autophagy termed *DRAM* (Crighton et al. 2006) further emphasizing the role of autophagy in cell death. Like many cell death regulators, DRAM levels are decreased in human cancers. In addition, a recent report shows that a human ovarian epithelial cell line that expresses oncogenic $H\text{-}Ras^{V12}$ undergoes caspase-independent autophagic cell death that relies on stress kinases, including MEK and ERK, Beclin-1 and Noxa (Elgendy et al. 2011). Interestingly, Debnath and colleagues identified a noncanonical Atg12- Atg3 complex that did not influence starvation-induced autophagy, but when disrupted resulted in increased mitochondrial mass and inhibition of cell death that is mediated by mitochondria

(Radoshevich et al. 2010). Although the association of Beclin-1 with Bcl-2 provided one of the first molecular connections between autophagy and cell death (Liang et al. 1998), much remains to be learned about the relationship(s) between autophagy and cell death regulatory pathways. Clearly, a mechanistic understanding of the relationship between autophagy and cell death is critical to the design of rationale therapies.

CONCLUSIONS

Here we have described the regulation and function of autophagy in cell survival and cell death, two important processes involved in health and disease. Autophagy is often considered a cell survival process (Levine and Kroemer 2009), and it is clear that under nutrient restriction and cell stress, autophagy is augmented to protect the cell and maintain homeostasis. However, accumulating evidence indicates that autophagy can promote cell death, and how autophagy influences cell death appears to depend on the type and context of the cell.

The connection of autophagy to the control of metabolism, stress, survival, and death suggests that organism-specific utilization of this catabolic process is likely to occur. Autophagy is widely used to provide an internal source of nutrients under starvation conditions in organisms as diverse as yeast and humans, but this process may be augmented under organism-specific situations. In developing mice, for example, essential embryonic nutrients are supplied by the mother through the placental interface. At birth, when this supply is terminated, neonates face severe acute starvation, and autophagy is induced until mice are fed (Kuma et al. 2004). Similarly, *C. elegans* enter an alternative dauer larval form during unfavorable environmental conditions. Autophagy is elevated in dauer larvae, and decreased function of *beclin-1*, *unc-51* (*Atg1* in worms), *Atg7*, *lgg-1*(*Atg8a* in worms) and *Atg18* inhibits the completion of dauer development (Melendez et al. 2003). Autophagy may also function to promote homeostasis by maintaining the health of stem cells. Indeed, a recent study showed that Atg7 plays a crucial role in adult mouse hematopoietic stem cell maintenance and survival by regulating mitochondrial quantity and quality (Mortensen et al. 2011). These studies show that autophagy is used in organism-specific biological programs to promote homeostasis. Although these programs appear to use conserved autophagy programs, it is possible that the stimuli that trigger autophagy are specialized. As mentioned above, steroids trigger autophagy in a stage and tissue-specific manner during *Drosophila* development. Furthermore, at least one factor Draper has been identified that is specifically required for autophagy in dying cells, but not during starvation-induced autophagy in *Drosophila* (McPhee et al. 2010). This study highlights the potential different roles and regulatory signaling mechanisms in different cell types. In addition, it is important to consider that differences in cell cargoes may also influence cell fates, including survival and death, as the depletion of survival factors is another way to kill a cell (Yu et al. 2006; Nezis et al. 2010).

Numerous reports suggest a role for autophagy in human diseases, including cancer, neurodegeneration and other disorders. These are age-associated disorders, and aging is associated with the accumulation of by-products of metabolism, cell damage and the inefficient function of the machinery that degrades damaged cell material. In this context, *Beclin-1*, *Atg7*, *Atg8*, and *Atg12* have been shown to be involved in lifespan of worms and flies (Hars et al. 2007; Juhász et al. 2007; Simonsen et al. 2007), which is consistent with studies in mammals (reviewed in Cuervo 2008). Thus, either inhibitors or inducers of autophagy might play a prominent role as therapeutics in combating diseases associated with autophagy (Fleming et al. 2011). Promise exists to support autophagy as a therapeutic target, but caution is prudent when designing drugs that influence a fundamental catabolic process that appears to be involved in the health of all cells, particularly because of its cell context-specific functions in survival and death.

The multiple functions of autophagy are supported by complexity of disease phenotypes. For example, autophagy was first recognized as a potential tumor suppressor mechanism based on mono-allelic loss of *Beclin1* in human tumors (Liang et al. 1999), and this is consistent in murine models (Qu et al. 2003; Yue et al. 2003). However, the mechanistic role of autophagy in tumor suppression is not completely clear. Although loss of autophagy can promote aneuploidy and the development of the transformed phenotype in cell lines (Mathew et al. 2009), it has also been implicated in tumor cell survival (Degenhardt et al. 2006). Significantly, loss of *Atg5* leads to benign adenomas in livers, but this phenotype is not observed in other tissues (Takamura et al. 2011). In addition, the failure of these benign adenomas to cause cancer suggests that autophagy is required for tumor progression. These results are consistent with studies showing that both pancreatic and mammary tumors require autophagy for maintenance of tumorigenesis (Wei et al. 2011; Yang et al. 2011), but differ from models in which autophagy activation facilitates tumor cell killing by multiple agents (Martin et al. 2009; Hamed et al. 2010). Thus, the diametrically opposite roles of autophagy in tumor progression warrants further consideration for the development of rationale cancer therapies (Mah and Ryan 2012).

The complex roles of autophagy in survival and death should also be considered when designing therapies for other disorders. Autophagy promotes the clearance of protein aggregates (Hara et al. 2006; Komatsu et al. 2006) and has an important neuroprotective role in several neurodegenerative disease models, including Alzheimer's and Huntington's (Menzies et al. 2011). In addition, recent evidence suggests that mitochondrial autophagy plays an important role in the pathogenesis of Parkinson's disease (Nixon and Yang 2012). Although the promotion of autophagy in neurodegenerative disease models results in healthier individuals, it is also possible that too much autophagy could have deleterious effects, including problems with bioenergetics or even worse killing the cells while trying to protect them. Future work should not only consider how autophagy may promote cell survival or death, but what the impact of modulating autophagy may have on the health of the test subject and patient.

ACKNOWLEDGMENTS

We apologize to the many authors who were not cited because of space limitations. Research on this subject is supported by NIH grants GM079431 and CA159314 to EHB.

REFERENCES

*Reference is also in this collection.

Ashford TP, Porter KR. 1962. Cytoplasmic components in hepatic cell lysosomes. *J Cell Biol* **12:** 198–202.

Axe EL, Walker SA, Manifava M, Chandra P, Roderick HL, Habermann A, Griffiths G, Ktistakis NT. 2008. Autophagosome formation from membrane compartments enriched in phosphatidylinositol 3-phosphate and dynamically connected to the endoplasmic reticulum. *J Cell Biol* **182:** 685–701.

Baehrecke EH. 2005. Autophagy: Dual roles in life and death? *Nat Rev Mol Cell Biol* **6:** 505–510.

Behrends C, Sowa ME, Gygi SP, Harper JW. 2010. Network organization of the human autophagy system. *Nature* **466:** 68–76.

Berry DL, Baehrecke EH. 2007. Growth arrest and autophagy are required for salivary gland cell degradation in *Drosophila*. *Cell* **131:** 1137–1148.

Bjørkøy G, Lamark T, Brech A, Outzen H, Perander M, Overvatn A, Stenmark H, Johansen T. 2005. p62/SQSTM1 forms protein aggregates degraded by autophagy and has a protective effect on huntingtin-induced cell death. *J Cell Biol* **171:** 603–614.

Boya P, Gonzalez-Polo RA, Casares N, Perfettini JL, Dessen P, Larochette N, Metivier D, Meley D, Souquere S, Yoshimori T, et al. 2005. Inhibition of macroautophagy triggers apoptosis. *Mol Cell Biol* **25:** 1025–1040.

Bursch W, Ellinger A, Kienzl H, Torok L, Pandey S, Sikorska M, Walker R, Hermann RS. 1996. Active cell death induced by the anti-estrogens tamoxifen and ICI164384 in human mammary carcinoma cells (MCF-7) in culture: The role of autophagy. *Carcinogenesis* **17:** 1595–1607.

Byfield MP, Murray JT, Backer JM. 2005. hVps34 is a nutrient-regulated lipid kinase required for activation of p70 S6 kinase. *J Biol Chem* **280:** 33076–33082.

Chan EY, Longatti A, McKnight NC, Tooze SA. 2009. Kinase-inactivated ULK proteins inhibit autophagy via their conserved C-terminal domains using an Atg13-independent mechanism. *Mol Cell Biol* **29:** 157–171.

Clark SLJ. 1957. Cellular differentiation in the kidneys of newborn mice studied with the electron microscope. *J Biophys Biochem Cytol* **3:** 349–362.

Clarke PGH. 1990. Developmental cell death: Morphological diversity and multiple mechanisms. *Anat Embryol* **181:** 195–213.

Crighton D, Wilkinson S, O'Prey J, Syed N, Smith P, Harrison PR, Gasco M, Garrone O, Crook T, Ryan KM. 2006. DRAM, a p53–induced modulator of autophagy, is critical for apoptosis. *Cell* **126:** 121–134.

Cuervo AM. 2008. Autophagy and aging: Keeping that old broom working. *Trends Genet* **24:** 604–612.

de Duve C, Wattiaux R. 1966. Functions of lysosomes. *Annu Rev Physiol* **28:** 435–492.

Degenhardt K, Mathew R, Beaudoin B, Bray K, Anderson D, Chen G, Mukherjee C, Shi Y, Gélinas C, Fan Y, et al. 2006. Autophagy promotes tumor cell survival and restricts necrosis, inflammation, and tumorigenesis. *Cancer Cell* **10:** 51–64.

Denton D, Shravage B, Simin R, Mills K, Berry DL, Baehrecke EH, Kumar S. 2009. Autophagy, not apoptosis, is essential for midgut cell death in *Drosophila*. *Curr Biol* **19:** 1741–1746.

Deretic V. 2011. Autophagy in immunity and cell-autonomous defense against intracellular microbes. *Immunol Rev* **240:** 92–104.

Dunn WA J. 1990. Studies on the mechanisms of autophagy: Formation of the autophagic vacuole. *J Cell Biol* **110:** 1923–1933.

Dusetti NJ, Jiang Y, Vaccaro MI, Tomasini R, Azizi Samir A, Calvo EL, Ropolo A, Fiedler F, Mallo GV, Dagorn JC, et al. 2002. Cloning and expression of the rat vacuole membrane protein 1 (VMP1), a new gene activated in pancreas with acute pancreatitis, which promotes vacuole formation. *Biochem Biophys Res Commun* **290:** 641–649.

Egan DF, Shackelford DB, Mihaylova MM, Gelino S, Kohnz RA, Mair W, Vasquez DS, Joshi A, Gwinn DM, Taylor R, et al. 2011. Phosphorylation of ULK1 (hATG1) by AMP-activated protein kinase connects energy sensing to mitophagy. *Science* **331:** 456–461.

Elgendy M, Sheridan C, Brumatti G, Martin SJ. 2011. Oncogenic Ras-induced expression of Noxa and Beclin-1 promotes autophagic cell death and limits clonogenic survival. *Mol Cell* **42:** 23–35.

Elmore SP, Qian T, Grissom SF, Lemasters JJ. 2001. The mitochondrial permeability transition initiates autophagy in rat hepatocytes. *FASEB J* **15:** 2286–2287.

Filimonenko M, Isakson P, Finley KD, Anderson M, Jeong H, Melia TJ, Bartlett BJ, Myers KM, Birkeland HC, Lamark T, et al. 2010. The selective macroautophagic degradation of aggregated proteins requires the PI3P-binding protein Alfy. *Mol Cell* **38:** 265–279.

Fimia GM, Stoykova A, Romagnoli A, Giunta L, Di Bartolomeo S, Nardacci R, Corazzari M, Fuoco C, Ucar A, Schwartz P, et al. 2007. Ambra1 regulates autophagy and development of the nervous system. *Nature* **447:** 1121–1125.

Fleming A, Noda T, Yoshimori T, Rubinsztein DC. 2011. Chemical modulators of autophagy as biological probes and potential therapeutics. *Nat Chem Biol* **7:** 9–17.

Gu Z, Flemington C, Chittenden T, Zambetti GP. 2000. ei24, a p53 response gene involved in growth suppression and apoptosis. *Mol Cell Biol* **20:** 233–241.

Gwinn DM, Shackelford DB, Egan DF, Mihaylova MM, Mery A, Vasquez DS, Turk BE, Shaw RJ. 2008. AMPK phosphorylation of raptor mediates a metabolic checkpoint. *Mol Cell* **30:** 214–226.

Hailey DW, Rambold AS, Satpute-Krishnan P, Mitra K, Sougrat R, Kim PK, Lippincott-Schwartz J. 2010. Mitochondria supply membranes for autophagosome biogenesis during starvation. *Cell* **141:** 656–667.

Hamed HA, Yacoub A, Park MA, Eulitt P, Sarkar D, Dimitrie IP, Chen CS, Grant S, Curiel DT, Fisher PB, et al. 2010. OSU-03012 enhances Ad.7-induced GBM cell killing via ER stress and autophagy and by decreasing expression of mitochondrial protective proteins. *Cancer Biol Ther* **9:** 526–536.

Hara T, Nakamura K, Matsui M, Yamamoto A, Nakahara Y, Suzuki-Migishima R, Yokoyama M, Mishima K, Saito I, Okano H, et al. 2006. Suppression of basal autophagy in neural cells causes neurodegenerative disease in mice. *Nature* **441:** 885–889.

Harding TM, Morano KA, Scott SV, Klionsky DJ. 1995. Isolation and characterization of yeast mutants in the cytoplasm to vacuole protein targeting pathway. *J Cell Bio* **131:** 591–602.

Harding TM, Hefner-Gravink A, Thumm M, Klionsky DJ. 1996. Genetic and phenotypic overlap between autophagy and the cytoplasm to vacuole protein. *J Biol Chem* **271:** 17621–17624.

Hars ES, Qi H, Ryazanov AG, Jin S, Cai L, Hu C, Liu LF. 2007. Autophagy regulates ageing in *C. elegans*. *Autophagy* **3:** 93–95.

Hou YC, Chittaranjan S, Barbosa SG, McCall K, Gorski SM. 2008. Effector caspase Dcp-1 and IAP protein Bruce regulate starvation-induced autophagy during *Drosophila melanogaster* oogenesis. *J Cell Biol* **182:** 1127–1139.

Ichimura Y, Kirisako T, Takao T, Satomi Y, Shimonishi Y, Ishihara N, Mizushima N, Tanida I, Kominami E, et al. 2000. A ubiquitin-like system mediates protein lipidation. *Nature* **408:** 488–492.

Inoki K, Zhu T, Guan KL. 2003. TSC2 mediates cellular energy response to control cell growth and survival. *Cell* **115:** 577–590.

Jaber N, Dou Z, Chen JS, Catanzaro J, Jiang YP, Ballou LM, Selinger E, Ouyang X, Lin RZ, Zhang J, et al. 2012. Class III PI3K Vps34 plays an essential role in autophagy and in heart and liver function. *Proc Natl Acad Sci* **109:** 2003–2008.

Jiang C, Baehrecke EH, Thummel CS. 1997. Steroid regulated programmed cell death during *Drosophila* metamorphosis. *Development* **124:** 4673–4683.

Johansen T, Lamark T. 2011. Selective autophagy mediated by autophagic adapter proteins. *Autophagy* **7:** 279–296.

Juhász G, Erdi B, Sass M, Neufeld TP. 2007. Atg7-dependent autophagy promotes neuronal health, stress tolerance, and longevity but is dispensable for metamorphosis in *Drosophila*. *Genes Dev* **21:** 3061–3066.

Juhász G, Hill JH, Yan Y, Sass M, Baehrecke EH, Backer JM, Neufeld TP. 2008. The class III PI(3)K Vps34 promotes autophagy and endocytosis but not TOR signaling in *Drosophila*. *J Cell Biol* **181:** 2347–2360.

Kabeya Y, Mizushima N, Ueno T, Yamamoto A, Kirisako T, Noda T, Kominami E, Ohsumi Y, Yoshimori T. 2000. LC3, a mammalian homologue of yeast Apg8p, is localized in autophagosome membranes after processing. *EMBO J* **19:** 5720–5728.

Kamada Y, Funakoshi T, Shintani T, Nagano K, Ohsumi M, Ohsumi Y. 2000. Tor-mediated induction of autophagy via an Apg1 protein kinase complex. *J Cell Biol* **150:** 1507–1513.

Kang C, You YJ, Avery L. 2007. Dual roles of autophagy in the survival of *Caenorhabditis elegans* during starvation. *Genes Dev* **21:** 2161–2171.

Kanki T, Wang K, Cao Y, Baba M, Klionsky DJ. 2009. Atg32 is a mitochondrial protein that confers selectivity during mitophagy. *Dev Cell* **17:** 98–109.

Kim J, Kundu M, Viollet B, Guan KL. 2011. AMPK and mTOR regulate autophagy through direct phosphorylation of Ulk1. *Nat Cell Biol* **13:** 132–141.

Kirisako T, Ichimura Y, Okada H, Kabeya Y, Mizushima N, Yoshimori T, Ohsumi M, Takao T, Noda T, Ohsumi Y. 2000. The reversible modification regulates the membrane-binding state of Apg8/Aut7 essential for autophagy and the cytoplasm to vacuole targeting pathway. *J Cell Biol* **151:** 263–276.

Kirkin V, Lamark T, Sou YS, Bjørkøy G, Nunn JL, Bruun JA, Shvets E, McEwan DG, Clausen TH, Wild P, et al. 2009. A role for NBR1 in autophagosomal degradation of ubiquitinated substrates. *Mol Cell* **33:** 505–516.

Klionsky DJ. 2008. Autophagy revisited: A conversation with Christian de Duve. *Autophagy* **4:** 740–743.

Klionsky DJ, Cregg JM, Dunn WAJ, Emr SD, Sakai Y, Sandoval IV, Sibirny A, Subramani S, Thumm M, Veenhuis M, et al. 2003. A unified nomenclature for yeast autophagy-related genes. *Dev Cell* **5:** 539–545.

Komatsu M, Waguri S, Chiba T, Murata S, Iwata J, Tanida I, Ueno T, Koike M, Uchiyama Y, Kominami E, et al. 2006. Loss of autophagy in the central nervous system causes neurodegeneration in mice. *Nature* **441:** 880–884.

Kosta A, Roisin-Bouffay C, Luciani MF, Otto GP, Kessin RH, Golstein P. 2004. Autophagy gene disruption reveals a non-vacuolar cell death pathway in *Dictyostelium*. *J Biol Chem* **279:** 48404–48409.

Kraft C, Deplazes A, Sohrmann M, Peter M. 2008. Mature ribosomes are selectively degraded upon starvation by an autophagy pathway requiring the Ubp3p/Bre5p ubiquitin protease. *Nat Cell Biol* **10:** 602–610.

Kuma A, Mizushima N, Ishihara N, Ohsumi Y. 2002. Formation of the approximately 350-kDa Apg12-Apg5.Apg16 multimeric complex, mediated by Apg16 oligomerization, is essential for autophagy in yeast. *J Biol Chem* **277:** 18619–18625.

Kuma A, Hatano M, Matsui M, Yamamoto A, Nakaya H, Yoshimori T, Ohsumi Y, Tokuhisa T, Mizushima N. 2004. The role of autophagy during the early neonatal starvation period. *Nature* **432:** 1032–1036.

Lee C-Y, Baehrecke EH. 2001. Steroid regulation of autophagic programmed cell death during development. *Development* **128:** 1443–1455.

Lee C-Y, Cooksey BAK, Baehrecke EH. 2002. Steroid regulation of midgut cell death during *Drosophila* development. *Dev Biol* **250:** 101–111.

Lee C-Y, Clough EA, Yellon P, Teslovich TM, Stephan DA, Baehrecke EH. 2003. Genome-wide analyses of steroid- and radiation-triggered programmed cell death in *Drosophila*. *Curr Biol* **13:** 350–357.

Lee JW, Park S, Takahashi Y, Wang HG. 2010. The association of AMPK with ULK1 regulates autophagy. *PLoS ONE* **5:** e15394.

Levine B, Kroemer G. 2009. Autophagy in aging, disease and death: The true identity of a cell death impostor. *Cell Death Differ* **16:** 1–2.

Levine B, Yuan J. 2005. Autophagy in cell death: An innocent convict? *J Clin Invest* **115:** 2679–2688.

Li TR, White KP. 2003. Tissue-specific gene expression and ecdysone-regulated genomic networks in *Drosophila*. *Dev Cell* **5:** 59–72.

Liang XH, Kleeman LK, Jiang HH, Gordon G, Goldman JE, Berry G, Herman B, Levine B. 1998. Protection against fatal Sindbis virus encephalitis by beclin, a novel Bcl-2-interacting protein. *J Virol* **72:** 8586–8596.

Liang XH, Jackson S, Seaman M, Brown K, Kempkes B, Hibshoosh H, Levine B. 1999. Induction of autophagy and inhibition of tumorigenesis by *beclin 1*. *Nature* **402:** 672–676.

Lipinski MM, Hoffman G, Ng A, Zhou W, Py BF, Hsu E, Liu X, Eisenberg J, Liu J, Blenis J, et al. 2010. A genome-wide siRNA screen reveals multiple mTORC1 independent signaling pathways regulating autophagy under normal nutritional conditions. *Dev Cell* **18:** 1041–1052.

Lum JJ, DeBerardinis RJ, Thompson CB. 2005. Autophagy in metazoans: Cell survival in the land of plenty. *Nat Rev Mol Cell Biol* **6:** 439–448.

* Mah LY, Ryan KM. 2012. Autophagy and cancer. *Cold Spring Harb Perspect Biol* **4:** a008821.

Manjithaya R, Nazarko TY, Farré JC, Subramani S. 2010. Molecular mechanism and physiological role of pexophagy. *FEBS Lett* **584:** 1367–1373.

Martin AP, Mitchell C, Rahmani M, Nephew KP, Grant S, Dent P. 2009. Inhibition of MCL-1 enhances lapatinib toxicity and overcomes lapatinib resistance via BAK-dependent autophagy. *Cancer Biol Ther* **8:** 20842096.

Mathew R, Karp CM, Beaudoin B, Vuong N, Chen G, Chen HY, Bray K, Reddy A, Bhanot G, Gelinas C, et al. 2009. Autophagy suppresses tumorigenesis through elimination of p62. *Cell* **137:** 1062–1075.

Matsuura A, Tsukada M, Wada Y, Ohsumi Y. 1997. Apg1p, a novel protein kinase required for the autophagic process in *Saccharomyces cerevisiae*. *Gene* **192:** 245–250.

McPhee CK, Baehrecke EH. 2009. Autophagy in *Drosophila melanogaster*. *Biochim Biophys Acta* **1793:** 1452–1460.

McPhee CK, Logan MA, Freeman MR, Baehrecke EH. 2010. Activation of autophagy during cell death requires the engulfment receptor Draper. *Nature* **465:** 1093–1096.

Melendez A, Talloczy Z, Seaman M, Eskelinen EL, Hall DH, Levine B. 2003. Autophagy genes are essential for dauer development and life-span extension in *C. elegans*. *Science* **301:** 1387–1391.

Meley D, Bauvy C, Houben-Weerts JH, Dubbelhuis PF, Helmond MT, Codogno P, Meijer AJ. 2006. AMP-activated protein kinase and the regulation of autophagic proteolysis. *J Biol Chem* **281:** 34870–34879.

Menzies FM, Moreau K, Rubinsztein DC. 2011. Protein misfolding disorders and macroautophagy. *Curr Opin Cell Biol* **23:** 190–197.

Mizushima N, Komatsu M. 2011. Autophagy: Renovation of cells and tissues. *Cell* **147:** 728–741.

Mohseni N, McMillan SC, Chaudhary R, Mok J, Reed BH. 2009. Autophagy promotes caspase-dependent cell death during *Drosophila* development. *Autophagy* **5:** 329–338.

Mortensen M, Soilleux EJ, Djordjevic G, Tripp R, Lutteropp M, Sadighi-Akha E, Stranks AJ, Glanville J, Knight S, Jacobsen SE, et al. 2011. The autophagy protein Atg7 is essential for hematopoietic stem cell maintenance. *J Exp Med* **208:** 455–467.

Nair U, Jotwani A, Geng J, Gammoh N, Richerson D, Yen WL, Griffith J, Nag S, Wang K, Moss T, et al. 2011. SNARE proteins are required for macroautophagy. *Cell* **146:** 290–302.

Narendra D, Tanaka A, Suen DF, Youle RJ. 2008. Parkin is recruited selectively to impaired mitochondria and promotes their autophagy. *J Cell Biol* **183:** 795–803.

Narendra DP, Jin SM, Tanaka A, Suen DF, Gautier CA, Shen J, Cookson MR, Youle RJ. 2010. PINK1 is selectively stabilized on impaired mitochondria to activate Parkin. *PLoS Biol* **8:** e1000298.

Nezis IP, Lamark T, Velentzas AD, Rusten TE, Bjørkøy G, Johansen T, Papassideri IS, Stravopodis DJ, Margaritis LH, Stenmark H, et al. 2009. Cell death during *Drosophila melanogaster* early oogenesis is mediated through autophagy. *Autophagy* **5:** 298–302.

Nezis IP, Shravage BV, Sagona AP, Lamark T, Bjørkøy G, Johansen T, Rusten TE, Brech A, Baehrecke EH, Stenmark H. 2010. Autophagic degradation of dBruce controls DNA fragmentation in nurse cells during late *Drosophila melanogaster* oogenesis. *J Cell Biol* **190:** 523–531.

* Nixon RA, Yang D.-S. 2012. Autophagy and neuronal cell death in neurological disorders. *Cold Spring Harb Perspect Biol* doi: 10.1101/cshperspect.a008839.

Nobukuni T, Joaquin M, Roccio M, Dann SG, Kim SY, Gulati P, Byfield MP, Backer JM, Natt F, Bos JL, et al. 2005. Amino acids mediate mTOR/raptor signaling through activation of class 3 phosphatidylinositol 3OH-kinase. *Proc Natl Acad Sci* **102:** 14238–14243.

Novak I, Kirkin V, McEwan DG, Zhang J, Wild P, Rozenknop A, Rogov V, Löhr F, Popovic D, Occhipinti A, et al. 2010. Nix is a selective autophagy receptor for mitochondrial clearance. *EMBO Rep* **11:** 45–51.

Novikoff AB, Essner E. 1962. Cytolysomes and mitochondrial degeneration. *J Cell Biol* **15:** 140–146.

Ohsumi Y. 2001. Molecular dissection of autophagy: Two ubiquitin-like systems. *Nat Rev Mol Cell Biol* **2:** 211–216.

Okamoto K, Kondo-Okamoto N, Ohsumi Y. 2009. Mitochondria-anchored receptor Atg32 mediates degradation of mitochondria via selective autophagy. *Dev Cell* **17:** 87–97.

Otto GP, Wu MY, Kazgan N, Anderson OR, Kessin RH. 2003. Macroautophagy is required for multicellular development of the social amoeba *Dictyostelium discoideum*. *J Biol Chem* **278:** 17636–17645.

Pattingre S, Tassa A, Qu X, Garuti R, Liang XH, Mizushima N, Packer M, Schneider MD, Levine B. 2005. Bcl-2 antiapoptotic proteins inhibit Beclin 1-dependent autophagy. *Cell* **122:** 927–939.

Qu X, Yu J, Bhagat G, Furuya N, Hibshoosh H, Troxel A, Rosen J, Eskelinen EL, Mizushima N, Ohsumi Y, et al. 2003. Promotion of tumorigenesis by heterozygous disruption of the beclin 1 autophagy gene. *J Clin Invest* **112:** 1809–1820.

Qu X, Zou Z, Sun Q, Luby-Phelps K, Cheng P, Hogan RN, Gilpin C, Levine B. 2007. Autophagy gene-dependent clearance of apoptotic cells during embryonic development. *Cell* **128:** 931–946.

Radoshevich L, Murrow L, Chen N, Fernandez E, Roy S, Fung C, Debnath J. 2010. ATG12 conjugation to ATG3 regulates mitochondrial homeostasis and cell death. *Cell* **142:** 590–600.

Ravikumar B, Moreau K, Jahreiss L, Puri C, Rubinsztein DC. 2010. Plasma membrane contributes to the formation of pre-autophagosomal structures. *Nat Cell Biol* **12:** 747–757.

Ryoo HD, Baehrecke EH. 2010. Distinct death mechanisms in *Drosophila* development. *Curr Opin Cell Biol* **22:** 889–895.

Samari HR, Seglen PO. 1998. Inhibition of hepatocytic autophagy by adenosine, aminoimidazole-4-carboxamide riboside, and N6-mercaptopurine riboside. Evidence for involvement of amp-activated protein kinase. *J Biol Chem* **273:** 23758–23763.

Sancak Y, Bar-Peled L, Zoncu R, Markhard AL, Nada S, Sabatini DM. 2010. Ragulator-Rag complex targets mTORC1 to the lysosomal surface and is necessary for its activation by amino acids. *Cell* **141:** 290–303.

Schweichel J-U, Merker H-J. 1973. The morphology of various types of cell death in prenatal tissues. *Teratology* **7:** 253–266.

Scott RC, Schuldiner O, Neufeld TP. 2004. Role and regulation of starvation-induced autophagy in the *Drosophila* fat body. *Dev Cell* **7:** 167–178.

Scott RC, Juhász G, Neufeld TP. 2007. Direct induction of autophagy by Atg1 inhibits cell growth and induces apoptotic cell death. *Curr Biol* **17:** 1–11.

Shimizu S, Kanaseki T, Mizushima N, Mizuta T, Arakawa-Kobayashi S, Thompson CB, Tsujimoto Y. 2004. Role of Bcl-2 family proteins in a non-apoptotic programmed cell death dependent on autophagy genes. *Nat Cell Biol* **6:** 1221–1228.

Simonsen A, Tooze SA. 2009. Coordination of membrane events during autophagy by multiple class III PI3-kinase complexes. *J Cell Biol* **186:** 773–782.

Simonsen A, Cumming RC, Brech A, Isakson P, Schubert DR, Finley KD. 2007. Promoting basal levels of autophagy in the nervous system enhances longevity and oxidant resistance in adult *Drosophila*. *Autophagy* **4:** 176–184.

Singh R, Kaushik S, Wang Y, Xiang Y, Novak I, Komatsu M, Tanaka K, Cuervo AM, Czaja MJ. 2009. Autophagy regulates lipid metabolism. *Nature* **458:** 1131–1135.

Sjöblom T, Jones S, Wood LD, Parsons DW, Lin J, Barber TD, Mandelker D, Leary RJ, Ptak J, Silliman N, et al. 2006. The consensus coding sequences of human breast and colorectal cancers. *Science* **314:** 268–274.

Takamura A, Komatsu M, Hara T, Sakamoto A, Kishi C, Waguri S, Eishi Y, Hino O, Tanaka K, Mizushima N. 2011. Autophagy-deficient mice develop multiple liver tumors. *Genes Dev* **25:** 795–800.

Tanida I, Mizushima N, Kiyooka M, Ohsumi M, Ueno T, Ohsumi Y, Kominami E. 1999. Apg7p/Cvt2p: A novel protein-activating enzyme essential for autophagy. *Mol Biol Cell* **10:** 1367–1379.

Thumm M, Egner R, Koch B, Schlumpberger M, Straub M, Veenhuis M, Wolf DH. 1994. Isolation of autophagocytosis mutants of *Saccharomyces cerevisiae*. *FEBS Lett* **349:** 275–280.

Tian Y, Li Z, Hu W, Ren H, Tian E, Zhao Y, Lu Q, Huang X, Yang P, Li X, et al. 2010. *C. elegans* screen identifies autophagy genes specific to multicellular organisms. *Cell* **141:** 1042–1055.

Tsukada M, Ohsumi Y. 1993. Isolation and characterization of autophagy-defective mutants of *Saccharomyces cerevisiae*. *FEBS Lett* **333:** 169–174.

Vergne I, Roberts E, Elmaoued RA, Tosch V, Delgado MA, Proikas-Cezanne T, Laporte J, Deretic V. 2009. Control of autophagy initiation by phosphoinositide 3-phosphatase Jumpy. *EMBO J* **28:** 2244–2258.

Wang Z, Wilson WA, Fujino MA, Roach PJ. 2001. Antagonistic controls of autophagy and glycogen accumulation by Snf1p, the yeast homolog of AMP-activated protein kinase, and the cyclin-dependent kinase Pho85p. *Mol Cell Biol* **21:** 5742–5752.

Wei H, Wei S, Gan B, Peng X, Zou W, Guan JL. 2011. Suppression of autophagy by FIP200 deletion inhibits mammary tumorigenesis. *Genes Dev* **25:** 1510–1527.

Weidberg H, Shvets E, Elazar Z. 2010. Biogenesis and cargo selectivity of autophagosomes. *Annu Rev Biochem* **80:** 125–156.

Wullschleger S, Loewith R, Hall MN. 2006. TOR signaling in growth and metabolism. *Cell* **124:** 471–484.

Yang S, Wang X, Contino G, Liesa M, Sahin E, Ying H, Bause A, Li Y, Stommel JM, Dell'antonio G, et al. 2011. Pancreatic cancers require autophagy for tumor growth. *Genes Dev* **25:** 717–729.

You YJ, Kim J, Cobb M, Avery L. 2006. Starvation activates MAP kinase through the muscarinic acetylcholine pathway in *Caenorhabditis elegans* pharynx. *Cell Metab* **3:** 237–245.

Yu L, Alva A, Su H, Dutt P, Freundt E, Welsh S, Baehrecke EH, Lenardo MJ. 2004. Regulation of an ATG7-beclin 1 program of autophagic cell death by caspase-8. *Science* **304:** 1500–1502.

Yu L, Wan F, Dutta S, Welsh S, Liu Z, Freundt E, Baehrecke EH, Lenardo MJ. 2006. Autophagic programmed cell death by selective catalase degradation. *Proc Natl Acad Sci* **103:** 4952–4957.

Yu L, McPhee CK, Zheng L, Mardones GA, Rong Y, Peng J, Mi N, Zhao Y, Liu Z, Wan F, et al. 2010. Autophagy termination and lysosome reformation regulated by mTOR. *Nature* **465:** 942–946.

Yue Z, Jin S, Yang C, Levine AJ, Heintz N. 2003. Beclin 1, an autophagy gene essential for early embryonic development, is a haploinsufficient tumor suppressor. *Proc Natl Acad Sci* **100:** 15077–15082.

Autophagy and Cancer

Li Yen Mah and Kevin M. Ryan

Tumour Cell Death Laboratory, Beatson Institute for Cancer Research, Glasgow G61 1BD, United Kingdom

Correspondence: k.ryan@beatson.gla.ac.uk

(Macro)autophagy is a cellular membrane trafficking process that serves to deliver cytoplasmic constituents to lysosomes for degradation. At basal levels, it is critical for maintaining cytoplasmic as well as genomic integrity and is therefore key to maintaining cellular homeostasis. Autophagy is also highly adaptable and can be modified to digest specific cargoes to bring about selective effects in response to numerous forms of intracellular and extracellular stress. It is not a surprise, therefore, that autophagy has a fundamental role in cancer and that perturbations in autophagy can contribute to malignant disease. We review here the roles of autophagy in various aspects of tumor suppression including the response of cells to nutrient and hypoxic stress, the control of programmed cell death, and the connection to tumor-associated immune responses.

THE MOLECULAR MECHANISMS AND TYPES OF AUTOPHAGY

"Autophagy" is broadly defined as a mechanism by which intracellular and extracellular substrates are delivered to lysosomes for degradation. This process is required for the maintenance of cellular homeostasis (Mizushima et al. 2008), generation of amino acids for sustained viability during periods of starvation (Cuervo 2004; Ciechanover 2005), and enhanced protection against pathogens (Shoji-Kawata and Levine 2009). On the basis of the delivery route and cargo specificity, three different types of autophagy have been distinguished—macroautophagy, microautophagy, and chaperone-mediated autophagy (CMA) (Mizushima et al. 2008). Of these, macroautophagy, which is often simply (and hereafter) referred to as autophagy, is the most characterized form and has been extensively researched in yeast and mammals. It is defined by the sequestration of bulk cytoplasm and organelles in double-membrane organelles termed autophagosomes (Fig. 1) (Eskelinen and Saftig 2009). In contrast, microautophagy is characterized by the direct uptake of cytoplasmic substrates by the invagination of the lysosomal membrane, and CMA by the shuttling of soluble proteins into the lysosome via lysosomal chaperone proteins (Mizushima et al. 2008).

Autophagy regulators are conserved from yeast to mammals, and they are the products of AuTophaGy(*Atg*)-related genes (Xie and Klionsky 2007). The role of several Atg proteins in autophagy is highlighted in this article and is depicted in Figure 1, but for a more extensive review of these factors, see Das et al. (2012). An alternative form of macroautophagy has also been described that does not rely on the complete cascade of Atg signaling. This form of autophagy has been termed "non-canonical"

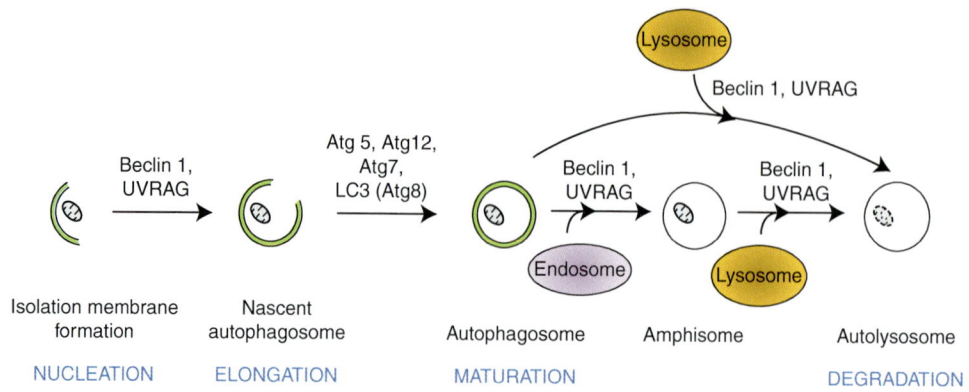

Figure 1. Cellular mechanism and molecular regulators of autophagy in eukaryotes. Nucleation: Beclin 1(Atg6) and UVRAG (UV irradiation resistance–associated gene) are required for the formation of the isolation membrane for sequestering the autophagic substrate. Elongation: The closure of the isolation membrane to form autophagosomes causes sequestration/entrapment of cytoplasmic constituents. This requires the conjugation of Atg5 and Atg12 and is catalyzed by Atg7 (E1-like enzyme). Meanwhile, pro-LC3 (Atg8) is cleaved to form LC3-I, which is then lipidated to form LC3-II. Maturation: Autophagosomes dock and fuse with lysosomes to form autolysosomes. Although the molecular mechanism is not fully understood, Beclin 1 and UVRAG are thought to mediate this process. Alternatively, autophagosomes can fuse with endosomal vesicles, such as endosomes and multivesicular bodies, to form amphisomes, which eventually dock with lysosomes. Degradation: After fusion with lysosomes, autolysosomes are generated where the sequestered materials are hydrolyzed. The inner membrane of autophagosomes and the cargoes are degraded by lysosomal enzymes, and breakdown products are released back into the cytosol.

or "alternative autophagy," and its relevance in relation to "classical" autophagy is currently a matter of debate (Scarlatti et al. 2008; Nishida et al. 2009; Klionsky and Lane 2010). One possibility is that alternative autophagy is adopted when cells fail to activate canonical autophagy owing to mutations in the cluster of *Atg* genes.

Although autophagy was initially considered a nonselective cellular process, the specific catabolism of cellular organelles like mitochondria, peroxisomes, endoplasmic reticulum, and ribosomes has been documented and is termed "mitophagy" (Kissova et al. 2004; Lemasters 2005), "pexophagy" (Sakai et al. 2006), "ER-phagy/reticulophagy" (Bernales et al. 2007), and "ribophagy" (Kraft et al. 2008), respectively.

In general, the role of autophagy to maintain cellular homeostasis requires the versatility to recognize a diverse range of substrates and the ability to regulate or respond to specific cellular pathways and stimuli. This reflects a complex signaling network in the regulation of autophagy. Collectively, this process is fundamental and indispensable such that in response to a block in the canonical signaling cascades, one can imagine that cells adopt alternative routes to activate autophagy in response to intracellular and extracellular cues that may not be equivalent, but are sufficient to sustain viability.

It is now well established that autophagy is connected to tumor development, although the exact roles played by the process at various stages of cancer progression are not yet clear and in some cases are contradictory. In the following sections, we outline the current knowledge regarding the regulation of autophagy in cancer and its impact on various processes that protect against malignant disease. Finally, we speculate as to new areas in cancer where autophagy may be important, and we discuss the possibility of targeting autophagy for cancer therapy.

AUTOPHAGY: FROM MOLECULES TO CANCER

The link between autophagy and cancer is now broad-based (Rosenfeldt and Ryan 2011) but was established based on two principal observations.

First, it was found that *BECN1*, the gene encoding Beclin 1 and the ortholog of yeast *Atg6*, is monoallelically deleted in breast, ovarian, and prostate cancers (Aita et al. 1999; Liang et al. 1999). In addition, ectopic overexpression of *BECN1* in MCF7 cells, which have extremely low levels of endogenous Beclin 1, resulted in activation of autophagy coincident with decreased proliferation and inhibition of tumorigenesis (Liang et al. 1999). Consistently, ectopic overexpression of *BECN1* in colon cancer cell lines with low expression of this gene results in growth inhibition (Koneri et al. 2007). Subsequent to these studies, mutations in other autophagy-related genes including *Atg2B*, *Atg5*, *Atg9B*, *Atg12*, and *UVRAG*, have been documented in gastric and colorectal cancers (Kim et al. 2008; Kang et al. 2009).

The second line of evidence, which consolidated a connection between autophagy and cancer, came from a series of studies involving genetically modified mouse models lacking autophagy regulators. It was found first in mice hemizygous for *BECN1*, and later in mice lacking Atg4C and BIF1, that a deficiency in these autophagic factors can lead to an increased incidence of tumor formation (Qu et al. 2003; Yue et al. 2003; Marino et al. 2007; Takahashi et al. 2007). Interestingly, however, in the case of mosaic deletion of Atg5 and liver-specific deletion of Atg7, only benign lesions were observed in the liver. Moreover, the mosaic loss of Atg5 in other tissues did not have any effect on tumor formation—either benign or malignant (Takamura et al. 2011). This implies that certain autophagy regulators—in this case, Atg5—may be redundant with respect to tumor suppression in many tissues and that in liver autophagy might prevent tumor initiation, but is required for tumor progression. Having said that, it should be pointed out that some autophagy regulators also control other cellular processes. For example, Beclin 1 also regulates endocytosis. Thus, the resulting tumor phenotype associated with mutation in these versatile regulators may be caused by autophagy-independent mechanisms or by synergistic effects on autophagy and other cellular mechanisms.

A plethora of genes, which are known to be perturbed in cancer, have now also been reported to modulate autophagy (for review, see Rosenfeldt and Ryan 2009). For example, the well-established tumor suppressor p53—the most frequently mutated gene in human cancer—has been reported to modulate autophagy both positively and negatively (Fig. 2) (Ryan 2011). Its positive effects on autophagy have been shown to occur via modulation of mTOR and through transcriptional up-regulation of the autophagy promoters Sestrin-2 and damage-regulated autophagy modulator-1 (DRAM-1) (Feng et al. 2005; Crighton et al. 2006, 2007; Maiuri et al. 2009). DRAM-1 has also been reported to be down-regulated in certain cancers indicating a direct role of this p53 target gene in tumor suppression (Crighton et al. 2006, 2007).

p53 has, however, also been shown to be a negative regulator of autophagy within the cytoplasm (Tasdemir et al. 2008). These dual effects on autophagy are not exclusive to p53. The potent oncogene Ras has also been shown to promote as well as inhibit autophagy (Furuta et al. 2004; Elgendy et al. 2011). Although these findings add to the connection between autophagy and cancer, the apparently conflicting reports may at first seem difficult to reconcile. It must be remembered, however, that multiple different forms of cellular stress occur during tumor development, and these observations may simply reflect different positive and negative effects of autophagy at different stages of the disease.

Additionally, many of these observations are conducted in transformed cell lines, which differ biologically and may not represent what happens in a genuine tumor microenvironment. As such, these results need to be consolidated through sophisticated and thorough in vivo genetically modified mouse models of cancer.

STIMULATING AUTOPHAGY IN CANCER

Autophagy occurs constitutively, but when cells are exposed to unfavorable conditions, autophagy is activated above basal levels to counteract "stress" and to promote cellular homeostasis. This switch from default housekeeping to a

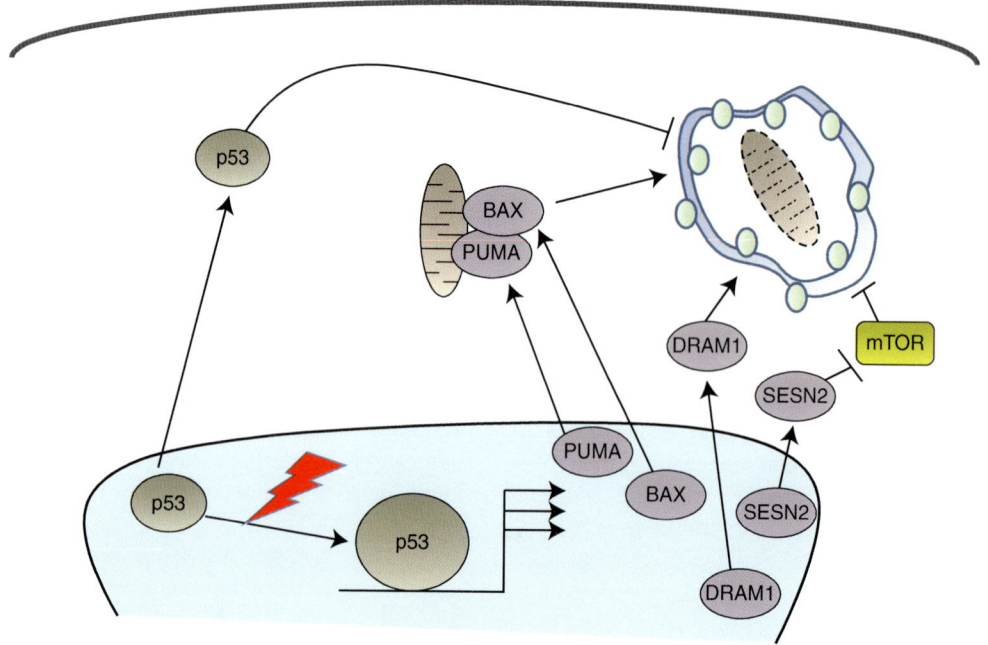

Figure 2. p53 modulates autophagy in multiple ways. Basal levels of p53 target repression of autophagy from within the cytoplasm. In response to cellular stress, the levels of p53 become elevated and accumulate in the nucleus. This results in activation of a series of target genes that positively regulate autophagy. PUMA and BAX also localize to mitochondria. DRAM-1, in contrast, localizes to lysosomes, and Sestrin-2 modulates autophagy via mTOR. SESN2, the gene encoding Sestrin-2.

specific cytoprotective role is triggered by various stimuli. In cancer, for example, a variety of adverse and hostile conditions that result in the production of reactive oxygen species (ROS) can lead to protein and DNA damage that destabilizes cellular homeostasis (Dewaele et al. 2010).

Under normal conditions, our cells are constantly breaking down macromolecules found abundantly in the surrounding environment for the synthesis of new building blocks and to provide energy to sustain their survival. However, under conditions in which nutrients are scarce, as occurs in the poorly vascularized regions of developing tumors, autophagy is activated as a mechanism to provide nutrients from within the cell in order to sustain viability for a limited period until external nutrients become available (Kuma et al. 2004; Lum et al. 2005a,b). In fact, several evolutionarily conserved nutrient sensors such as mTOR, AMPK, and Sirtuins are regulators of autophagy (Rosenfeldt and Ryan 2009, 2011; Kroemer et al. 2010; Mehrpour et al. 2010). In addition, cellular compartments have recently been identified that enable linkage between autophagy and mTOR (Narita et al. 2011). The importance of this response is exemplified by the fact that when cells are deprived of amino acids, mature autophagic vacuoles have been reported to occupy ∼1% of the cytoplasm (Mizushima et al. 2001). Likewise, under growth factor–limiting conditions, hematopoietic cells activate autophagy to produce ATP for survival (Lum et al. 2005a). Using similar measurements, glucose and serum withdrawal have also been shown to induce autophagy, although the net outcome in these situations often skews toward cell death (Aki et al. 2003; Steiger-Barraissoul and Rami 2009).

In addition to nutrient deprivation, metabolic stress is also caused by hypoxia. Solid tumors contain poorly vascularized regions that are very hypoxic. Studies have now shown that cells within these areas display high levels of macroautophagy (Degenhardt et al. 2006). It was concluded from this study that tumor cells rely heavily on autophagy to survive. The most documented hypoxic signaling pathway involves the activation of the hypoxia inducible factor 1 (HIF), which is composed of α and β subunits. The HIF1α subunit is affected by changes in oxygen concentration and modulates genes involved in regulating erythropoiesis, angiogenesis, energy metabolism, pH regulation, cell migration, and tumor invasion (Mazure and Pouysségur 2010). Another HIF family member, HIF2α, is also stimulated by hypoxia, and both HIF1α and HIF2α, as well as the autophagy regulator Beclin 1, have been shown to be required for prosurvival hypoxia-induced autophagy in normal and cancer cell lines (Zhang et al. 2008; Bellot et al. 2009; Wilkinson et al. 2009). In tumor cells, this response is also enhanced by autocrine growth factor signaling via PDGF receptors, leading to enhanced HIF activity and a robust and selective autophagic response to promote cell survival (Wilkinson and Ryan 2009; Wilkinson et al. 2009).

Several reports have also shown that hypoxia can stimulate autophagy in an HIF-independent manner (Pursiheimo et al. 2009). Hypoxia-induced autophagy removes the autophagy "adaptor" and signaling protein p62/SQSTM1, a well-known autophagic substrate, suggesting a link between p62/SQSTM1 and the regulation of hypoxic cancer cell survival responses (Wilkinson and Ryan 2009; Wilkinson et al. 2009). Furthermore, the lack of p62/SQSTM1 removal in autophagy-deficient cells has been shown to be a contributing factor to tumor development (Mathew et al. 2009).

In short, autophagy is activated following metabolite deprivation and hypoxia, but also in response to a multitude of factors encountered during the development of cancer that perturb the equilibrated cellular environment. In addition, autophagy is also activated in response to a variety of chemotherapeutic drugs (Kondo et al. 2005). Thus, this draws us to the tentative conclusion that most, if not all, forms of stress stimulate autophagy, which when taken together has been termed the "integrated stress response" (Kroemer et al. 2010).

AUTOPHAGY—NOT JUST CELL AUTONOMOUS IN CANCER

Tumor cells are thought to favor metabolism of glucose via glycolysis, because they display high levels of glucose uptake and lactate production, even when oxygen is abundant. This form of aerobic glycolysis is termed the "Warburg Effect" (Warburg 1956). Tumor cells that fail to keep up with energy demands can die by necrosis, resulting in the production and accumulation of ROS (Lin et al. 2004). This induces oxidative stress in the tumor microenvironment. As a result, bystander cells can switch on autophagy to remove ROS together with damaged cellular organelles, and tumor cells have hijacked this housekeeping program to fuel their own growth. Martinez-Outschoorn et al. (2011) showed that cancer cells transmit oxidative stress to neighboring fibroblasts to down-regulate Cav-1 (a marker often associated with early tumor recurrence, lymph node metastasis, and tamoxifen resistance). They also showed that cancer-associated fibroblasts are subsequently plagued with mitochondrial dysfunction, oxidative stress, and aerobic glycolysis. They proposed that these stromal cells activate autophagy to remove damaged organelles, and the resulting degradation products are fed to tumor cells in a manner similar to what has been dubbed the "reverse Warburg effect" (Pavlides et al. 2009). Thus, not only is autophagy exploited intrinsically by cancer cells for their benefit, but also signals for autophagy activation are transmitted to the surrounding untransformed cells, and their activity is used to fuel tumor cell growth.

AUTOPHAGY: GUARDIAN OF THE GENOME IN ADDITION TO THE PROTEOME

Maintenance of genome integrity is critical to avoid tumorigenesis (Bartek et al. 2007). One of the main consequences of metabolic stress is the accumulation of ROS, which can cause DNA damage by inducing DNA base changes and strand breaks. This leads to the inactivation of tumor-suppressor genes and enhanced expression and/or activation of proto-oncogenes (Wiseman and Halliwell 1996). Consistently, ROS scavenge and clean up debris during neurodegeneration in preparation for regenerative processes. The connection between autophagy and neuronal cell death reemerged in the 1970s from observations of Clarke and colleagues, who presented evidence that the developing brain deployed autophagy as a form of programmed neuronal cell death during which autophagy was massively up-regulated to eliminate cytoplasmic components, at once killing the neuron and reducing its cell mass for easy removal. Self-degradation was suggested as a more efficient elimination mechanism than apoptosis, which requires a large population of phagocytic cells and access of these cells to the dying region (Baehrecke 2005). Indeed, the best evidence for this process is in the context of massive cell death, as in metamorphosis and involutional states (Das et al. 2012).

Clarke proposed that autophagic cell death (ACD)—type 2 programmed cell death (PCD)— could be a relatively common alternative route to death distinct from apoptosis—type 1 PCD (Clarke 1990)—or caspase-independent cell death—type 3 PCD (Fig. 1). The distinguishing features of ACD are marked proliferation of AVs and progressive disappearance of organelles but relative preservation of cytoskeletal and nuclear integrity until late in the process (Schweichel and Merker 1973; Hornung et al. 1989). In this original concept of ACD or type 2 PCD, death is achieved by autophagic digestion of organelles and essential regulatory molecules and elimination of death inhibitory factors (Baehrecke 2005). With the advent of the molecular era of autophagy research in the 1990s, it became possible to verify the most important implication of ACD, namely, that the death could be prevented by inhibiting autophagy genetically or pharmacologically. Meanwhile, reports of prominent lysosomal/autophagic pathology in Alzheimer's disease (AD) (Cataldo et al. 1997; Nixon et al. 2000, 2005) and other neuropathic states (Anglade et al. 1997; Rubinsztein et al. 2005) raised important questions about whether autophagy pathology signifies a prodeath program or an attempt to maintain survival—a critical question for any potential therapy based on autophagy modulation. In this article, we will examine evidence for the various neuroprotective roles of autophagy and review our current understanding of how specific stages of autophagy may become disrupted and influence the neurodegenerative pattern seen in major adult-onset neurological diseases. We will particularly focus on how neurons regulate the balance between prosurvival autophagy and well-established cell death mechanisms in making life or death decisions.

THE STAGES OF AUTOPHAGY

Since its initial description, autophagy has referred to the process by which unwarranted constituents of the cell are sequestered within vacuoles and degraded by lysosomes (Klionsky et al. 2010). Lysosomal digestion of the cell's own cytoplasmic material is the common cardinal feature of all forms of autophagy. Different mechanisms for delivery of substrates to the lysosomal compartment, however, distinguish among the three major subtypes of autophagy in mammalian cells (Fig. 2). The least well understood of these, microautophagy, is a constitutive and sometimes selective process by which proteins or organelles may be engulfed by invagination of the lysosomal or endosomal membrane and scission of the cargo-containing vesicle inside the lumen. Cargo selection and delivery to endosomes is facilitated by the chaperone Hsc70 and internalization relies on the same endosomal sorting complex required for transport (ESCRT) I and III complexes involved in multivesicular body (MVB) formation (Sahu et al. 2011). In chaperone-mediated autophagy (CMA), cytosolic proteins containing a KFERQ motif are delivered to a LAMP2A-containing complex on the lysosomal membrane via a complex of chaperones, including Hsc70 and HSP70, and cochaperones (EGHSP40, HSP90, HIP, HOP, and BAG1). The substrate is subsequently unfolded and degraded with the help of lysosomal Hsc70 and a complex of proteins that disassembles the transmembrane translocation complex (Bandyopadhyay et al. 2008, 2010). CMA activity is constitutive in most cells but up-regulated under conditions of stress or nutrient deprivation.

macroautophagy (Degenhardt et al. 2006). It was concluded from this study that tumor cells rely heavily on autophagy to survive. The most documented hypoxic signaling pathway involves the activation of the hypoxia inducible factor 1 (HIF), which is composed of α and β subunits. The HIF1α subunit is affected by changes in oxygen concentration and modulates genes involved in regulating erythropoiesis, angiogenesis, energy metabolism, pH regulation, cell migration, and tumor invasion (Mazure and Pouyssegur 2010). Another HIF family member, HIF2α, is also stimulated by hypoxia, and both HIF1α and HIF2α, as well as the autophagy regulator Beclin 1, have been shown to be required for prosurvival hypoxia-induced autophagy in normal and cancer cell lines (Zhang et al. 2008; Bellot et al. 2009; Wilkinson et al. 2009). In tumor cells, this response is also enhanced by autocrine growth factor signaling via PDGF receptors, leading to enhanced HIF activity and a robust and selective autophagic response to promote cell survival (Wilkinson and Ryan 2009; Wilkinson et al. 2009).

Several reports have also shown that hypoxia can stimulate autophagy in an HIF-independent manner (Pursiheimo et al. 2009). Hypoxia-induced autophagy removes the autophagy "adaptor" and signaling protein p62/SQSTM1, a well-known autophagic substrate, suggesting a link between p62/SQSTM1 and the regulation of hypoxic cancer cell survival responses (Wilkinson and Ryan 2009; Wilkinson et al. 2009). Furthermore, the lack of p62/SQSTM1 removal in autophagy-deficient cells has been shown to be a contributing factor to tumor development (Mathew et al. 2009).

In short, autophagy is activated following metabolite deprivation and hypoxia, but also in response to a multitude of factors encountered during the development of cancer that perturb the equilibrated cellular environment. In addition, autophagy is also activated in response to a variety of chemotherapeutic drugs (Kondo et al. 2005). Thus, this draws us to the tentative conclusion that most, if not all, forms of stress stimulate autophagy, which when taken together has been termed the "integrated stress response" (Kroemer et al. 2010).

AUTOPHAGY—NOT JUST CELL AUTONOMOUS IN CANCER

Tumor cells are thought to favor metabolism of glucose via glycolysis, because they display high levels of glucose uptake and lactate production, even when oxygen is abundant. This form of aerobic glycolysis is termed the "Warburg Effect" (Warburg 1956). Tumor cells that fail to keep up with energy demands can die by necrosis, resulting in the production and accumulation of ROS (Lin et al. 2004). This induces oxidative stress in the tumor microenvironment. As a result, bystander cells can switch on autophagy to remove ROS together with damaged cellular organelles, and tumor cells have hijacked this housekeeping program to fuel their own growth. Martinez-Outschoorn et al. (2011) showed that cancer cells transmit oxidative stress to neighboring fibroblasts to down-regulate Cav-1 (a marker often associated with early tumor recurrence, lymph node metastasis, and tamoxifen resistance). They also showed that cancer-associated fibroblasts are subsequently plagued with mitochondrial dysfunction, oxidative stress, and aerobic glycolysis. They proposed that these stromal cells activate autophagy to remove damaged organelles, and the resulting degradation products are fed to tumor cells in a manner similar to what has been dubbed the "reverse Warburg effect" (Pavlides et al. 2009). Thus, not only is autophagy exploited intrinsically by cancer cells for their benefit, but also signals for autophagy activation are transmitted to the surrounding untransformed cells, and their activity is used to fuel tumor cell growth.

AUTOPHAGY: GUARDIAN OF THE GENOME IN ADDITION TO THE PROTEOME

Maintenance of genome integrity is critical to avoid tumorigenesis (Bartek et al. 2007). One of the main consequences of metabolic stress is the accumulation of ROS, which can cause DNA damage by inducing DNA base changes and strand breaks. This leads to the inactivation of tumor-suppressor genes and enhanced expression and/or activation of proto-oncogenes (Wiseman and Halliwell 1996). Consistently, ROS has long been associated with human cancer development (Wiseman and Halliwell 1996), and the role of autophagy in lowering ROS levels has been confirmed by many studies (Rouschop et al. 2009). Given that ROS stimulate autophagy, we can summarize that when ROS are present at high levels, autophagy is activated to scavenge ROS, thus preventing DNA damage and tumorigenesis.

Despite being more sensitive to metabolic stress, autophagy-deficient cancer cells are more likely to accumulate genomic damage. Mathew et al. (2007) showed that $BECN1^{+/-}$ and $Atg5^{-/-}$ immortalized baby mouse kidney (iBMK) cells accumulate higher levels of mutation, chromosomal instability, and accelerated progression into aneuploidy (Mathew et al. 2007). This finding is confirmed in an in vivo mammary model. When immortalized $BECN1^{+/+}$ and $BECN1^{+/-}$ mouse mammary epithelial cells (iMMEC) were injected into mice to form xenografts, $BECN1$ hemizygosity resulted in genome damage under metabolic stress and gene amplification (Karantza-Wadsworth et al. 2007).

The role of autophagy as the guardian of the genome is multifaceted. The decision whether to die via apoptosis or necrosis, or to stay alive with unresolved damage, is not yet completely understood. These variations may be governed by the extent of damage, status of the affected cells, or the type of drug, which might also stimulate other responses apart from DNA damage. Either way, a defect in autophagy can be perceived as a road not only to impaired cytoplasmic homeostasis, but also to replication of cells containing DNA damage that may be predisposed to tumor development.

AUTOPHAGY PROGRAMS CELL DEATH

Apoptosis is long established as a form of programmed cell death and is known to be a key component of tumor suppression. The apoptotic pathways are well studied, and the role of autophagy in modulating apoptosis has been convincingly documented in vivo. Autophagy has been shown to be required for cell death in salivary glands during *Drosophila* development (Berry and Baehrecke 2007)—we refer the reader to the accompanying article in this collection by Das et al. (2012). Similarly, overexpression of wild-type *Atg1* in *Drosophila* elicits a strong autophagic response, and cells that have high levels of this protein are selectively and rapidly eliminated (Scott et al. 2007).

Although the exact mechanism(s) connecting autophagy and apoptosis as partners in cell killing is still unclear, the cross talk between autophagy and apoptosis has been dissected by many researchers (Fig. 3). Classical apoptotic regulators such as the antiapoptotic Bcl-2 family members Bcl-X_L and Bcl-2 have been shown to regulate autophagy, and, reciprocally, a cleaved form of the essential autophagy protein Atg5 has been shown to induce apoptosis directly at mitochondria (Pattingre et al. 2005; Yousefi et al. 2006).

The coactivation of autophagy and apoptosis has also been shown in cancer cells. For example, DRAM-1 was found to have both proautophagic and proapoptotic roles downstream from p53 (Crighton et al. 2006, 2007). Similarly, Yee et al. (2009) showed that another p53 target gene, the potent proapoptotic Puma, induces mitophagy and the subsequent release of cytochrome *c*, leading to apoptosis. The inhibition of PUMA-induced autophagy diminishes the apoptotic response, thus highlighting a synergy between autophagy and apoptosis in this cell death response.

Although apoptosis and autophagy can occur in a cooperative manner to elicit cell death, these processes are sometimes mutually antagonistic (Fig. 3) (Boya et al. 2005). For example, caspases (the effectors of the apoptotic cell death) have been shown to cleave and inactivate Atg6/Beclin 1, and the suppression of Atg6 function increases apoptotic cell death (Cho et al. 2009). These lines of evidence imply that apoptosis is either inhibited or delayed when autophagy is present, with the probable conclusion that autophagy is activated to protect cells from dying (Fig. 3).

Other contrasting behaviors between apoptosis and autophagy have been attributed to the status of the cell, or stage of transformation. Dominant-negative FADD invokes a death

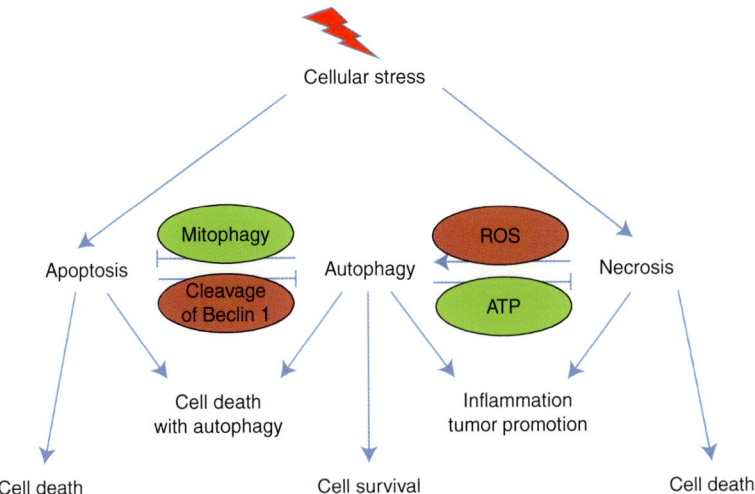

Figure 3. Autophagy and the control of cell death. There is cross talk between apoptosis and autophagy in the control of cell death. The two pathways also repress one another, fighting for either cell survival or cell death. Autophagy also regulates necrosis. Both pathways can regulate each other, and the products of inflammation can positively feed back to enhance the amplitude of these responses. Examples are shown to indicate how apoptosis and autophagy and necrosis and autophagy are connected.

stimulus involving autophagy in healthy cells but not in cancer cells and induces varying amplitudes of death responses at different stages of cancer progression (Thorburn et al. 2005). In addition, oncogenic Ras has also recently been shown to cause autophagic cell death in the absence of apoptosis (Elgendy et al. 2011), even though other reports have indicated that autophagy is required for Ras-driven tumor growth (Guo et al. 2011; Yang et al. 2011).

These lines of evidence suggest that the decision to activate, repress, or simply not to manipulate autophagy, may be governed by the nature of the stimuli or the upstream regulator of an autophagic and/or apoptotic protein, as well as the health status of cells, and the net outcome of these three possibilities is either to promote or repress tumor survival, or both.

AUTOPHAGY AND NECROSIS

Necrosis is a form of cellular demise characterized by several features such as ATP depletion, the loss of cellular osmolarity, and release of various factors such as HMGB1 and cell lysis, which all lead to a strong inflammatory response (Edinger and Thompson 2004). Although necrosis is often perceived as an unregulated form of cell death, emerging evidence reveals that it can also occur as a form of caspase-independent programmed cell death and is specifically termed "necroptosis" (Vandenabeele et al. 2010). Programmed necrotic cell death is also known to be the mediator of cell death in response to certain classes of chemotherapeutic drugs (Zong et al. 2004).

As if contradicting the role of cell death in tumor suppression, necrosis is generally considered to be a tumor promoter and is often associated with poor prognosis (Swinson et al. 2002). Necrotic cells in vivo cause a strong inflammatory response, accompanied by the production of cytokines, chemokines, and other inflammatory enzymes, which can, in turn, positively feed back to cause further damage in surviving cells with enhanced tumorigenic potential (Balkwill et al. 2005).

Necrosis and autophagy, like apoptosis and autophagy, often occur coincidently (Fig. 3). Therapeutic treatment of cancer cells triggers necrosis due to bioenergetic compromise. Tumor cells can evade this ATP-limiting demise by activating the energy sensor LKB1/AMPK complex, which, in turn, inhibits the mammalian target of rapamycin (mTOR), leading to activation of autophagy (Amaravadi and Thompson 2007). Although this seems to favor the viability of cancer cells, the major role for autophagy activation in promoting tumor development lies in the repression of necrosis-associated inflammatory responses. These responses can promote the release of prometastatic immune-modulatory factors such as HMBG1, and this may lead to increased metastasis (Thorburn et al. 2009a,b). Additionally, necrosis has been shown to be activated by AKT and Ras oncogenic signaling and is up-regulated when autophagy is compromised by monoallelic deletion of *BECN1* (Degenhardt et al. 2006). Thus, autophagy is perceived to prevent necrosis in order to limit further cellular damage that may promote tumorigenesis and metastasis. However, the situation is not entirely straightforward because tumors that survive necrotic stress via autophagy could obtain an advantage to thrive under nutrient-limiting conditions and acquire mutations that cause resistance to cell death.

AUTOPHAGY AND SENESCENCE: A BARRIER TO TUMOR PROGRESSION?

Cellular senescence is a process in which cells enter a state of irreversible cell cycle arrest (Krizhanovsky et al. 2008). The permanent withdrawal from the cell cycle can restrict tumor cell proliferation and is therefore considered an important mechanism of tumor suppression. Senescent cells do, however, secrete a spectrum of proinflammatory cytokines, termed the senescence-associated secretory phenotype (SASP), causing inflammation that is, in turn, a vector for tumor cell growth (Davalos et al. 2010).

Recently, Young et al. (2009) reported that several *Atg* genes—*ULK3*, *Atg7*, and *LC3* (*Atg8*)—are up-regulated during oncogene-induced senescence. They also showed that autophagy is required for the transition of mitotic to senescent phase, and genetic ablation of autophagy delays the onset of senescence. Since then, much thought has been put into the "hows" and "whys" of the autophagy-regulated senescence program. Some researchers believe that autophagy is evoked to break down specific cellular components to enable physical remodelling associated with senescence (White and Lowe 2009). Additionally, autophagy may supplement senescent cells with the monomers that are required for the production and secretion of a plethora of growth factors, as well as for restructuring the cellular cytoskeleton (Adams 2009).

AUTOPHAGY AND TUMOR IMMUNITY

Cellular immunity can be generally classified into two categories; innate and adaptive. Innate immunity, analogous to the body's first line of defense, is almost always engaged in triggering the complement system and inflammation. Adaptive immunity, on the other hand, results in a more specific and stronger response. This involves the surveillance, capture, and presentation of pathogens or pathogenic/non-self peptides by antigen-presenting cells (APCs) such as macrophages, B cells, and dendritic cells (DCs), which then stimulate T-lymphocytes to invoke cell death, among other responses. Autophagy and immunity have long been considered as two inseparable entities (Schmid et al. 2007; Shoji-Kawata and Levine 2009). The roles of autophagy in regulating inflammation and adaptive immunity, ranging from lymphocyte development (Nedjic et al. 2008), pathogen recognition and destruction, to antigen presentation have been shown (Schmid et al. 2007).

Over the last few decades, neoplastic cells have been shown to express a panel of unique antigens that are recognized by T cells (Sensi and Anichini 2006). Hence, tumor antigen presentation is an important aspect of antitumor responses, and a defect in this system can result in tumor escape from immune surveillance—a facet often associated with cancer progression (Garcia-Lora et al. 2003; Hanahan and Weinberg 2011). DCs are one of the most effective professional APCs. One of the many contributors that impede DC function in tumor defense is the buildup of ROS, which are present abundantly in the tumor microenvironment (Fricke and Gabrilovich 2006). High levels of ROS have been shown to induce oxidative stress, resulting in JNK-mediated denditric cell death (Handley et al. 2005). Because autophagy has a major role in buffering ROS, a defect in this lysosomal degradation pathway would handicap DCs due to the accumulation of ROS.

Autophagy is also intrinsically connected to the process of antigen presentation. Classically, antigens derived from outside the cell are degraded in lysosomes, and autophagy has important roles in trafficking antigens destined for degradation, as well as trafficking peptides from degraded antigen back to the cell surface for presentation on the class II major histocompatibility complex (MHC). Most tumor antigens are, of course, derived from within the cell, and degradation in this context is primarily via proteasomes, resulting in presentation on class I MHC. Autophagy, however, stills plays a major role in the presentation of tumor antigens on APCs directly on class II MHC or via a process termed "cross-presentation" on class I MHC (Munz 2010). With respect to this, priming of $CD8^+$ T cells APCs presenting the melanocyte-derived tumor antigen gp100 is greatly enhanced when autophagy is pharmacologically induced in the melanocytes with rapamycin, and this is reversed when autophagy is blocked with 3-methyladenine (3-MA) (Li et al. 2008). The investigators also found that blocking autophagy at later stages by inhibiting autophagosome turnover facilitated antigen cross-presentation, subsequently revealing that autophagosomes are efficient transporters of antigen from antigen-presenting cells to T cells. This suggests that the stabilization of autophagosomes, rather than the initiation and completion of autophagy itself, is required for effective priming of cytotoxic T cells. Elegant in vivo studies by Lee et al. (2010) have confirmed that autophagy is particularly required to enhance class I antigen presentation by DCs. They showed that $Atg5^{-/-}$ DCs present antigen on class II molecules but have a reduced ability to activate $CD4^+$ lymphocytes by promoting antigen degradation in autolysosomes. No differences were observed in DC migration or DC antigen capture via endocytosis and phagocytosis, as well as cross-presentation on class I. They also discovered that autophagy induced by starvation and rapamycin treatment decreases class II presentation.

Albeit varying in the mode of action, autophagy initiation is required for both class I and II MHC processing. Evidence indicates that the synthesis, but not degradation, of autophagosomes is the determining factor for class I, whereas for class II, both synthesis and degradation of autophagosomes are crucial (Dorfel et al. 2005). Aside from macroautophagy, chaperone-mediated autophagy has also been shown to play a role in antigen presentation, particularly in the presentation of endogenous antigen of cytoplasmic origin (Deretic 2005). It is conceivable, therefore, that more than one form of autophagy operates in antigen processing, and the dissection of this network, particularly in the context of tumor antigen presentation, may represent an encouraging step toward new, rationally designed tumor therapies.

CONCLUSIONS AND PERSPECTIVES

Autophagy is not only a critical degradation process for housekeeping purposes in normal tissues, but it also affects various cellular mechanisms that are critical, or altered in cancer cells (Fig. 4). There may, however, be additional mechanisms relevant to tumor development where a role for autophagy is yet to be defined. For example, in addition to genetic alterations, cancer cells usually have tremendous changes in their epigenetic landscape (Sharma et al. 2010). Given that autophagosomes can sequester chromosomes (Sit et al. 1996) and autophagic breakdown of the nucleus is well documented in yeast (Kvam and Goldfarb 2007), it might be possible for autophagy to regulate epigenetic regulatory factors such as histone methyltransferases and deacetylases in the nucleus of mammalian cells. Could this be an additional role for autophagy in cancer?

It is a defining feature of cancer that tumor cells show uncontrolled proliferation compared with normal cells. Because autophagy is often deregulated in transformed cells, is there a link between autophagy and cell cycle deregulation?

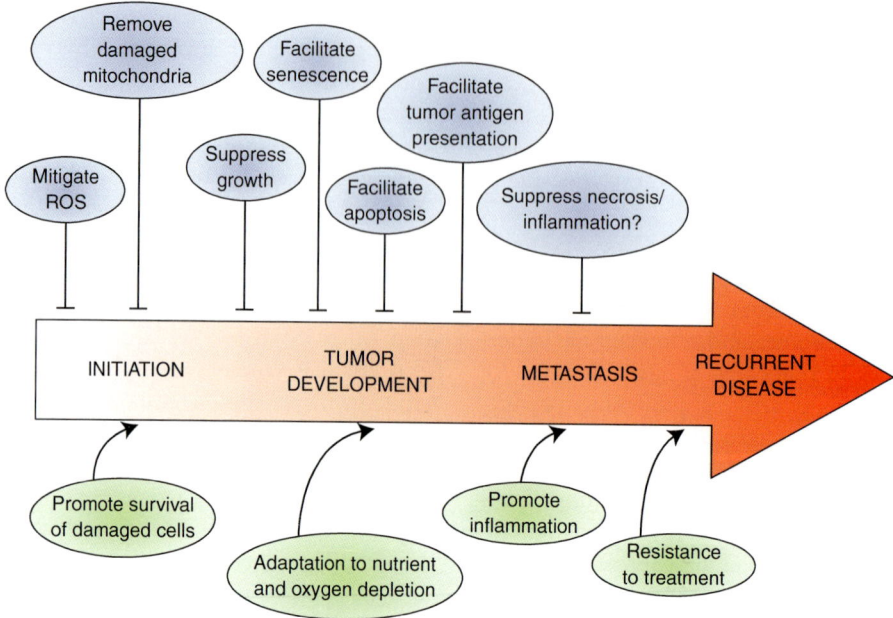

Figure 4. The contrasting roles of autophagy in cancer. Activating autophagy at different stages of cancer yields multiple opposing effects. In healthy cells, autophagy prevents cellular transformation by removing ROS and damaged mitochondria. However, following transformation, activation of autophagy can promote and suppress cancer progression, depending on the timing or stage of disease. Autophagy either mediates its effects directly or "communicates" with other cellular pathways such as senescence, apoptosis, necrosis, and inflammation.

In this regard, autophagy has been reported to be activated in response to TGF-β, resulting in enhancement of TGF-β-mediated growth inhibition in human hepatocellular carcinoma cells (Kiyono et al. 2009). Would this imply that autophagy could attenuate the cell cycle and limit cancer cell proliferation?

Because of these extensive links, autophagy is a very attractive target for cancer therapeutics. In fact, most existing drugs designed to kill cancer cells also induce autophagy (Kondo et al. 2005). Whether autophagy is activated to enhance cell killing or as a counterstress mechanism is governed by many factors, ranging from the nature of the stimulus to the health status of the cell. Recently, the role of autophagy in the interplay of stromal cells within the tumor niche has also been acknowledged (Martinez-Outschoorn et al. 2010, 2011). This adds another level of complexity to the existing theory whereby the induction of autophagy at different stages of cancer yields different effects (Fig. 4).

With this knowledge, treatments, or at least adjuvants that modulate autophagy, could be incorporated into the cancer treatment regime (Amaravadi and Thompson 2007). Perhaps the delivery of autophagic modulators either positive or negative should also be tailored to the stage and type of cancer. In addition, would the modulation of autophagy be a strategy for both initial treatment and for treatment of relapsed disease?

These are all questions that remain to be answered. Nevertheless, the roles of autophagy in modulating various cellular processes involved in cancer progression are now without question. As a result, much excitement currently surrounds the possibility of targeting autophagy for tumor therapy. The issue, however, is not straightforward. As we have highlighted here, autophagy can have both positive and negative effects on tumor development, thus making it difficult to know whether to positively or negatively modulate autophagy in any given scenario. Moreover, it must be remembered that autophagy is not only serving to protect us against cancer, but also has major roles in protecting us against many other forms of disease (Ravikumar et al. 2010). Systemic modulation of autophagy, although beneficial for tumor therapy, may therefore have detrimental roles in normal tissues. Ultimately, however, with an optimistic view, it may be that transient modulation of autophagy may be sufficient for therapy without impact in normal tissues or that ways to selectively target tumor-associated autophagy may be devised. Numerous laboratories are currently addressing these issues, and the exciting answers should be revealed in the not-too-distant future.

ACKNOWLEDGMENTS

We apologize to those researchers whose studies on autophagy we were unable to cite because of the length of this article. We thank Simon Milling and members of the Tumour Cell Death Laboratory for critical reading of the manuscript. Work in the Tumour Cell Death Laboratory is supported by Cancer Research UK and the Association for International Cancer Research.

CONFLICT OF INTEREST STATEMENT

The authors declare that they have no conflicts of interest in relation to this article.

REFERENCES

*Reference is also in this collection.

Adams PD. 2009. Healing and hurting: Molecular mechanisms, functions, and pathologies of cellular senescence. *Mol Cell* **36:** 2–14.

Aita VM, Liang XH, Murty VV, Pincus DL, Yu W, Cayanis E, Kalachikov S, Gilliam TC, Levine B. 1999. Cloning and genomic organization of *beclin 1*, a candidate tumor suppressor gene on chromosome 17q21. *Genomics* **59:** 59–65.

Aki T, Yamaguchi K, Fujimiya T, Mizukami Y. 2003. Phosphoinositide 3-kinase accelerates autophagic cell death during glucose deprivation in the rat cardiomyocyte-derived cell line H9c2. *Oncogene* **22:** 8529–8535.

Amaravadi RK, Thompson CB. 2007. The roles of therapy-induced autophagy and necrosis in cancer treatment. *Clin Cancer Res* **13:** 7271–7279.

Balkwill F, Charles KA, Mantovani A. 2005. Smoldering and polarized inflammation in the initiation and promotion of malignant disease. *Cancer Cell* **7:** 211–217.

Bartek J, Bartkova J, Lukas J. 2007. DNA damage signalling guards against activated oncogenes and tumour progression. *Oncogene* **26:** 7773–7779.

Bellot G, Garcia-Medina R, Gounon P, Chiche J, Roux D, Pouyssegur J, Mazure NM. 2009. Hypoxia-induced autophagy is mediated through hypoxia-inducible factor induction of BNIP3 and BNIP3L via their BH3 domains. *Mol Cell Biol* **29:** 2570–2581.

Bernales S, Schuck S, Walter P. 2007. ER-phagy: Selective autophagy of the endoplasmic reticulum. *Autophagy* **3:** 285–287.

Berry DL, Baehrecke EH. 2007. Growth arrest and autophagy are required for salivary gland cell degradation in *Drosophila*. *Cell* **131:** 1137–1148.

Boya P, Gonzalez-Polo RA, Casares N, Perfettini JL, Dessen P, Larochette N, Metivier D, Meley D, Souquere S, Yoshimori T, et al. 2005. Inhibition of macroautophagy triggers apoptosis. *Mol Cell Biol* **25:** 1025–1040.

Cho DH, Jo YK, Hwang JJ, Lee YM, Roh SA, Kim JC. 2009. Caspase-mediated cleavage of ATG6/Beclin-1 links apoptosis to autophagy in HeLa cells. *Cancer Lett* **274:** 95–100.

Ciechanover A. 2005. Proteolysis: From the lysosome to ubiquitin and the proteasome. *Nat Rev Mol Cell Biol* **6:** 79–87.

Crighton D, Wilkinson S, O'Prey J, Syed N, Smith P, Harrison PR, Gasco M, Garrone O, Crook T, Ryan KM. 2006. DRAM, a p53-induced modulator of autophagy, is critical for apoptosis. *Cell* **126:** 121–134.

Crighton D, Wilkinson S, Ryan KM. 2007. DRAM links autophagy to p53 and programmed cell death. *Autophagy* **3:** 72–74.

Cuervo AM. 2004. Autophagy: In sickness and in health. *Trends Cell Biol* **14:** 70–77.

*Das G, Shravage BV, Baehrecke EH. 2012. Regulation and function of autophagy during cell survival and cell death. *Cold Spring Harb Perspect Biol* **4:** a008813.

Davalos AR, Coppe JP, Campisi J, Desprez PY. 2010. Senescent cells as a source of inflammatory factors for tumor progression. *Cancer Metastasis Rev* **29:** 273–283.

Degenhardt K, Mathew R, Beaudoin B, Bray K, Anderson D, Chen G, Mukherjee C, Shi Y, Gelinas C, Fan Y, et al. 2006. Autophagy promotes tumor cell survival and restricts necrosis, inflammation, and tumorigenesis. *Cancer Cell* **10:** 51–64.

Deretic V. 2005. Autophagy in innate and adaptive immunity. *Trends Immunol* **26:** 523–528.

Dewaele M, Maes H, Agostinis P. 2010. ROS-mediated mechanisms of autophagy stimulation and their relevance in cancer therapy. *Autophagy* **6:** 838–854.

Dorfel D, Appel S, Grunebach F, Weck MM, Muller MR, Heine A, Brossart P. 2005. Processing and presentation of HLA class I and II epitopes by dendritic cells after transfection with in vitro–transcribed MUC1 RNA. *Blood* **105:** 3199–3205.

Edinger AL, Thompson CB. 2004. Death by design: Apoptosis, necrosis and autophagy. *Curr Opin Cell Biol* **16:** 663–669.

Elgendy M, Sheridan C, Brumatti G, Martin SJ. 2011. Oncogenic ras-induced expression of Noxa and Beclin-1 promotes autophagic cell death and limits clonogenic survival. *Mol Cell* **42**: 23–35.

Eskelinen EL, Saftig P. 2009. Autophagy: A lysosomal degradation pathway with a central role in health and disease. *Biochim Biophys Acta* **1793**: 664–673.

Feng Z, Zhang H, Levine AJ, Jin S. 2005. The coordinate regulation of the p53 and mTOR pathways in cells. *Proc Natl Acad Sci* **102**: 8204–8209.

Fricke I, Gabrilovich DI. 2006. Dendritic cells and tumor microenvironment: A dangerous liaison. *Immunol Invest* **35**: 459–483.

Furuta S, Hidaka E, Ogata A, Yokota S, Kamata T. 2004. Ras is involved in the negative control of autophagy through the class I PI3-kinase. *Oncogene* **23**: 3898–3904.

Garcia-Lora A, Algarra I, Garrido F. 2003. MHC class I antigens, immune surveillance, and tumor immune escape. *J Cell Physiol* **195**: 346–355.

Guo JY, Chen HY, Mathew R, Fan J, Strohecker AM, Karsli-Uzunbas G, Kamphorst JJ, Chen G, Lemons JM, Karantza V, et al. 2011. Activated Ras requires autophagy to maintain oxidative metabolism and tumorigenesis. *Genes Dev* **25**: 460–470.

Hanahan D, Weinberg RA. 2011. Hallmarks of cancer: The next generation. *Cell* **144**: 646–674.

Handley ME, Thakker M, Pollara G, Chain BM, Katz DR. 2005. JNK activation limits dendritic cell maturation in response to reactive oxygen species by the induction of apoptosis. *Free Radic Biol Med* **38**: 1637–1652.

Kang MR, Kim MS, Oh JE, Kim YR, Song SY, Kim SS, Ahn CH, Yoo NJ, Lee SH. 2009. Frameshift mutations of autophagy-related genes *ATG2B*, *ATG5*, *ATG9B* and *ATG12* in gastric and colorectal cancers with microsatellite instability. *J Pathol* **217**: 702–706.

Karantza-Wadsworth V, Patel S, Kravchuk O, Chen G, Mathew R, Jin S, White E. 2007. Autophagy mitigates metabolic stress and genome damage in mammary tumorigenesis. *Genes Dev* **21**: 1621–1635.

Kim MS, Jeong EG, Ahn CH, Kim SS, Lee SH, Yoo NJ. 2008. Frameshift mutation of UVRAG, an autophagy-related gene, in gastric carcinomas with microsatellite instability. *Hum Pathol* **39**: 1059–1063.

Kissova I, Deffieu M, Manon S, Camougrand N. 2004. Uth1p is involved in the autophagic degradation of mitochondria. *J Biol Chem* **279**: 39068–39074.

Kiyono K, Suzuki HI, Matsuyama H, Morishita Y, Komuro A, Kano MR, Sugimoto K, Miyazono K. 2009. Autophagy is activated by TGF-β and potentiates TGF-β-mediated growth inhibition in human hepatocellular carcinoma cells. *Cancer Res* **69**: 8844–8852.

Klionsky DJ, Lane JD. 2010. Alternative macroautophagy. *Autophagy* **6**: 201.

Kondo Y, Kanzawa T, Sawaya R, Kondo S. 2005. The role of autophagy in cancer development and response to therapy. *Nat Rev Cancer* **5**: 726–734.

Koneri K, Goi T, Hirono Y, Katayama K, Yamaguchi A. 2007. Beclin 1 gene inhibits tumor growth in colon cancer cell lines. *Anticancer Res* **27**: 1453–1457.

Kraft C, Deplazes A, Sohrmann M, Peter M. 2008. Mature ribosomes are selectively degraded upon starvation by an autophagy pathway requiring the Ubp3p/Bre5p ubiquitin protease. *Nat Cell Biol* **10**: 602–610.

Krizhanovsky V, Xue W, Zender L, Yon M, Hernando E, Lowe SW. 2008. Implications of cellular senescence in tissue damage response, tumor suppression, and stem cell biology. *Cold Spring Harb Symp Quant Biol* **73**: 513–522.

Kroemer G, Marino G, Levine B. 2010. Autophagy and the integrated stress response. *Mol Cell* **40**: 280–293.

Kuma A, Hatano M, Matsui M, Yamamoto A, Nakaya H, Yoshimori T, Ohsumi Y, Tokuhisa T, Mizushima N. 2004. The role of autophagy during the early neonatal starvation period. *Nature* **432**: 1032–1036.

Kvam E, Goldfarb DS. 2007. Nucleus–vacuole junctions and piecemeal microautophagy of the nucleus in S. cerevisiae. *Autophagy* **3**: 85–92.

Lee HK, Mattei LM, Steinberg BE, Alberts P, Lee YH, Chervonsky A, Mizushima N, Grinstein S, Iwasaki A. 2010. In vivo requirement for Atg5 in antigen presentation by dendritic cells. *Immunity* **32**: 227–239.

Lemasters JJ. 2005. Selective mitochondrial autophagy, or mitophagy, as a targeted defense against oxidative stress, mitochondrial dysfunction, and aging. *Rejuvenation Res* **8**: 3–5.

Li Y, Wang LX, Yang G, Hao F, Urba WJ, Hu HM. 2008. Efficient cross-presentation depends on autophagy in tumor cells. *Cancer Res* **68**: 6889–6895.

Liang XH, Jackson S, Seaman M, Brown K, Kempkes B, Hibshoosh H, Levine B. 1999. Induction of autophagy and inhibition of tumorigenesis by *beclin 1*. *Nature* **402**: 672–676.

Lin Y, Choksi S, Shen HM, Yang QF, Hur GM, Kim YS, Tran JH, Nedospasov SA, Liu ZG. 2004. Tumor necrosis factor–induced nonapoptotic cell death requires receptor-interacting protein-mediated cellular reactive oxygen species accumulation. *J Biol Chem* **279**: 10822–10828.

Lum JJ, Bauer DE, Kong M, Harris MH, Li C, Lindsten T, Thompson CB. 2005a. Growth factor regulation of autophagy and cell survival in the absence of apoptosis. *Cell* **120**: 237–248.

Lum JJ, DeBerardinis RJ, Thompson CB. 2005b. Autophagy in metazoans: Cell survival in the land of plenty. *Nat Rev Mol Cell Biol* **6**: 439–448.

Maiuri MC, Malik SA, Morselli E, Kepp O, Criollo A, Mouchel PL, Carnuccio R, Kroemer G. 2009. Stimulation of autophagy by the p53 target gene *Sestrin2*. *Cell Cycle* **8**: 1571–1576.

Marino G, Salvador-Montoliu N, Fueyo A, Knecht E, Mizushima N, Lopez-Otin C. 2007. Tissue-specific autophagy alterations and increased tumorigenesis in mice deficient in Atg4C/autophagin-3. *J Biol Chem* **282**: 18573–18583.

Martinez-Outschoorn UE, Balliet RM, Rivadeneira DB, Chiavarina B, Pavlides S, Wang C, Whitaker-Menezes D, Daumer KM, Lin Z, Witkiewicz AK, et al. 2010. Oxidative stress in cancer associated fibroblasts drives tumor-stroma co-evolution: A new paradigm for understanding tumor metabolism, the field effect and genomic instability in cancer cells. *Cell Cycle* **9**: 3256–3276.

Martinez-Outschoorn UE, Whitaker-Menezes D, Lin Z, Flomenberg N, Howell A, Pestell RG, Sotgia F, Lisanti MP. 2011. Cytokine production and inflammation drive autophagy in the tumor microenvironment: Role of stromal caveolin-1 as a key regulator. *Cell Cycle* **10:** 1784–1793.

Mathew R, Kongara S, Beaudoin B, Karp CM, Bray K, Degenhardt K, Chen G, Jin S, White E. 2007. Autophagy suppresses tumor progression by limiting chromosomal instability. *Genes Dev* **21:** 1367–1381.

Mathew R, Karp CM, Beaudoin B, Vuong N, Chen G, Chen HY, Bray K, Reddy A, Bhanot G, Gelinas C, et al. 2009. Autophagy suppresses tumorigenesis through elimination of p62. *Cell* **137:** 1062–1075.

Mazure NM, Pouyssegur J. 2010. Hypoxia-induced autophagy: Cell death or cell survival? *Curr Opin Cell Biol* **22:** 177–180.

Mehrpour M, Esclatine A, Beau I, Codogno P. 2010. Overview of macroautophagy regulation in mammalian cells. *Cell Res* **20:** 748–762.

Mizushima N, Yamamoto A, Hatano M, Kobayashi Y, Kabeya Y, Suzuki K, Tokuhisa T, Ohsumi Y, Yoshimori T. 2001. Dissection of autophagosome formation using Apg5-deficient mouse embryonic stem cells. *J Cell Biol* **152:** 657–668.

Mizushima N, Levine B, Cuervo AM, Klionsky DJ. 2008. Autophagy fights disease through cellular self-digestion. *Nature* **451:** 1069–1075.

Munz C. 2010. Antigen processing via autophagy—not only for MHC class II presentation anymore? *Curr Opin Immunol* **22:** 89–93.

Narita M, Young AR, Arakawa S, Samarajiwa SA, Nakashima T, Yoshida S, Hong S, Berry LS, Reichelt S, Ferreira M, et al. 2011. Spatial coupling of mTOR and autophagy augments secretory phenotypes. *Science* **332:** 966–970.

Nedjic J, Aichinger M, Emmerich J, Mizushima N, Klein L. 2008. Autophagy in thymic epithelium shapes the T-cell repertoire and is essential for tolerance. *Nature* **455:** 396–400.

Nishida Y, Arakawa S, Fujitani K, Yamaguchi H, Mizuta T, Kanaseki T, Komatsu M, Otsu K, Tsujimoto Y, Shimizu S. 2009. Discovery of Atg5/Atg7-independent alternative macroautophagy. *Nature* **461:** 654–658.

Pattingre S, Tassa A, Qu X, Garuti R, Liang XH, Mizushima N, Packer M, Schneider MD, Levine B. 2005. Bcl-2 antiapoptotic proteins inhibit Beclin 1-dependent autophagy. *Cell* **122:** 927–939.

Pavlides S, Whitaker-Menezes D, Castello-Cros R, Flomenberg N, Witkiewicz AK, Frank PG, Casimiro MC, Wang C, Fortina P, Addya S, et al. 2009. The reverse Warburg effect: Aerobic glycolysis in cancer associated fibroblasts and the tumor stroma. *Cell Cycle* **8:** 3984–4001.

Pursiheimo JP, Rantanen K, Heikkinen PT, Johansen T, Jaakkola PM. 2009. Hypoxia-activated autophagy accelerates degradation of SQSTM1/p62. *Oncogene* **28:** 334–344.

Qu X, Yu J, Bhagat G, Furuya N, Hibshoosh H, Troxel A, Rosen J, Eskelinen EL, Mizushima N, Ohsumi Y, et al. 2003. Promotion of tumorigenesis by heterozygous disruption of the *beclin 1* autophagy gene. *J Clin Invest* **112:** 1809–1820.

Ravikumar B, Sarkar S, Davies JE, Futter M, Garcia-Arencibia M, Green-Thompson ZW, Jimenez-Sanchez M, Korolchuk VI, Lichtenberg M, Luo R, et al. 2010. Regulation of mammalian autophagy in physiology and pathophysiology. *Physiol Rev* **90:** 1383–1435.

Rosenfeldt MT, Ryan KM. 2009. The role of autophagy in tumour development and cancer therapy. *Expert Rev Mol Med* **11:** e36.

Rosenfeldt MT, Ryan KM. 2011. The multiple roles of autophagy in cancer. *Carcinogenesis* **32:** 955–963.

Rouschop KM, Ramaekers CH, Schaaf MB, Keulers TG, Savelkouls KG, Lambin P, Koritzinsky M, Wouters BG. 2009. Autophagy is required during cycling hypoxia to lower production of reactive oxygen species. *Radiother Oncol* **92:** 411–416.

Ryan KM. 2011. p53 and autophagy in cancer: Guardian of the genome meets guardian of the proteome. *Eur J Cancer* **47:** 44–50.

Sakai Y, Oku M, van der Klei IJ, Kiel JA. 2006. Pexophagy: Autophagic degradation of peroxisomes. *Biochim Biophys Acta* **1763:** 1767–1775.

Scarlatti F, Maffei R, Beau I, Ghidoni R, Codogno P. 2008. Non-canonical autophagy: An exception or an underestimated form of autophagy? *Autophagy* **4:** 1083–1085.

Schmid D, Pypaert M, Munz C. 2007. Antigen-loading compartments for major histocompatibility complex class II molecules continuously receive input from autophagosomes. *Immunity* **26:** 79–92.

Scott RC, Juhasz G, Neufeld TP. 2007. Direct induction of autophagy by Atg1 inhibits cell growth and induces apoptotic cell death. *Curr Biol* **17:** 1–11.

Sensi M, Anichini A. 2006. Unique tumor antigens: Evidence for immune control of genome integrity and immunogenic targets for T cell-mediated patient-specific immunotherapy. *Clin Cancer Res* **12:** 5023–5032.

Sharma S, Kelly TK, Jones PA. 2010. Epigenetics in cancer. *Carcinogenesis* **31:** 27–36.

Shoji-Kawata S, Levine B. 2009. Autophagy, antiviral immunity, and viral countermeasures. *Biochim Biophys Acta* **1793:** 1478–1484.

Sit KH, Paramananthan R, Bay BH, Chan HL, Wong KP, Thong P, Watt F. 1996. Sequestration of mitotic (M-phase) chromosomes in autophagosomes: Mitotic programmed cell death in human Chang liver cells induced by an OH* burst from vanadyl(4). *Anat Rec* **245:** 1–8.

Steiger-Barraissoul S, Rami A. 2009. Serum deprivation induced autophagy and predominantly an AIF-dependent apoptosis in hippocampal HT22 neurons. *Apoptosis* **14:** 1274–1288.

Swinson DE, Jones JL, Richardson D, Cox G, Edwards JG, O'Byrne KJ. 2002. Tumour necrosis is an independent prognostic marker in non-small cell lung cancer: Correlation with biological variables. *Lung Cancer* **37:** 235–240.

Takahashi Y, Coppola D, Matsushita N, Cualing HD, Sun M, Sato Y, Liang C, Jung JU, Cheng JQ, Mule JJ, et al. 2007. Bif-1 interacts with Beclin 1 through UVRAG and regulates autophagy and tumorigenesis. *Nat Cell Biol* **9:** 1142–1151.

Takamura A, Komatsu M, Hara T, Sakamoto A, Kishi C, Waguri S, Eishi Y, Hino O, Tanaka K, Mizushima N. 2011. Autophagy-deficient mice develop multiple liver tumors. *Genes Dev* **25:** 795–800.

Tasdemir E, Maiuri MC, Galluzzi L, Vitale I, Djavaheri-Mergny M, D'Amelio M, Criollo A, Morselli E, Zhu C, Harper F, et al. 2008. Regulation of autophagy by cytoplasmic p53. *Nat Cell Biol* **10:** 676–687.

Thorburn J, Moore F, Rao A, Barclay WW, Thomas LR, Grant KW, Cramer SD, Thorburn A. 2005. Selective inactivation of a Fas-associated death domain protein (FADD)–dependent apoptosis and autophagy pathway in immortal epithelial cells. *Mol Biol Cell* **16:** 1189–1199.

Thorburn J, Frankel AE, Thorburn A. 2009a. Regulation of HMGB1 release by autophagy. *Autophagy* **5:** 247–249.

Thorburn J, Horita H, Redzic J, Hansen K, Frankel AE, Thorburn A. 2009b. Autophagy regulates selective HMGB1 release in tumor cells that are destined to die. *Cell Death Differ* **16:** 175–183.

Vandenabeele P, Galluzzi L, Vanden Berghe T, Kroemer G. 2010. Molecular mechanisms of necroptosis: An ordered cellular explosion. *Nat Rev Mol Cell Biol* **11:** 700–714.

Warburg O. 1956. On the origin of cancer cells. *Science* **123:** 309–314.

White E, Lowe SW. 2009. Eating to exit: Autophagy-enabled senescence revealed. *Genes Dev* **23:** 784–787.

Wilkinson S, Ryan KM. 2009. Growth factor signaling permits hypoxia-induced autophagy by a HIF1α-dependent, BNIP3/3L-independent transcriptional program in human cancer cells. *Autophagy* **5:** 1068–1069.

Wilkinson S, O'Prey J, Fricker M, Ryan KM. 2009. Hypoxia-selective macroautophagy and cell survival signaled by autocrine PDGFR activity. *Genes Dev* **23:** 1283–1288.

Wiseman H, Halliwell B. 1996. Damage to DNA by reactive oxygen and nitrogen species: Role in inflammatory disease and progression to cancer. *Biochem J* **313:** 17–29.

Xie Z, Klionsky DJ. 2007. Autophagosome formation: Core machinery and adaptations. *Nat Cell Biol* **9:** 1102–1109.

Yang S, Wang X, Contino G, Liesa M, Sahin E, Ying H, Bause A, Li Y, Stommel JM, Dell'antonio G, et al. 2011. Pancreatic cancers require autophagy for tumor growth. *Genes Dev* **25:** 717–729.

Yee KS, Wilkinson S, James J, Ryan KM, Vousden KH. 2009. PUMA- and Bax-induced autophagy contributes to apoptosis. *Cell Death Differ* **16:** 1135–1145.

Young AR, Narita M, Ferreira M, Kirschner K, Sadaie M, Darot JF, Tavare S, Arakawa S, Shimizu S, Watt FM. 2009. Autophagy mediates the mitotic senescence transition. *Genes Dev* **23:** 798–803.

Yousefi S, Perozzo R, Schmid I, Ziemiecki A, Schaffner T, Scapozza L, Brunner T, Simon HU. 2006. Calpain-mediated cleavage of Atg5 switches autophagy to apoptosis. *Nat Cell Biol* **8:** 1124–1132.

Yue Z, Jin S, Yang C, Levine AJ, Heintz N. 2003. Beclin 1, an autophagy gene essential for early embryonic development, is a haploinsufficient tumor suppressor. *Proc Natl Acad Sci* **100:** 15077–15082.

Zhang H, Bosch-Marce M, Shimoda LA, Tan YS, Baek JH, Wesley JB, Gonzalez FJ, Semenza GL. 2008. Mitochondrial autophagy is an HIF-1-dependent adaptive metabolic response to hypoxia. *J Biol Chem* **283:** 10892–10903.

Zong WX, Ditsworth D, Bauer DE, Wang ZQ, Thompson CB. 2004. Alkylating DNA damage stimulates a regulated form of necrotic cell death. *Genes Dev* **18:** 1272–1282.

Autophagy and Neuronal Cell Death in Neurological Disorders

Ralph A. Nixon[1,2,3] and Dun-Sheng Yang[1,2]

[1]Center for Dementia Research, Nathan S. Kline Institute, Orangeburg, New York 10962
[2]Department of Psychiatry, New York University Langone Medical Center, New York, New York 10016
[3]Department of Cell Biology, New York University Langone Medical Center, New York, New York 10016

Correspondence: nixon@nki.rfmh.org

Autophagy is implicated in the pathogenesis of major neurodegenerative disorders, although concepts about how it influences these diseases are still evolving. Once proposed to be mainly an alternative cell death pathway, autophagy is now widely viewed as both a vital homeostatic mechanism in healthy cells and as an important cytoprotective response mobilized in the face of aging- and disease-related metabolic challenges. In Alzheimer's, Parkinson's, Huntington's, amyotrophic lateral sclerosis, and other diseases, impairment at different stages of autophagy leads to the buildup of pathogenic proteins and damaged organelles, while defeating autophagy's crucial prosurvival and antiapoptotic effects on neurons. The differences in the location of defects within the autophagy pathway and their molecular basis influence the pattern and pace of neuronal cell death in the various neurological disorders. Future therapeutic strategies for these disorders will be guided in part by understanding the manifold impact of autophagy disruption on neurodegenerative diseases.

Soon after the discovery of lysosomes by de Duve in the 1950s, electron microscopists recognized the presence of cytoplasmic organelles within membrane-limited vacuoles (Clark 1957) and observed what appeared to be the progressive breakdown of these contents (Ashford and Porter 1962). Proposing that "prelysosomes" containing sequestered cytoplasm matured to autolysosomes by fusion with primary lysosomes, de Duve and colleagues (de Duve 1963; de Duve and Wattiaux 1966) named this process "autophagy" (self-eating). Neurons, as cells particularly rich in acid phosphatase-positive lysosomes, were a preferred model in the initial investigations of autophagy. Early studies of pathologic states such as neuronal chromatolysis (Holtzman and Novikoff 1965; Holtzman et al. 1967) linked neurodegenerative phenomena to robust proliferation of autophagic vacuoles (AVs) and lysosomes. Although de Duve appreciated the importance of lysosomes for maintaining cell homeostasis, he was especially intrigued with their potential as "suicide bags" capable of triggering cell death by releasing proteases into the cytoplasm. Despite some support for this notion (Brunk and Brun 1972), the concept was not significantly embraced until many decades later. Instead, for many years, lysosomes and autophagy were mainly considered to perform cellular housekeeping and to scavenge and clean up debris during neurodegeneration in preparation for regenerative processes. The connection between autophagy and neuronal cell death reemerged in the 1970s from observations of Clarke and colleagues, who presented evidence that the developing brain deployed autophagy as a form of programmed neuronal cell death during which autophagy was massively up-regulated to eliminate cytoplasmic components, at once killing the neuron and reducing its cell mass for easy removal. Self-degradation was suggested as a more efficient elimination mechanism than apoptosis, which requires a large population of phagocytic cells and access of these cells to the dying region (Baehrecke 2005). Indeed, the best evidence for this process is in the context of massive cell death, as in metamorphosis and involutional states (Das et al. 2012).

Clarke proposed that autophagic cell death (ACD)—type 2 programmed cell death (PCD)—could be a relatively common alternative route to death distinct from apoptosis—type 1 PCD (Clarke 1990)—or caspase-independent cell death—type 3 PCD (Fig. 1). The distinguishing features of ACD are marked proliferation of AVs and progressive disappearance of organelles but relative preservation of cytoskeletal and nuclear integrity until late in the process (Schweichel and Merker 1973; Hornung et al. 1989). In this original concept of ACD or type 2 PCD, death is achieved by autophagic digestion of organelles and essential regulatory molecules and elimination of death inhibitory factors (Baehrecke 2005). With the advent of the molecular era of autophagy research in the 1990s, it became possible to verify the most important implication of ACD, namely, that the death could be prevented by inhibiting autophagy genetically or pharmacologically. Meanwhile, reports of prominent lysosomal/autophagic pathology in Alzheimer's disease (AD) (Cataldo et al. 1997; Nixon et al. 2000, 2005) and other neuropathic states (Anglade et al. 1997; Rubinsztein et al. 2005) raised important questions about whether autophagy pathology signifies a prodeath program or an attempt to maintain survival—a critical question for any potential therapy based on autophagy modulation. In this article, we will examine evidence for the various neuroprotective roles of autophagy and review our current understanding of how specific stages of autophagy may become disrupted and influence the neurodegenerative pattern seen in major adult-onset neurological diseases. We will particularly focus on how neurons regulate the balance between prosurvival autophagy and well-established cell death mechanisms in making life or death decisions.

THE STAGES OF AUTOPHAGY

Since its initial description, autophagy has referred to the process by which unwarranted constituents of the cell are sequestered within vacuoles and degraded by lysosomes (Klionsky et al. 2010). Lysosomal digestion of the cell's own cytoplasmic material is the common cardinal feature of all forms of autophagy. Different mechanisms for delivery of substrates to the lysosomal compartment, however, distinguish among the three major subtypes of autophagy in mammalian cells (Fig. 2). The least well understood of these, microautophagy, is a constitutive and sometimes selective process by which proteins or organelles may be engulfed by invagination of the lysosomal or endosomal membrane and scission of the cargo-containing vesicle inside the lumen. Cargo selection and delivery to endosomes is facilitated by the chaperone Hsc70 and internalization relies on the same endosomal sorting complex required for transport (ESCRT) I and III complexes involved in multivesicular body (MVB) formation (Sahu et al. 2011). In chaperone-mediated autophagy (CMA), cytosolic proteins containing a KFERQ motif are delivered to a LAMP2A-containing complex on the lysosomal membrane via a complex of chaperones, including Hsc70 and HSP70, and cochaperones (EGHSP40, HSP90, HIP, HOP, and BAG1). The substrate is subsequently unfolded and degraded with the help of lysosomal Hsc70 and a complex of proteins that disassembles the transmembrane translocation complex (Bandyopadhyay et al. 2008, 2010). CMA activity is constitutive in most cells but up-regulated under conditions of stress or nutrient deprivation.

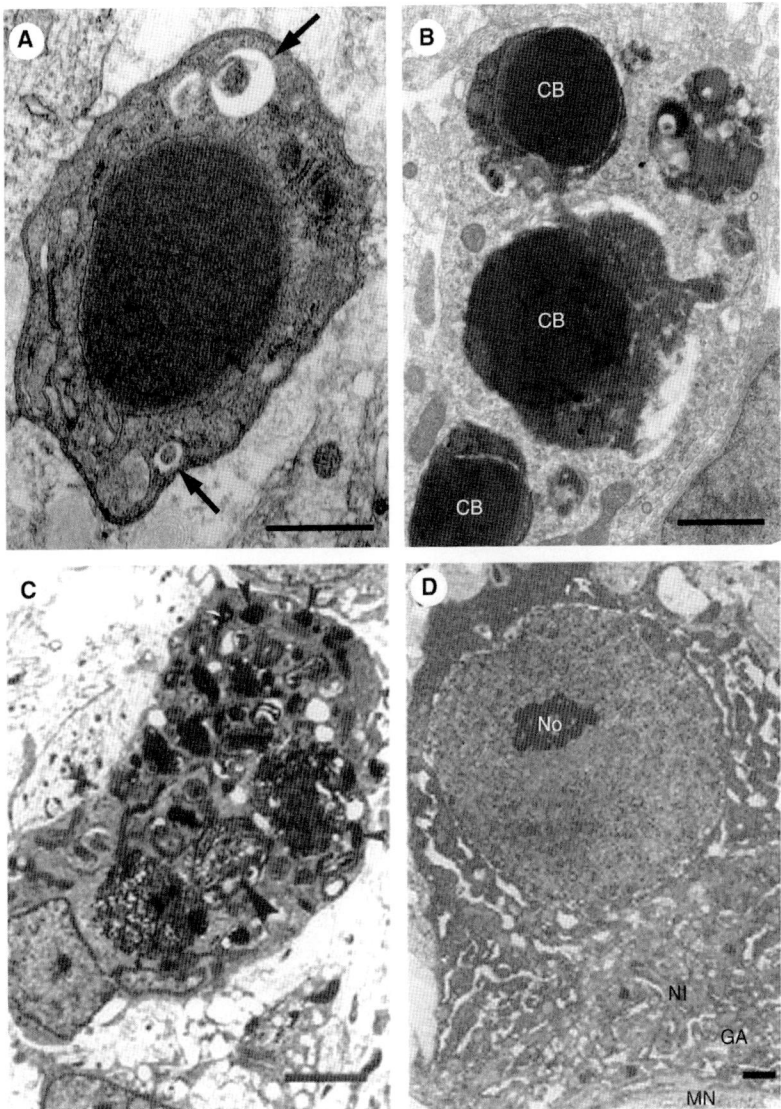

Figure 1. Neuronal cell death: three general morphological types of dying cells in the developing nervous system, as initially classified by Schweichel and Merker (1973) and later Clarke (1990). (*A,B*) Type 1 ("apoptotic") cell death: (*A*) A neuron, from the brain of a postnatal day 6 mouse pup, in the middle of apoptotic degeneration showing cell shrinkage, cytoplasmic condensation, ruffled plasma membrane, and a highly electron-dense nucleus. Endoplasmic reticulum (ER) is still recognizable and some are dilated. A small number of autophagic vacuoles (AVs) can be seen (arrows). (*B*) A late-stage apoptotic neuron displaying electron-dense chromatin balls (CB), each surrounded by a small amount of highly condensed cytoplasm. (Panel from Yang et al. 2008; reprinted, with permission, from the American Association of Pathologists and Bacteriologists.) (*C*) Type 2 ("autophagic") cell death: a deafferented isthmo-optic neuron in developing chick brain after uptake of horseradish peroxidase to highlight (electron dense) endocytic and autophagic compartments. The cell death pattern features pyknosis, abundant AVs, and sometimes dilated ER and mitochondria. (Panel from Hornung et al. 1989; reproduced, with permission, from John Wiley & Sons.) (*D*) Type 3 ("cytoplasmic, nonlysosomal") cell death: a motoneuron displaying markedly dilated rough ER, Golgi, and nuclear envelope, late vacuolization, and increased chromatin granularity. (Panel from Chu-Wang and Oppenheim 1978; reproduced, with permission, from John Wiley & Sons.) Scale bars, 1 μm (*A,B*); 2 μm (*C,D*).

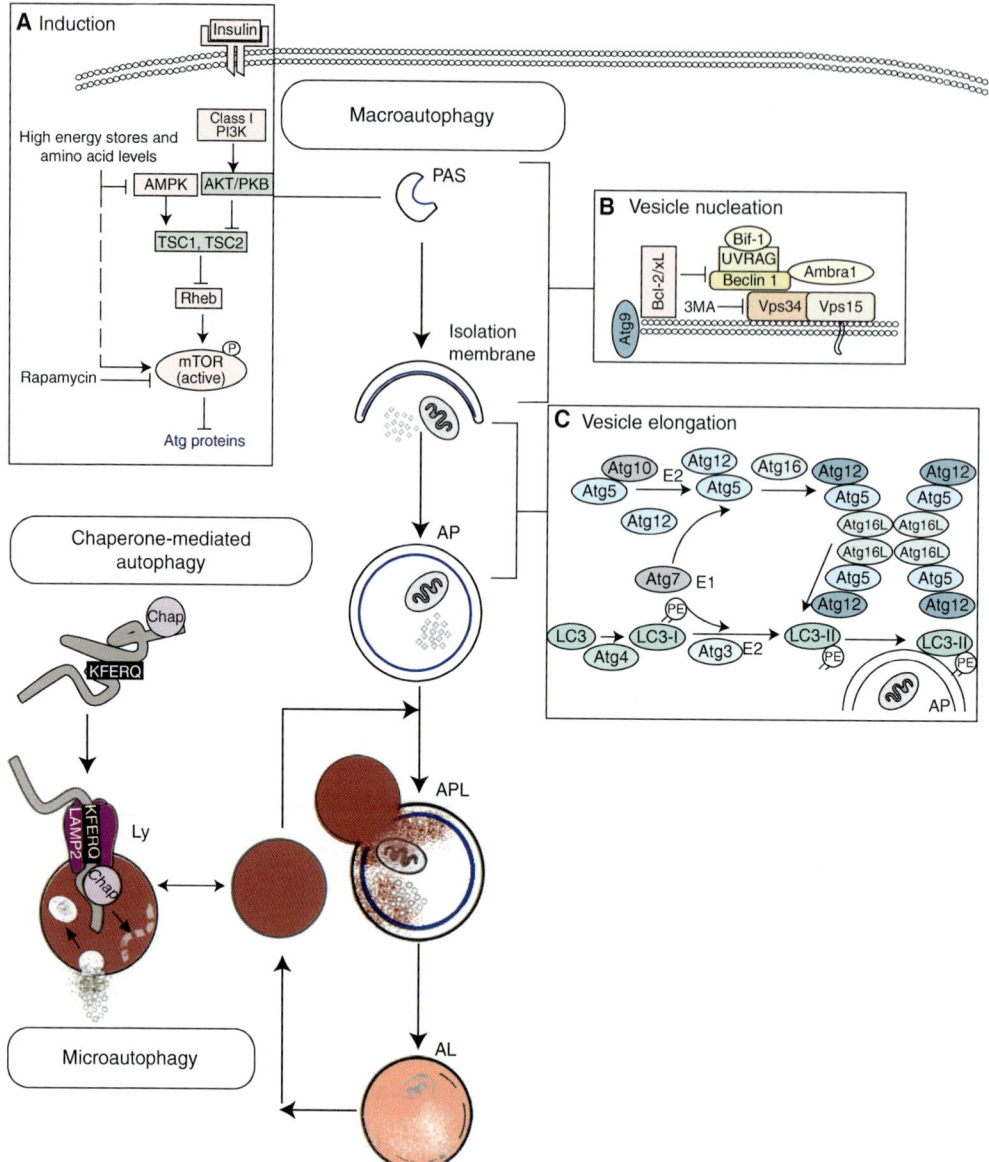

Figure 2. Schematic for the three major subtypes of autophagy—macroautophagy, chaperone-mediated autophagy, and microautophagy. The mammalian target of rapamycin (mTOR) kinase, the principal regulator of macroautophagy, is activated by inhibiting the tuberous sclerosis complex (TSC1) and TSC2, thereby increasing the function of the GTP-binding protein Rheb (A). Insulin or growth factors suppress autophagy by activating the class I phosphatidylinositol 3 kinase (PI3K)–Akt/protein kinase B (PKB) pathway. Abundant intracellular stores of amino acids and ATP suppress autophagy by inhibiting AMP-activated protein kinase (AMPK), an activator of the TSC complex. Macroautophagy is orchestrated by complexes composed of autophagy gene (Atg)-related proteins, which coordinate specific steps in autophagy induction and sequestration as described in the text. The process is initiated when an "isolation" membrane is created from a preautophagosomal structure (PAS) under the direction of the class III PI3K complex and Atg proteins, including Beclin 1 (Atg6) (B). Two ubiquitin-like protein conjugation pathways (C) direct the expansion of the isolation membrane as it sequesters a region of cytoplasm and organelles into a double-membrane-limited autophagosome (AP). (*See facing page for legend.*)

A third form of autophagy, macroautophagy, mediates either bulk or selective degradation of cytoplasmic constituents and is arguably the mechanism with the highest degradative capacity. In this process, a double-membrane structure, termed the "phagophore," originates from several possible sources of cellular membrane and becomes specialized as a preautophagosomal structure (PAS) (Fig. 2B). Under the coordination of two ubiquitin-like protein conjugation pathways (Fig. 2C) (Ohsumi 2001), the PAS elongates into an "isolation" membrane that envelops a region of cytoplasm ultimately forming a double-membrane-limited vesicle, the autophagosome. The outer membrane of the autophagosome fuses with a lysosome or late endosome, creating an autolysosome or an amphisome, respectively. Amphisome formation predominates in axons in which autophagosomes are actively formed and fuse rapidly with late endosomes before being delivered by retrograde transport to the lysosomes concentrated mainly at proximal axon levels and in the perikaryon (Lee et al. 2011). The digestion of sequestered material by dozens of acidic hydrolases (Gordon and Seglen 1988; Liou et al. 1997; Eskelinen 2005; Fader and Colombo 2009; Noda et al. 2009) is promoted when the autolysosome is acidified by vacuolar ATPase (vATPase), a proton pump on the lysosomal membrane, and cathepsins become fully activated (Yoshimori et al. 1991). Autophagy induction is regulated by activity of the mammalian target of rapamycin (mTOR) kinase, which is controlled by growth factor supply and signaling via the PI3 kinase-AKT pathway and by specific amino acids and ATP through AMP kinase (Fig. 2A) (He and Klionsky 2009; Ravikumar et al. 2010). Lysosomes are reformed from autolysosomes through a process also regulated in part by mTOR (Yu et al. 2010).

NEUROPROTECTIVE ACTIONS OF AUTOPHAGY

Macroautophagy is constitutive in neurons and nonselective under nutrient deprivation conditions but can selectively target an organelle that is damaged and exposes a molecular signal that promotes sequestration (Dikic et al. 2010; Weidberg et al. 2011; Youle and Narendra 2011). Recognition of this selective form of macroautophagy has highlighted the working relationship between autophagy and the ubiquitin-proteasome system (UPS) in which a decline in UPS activity may up-regulate autophagy (Fortun et al. 2003; Korolchuk et al. 2009). Certain proteins play regulatory roles in both processes, such as p62, an adaptor protein for autophagy, VCP/P97, which acts through p62 and ubiquitin (Tresse et al. 2010), and Parkin, an E3 ubiquitin ligase implicated in Parkinson's disease (PD) (Yoshii et al. 2011).

It is widely believed that autophagy exerts mainly prosurvival influences on normal cells and its increased activity has been linked to enhanced longevity in animal models (Cavallini et al. 2008; Morselli et al. 2010; Rubinsztein et al. 2011). Autophagy protects cells under stress from nutrient deprivation (Levine and Yuan 2005), loss of energy, or states of protein aggregation by breaking down less essential cellular constituents to recover amino acids, lipids, and other metabolites needed to maintain bioenergetics and synthesize more adaptive proteins, thereby forestalling apoptosis. Another important protective strategy in neurodegenerative disease states is the accelerated elimination by

Figure 2. (*Continued*) Microtubule-associated protein light chain 3-II (LC3-II), formed by phosphoethanolamine conjugation of LC3-I, translocates to the autophagosome membrane (*C*). Digestion of the sequestered cytoplasmic cargo is initiated when a lysosome (Ly) fuses with the outer membrane of the autophagosome to form an autophagolysosome (APL) and lysosomal hydrolases are introduced. The completion of substrate digestion within autolysosomes (AL) ultimately results in lysosome retrieval. In chaperone-mediated autophagy, cytosolic proteins containing a KFERQ motif are selectively chaperoned (Chap) and delivered to the lysosomal lumen for degradation through the binding to LAMP2 located on the lysosomal membrane. In microautophagy, small quantities of proteins or organelles directly enter lysosomes by invagination of the lysosomal or endosomal membrane.

autophagy of damaged and potentially toxic proteins as well as proapoptotic molecules. For example, activated caspase-3 appears in affected neurons in mouse models of AD but is mainly sequestered within autophagosomes (Yang et al. 2008). Similarly, mitophagy, the selective removal of damaged mitochondria by autophagy, is a crucial line of defense against apoptosis believed to fail in some neurodegenerative diseases. In mitophagy, the E3-ubiquitin ligase Parkin, or a related protein such as Nix (Bnip3L) (Sandoval et al. 2008), is recruited to the compromised mitochondrion by certain mitochondrial proteins exposed by the injury. The ubiquitinated protein(s) serve as a binding site for p62 and light chain 3 (LC3) to induce sequestration (Narendra et al. 2008).

AUTOPHAGY AND CELL DEATH PATHWAYS

The cross-regulation of autophagy and apoptosis provides insight into how autophagy protects against apoptotic cell death and conversely how the apoptotic program subverts autophagy to achieve its goal. Central to this cross talk is the interaction between the key apoptosis regulators, Bcl-2 and Bcl-xL, and beclin 1, a component of two separate protein complexes regulating autophagosome formation and endosome maturation, respectively. As a Bcl-2 homology (BH)–3 domain-only protein (Oberstein et al. 2007), beclin 1 may interact with Bcl-2 and Bcl-xL (Pattingre et al. 2005), which diminishes its interactions with these complexes and deprives cells of autophagy's antiapoptotic actions. Mutating the BH3 domain of beclin or the BH2 receptor domain of Bcl-2/Bcl-xL abolishes this capacity to inhibit beclin-dependent autophagy (Maiuri et al. 2007). Conversely, dissociation of beclin 1 and Bcl-2/Bcl-xL, resulting in increased autophagy and antiapoptotic effects, can be achieved in several ways. These include competitive displacement of either partner by interacting proteins (Bad, Bnip3, or Tbid), which include components of the beclin 1 complexes that support autophagy (e.g., UVRAG, Atg14L/Barkor, and Hmgb1). The interaction is also blocked by beclin phosphorylation via death-associated protein kinase (DAPK) (Zalckvar et al. 2009), MAP kinase (ERK1 or JNK1)-mediated phosphorylation of Bcl-2 (Wei et al. 2008; Tang et al. 2010), or Traf6-mediated ubiquitination of beclin 1 (Shi and Kehrl 2010).

Pathological activation of calpains, a family of calcium-activated neutral proteases, can also inhibit autophagy by truncating Atg5, which then translocates from the cytosol to mitochondria where it associates with Bcl-xL and can trigger cytochrome c release (Yousefi 2006). Calpains can also inactivate caspases and potentially convert apoptotic death to necrotic death (Lankiewicz et al. 2000; Syntichaki and Tavernarakis 2002). Whether apoptosis or necrosis ensues depends on various factors, including the developmental status of the neuron, extracellular glutamate levels, cytosolic and mitochondrial calcium levels, mitochondrial membrane potential, ATP levels, and calpain activity (Pang and Geddes 1997; Nasr et al. 2003; Pang et al. 2003).

AUTOPHAGY IN NEURONS AND NEURODEGENERATION

The paucity of AVs in healthy neurons (Mizushima et al. 2004; Nixon, et al. 2005) initially led investigators to suggest that autophagy activity in neurons is relatively low unless induced by some form of stress (Holtzman et al. 1967). Recent evidence has shown, however, that neuronal autophagy is constitutive and, in fact, may be quite active, although AVs remain scarce because lysosomal clearance of these intermediates is very efficient (Boland et al. 2008). One important implication is that, although neurons seem able to accommodate high autophagy activity with minimal ill effects, they are unusually vulnerable to any impairment of autolysosomal clearance. The special reliance of neurons on autophagy has long been suspected from observations that many human primary lysosomal diseases (e.g., lysosomal storage disorders) preferentially affect the brain. This vulnerability of neurons to autophagic/lysosomal dysfunction is not surprising when it is considered that neurons must survive throughout the life of the organism and lose the capability for mitosis, which helps mitotic cells cope with accumulating waste materials by diluting this burden through cell division. Also, neurons must maintain especially large volumes of membrane and cytoplasm associated with long axons and dendrites and must continually traffic autophagy-related compartments long distances back to the cell body where substrate clearance by lysosomes is most active (Lee et al. 2011). Even briefly inhibiting lysosomal proteolytic activity disrupts transport of AVs and causes them to selectively accumulate in dystrophic swelling (Lee et al. 2011) and in perikarya (Ivy et al. 1984; Bednarski et al. 1997).

Investigating cell death pathways in neurons and autophagy is complicated by the slow evolution of neurodegenerative diseases in which prodeath disease and prosurvival factors within a given neuron may battle over many months or years and proteolytic systems may cross talk extensively, yielding in the end a complex picture that defies classification as a distinct pattern of cell death. In AD, for example, neurons with dystrophic axons containing activated caspases, activated calpains, and massive AV accumulations may persist for months or years (Coleman 2005; Rao et al. 2008; Adalbert et al. 2009). Moreover, in most chronic neurodegenerative diseases, only a tiny percentage of the neurons may be degenerating at any given time and may be at different stages of the degenerative process when analyzed. Moreover, cell death patterns may vary across neuronal cell types even in the same disease model, whereas nonneuronal cells in brain may be mounting quite different biochemical responses. Analysis at the single cell level is, therefore, the optimal approach to cell death investigations in brain.

AUTOPHAGY OVERACTIVATION—AN UNCOMMON CELL DEATH PATHWAY IN NEURONS

Although autophagy is generally neuroprotective, can too much of a good thing at the wrong time be deleterious? ACD in the context of involution and metamorphosis during invertebrate development may be among the rare instances in which ACD, as originally defined by Clarke, takes place, and the phenomenon in this situation seems to serve the physiological purpose of eliminating entire cell populations. In mammalian cells, however, ACD strictly defined as death executed by the autocannibalism of constituents essential for cell survival and dependent on autophagy genes is uncommon. An autophagic (type 2 PCD) pattern of death blocked by Atg-gene deletion has been observed in cells in which apoptosis is prevented by deleting Bcl-2 family members, Bax, and Bak or by inhibiting caspases (Shimizu et al. 2004; Yu et al. 2004). Even in this specialized situation, however, it is not entirely clear that autophagy is sufficient to execute death without help from proteases that mediate necrosis. In cells that have competent caspases, additional cell death pathways are invariably activated during cell death showing the type 2 PCD morphological pattern, although autophagy may serve as the upstream trigger (Pyo et al. 2005; Nixon 2006). In some cases, AV proliferation occurs in the context of cell death executed by caspases and may facilitate execution but is not essential for death (Nixon 2006).

Autophagy inhibition by 3-methyl adenine (3MA) has been used to implicate autophagy in cell death execution by showing blocked or delayed cell death after this treatment, although the interpretation of protection via autophagy inhibition should be qualified because this compound inhibits not only the class III PI3 kinase regulating autophagy but also class I PI3 kinase involved in endocytosis. Also, in most of these cases, cytoprotection is not absolute and death ultimately ensues via cytochrome c release and caspase-3 activation (Uchiyama 2001; Canu et al. 2005; Kaasik et al. 2005), indicating that an apoptotic pathway may be operating in parallel. Furthermore, the inhibition of a particular cathepsin (Uchiyama 2001; Canu et al. 2005; Kaasik et al. 2005) also delays or blocks cell death in many of the same models, further supporting the idea that lysosomal destabilization and cathepsin release ultimately triggers apoptosis.

Inhibiting autophagy has been reported to be beneficial in several neurodegenerative states, the clearest example being hypoxic ischemic

(HI) brain injury. In the most compelling of these studies, Atg deletion reduced autophagy-related pathology and nearly completely prevented severe hippocampal damage after HI brain injury in neonatal or adult mice (Koike et al. 2008). In both mouse models, however, effects of autophagy modulation were upstream of the neuronal cell death executioner pathways that were also activated in these models—both caspase-3-dependent and -independent pathways, in the case of the neonatal model, and the caspase-3-independent pathway only in the adult model. Similar observations on this injury model have been made using 3MA to assess the involvement of autophagy in neuronal death. (Wen et al. 2008; Puyal et al. 2009; Piras et al. 2011; Wang et al. 2011; Xin et al. 2011; Zhang 2012). Beneficial effects of autophagic inhibition have also been seen in several models of chronic neurodegenerative diseases. In 6-hydroxydopamine (6-OHDA) injured rat substantia nigra neurons, up-regulated autophagy (as evidenced by AV accumulation) and cell death were prevented by 3MA pretreatment (Li et al. 2011). Neuronal death induced by dysfunctional ESCRT-III, which is associated with frontotemporal dementia linked to chromosome 3, could be delayed by 3MA administration or by knocking down Atg5 and Atg7, suggesting that autophagic stress by excess accumulation of autophagosomes is detrimental to neuronal survival (Lee et al. 2007; Lee and Gao 2009).

When autophagy *inhibition* appears to be cytoprotective, the particular step in autophagy involved is not straightforward. Autophagy induction is often accompanied by increased lysosome biogenesis and up-regulated hydrolase expression (Dehay et al. 2010; Palmieri et al. 2011; Settembre et al. 2011). In situations in which lysosomal function is also compromised or lysosomes are destabilized, it is possible that autophagy inhibitors relieve autophagic stress on compromised lysosomes by slowing delivery of autophagic substrates rather than by attenuating an overaggressive autocannibalistic process. Indeed, healthy neurons in culture seem to tolerate strong autophagy induction (Lee et al. 2011) unless lysosomal function is also impaired. In situations in which blocking autophagy is protective, fending off lysosome destabilization could be more germane to the mechanism of cytoprotection than is the blockade of authentic ACD.

NEURONAL CELL DEATH ASSOCIATED WITH FAILURE OF AUTOPHAGY

In contrast to autophagy overactivation, autophagy failure is commonly linked to a lysosome-dependent form of cell death (Boya 2011), which is relevant to the loss of neurons in various neurodegenerative diseases (Nixon et al. 2008; Settembre et al. 2008a,b; Dehay et al. 2010; Rubinsztein et al. 2011). A neuronal cell death with type 2 morphology is seen in lysosomal storage disorders owing to defects at autolysosomal stages of autophagy—loss of function of lysosomal hydrolases or structural proteins (Hartmann et al. 2000; Koike et al. 2000; Ko et al. 2005; Willenborg et al. 2005; Nixon et al. 2008; Settembre et al. 2008a; Wolfe and Nixon 2012). For example, cathepsin D is ubiquitously expressed, yet mutations that markedly lower cathepsin D activity affect the brain disproportionately causing profuse AV accumulation in neurites and progressive neuronal loss in the neocortex and hippocampus (Tyynela et al. 2000). Lysosomal dysfunction leading to the destabilization of lysosomal membranes and death of nonneuronal cells is well established (Brunk and Svensson 1999; Boya and Kroemer 2008; Boya 2011) and likely relevant to neurons in neurological disorders (Nixon et al. 2008; Dehay et al. 2010), although the evidence is mostly in vitro at present. Autophagy failure at early stages of this pathway may also induce cell death although possibly with less florid autophagic pathology. Deleting factors critical for autophagy induction or autophagosome formation such as FIP200, Atg5, or Atg7 has been shown to induce delayed neuronal cell death and cytoplasmic accumulation of ubiquitinated proteins or organelles (Hara et al. 2006; Komatsu et al. 2006, 2007; Liang et al. 2010).

Autophagy failure, depending on where the defect is along the pathway, can trigger neuronal cell death in several ways. When proteolytic clearance steps are compromised, autolysosomes/lysosomes accumulate mutant and oxidized proteins, protein oligomers and aggregates, damaged organelles, and other incompletely digested products, certain of which increase the permeability of lysosomal membranes causing hydrolases to be released into the cytoplasm, in some cases even from otherwise intact lysosomes (Kroemer and Jaattela 2005). A long list of exogenous agents, including many anticancer drugs (Erdal et al. 2005), are able to disrupt lysosomal membrane integrity directly and induce rapid lysosome-dependent cell death. Similarly, the list of endogenous factors able to induce lysosomal membrane permeabilization (LMP) is growing and includes ceramide, sphingosine, oxidized lipids or lipoproteins, reactive oxygen species (ROS), calpains, certain caspases, and a few proteins/peptides implicated in AD, such as Aβ and ApoE (Brunk and Svensson 1999; Yuan and Yankner 2000; Boya and Kroemer 2008; Johansson et al. 2010; Repnik et al. 2012). Agents that cause cataclysmic disruption of lysosomal membranes are likely to induce rapid necrosis during which released hydrolases participate as both a trigger and as executioners along with caspases that are activated by cathepsin-mediated cleavage (Hayashida et al. 1993; Werneburg et al. 2007). Slower release of cathepsins from lysosomes may first activate apoptotic cascades via the caspase-mediated cleavage of BID releasing of mitochondrial cytochrome c (Guicciardi et al. 2000), degradation of antiapoptotic Bcl-2 homologs, and activation of Bax that releases mitochondrial apoptosis-initiating factor (AIF) (Bidere et al. 2003) and can also induce LMP (Werneburg et al. 2004; Kilinc et al. 2010). In a pathological situation in which either autophagy induction, substrate recognition, or sequestration are impaired, the resultant increase in numbers of damaged mitochondria can trigger apoptosis through the intrinsic pathway and via ROS generation that oxidizes membrane lipids and destabilizes the lysosome membrane. Reduced autophagic elimination of other proapoptotic factors, such as activated caspases, may also accelerate apoptosis under these conditions (Yang et al. 2008).

AUTOPHAGY IN MAJOR NEURODEGENERATIVE DISEASES

Autophagy-related pathology has been noted in late-onset neurodegenerative diseases including AD (Cataldo et al. 1997; Nixon et al. 2005; Nixon and Cataldo 2006), PD (Anglade et al. 1997), amyotrophic lateral sclerosis (ALS) (Hart et al. 1977; Nakano et al. 1993; Sasaki 2011), Huntington's disease (HD) (Roos et al. 1985; Rudnicki et al. 2008), and several others (Liberski et al. 1995; Yue et al. 2002; Rudnicki et al. 2008). The pathology in different disease states has so far been rarely assessed quantitatively. The extent of autophagy waste "storage" in perikarya and neuronal processes and the nature of the AV subtypes that accumulate may differ significantly across these diseases, which in some cases provide useful clues to the underlying disturbance of autophagy. Combined with biochemical analyses of cell and animal models, a picture is emerging from studies of different disorders that autophagy may be altered at early stages (e.g., induction and autophagosome formation) or at late stages (e.g., proteolytic clearance of autophagosomes by lysosomes), or both, which has a different impact on the pattern of neurodegeneration.

PD

In PD, the death mainly of dopaminergic neurons in the substantia nigra is associated with accumulation of the cytosolic protein α-synuclein within inclusions called Lewy bodies. The PD cell death pattern shows features of apoptosis and necrosis as well as increased numbers of autophagosomelike structures (Fig. 3) (Stefanis 2005). Autophagy defects at early stages of autophagy as well as at later lysosomal clearance stages contribute to this complex pattern. Mutations of Parkin (Park2) and PINK1 account for the majority of autosomal recessive cases of parkinsonism (Gasser 2009). Both proteins operate in a pathway that controls mitochondrial fusion-fission events (Yao and Wood 2009) but also play roles in mitophagy that are critical to PD pathogenesis. PINK1, a serine/threonine kinase, localizes to the outer mitochondrial

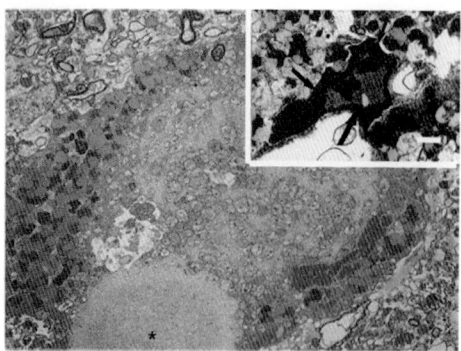

Figure 3. Neuropathology in PD. A neuron of the substantia nigra from a case of juvenile parkinsonism. The pale central area of the perikaryon, next to a Lewy body (asterisk), shows disappearance of rough ER and mitochondria accumulation, which may imply impaired mitophagy. (Panel from Hayashida et al. 1993; reproduced, with permission, from Springer Science + Business Media.) (*Inset*) A dopaminergic neuron from a PD case, filled with neuromelanin-containing vesicles, shows a shrunken nucleus with highly condensed, marginalized chromatin (arrows). Scale bar, 1 μm. (Panel from Hartmann et al. 2000; reproduced, with permission, from the National Academy of Sciences.)

membrane and serves as a sensor of mitochondrial membrane polarization. On normal mitochondria, PINK1 is proteolytically cleaved constitutively (Abeliovich 2010), but it becomes selectively stabilized on the outer mitochondrial membrane when mitochondrial function is compromised. This enables Parkin (a cytoplasmic E3 ubiquitin ligase) to bind and ubiquitinate certain exposed membrane proteins, which recruits p62 and LC3, that initiate mitophagy of the damaged organelle (Geisler et al. 2010; Narendra et al. 2010; Wild and Dikic 2010). It is believed that PD-causing mutations of PINK and Parkin impede mitophagy, causing damaged mitochondria to accumulate and potentially initiate apoptotic events. Additional influences of defective PINK on mitophagy and other mitochondrial dynamics, such as fission–fusion events and ROS production (Wood-Kaczmar et al. 2008; Dagda et al. 2009; Michiorri et al. 2010), may add complexity to the net role mitophagy failure plays in PD. Mutations of α-synuclein in familial PD also disrupt substrate access during CMA by binding abnormally tightly to LAMP2A, thereby blocking not only its own uptake into lysosomes but that of other CMA substrates (Cuervo et al. 2004). Wild-type α-synuclein modified by dopamine, seems to cause similar CMA dysfunction suggesting that this form of α-synuclein toxicity applies to sporadic as well as familial forms of PD (Xilouri et al. 2009).

α-Synuclein is also a substrate for macroautophagy (Webb et al. 2003), which may be recruited as a compensatory mechanism when CMA is impaired. Macroautophagy stimulation by beclin 1 gene transfer, ATG overexpression, or rapamycin significantly ameliorates pathology in α-synuclein PD models (Spencer et al. 2009; Yu et al. 2009). Certain PD-related disease factors, however, may also disrupt lysosomal function and the clearance of substrates by macroautophagy. One example is 1-methyl-4-phenyl-1,2,3,6-tetrahydropyridine (MPTP), a well-characterized neurotoxin model of PD induced by mitochondrial dysfunction and oxidative stress (Yokoyama et al. 2008; Bezard and Przedborski 2011). Lysosome number drops in dopaminergic neurons very early after MPTP exposure, related in part to LMP and cytosolic release of cathepsins—events that are partially prevented by cathepsin inhibitors. Autophagosome accumulation accompanies lysosomal depletion. All of these events and neurodegeneration are attenuated by genetic or pharmacological activation of transcription factor EB (TFEB) or rapamycin, both of which increase lysosomal biogenesis (Dehay et al. 2010) in addition to inducing autophagy. Other factors contributing to general lysosomal dysfunction in some forms of PD include loss-of-function mutations of the lysosomal ATPase, ATP 13A2, a rare cause of PD (Ramirez et al. 2006; Di Fonzo et al. 2007), and mutations in the lysosomal hydrolase glucocerebrosidase that cause the lysosomal storage disorder Gaucher disease and also increase the risk of PD (Sidransky et al. 2009). The neuroprotective effects of enhancing autophagy in most PD models (Dagda et al. 2009; Dadakhujaev et al. 2010), but not all (Zhu et al. 2007), supports the pathogenic importance of autophagy defects observed in PD and PD mouse models.

Polyglutamine Expansion Diseases

Polyglutamine disorders are adult-onset progressive neurodegenerative diseases caused by expansion of a cytosine-adenine-guanine (CAG) repeat motif within the coding region of various genes. The respective proteins containing the long polyglutamine tracts typically form aggregates in the cytoplasm. Generally the age at disease onset correlates with the increased number of CAG repeats. These aggregate-prone proteins tend to be good macroautophagy substrates, and blocking autophagy leads to their increased aggregation and neurotoxicity when the protein is overexpressed (Ravikumar et al. 2004). Besides HD, the first repeat expansion disease discovered to be associated with defective autophagy as discussed below, AVs are increased in animal models of spinocerebellar ataxia 1 and 7, and spinal and bulbar muscular atrophy and neuroprotective effects of autophagic enhancement indirectly imply a role of autophagy in the disease pathogenesis (La Spada and Taylor 2010).

HD, the best known of the polyglutamine disorders, is caused by mutations of the Huntingtin (Htt) gene, encoding a 350-kDa protein that is ubiquitously expressed even though HD principally affects the brain and particularly the striatum. Although the relative neurotoxicity of aggregated versus nonaggregated Htt remains controversial, soluble forms of mutant Htt are increasingly implicated as the toxic species (Dunah et al. 2002; Arrasate et al. 2004; Cui et al. 2006). Several caspases and calpains, known to be activated in HD brain, generate toxic amino-terminal fragments of Htt most closely linked to HD pathogenesis (Ona et al. 1999; Sanchez et al. 1999; Gafni and Ellerby 2002; Graham et al. 2006). Caspase-6 cleavage of mutant Htt seems particularly important in mediating the sensitivity of HD striatal neurons to excitotoxicity (Graham et al. 2006), although polyglutamine expansion also impedes various antiapoptotic effects of Htt including inhibitory effects of caspase-3 activation (Gervais et al. 2002; Zhang et al. 2006). Although activation of apoptotic pathways is likely, the typical morphologic pattern of apoptosis is incomplete (Hickey and Chesselet 2003).

Autophagy was initially implicated in HD from the presence of abnormal endosomes and AV accumulations in HD neurons (Tellez-Nagel et al. 1974) and by the similar morphologies seen in cell models after mutant Htt is overexpressed (Kegel et al. 2000). Htt aggregates, which are poor UPS substrates (Jana et al. 2001), can be cleared by autophagy. Autophagy inducers accelerate their clearance while improving toxicity phenotypes in cell and animal models of HD, whereas inhibiting autophagosome formation or fusion with lysosomes promotes inclusion formation and cell death (Ravikumar et al. 2004). Aggregate-prone Htt seems to be a selective macroautophagy target because autophagy modulation does not affect wild-type Htt levels. Interestingly, acetylation of mutant Htt targets it to autophagosomes (Jeong et al. 2009) and selective clearance of mutant Htt can be achieved by hyperacetylation of the mutant protein with HDAC (histone deacetylase) inhibitors, which are neuroprotective in various HD models (Kazantsev and Thompson 2008).

The efficacy of autophagy inducers in clearing overexpressed mutant Htt from autophagosomes suggests that lysosomal clearance mechanisms are not the primary site of autophagy disruption in HD. Evidence from HD tissue and disease models points instead to defective recognition of cargoes during early sequestration steps. In several HD models, autophagosomes form properly and are cleared even though protein turnover by macroautophagy is paradoxically slow (Martinez-Vicente et al. 2010). The autophagosomes in these cells, however, appear relatively devoid of cargo, suggesting a possible failure of autophagosome membranes to properly engage the substrates during sequestration. In this regard, mutant Htt aggregates sequester beclin 1, which can interfere with autophagosome membrane nucleation (Shibata et al. 2006) and perhaps promote derangement of the sequestration process. Deficient autophagy would be expected to compound effects of Htt-related alterations of mitochondrial function, including loss of antiapoptotic functions of Htt when mutated in HD (Gil and Rego 2008).

ALS and Related Disorders

ALS, the prototypical degenerative disease of motor neurons, is sporadic in approximately 90% of cases but the remaining cases are familial, caused by mutations of at least three different genes, namely, the Cu/Zn superoxide dismutase-1 gene (SOD-1), the TAR DNA-binding protein 43 gene (TARDBP or TDP-43), and the fused in sarcoma/translation in liposarcoma gene (FUS/TLS) (Liscic and Breljak 2011). The recent finding that ALS also arises from mutations of TDP-43, a major cause of frontotemporal dementia (FTD), has highlighted the overlapping clinicopathological features of ALS, FTD-MND (FTD with motor neuron disease), or FTD-U (FTD with ubiquitin-immunoreactive inclusions). In the latter disorders, ubiquitinated inclusions appear in the cytoplasm replacing the normal nuclear staining of TDP-43 (Hirano 1996; Nakano 2000; Mackenzie 2007; Strong 2008; Strong et al. 2009; Geser et al. 2010).

A significant autophagy response occurs very early in ALS. Autophagosomes are frequently seen in otherwise normal-appearing perikarya of motor neurons in sporadic ALS patients (Sasaki 2011) and significantly are elevated by the time clinical symptoms appear (Venkatachalam et al. 2008; Pasquali et al. 2009). Autophagosomes and autolysosomes appear in close association with p62-positive inclusions and both become more frequent in degenerating neurons (Sasaki 2011). The high number of autophagosomes may imply elevated autophagy induction given the decreased mTOR phosphorylation seen in several genetic ALS models (Morimoto et al. 2007; Li et al. 2008; Hetz et al. 2009). Although damaged mitochondria increase in motor neurons in SOD1 mouse ALS models (Liu et al. 2004; Martin et al. 2007; Vande Velde et al. 2008), these are frequently in AVs, suggesting that mitophagy is competent (Wong et al. 1995). Investigations to date have not identified a significant impairment of cargo recognition, sequestration, or autophagosome formation.

Alternatively, or in addition to the possibility of increased autophagy induction, defective autophagosome clearance is also consistent with this morphological pattern and with other data, especially given that neurons usually clear autophagosomes efficiently (Boland et al. 2008). Indeed, growing evidence points to a defect in autophagosome clearance owing to impaired fusion of endosomes, MVP, or lysosomes in some forms of ALS. Defects in the ESCRT machinery that regulates fusion of endosomes/MVB with autophagosomes to produce amphisomes (Filimonenko et al. 2007) have been implicated in frontotemporal dementia linked to chromosome 3 (FTD3) (Skibinski et al. 2005) and ALS (Momeni et al. 2006; Parkinson et al. 2006). Cell models of these diseases have shown that loss of function of various ESCRT genes disrupts autophagosome maturation by impeding autophagosome fusion with lysosomes or endosomes. For example, mutations in charged multivesicular body protein-2B (CHMP2B) in familial ALS disrupt ESCRT pathway function causing aggregates of ubiquitinated proteins and p62 to accumulate (Filimonenko et al. 2007; Lee et al. 2007). Loss-of-function mutations of dynactin, a component of the dynein complex regulating retrograde axonal transport, impair autophagy by disrupting autophagosome retrograde transport and fusion with lysosomes (Ravikumar et al. 2005; Laird et al. 2008).

AD

AD is defined by the presence of "senile or neuritic plaques" consisting of β-amyloid deposits surrounding foci of degenerating or dystrophic axons and dendrites and intraneuronal filamentous aggregates of the microtubule-associated protein tau. AD leads to widespread death of many subtypes of neurons beginning in the hippocampus and spreading progressively to the cortex in a circuit-selective pattern. Neuronal death in AD does not conform to a conventional cell death pattern and may vary in its features among neuron subtypes, showing in various proportions the features of apoptosis (e.g., activation of multiple caspases), necrosis (e.g., calpain activation and cathepsin up-regulation), and florid autophagic/lysosomal pathology.

Abnormalities of the lysosomal system in AD (Fig. 4) are a continuum that includes endosome anomalies associated with endocytic

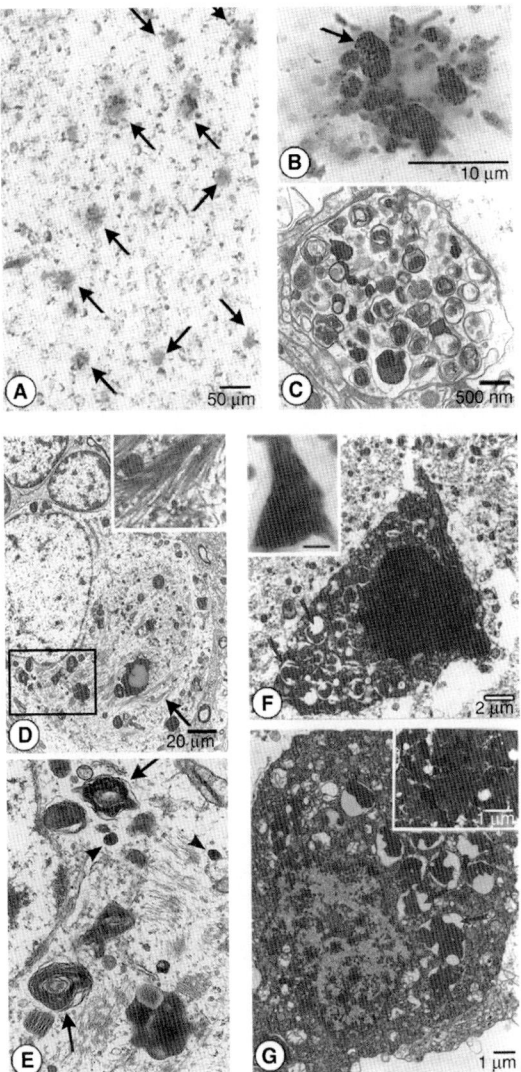

Figure 4. Autophagic pathology and neurodegeneration in AD brains. (A) Cathepsin D (Cat D) immunocytochemistry of AD neocortex reveals that many amyloid plaques are cathepsin D positive (arrows). (B) A plaque of higher magnification depicts numerous Cat D-positive dystrophic neurites. (C) Electron microscopy shows that dystrophic neurites are filled mainly with autophagosomes and autolysosomes containing undigested materials. (D) A tangle-bearing neuron showing scattered bundles of paired helical filaments (arrow and *inset*) and a peripherally displaced but otherwise normal nucleus. (E) The boxed area of D is shown at higher magnification. The cytoplasm contains numerous AVs including double-membrane dense structures and multilamellar bodies (arrows) as well as many small dense bodies or lysosomes (arrowheads). (Panels C–E from Nixon et al. 2005; reproduced, with permission, from the author.) (F,G) Neurons in AD neocortex immunolabeled with Cat D antibodies depicting a later stage of degeneration with altered nuclear morphology and accumulation of hydrolase-positive lipofuscin or lipofuscin-containing autolysosomes (arrows). (F) The same neocortical pyramidal neuron showing atrophic changes by Nissl staining (F, *inset*; scale bar, 10 μm) contains AVs and many lipofuscin granules when examined ultrastructurally. The nucleus is highly electron dense and the morphology of the whole cell suggests dark-neuron degeneration, whereas the nucleus of a neuron in G shows chromatin clumping and marginalization, suggesting ongoing apoptotic changes. G (*inset*) shows hydrolase-positive aggregates of lipofuscin granules. (Panels F and G from Cataldo et al. 1994; reprinted, with permission, from Elsevier.)

pathway up-regulation and increased lysosome biogenesis—the earliest appearing disease-specific cellular pathologies in the disease (Nixon et al. 2006). AVs progressively accumulate profusely in affected neurons and become the predominant organelles within enormously swollen dystrophic neurites, a hallmark of AD neuropathology. The selective accumulation of AVs of all types within the dystrophic swellings likely reflects selective and specific impairment of the axonal transport of autophagy/lysosomal-related compartments (Lee et al. 2011). Neuritic dystrophy in AD is much more extensive than that seen in other aging-related neurodegenerative diseases, and the huge mass of accumulated waste protein within AVs is reminiscent of that seen in some primary lysosomal storage disorders (Nixon et al. 2008).

Current evidence indicates that autophagy is principally defective at the stage of autolysosomal proteolysis in AD. The presence of abundant mature autophagosomes containing partially digested substrates and cathepsins (Nixon et al. 2005) contrasts with the exceptional efficiency of AV clearance in normal neurons (Boland et al. 2008). Neuritic dystrophy and selective AV accumulation can be reproduced by blocking lysosomal proteolysis pharmacologically or genetically in in vitro and in vivo models (reviewed in Nixon and Yang 2011), whereas they are not seen when autophagy is strongly induced in otherwise healthy neurons (Boland et al. 2008; Lee et al. 2011). Lysosomal proteolysis failure is further supported by the recently identified role of presenilin 1 (PS1) as a chaperone essential for the delivery of the proton pump, vacuolar vATPase, to lysosomes, which is essential for lysosome acidification and protease activation. Mutations of PS1, the most common cause of early-onset familial AD (Sherrington et al. 1995), lead to markedly defective lysosomal acidification and autolysosomal maturation (Lee et al. 2010), explaining why PS1 mutations potentiate autophagic/lysosomal, amyloid, and tau pathologies as well as accelerate neuronal cell death in patients with PS-FAD or mouse models (Cataldo et al. 2004).

Inheritance of the ε4 allele of the apolipoprotein E gene (ApoE4) is the strongest genetic risk factor for late-onset AD. ApoE4 has numerous biological activities but prominent among them are allele-specific effects on cholesterol regulation that promote the specific endocytic dysfunction arising in early AD (Cataldo et al. 1996, 2000). Moreover, a unique proteolytically processing pattern of ApoE4 in lysosomes yields a "molten globule" structure that induces reactive intermediates, which destabilize lysosomal membranes leading to lysosomal leakage and apoptosis (Ji et al. 2002, 2006; Mahley and Huang 2006). The expression of ApoE4, but not the ApoE3 allele, increases levels of intracellular Aβ peptide (Aβ), enlarges lysosomes and alters their morphology in a mouse AD model, and causes neurodegeneration of neurons typically vulnerable in AD (Belinson et al. 2008). Similarly, overexpression of human Aβ 42, but not Aβ 40, in Drosophila neurons induces age-related autophagic/lysosomal dysfunction and neurotoxicity (Ling et al. 2009) believed to arise from lysosomal membrane destabilization mediated directly by Aβ (Yang et al. 1998; Glabe 2001) or by oxidized, incompletely degraded autophagic substrates (Terman and Brunk 2006; Kurz et al. 2008).

The prominent defects in clearance of autophagic substrates by lysosomes may be compounded by an increased induction of autophagy in AD, which is known to accelerate autophagy pathology (Boland et al. 2008) and increase neurotoxicity. Lysosome biogenesis in AD brain and AD mouse models (Cataldo et al. 1995, 1996, 2004; Ginsberg et al. 2010) is up-regulated as is the transcription of a wide range of autophagy-related genes in the AD cortex (Lipinski et al. 2010) and in CA1 hippocampal neurons specifically (S Ginsberg and R Nixon, unpubl.). Some analyses of mTOR signaling, although not all, are consistent with increased autophagy induction (Bhaskar et al. 2009; Caccamo et al. 2010; Nixon and Yang 2011). Although an initial report of lowered beclin 1 levels in brain suggested that early steps of autophagy may be impaired (Pickford et al. 2008; Jaeger et al. 2010), this finding has not been confirmed in a later more detailed analysis (R Nixon, P Mohan, and E Masliah, unpubl.).

The significance of autophagy failure to the pathogenesis of AD is underscored by the efficacy of autophagy enhancement in ameliorating AD-related pathologies (e.g., Aβ deposition and tau pathology) and synaptic and cognitive deficits in AD models (Karuppagounder et al. 2009; Ahmed et al. 2010; Greco et al. 2010; Spilman et al. 2010; Majumder et al. 2011; Tian et al. 2011; Yang et al. 2011). Interestingly, in one recent study, rapamycin administration before development of AD pathology and presumably before lysosomal failure delayed the onset and diminished the severity of the AD phenotype, whereas administration after the appearance of pathology had little beneficial effect, highlighting the importance of lysosomal failure (Majumder et al. 2011). The therapeutic efficacy of genetic manipulations selectively targeting the lysosome to increase its proteolytic efficiency (Sun et al. 2008; Yang et al. 2011) has established the particular importance of lysosomal proteolysis defects in promoting disease progression.

CONCLUDING REMARKS

Investigations of autophagy in neurodegenerative disease have greatly intensified since the first reviews appeared on this subject a decade ago (Nixon et al. 1995). Autophagy is now firmly established as a vital homeostatic and quality control mechanism in healthy neurons and as a cytoprotective response when further induced in chronic neurodegenerative diseases. Autophagy protects neurons against metabolic challenges by generating supplies for energy production and cell repair, eliminating toxic damaged proteins and organelles, and suppressing apoptotic signaling. With increasing cellular age, autophagy becomes less efficient despite a greater need to eliminate mounting levels of damaged proteins, organelles, and other apoptosis-inducing stimuli. In AD, PD, and likely other disorders, progressive autophagy dysfunction, including impairment driven by causative gene mutations or other disease risk factors, compounds the adverse effects of aging on autophagy in promoting neuronal cell death. Defects in autophagy that impair the early stages of this process—induction or cargo recognition and sequestration—cause autophagic substrates, including mitochondria, to accumulate in the cytoplasm and trigger apoptosis, as in PD. Failure at later stages of autophagy related to the efficient digestion of sequestered substrates leads to the buildup of partially digested toxic products in autolysosomes and lysosomes, as seen in AD. In this

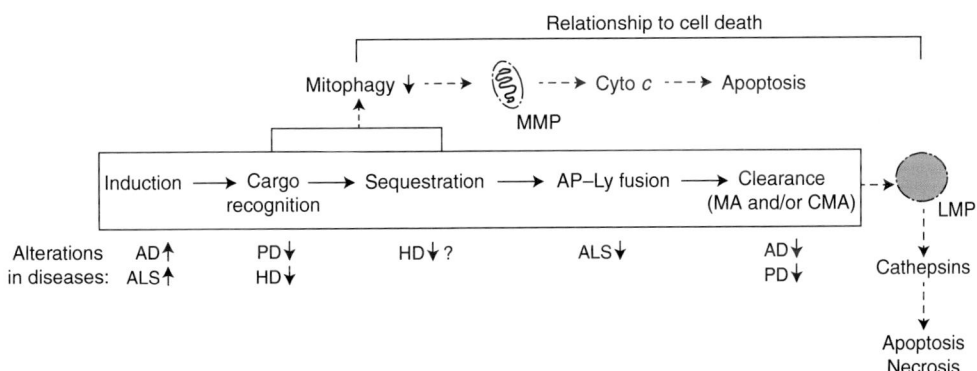

Figure 5. Possible states of impairment in the autophagy pathway in different neurodegenerative diseases and possible links to neuronal cell death. The autophagy defect may conceivably promote neuronal cell death via two possible mechanisms: (1) impaired mitophagy resulting in accumulation of damaged mitochondria and mitochondrial membrane permeability (MMP) leading to cytochrome c release and apoptotic cell death; and (2) impaired lysosomal clearance of autophagy substrate leading to lysosomal membrane permeabilization (LMP) and cathepsin release into cytosol, thereby inducing either apoptotic or necrotic cell death. Increased autophagy induction in the face of a downstream block in the pathway may be a counterproductive response. AP, autophagosome; CMA, chaperone-mediated autophagy; Cyto c, cytochrome c; Ly, lysosome; MA, macroautophagy.

situation, consequent injury and destabilization of lysosomal membranes and leakage of cathepsins can trigger either apoptotic or necrotic cascades (Fig. 5). Extensive cross talk between regulators of autophagy and apoptosis has provided insight into how autophagy protects against apoptosis and how the balance between these opposing forces contributes to the complex patterns of neuronal cell death seen in nervous system diseases. Although much more work is needed to clarify these mechanisms, recent preclinical investigations in neurodegenerative disease models have provided glimpses that autophagy modulation may be a fruitful therapeutic strategy.

ACKNOWLEDGMENTS

We are grateful to Nicole Piorkowski for assistance with manuscript preparation, and to Corrinne Peterhoff for assistance with figure preparation and artwork. Studies from our laboratories are supported by the National Institute on Aging and the Alzheimer's Association.

REFERENCES

*Reference is also in this collection.

Abeliovich A. 2010. Parkinson's disease: Mitochondrial damage control. *Nature* **463:** 744–745.

Adalbert R, Nogradi A, Babetto E, Janeckova L, Walker SA, Kerschensteiner M, Misgeld T, Coleman MP. 2009. Severely dystrophic axons at amyloid plaques remain continuous and connected to viable cell bodies. *Brain* **132:** 402–416.

Ahmed T, Enam SA, Gilani AH. 2010. Curcuminoids enhance memory in an amyloid-infused rat model of Alzheimer's disease. *Neuroscience* **169:** 1296–1306.

Anglade P, Vyas S, Javoy-Agid F, Herrero MT, Michel PP, Marquez J, Mouatt-Prigent A, Ruberg M, Hirsch EC, Agid Y. 1997. Apoptosis and autophagy in nigral neurons of patients with Parkinson's disease. *Histol Histopathol* **12:** 25–31.

Arrasate M, Mitra S, Schweitzer ES, Segal MR, Finkbeiner S. 2004. Inclusion body formation reduces levels of mutant huntingtin and the risk of neuronal death. *Nature* **431:** 805–810.

Ashford TP, Porter KR. 1962. Cytoplasmic components in hepatic cell lysosomes. *J Cell Biol* **12:** 198–202.

Baehrecke EH. 2005. Autophagy: Dual roles in life and death? *Nat Rev Mol Cell Biol* **6:** 505–510.

Bandyopadhyay U, Kaushik S, Varticovski L, Cuervo AM. 2008. The chaperone-mediated autophagy receptor organizes in dynamic protein complexes at the lysosomal membrane. *Mol Cell Biol* **28:** 5747–5763.

Bandyopadhyay U, Sridhar S, Kaushik S, Kiffin R, Cuervo AM. 2010. Identification of regulators of chaperone-mediated autophagy. *Mol Cell* **39:** 535–547.

Bednarski E, Ribak CE, Lynch G. 1997. Suppression of cathepsins B and L causes a proliferation of lysosomes and the formation of meganeurites in hippocampus. *J Neurosci* **17:** 4006–4021.

Belinson H, Lev D, Masliah E, Michaelson DM. 2008. Activation of the amyloid cascade in apolipoprotein E4 transgenic mice induces lysosomal activation and neurodegeneration resulting in marked cognitive deficits. *J Neurosci* **28:** 4690–4701.

Bezard E, Przedborski S. 2011. A tale on animal models of Parkinson's disease. *Mov Disord* **26:** 993–1002.

Bhaskar K, Miller M, Chludzinski A, Herrup K, Zagorski M, Lamb BT. 2009. The PI3K-Akt-mTOR pathway regulates Aβ oligomer induced neuronal cell cycle events. *Mol Neurodegener* **4:** 14.

Bidere N, Lorenzo HK, Carmona S, Laforge M, Harper F, Dumont C, Senik A. 2003. Cathepsin D triggers Bax activation, resulting in selective apoptosis-inducing factor (AIF) relocation in T lymphocytes entering the early commitment phase to apoptosis. *J Biol Chem* **278:** 31401–31411.

Boland B, Kumar A, Lee S, Platt FM, Wegiel J, Yu WH, Nixon RA. 2008. Autophagy induction and autophagosome clearance in neurons: Relationship to autophagic pathology in Alzheimer's disease. *J Neurosci* **28:** 6926–6937.

Boya P. 2011. Lysosomal function and dysfunction: Mechanism and disease. *Antioxid Redox Signal* **17:** 766–774.

Boya P, Kroemer G. 2008. Lysosomal membrane permeabilization in cell death. *Oncogene* **27:** 6434–6451.

Brunk U, Brun A. 1972. The effect of aging on lysosomal permeability in nerve cells of the central nervous system. An enzyme histochemical study in rat. *Histochemie* **30:** 315–324.

Brunk UT, Svensson I. 1999. Oxidative stress, growth factor starvation and Fas activation may all cause apoptosis through lysosomal leak. *Redox Rep* **4:** 3–11.

Caccamo A, Majumder S, Richardson A, Strong R, Oddo S. 2010. Molecular interplay between mammalian target of rapamycin (mTOR), amyloid-β, and Tau: Effects on cognitive impairments. *J Biol Chem* **285:** 13107–13120.

Canu N, Tufi R, Serafino AL, Amadoro G, Ciotti MT, Calissano P. 2005. Role of the autophagic-lysosomal system on low potassium-induced apoptosis in cultured cerebellar granule cells. *J Neurochem* **92:** 1228–1242.

Cataldo AM, Hamilton DJ, Nixon RA. 1994. Lysosomal abnormalities in degenerating neurons link neuronal compromise to senile plaque development in Alzheimer disease. *Brain Res* **640:** 68–80.

Cataldo AM, Barnett JL, Berman SA, Li J, Quarless S, Bursztajn S, Lippa C, Nixon RA. 1995. Gene expression and cellular content of cathepsin D in Alzheimer's disease brain: Evidence for early up-regulation of the endosomal-lysosomal system. *Neuron* **14:** 671–680.

Cataldo AM, Hamilton DJ, Barnett JL, Paskevich PA, Nixon RA. 1996. Properties of the endosomal-lysosomal system in the human central nervous system: Disturbances mark most neurons in populations at risk to degenerate in Alzheimer's disease. *J Neurosci* **16:** 186–199.

Cataldo AM, Barnett JL, Pieroni C, Nixon RA. 1997. Increased neuronal endocytosis and protease delivery to early endosomes in sporadic Alzheimer's disease: Neuropathologic evidence for a mechanism of increased β-amyloidogenesis. *J Neurosci* **17:** 6142–6151.

Cataldo AM, Peterhoff CM, Troncoso JC, Gomez-Isla T, Hyman BT, Nixon RA. 2000. Endocytic pathway abnormalities precede amyloid β deposition in sporadic Alzheimer's disease and Down syndrome: Differential effects of APOE genotype and presenilin mutations. *Am J Pathol* **157:** 277–286.

Cataldo AM, Peterhoff CM, Schmidt SD, Terio NB, Duff K, Beard M, Mathews PM, Nixon RA. 2004. Presenilin mutations in familial Alzheimer disease and transgenic mouse models accelerate neuronal lysosomal pathology. *J Neuropathol Exp Neurol* **63:** 821–830.

Cavallini G, Donati A, Gori Z, Bergamini E. 2008. Towards an understanding of the anti-aging mechanism of caloric restriction. *Curr Aging Sci* **1:** 4–9.

Chu-Wang IW, Oppenheim RW. 1978. Cell death of motoneurons in the chick embryo spinal cord. I. A light and electron microscopic study of naturally occurring and induced cell loss during development. *J Comp Neurol* **177:** 33–57.

Clark SL Jr. 1957. Cellular differentiation in the kidneys of newborn mice studies with the electron microscope. *J Biophys Biochem Cytol* **3:** 349–362.

Clarke PG. 1990. Developmental cell death: Morphological diversity and multiple mechanisms. *Anat Embryol* **181:** 195–213.

Coleman M. 2005. Axon degeneration mechanisms: Commonality amid diversity. *Nat Rev Neurosci* **6:** 889–898.

Cuervo AM, Stefanis L, Fredenburg R, Lansbury PT, Sulzer D. 2004. Impaired degradation of mutant α-synuclein by chaperone-mediated autophagy. *Science* **305:** 1292–1295.

Cui L, Jeong H, Borovecki F, Parkhurst CN, Tanese N, Krainc D. 2006. Transcriptional repression of PGC-1α by mutant huntingtin leads to mitochondrial dysfunction and neurodegeneration. *Cell* **127:** 59–69.

Dadakhujaev S, Noh HS, Jung EJ, Cha JY, Baek SM, Ha JH, Kim DR. 2010. Autophagy protects the rotenone-induced cell death in α-synuclein overexpressing SH-SY5Y cells. *Neurosci Lett* **472:** 47–52.

Dagda RK, Cherra SJ 3rd, Kulich SM, Tandon A, Park D, Chu CT. 2009. Loss of PINK1 function promotes mitophagy through effects on oxidative stress and mitochondrial fission. *J Biol Chem* **284:** 13843–13855.

* Das G, Shravage BV, Baehrecke EH. 2012. Regulation and function of autophagy during cell survival and cell death. *Cold Spring Harb Perspect Biol* **4:** a008813.

de Duve C. 1963. The lysosome. *Sci Am* **208:** 64–72.

de Duve C, Wattiaux R. 1966. Functions of lysosomes. *Annu Rev Physiol* **28:** 435–492.

Dehay B, Bove J, Rodriguez-Muela N, Perier C, Recasens A, Boya P, Vila M. 2010. Pathogenic lysosomal depletion in Parkinson's disease. *J Neurosci* **30:** 12535–12544.

Di Fonzo A, Chien HF, Socal M, Giraudo S, Tassorelli C, Iliceto G, Fabbrini G, Marconi R, Fincati E, Abbruzzese G, et al. 2007. ATP13A2 missense mutations in juvenile parkinsonism and young onset Parkinson disease. *Neurology* **68:** 1557–1562.

Dikic I, Johansen T, Kirkin V. 2010. Selective autophagy in cancer development and therapy. *Cancer Res* **70:** 3431–3434.

Dunah AW, Jeong H, Griffin A, Kim YM, Standaert DG, Hersch SM, Mouradian MM, Young AB, Tanese N, Krainc D. 2002. Sp1 and TAFII130 transcriptional activity disrupted in early Huntington's disease. *Science* **296:** 2238–2243.

Erdal H, Berndtsson M, Castro J, Brunk U, Shoshan MC, Linder S. 2005. Induction of lysosomal membrane permeabilization by compounds that activate p53-independent apoptosis. *Proc Natl Acad Sci* **102:** 192–197.

Eskelinen EL. 2005. Maturation of autophagic vacuoles in Mammalian cells. *Autophagy* **1:** 1–10.

Eskelinen EL, Saftig P. 2009. Autophagy: A lysosomal degradation pathway with a central role in health and disease. *Biochim Biophys Acta* **1793:** 664–673.

Fader CM, Colombo MI. 2009. Autophagy and multivesicular bodies: Two closely related partners. *Cell Death Differ* **16:** 70–78.

Filimonenko M, Stuffers S, Raiborg C, Yamamoto A, Malerod L, Fisher EMC, Isaacs A, Brech A, Stenmark H, Simonsen A. 2007. Functional multivesicular bodies are required for autophagic clearance of protein aggregates associated with neurodegenerative disease. *J Cell Biol* **179:** 485–500.

Fortun J, Dunn WA Jr, Joy S, Li J, Notterpek L. 2003. Emerging role for autophagy in the removal of aggresomes in Schwann cells. *J Neurosci* **23:** 10672–10680.

Gafni J, Ellerby LM. 2002. Calpain activation in Huntington's disease. *J Neurosci* **22:** 4842–4849.

Gasser T. 2009. Molecular pathogenesis of Parkinson disease: Insights from genetic studies. *Expert Rev Mol Med* **11:** e22.

Geisler S, Holmstrom KM, Skujat D, Fiesel FC, Rothfuss OC, Kahle PJ, Springer W. 2010. PINK1/Parkin-mediated mitophagy is dependent on VDAC1 and p62/SQSTM1. *Nat Cell Biol* **12:** 119–131.

Gervais FG, Singaraja R, Xanthoudakis S, Gutekunst CA, Leavitt BR, Metzler M, Hackam AS, Tam J, Vaillancourt JP, Houtzager V, et al. 2002. Recruitment and activation of caspase-8 by the Huntingtin-interacting protein Hip-1 and a novel partner Hippi. *Nat Cell Biol* **4:** 95–105.

Geser F, Lee VM, Trojanowski JQ. 2010. Amyotrophic lateral sclerosis and frontotemporal lobar degeneration: A spectrum of TDP-43 proteinopathies. *Neuropathology* **30:** 103–112.

Gil JM, Rego AC. 2008. Mechanisms of neurodegeneration in Huntington's disease. *Eur J Neurosci* **27:** 2803–2820.

Ginsberg SD, Alldred MJ, Counts SE, Cataldo AM, Neve RL, Jiang Y, Wuu J, Chao MV, Mufson EJ, Nixon RA, et al. 2010. Microarray analysis of hippocampal CA1 neurons implicates early endosomal dysfunction during Alzheimer's disease progression. *Biol Psychiatry* **68:** 885–893.

Glabe C. 2001. Intracellular mechanisms of amyloid accumulation and pathogenesis in Alzheimer's disease. *J Mol Neurosci* **17:** 137–145.

Gordon PB, Seglen PO. 1988. Prelysosomal convergence of autophagic and endocytic pathways. *Biochem Biophys Res Commun* **151:** 40–47.

Graham RK, Deng Y, Slow EJ, Haigh B, Bissada N, Lu G, Pearson J, Shehadeh J, Bertram L, Murphy Z, et al. 2006. Cleavage at the caspase-6 site is required for neuronal dysfunction and degeneration due to mutant huntingtin. *Cell* **125:** 1179–1191.

Greco SJ, Bryan KJ, Sarkar S, Zhu X, Smith MA, Ashford JW, Johnston JM, Tezapsidis N, Casadesus G. 2010. Leptin reduces pathology and improves memory in a transgenic mouse model of Alzheimer's disease. *J Alzheimers Dis* **19:** 1155–1167.

Guicciardi ME, Deussing J, Miyoshi H, Bronk SF, Svingen PA, Peters C, Kaufmann SH, Gores GJ. 2000. Cathepsin B contributes to TNF-α-mediated hepatocyte apoptosis by promoting mitochondrial release of cytochrome c. *J Clin Invest* **106:** 1127–1137.

Hara T, Nakamura K, Matsui M, Yamamoto A, Nakahara Y, Suzuki-Migishima R, Yokoyama M, Mishima K, Saito I, Okano H, et al. 2006. Suppression of basal autophagy in neural cells causes neurodegenerative disease in mice. *Nature* **441:** 885–889.

Hart MN, Cancilla PA, Frommes S, Hirano A. 1977. Anterior horn cell degeneration and Bunina-type inclusions associated with dementia. *Acta Neuropathol* **38:** 225–228.

Hartmann A, Hunot S, Michel PP, Muriel MP, Vyas S, Faucheux BA, Mouatt-Prigent A, Turmel H, Srinivasan A, Ruberg M, et al. 2000. Caspase-3: A vulnerability factor and final effector in apoptotic death of dopaminergic neurons in Parkinson's disease. *Proc Natl Acad Sci* **97:** 2875–2880.

Hayashida K, Oyanagi S, Mizutani Y, Yokochi M. 1993. An early cytoplasmic change before Lewy body maturation: An ultrastructural study of the substantia nigra from an autopsy case of juvenile parkinsonism. *Acta Neuropathol* **85:** 445–448.

He C, Klionsky DJ. 2009. Regulation mechanisms and signaling pathways of autophagy. *Annu Rev Genet* **43:** 67–93.

Hetz C, Thielen P, Matus S, Nassif M, Court F, Kiffin R, Martinez G, Cuervo AM, Brown RH, Glimcher LH. 2009. XBP-1 deficiency in the nervous system protects against amyotrophic lateral sclerosis by increasing autophagy. *Genes Dev* **23:** 2294–2306.

Hickey MA, Chesselet MF. 2003. Apoptosis in Huntington's disease. *Prog Neuropsychopharmacol Biol Psychiatry* **27:** 255–265.

Hirano A. 1996. Neuropathology of ALS: An overview. *Neurology* **47:** S63–S66.

Holtzman E, Novikoff AB. 1965. Lysomes in the rat sciatic nerve following crush. *J Cell Biol* **27:** 651–669.

Holtzman E, Novikoff AB, Villaverde H. 1967. Lysosomes and GERL in normal and chromatolytic neurons of the rat ganglion nodosum. *J Cell Biol* **33:** 419–435.

Hornung JP, Koppel H, Clarke PG. 1989. Endocytosis and autophagy in dying neurons: An ultrastructural study in chick embryos. *J Comp Neurol* **283:** 425–437.

Ivy GO, Schottler F, Wenzel J, Baudry M, Lynch G. 1984. Inhibitors of lysosomal enzymes: Accumulation of lipofuscin-like dense bodies in the brain. *Science* **226:** 985–987.

Jaeger PA, Pickford F, Sun CH, Lucin KM, Masliah E, Wyss-Coray T. 2010. Regulation of amyloid precursor protein processing by the Beclin 1 complex. *PLoS ONE* **5:** e11102.

Jana NR, Zemskov EA, Wang G, Nukina N. 2001. Altered proteasomal function due to the expression of polyglutamine-expanded truncated N-terminal huntingtin induces apoptosis by caspase activation through mitochondrial cytochrome c release. *Hum Mol Genet* **10:** 1049–1059.

Jeong H, Then F, Melia TJ Jr, Mazzulli JR, Cui L, Savas JN, Voisine C, Paganetti P, Tanese N, Hart AC, et al. 2009. Acetylation targets mutant huntingtin to autophagosomes for degradation. *Cell* **137:** 60–72.

Ji ZS, Miranda RD, Newhouse YM, Weisgraber KH, Huang Y, Mahley RW. 2002. Apolipoprotein E4 potentiates amyloid β peptide-induced lysosomal leakage and apoptosis in neuronal cells. *J Biol Chem* **277:** 21821–21828.

Ji ZS, Mullendorff K, Cheng IH, Miranda RD, Huang Y, Mahley RW. 2006. Reactivity of apolipoprotein E4 and amyloid β peptide: Lysosomal stability and neurodegeneration. *J Biol Chem* **281:** 2683–2692.

Johansson AC, Appelqvist H, Nilsson C, Kagedal K, Roberg K, Ollinger K. 2010. Regulation of apoptosis-associated lysosomal membrane permeabilization. *Apoptosis* **15:** 527–540.

Kaasik A, Rikk T, Piirsoo A, Zharkovsky T, Zharkovsky A. 2005. Up-regulation of lysosomal cathepsin L and autophagy during neuronal death induced by reduced serum and potassium. *Eur J Neurosci* **22:** 1023–1031.

Karuppagounder SS, Pinto JT, Xu H, Chen HL, Beal MF, Gibson GE. 2009. Dietary supplementation with resveratrol reduces plaque pathology in a transgenic model of Alzheimer's disease. *Neurochem Int* **54:** 111–118.

Kazantsev AG, Thompson LM. 2008. Therapeutic application of histone deacetylase inhibitors for central nervous system disorders. *Nat Rev Drug Discov* **7:** 854–868.

Kegel KB, Kim M, Sapp E, McIntyre C, Castano JG, Aronin N, DiFiglia M. 2000. Huntingtin expression stimulates endosomal-lysosomal activity, endosome tubulation, and autophagy. *J Neurosci* **20:** 7268–7278.

Kilinc M, Gursoy-Ozdemir Y, Gurer G, Erdener SE, Erdemli E, Can A, Dalkara T. 2010. Lysosomal rupture, necroapoptotic interactions and potential crosstalk between cysteine proteases in neurons shortly after focal ischemia. *Neurobiol Dis* **40:** 293–302.

Klionsky DJ, Codogno P, Cuervo AM, Deretic V, Elazar Z, Fueyo-Margareto J, Gewirtz DA, Kroemer G, Levine B, Mizushima N, et al. 2010. A comprehensive glossary of autophagy-related molecules and processes. *Autophagy* **6:** 438–448.

Ko DC, Milenkovic L, Beier SM, Manuel H, Buchanan J, Scott MP. 2005. Cell-autonomous death of cerebellar purkinje neurons with autophagy in Niemann-Pick type C disease. *PLoS Genet* **1:** 81–95.

Koike M, Nakanishi H, Saftig P, Ezaki J, Isahara K, Ohsawa Y, Schulz-Schaeffer W, Watanabe T, Waguri S, Kametaka S, et al. 2000. Cathepsin D deficiency induces lysosomal storage with ceroid lipofuscin in mouse CNS neurons. *J Neurosci* **20:** 6898–6906.

Koike M, Shibata M, Tadakoshi M, Gotoh K, Komatsu M, Waguri S, Kawahara N, Kuida K, Nagata S, Kominami E, et al. 2008. Inhibition of autophagy prevents hippocampal pyramidal neuron death after hypoxic-ischemic injury. *Am J Pathol* **172:** 454–469.

Komatsu M, Waguri S, Chiba T, Murata S, Iwata J, Tanida I, Ueno T, Koike M, Uchiyama Y, Kominami E, et al. 2006. Loss of autophagy in the central nervous system causes neurodegeneration in mice. *Nature* **441:** 880–884.

Komatsu M, Wang QJ, Holstein GR, Friedrich VL Jr, Iwata J, Kominami E, Chait BT, Tanaka K, Yue Z. 2007. Essential role for autophagy protein Atg7 in the maintenance of axonal homeostasis and the prevention of axonal degeneration. *Proc Natl Acad Sci* **104:** 14489–14494.

Korolchuk VI, Mansilla A, Menzies FM, Rubinsztein DC. 2009. Autophagy inhibition compromises degradation of ubiquitin-proteasome pathway substrates. *Mol Cell* **33:** 517–527.

Kroemer G, Jaattela M. 2005. Lysosomes and autophagy in cell death control. *Nat Rev Cancer* **5:** 886–897.

Kurz T, Terman A, Gustafsson B, Brunk UT. 2008. Lysosomes and oxidative stress in aging and apoptosis. *Biochim Biophys Acta* **1780:** 1291–1303.

Laird FM, Farah MH, Ackerley S, Hoke A, Maragakis N, Rothstein JD, Griffin J, Price DL, Martin LJ, Wong PC. 2008. Motor neuron disease occurring in a mutant dynactin mouse model is characterized by defects in vesicular trafficking. *J Neurosci* **28:** 1997–2005.

Lankiewicz S, Marc Luetjens C, Truc Bui N, Krohn AJ, Poppe M, Cole GM, Saido TC, Prehn JH. 2000. Activation of calpain I converts excitotoxic neuron death into a caspase-independent cell death. *J Biol Chem* **275:** 17064–17071.

La Spada AR, Taylor JP. 2010. Repeat expansion disease: Progress and puzzles in disease pathogenesis. *Nat Rev Genet* **11:** 247–258.

Lee JA, Gao FB. 2009. Inhibition of autophagy induction delays neuronal cell loss caused by dysfunctional ESCRT-III in frontotemporal dementia. *J Neurosci* **29:** 8506–8511.

Lee JA, Beigneux A, Ahmad ST, Young SG, Gao FB. 2007. ESCRT-III dysfunction causes autophagosome accumulation and neurodegeneration. *Curr Biol* **17:** 1561–1567.

Lee JH, Yu WH, Kumar A, Lee S, Mohan PS, Peterhoff CM, Marinez-Vicente M, Massey AG, Sovak G, Uchiyama Y, et al. 2010. Lysosomal proteolysis and autophagy require presenilin 1 and are disrupted by Alzheimer-related PS1 mutations. *Cell* **141:** 1146–1158.

Lee S, Sato Y, Nixon RA. 2011. Lysosomal proteolysis inhibition selectively disrupts axonal transport of degradative organelles and causes an Alzheimer's-like axonal dystrophy. *J Neurosci* **31:** 7817–7830.

Levine B, Yuan J. 2005. Autophagy in cell death: An innocent convict? *J Clin Invest* **115:** 2679–2688.

Li L, Zhang X, Le W. 2008. Altered macroautophagy in the spinal cord of SOD1 mutant mice. *Autophagy* **4:** 290–293.

Li L, Wang X, Fei X, Xia L, Qin Z, Liang Z. 2011. Parkinson's disease involves autophagy and abnormal distribution of cathepsin L. *Neurosci Lett* **489:** 62–67.

Liang CC, Wang C, Peng X, Gan B, Guan JL. 2010. Neural-specific deletion of FIP200 leads to cerebellar degeneration caused by increased neuronal death and axon degeneration. *J Biol Chem* **285:** 3499–3509.

Liberski PP, Budka H, Yanagihara R, Gajdusek DC. 1995. Neuroaxonal dystrophy in experimental Creutzfeldt-Jakob disease: Electron microscopical and immunohistochemical demonstration of neurofilament accumulations within affected neurites. *J Comp Pathol* **112:** 243–255.

Ling D, Song HJ, Garza D, Neufeld TP, Salvaterra PM. 2009. Aβ42-induced neurodegeneration via an age-dependent autophagic-lysosomal injury in Drosophila. *PLoS ONE* **4:** e4201.

Liou W, Geuze HJ, Geelen MJ, Slot JW. 1997. The autophagic and endocytic pathways converge at the nascent autophagic vacuoles. *J Cell Biol* **136:** 61–70.

Lipinski MM, Zheng B, Lu T, Yan Z, Py BF, Ng A, Xavier RJ, Li C, Yankner BA, Scherzer CR, et al. 2010. Genome-wide analysis reveals mechanisms modulating autophagy in normal brain aging and in Alzheimer's disease. *Proc Natl Acad Sci* **107:** 14164–14169.

Liscic RM, Breljak D. 2011. Molecular basis of amyotrophic lateral sclerosis. *Prog Neuropsychopharmacol Biol Psychiatry* **35:** 370–372.

Liu J, Lillo C, Jonsson PA, Vande Velde C, Ward CM, Miller TM, Subramaniam JR, Rothstein JD, Marklund S, Andersen PM, et al. 2004. Toxicity of familial ALS-linked SOD1 mutants from selective recruitment to spinal mitochondria. *Neuron* **43:** 5–17.

Mackenzie IR. 2007. The neuropathology of FTD associated with ALS. *Alzheimer Dis Assoc Disord* **21:** S44–S49.

Mahley RW, Huang Y. 2006. Apolipoprotein (apo) E4 and Alzheimer's disease: Unique conformational and biophysical properties of apoE4 can modulate neuropathology. *Acta Neurol Scand Suppl* **185:** 8–14.

Maiuri MC, Zalckvar E, Kimchi A, Kroemer G. 2007. Self-eating and self-killing: Crosstalk between autophagy and apoptosis. *Nat Rev Mol Cell Biol* **8:** 741–752.

Majumder S, Richardson A, Strong R, Oddo S. 2011. Inducing autophagy by rapamycin before, but not after, the formation of plaques and tangles ameliorates cognitive deficits. *PLoS ONE* **6:** e25416.

Martin DN, Balgley B, Dutta S, Chen J, Rudnick P, Cranford J, Kantartzis S, DeVoe DL, Lee C, Baehrecke EH. 2007. Proteomic analysis of steroid-triggered autophagic programmed cell death during Drosophila development. *Cell Death Differ* **14:** 916–923.

Martinez-Vicente M, Talloczy Z, Wong E, Tang G, Koga H, Kaushik S, de Vries R, Arias E, Harris S, Sulzer D, et al. 2010. Cargo recognition failure is responsible for inefficient autophagy in Huntington's disease. *Nat Neurosci* **13:** 567–576.

Michiorri S, Gelmetti V, Giarda E, Lombardi F, Romano F, Marongiu R, Nerini-Molteni S, Sale P, Vago R, Arena G, et al. 2010. The Parkinson-associated protein PINK1 interacts with Beclin1 and promotes autophagy. *Cell Death Differ* **17:** 962–974.

Tang D, Kang R, Livesey KM, Cheh CW, Farkas A, Loughran P, Hoppe G, Bianchi ME, Tracey KJ, Zeh HJ 3rd, et al. 2010. Endogenous HMGB1 regulates autophagy. *J Cell Biol* **190:** 881–892.

Tellez-Nagel I, Johnson AB, Terry RD. 1974. Studies on brain biopsies of patients with Huntington's chorea. *J Neuropathol Exp Neurol* **33:** 308–332.

Terman A, Brunk UT. 2006. Oxidative stress, accumulation of biological "garbage," and aging. *Antioxid Redox Signal* **8:** 197–204.

Tian Y, Bustos V, Flajolet M, Greengard P. 2011. A small-molecule enhancer of autophagy decreases levels of Aβ and APP-CTF via Atg5-dependent autophagy pathway. *FASEB J* **25:** 1934–1942.

Tresse E, Salomons FA, Vesa J, Bott LC, Kimonis V, Yao TP, Dantuma NP, Taylor JP. 2010. VCP/p97 is essential for maturation of ubiquitin-containing autophagosomes and this function is impaired by mutations that cause IBMPFD. *Autophagy* **6:** 217–227.

Tyynela J, Sohar I, Sleat DE, Gin RM, Donnelly RJ, Baumann M, Haltia M, Lobel P. 2000. A mutation in the ovine cathepsin D gene causes a congenital lysosomal storage disease with profound neurodegeneration. *Embo J* **19:** 2786–2792.

Uchiyama Y. 2001. Autophagic cell death and its execution by lysosomal cathepsins. *Arch Histol Cytol* **64:** 233–246.

Vande Velde C, Miller TM, Cashman NR, Cleveland DW. 2008. Selective association of misfolded ALS-linked mutant SOD1 with the cytoplasmic face of mitochondria. *Proc Natl Acad Sci* **105:** 4022–4027.

Venkatachalam K, Long AA, Elsaesser R, Nikolaeva D, Broadie K, Montell C. 2008. Motor deficit in a *Drosophila* model of mucolipidosis type IV due to defective clearance of apoptotic cells. *Cell* **135:** 838–851.

Wang JY, Xia Q, Chu KT, Pan J, Sun LN, Zeng B, Zhu YJ, Wang Q, Wang K, Luo BY. 2011. Severe global cerebral ischemia-induced programmed necrosis of hippocampal CA1 neurons in rat is prevented by 3-methyladenine: A widely used inhibitor of autophagy. *J Neuropathol Exp Neurol* **70:** 314–322.

Webb JL, Ravikumar B, Atkins J, Skepper JN, Rubinsztein DC. 2003. α-Synuclein is degraded by both autophagy and the proteasome. *J Biol Chem* **278:** 25009–25013.

Wei Y, Pattingre S, Sinha S, Bassik M, Levine B. 2008. JNK1-mediated phosphorylation of Bcl-2 regulates starvation-induced autophagy. *Mol Cell* **30:** 678–688.

Weidberg H, Shvets E, Elazar Z. 2011. Biogenesis and cargo selectivity of autophagosomes. *Annu Rev Biochem* **80:** 125–156.

Wen YD, Sheng R, Zhang LS, Han R, Zhang X, Zhang XD, Han F, Fukunaga K, Qin ZH. 2008. Neuronal injury in rat model of permanent focal cerebral ischemia is associated with activation of autophagic and lysosomal pathways. *Autophagy* **4:** 762–769.

Werneburg N, Guicciardi ME, Yin XM, Gores GJ. 2004. TNF-α-mediated lysosomal permeabilization is FAN and caspase 8/Bid dependent. *Am J Physiol Gastrointest Liver Physiol* **287:** G436–G443.

Werneburg NW, Guicciardi ME, Bronk SF, Kaufmann SH, Gores GJ. 2007. Tumor necrosis factor-related apoptosis-inducing ligand activates a lysosomal pathway of apoptosis that is regulated by Bcl-2 proteins. *J Biol Chem* **282:** 28960–28970.

Wild P, Dikic I. 2010. Mitochondria get a Parkin' ticket. *Nat Cell Biol* **12:** 104–106.

Willenborg M, Schmidt CK, Braun P, Landgrebe J, von Figura K, Saftig P, Eskelinen EL. 2005. Mannose 6-phosphate receptors, Niemann-Pick C2 protein, and lysosomal cholesterol accumulation. *J Lipid Res* **46:** 2559–2569.

Wolfe DM, Nixon RA. 2012. Autophagy failure in Alzheimer's disease and lysosomal storage disorders: A common pathway to neurodegeneration. In *Autophagy of the nervous system: Cellular self-digestion in neurons and neurological diseases* (ed. Yue Z, Chu CT). World Scientific, Hackensack, NJ.

Wong PC, Pardo CA, Borchelt DR, Lee MK, Copeland NG, Jenkins NA, Sisodia SS, Cleveland DW, Price DL. 1995. An adverse property of a familial ALS-linked SOD1 mutation causes motor neuron disease characterized by vacuolar degeneration of mitochondria. *Neuron* **14:** 1105–1116.

Wood-Kaczmar A, Gandhi S, Yao Z, Abramov AY, Miljan EA, Keen G, Stanyer L, Hargreaves I, Klupsch K, Deas E, et al. 2008. PINK1 is necessary for long term survival and mitochondrial function in human dopaminergic neurons. *PLoS ONE* **3:** e2455.

Xilouri M, Vogiatzi T, Vekrellis K, Park D, Stefanis L. 2009. Abberant α-synuclein confers toxicity to neurons in part through inhibition of chaperone-mediated autophagy. *PLoS ONE* **4:** e5515.

Xin XY, Pan J, Wang XQ, Ma JF, Ding JQ, Yang GY, Chen SD. 2011. 2-Methoxyestradiol attenuates autophagy activation after global ischemia. *Can J Neurol Sci* **38:** 631–638.

Yamashima T, Oikawa S. 2009. The role of lysosomal rupture in neuronal death. *Prog Neurobiol* **89:** 343–358.

Yang AJ, Chandswangbhuvana D, Margol L, Glabe CG. 1998. Loss of endosomal/lysosomal membrane impermeability is an early event in amyloid Aβ1–42 pathogenesis. *J Neurosci Res* **52:** 691–698.

Yang DS, Kumar A, Stavrides P, Peterson J, Peterhoff CM, Pawlik M, Levy E, Cataldo AM, Nixon RA. 2008. Neuronal apoptosis and autophagy cross talk in aging PS/APP mice, a model of Alzheimer's disease. *Am J Pathol* **173:** 665–681.

Yang DS, Stavrides P, Mohan PS, Kaushik S, Kumar A, Ohno M, Schmidt SD, Wesson D, Bandyopadhyay U, Jiang Y, et al. 2011. Reversal of autophagy dysfunction in the TgCRND8 mouse model of Alzheimer's disease ameliorates amyloid pathologies and memory deficits. *Brain* **134:** 258–277.

Yao Z, Wood NW. 2009. Cell death pathways in Parkinson's disease: Role of mitochondria. *Antioxid Redox Signal* **11:** 2135–2149.

Yokoyama H, Kuroiwa H, Yano R, Araki T. 2008. Targeting reactive oxygen species, reactive nitrogen species and inflammation in MPTP neurotoxicity and Parkinson's disease. *Neurol Sci* **29:** 293–301.

Yoshii SR, Kishi C, Ishihara N, Mizushima N. 2011. Parkin mediates proteasome-dependent protein degradation and rupture of the outer mitochondrial membrane. *J Biol Chem* **286:** 19630–19640.

Yoshimori T, Yamamoto A, Moriyama Y, Futai M, Tashiro Y. 1991. Bafilomycin A1, a specific inhibitor of vacuolar-type H^+-ATPase, inhibits acidification and protein degradation in lysosomes of cultured cells. *J Biol Chem* **266**: 17707–17712.

Youle RJ, Narendra DP. 2011. Mechanisms of mitophagy. *Nat Rev Mol Cell Biol* **12**: 9–14.

Yousefi S. 2006. Calpain-mediated cleavage of Atg5 switches autophagy to apoptosis. *Nat Cell Biol* **8**: 1124–1132.

Yu L, Alva A, Su H, Dutt P, Freundt E, Welsh S, Baehrecke EH, Lenardo MJ. 2004. Regulation of an ATG7-beclin 1 program of autophagic cell death by caspase-8. *Science* **304**: 1500–1502.

Yu WH, Dorado B, Figueroa HY, Wang L, Planel E, Cookson MR, Clark LN, Duff KE. 2009. Metabolic activity determines efficacy of macroautophagic clearance of pathological oligomeric α-synuclein. *Am J Pathol* **175**: 736–747.

Yu L, McPhee CK, Zheng L, Mardones GA, Rong Y, Peng J, Mi N, Zhao Y, Liu Z, Wan F, et al. 2010. Termination of autophagy and reformation of lysosomes regulated by mTOR. *Nature* **465**: 942–946.

Yuan J, Yankner BA. 2000. Apoptosis in the nervous system. *Nature* **407**: 802–809.

Yue Z, Horton A, Bravin M, DeJager PL, Selimi F, Heintz N. 2002. A novel protein complex linking the δ2 glutamate receptor and autophagy: Implications for neurodegeneration in lurcher mice. *Neuron* **35**: 921–933.

Zalckvar E, Berissi H, Mizrachy L, Idelchuk Y, Koren I, Eisenstein M, Sabanay H, Pinkas-Kramarski R, Kimchi A. 2009. DAP-kinase-mediated phosphorylation on the BH3 domain of beclin 1 promotes dissociation of beclin 1 from Bcl-XL and induction of autophagy. *EMBO Rep* **10**: 285–292.

Zhang Y, Leavitt BR, van Raamsdonk JM, Dragatsis I, Goldowitz D, MacDonald ME, Hayden MR, Friedlander RM. 2006. Huntingtin inhibits caspase-3 activation. *EMBO J* **25**: 5896–5906.

Zhang J, Zhang Y, Li J, Xing S, Li C, Li Y, Dang C, Fan Y, Yu J, Pei Z, et al. 2012. Autophagosomes accumulation is associated with β-amyloid deposits and secondary damage in the thalamus after focal cortical infarction in hypertensive rats. *J Neurochem* **120**: 564–573.

Zhu JH, Horbinski C, Guo F, Watkins S, Uchiyama Y, Chu CT. 2007. Regulation of autophagy by extracellular signal-regulated protein kinases during 1-methyl-4-phenylpyridinium-induced cell death. *Am J Pathol* **170**: 75–86.

Index

A

ABCA1, 208, 210
ABCA7, 208
ABI-1, 208
ABL-1, engulfment signaling, 208
ACL. See ATP citrate lyase
Activity-based probes. See Caspases
AD. See Alzheimer's disease
AIM2, inflammasome formation, 21–22
Akt
    autophagy regulation, 234–237
    long-term depression role, 74
    mTORC regulation, 251–252, 256–258
    oncogenic activity, 272
ALPS. See Autoimmune lymphoproliferative syndrome
ALS. See Amyotrophic lateral sclerosis
Alzheimer's disease (AD)
    caspase studies, 28–29
    neuron autophagy, 360–363
AMPK
    autophagy regulation, 236–237, 239–240, 255, 325
    TSC2 activation, 253
Amyloid-β, 362
Amyotrophic lateral sclerosis (ALS), neuron autophagy, 360
Apaf-1
    apoptosome formation and caspase-9 activation, 127
    homologs, 2, 7–8, 296
    intrinsic pathway of apoptosis, 16, 60
    olfactory sensory neuron function, 74
    transcriptional regulation, 73
APO-1. See Fas
Apo2 L. See TRAIL
ApoE3, 362
ApoE4, 362
Apoptosis
    evolution. See Evolution, apoptosis
    modulators. See specific proteins
ARF, 270–271
Ark, 296
ARK5, 70–71
Atg genes. See Autophagy
ATP, phagocyte chemotaxis, 202
ATP citrate lyase (ACL), 241
Autoimmune lymphoproliferative syndrome (ALPS), caspase studies, 29

Autophagy
    Atg genes, 322–329, 335–336
    Bcl-2 regulation, 170–171
    cancer association
        genomic integrity maintenance, 339–340
        overview, 336–337
        prospects for study, 343–345
        senescence as barrier to tumor progression, 342
        stimulation mechanisms, 337–339
        tumor immunity studies, 342–343
    cell death role, 326–329, 340–342, 350
    cell survival and nutrient utilization role, 325–326
    lysosomes in resistance, 228
    metabolic stress control, 236–238
    neuron studies
        apoptosis cross-regulation, 354
        neurodegeneration studies
            Alzheimer's disease, 360–363
            amyotrophic lateral sclerosis, 360
            autophagy failure, 356–363
            autophagy overactivation, 355–356
            Huntington's disease, 359
            overview, 354–355
            Parkinson's disease, 357–358
        neuroprotection, 353–354
        prospects for study, 363–364
    overview, 321–322
    pathology, 329–330
    phosphatidylinositol-3-kinase/Akt/mTORC1 pathway in regulation, 234–237, 254–256
    stages, 350, 353
    types, 335–336, 350, 352–353

B

BAI1, apoptotic cell engulfment receptor, 205
Bak
    activation mechanisms, 143–144
    calcium homeostasis role, 168
    functional overview, 140, 162–163
    lysosomal pathway of apoptosis, 227–228
    mitochondrial outer membrane permeabilization role, 121, 123, 140, 142–144
BALF1, 169
Bax
    activation mechanisms, 143–144
    calcium homeostasis role, 167–168

Bax (*Continued*)
  functional overview, 140, 162–163
  lysosomal pathway of apoptosis, 227–228
  mitochondrial outer membrane permeabilization role, 121, 123, 140, 142–144
B-cell lymphoma, genomic rearrangements, 268–269
Bcl-2
  antiapoptotic family members. *See also* Bcl-XL
    comparison of proteins, 148, 150
    protein/protein interactions, 146
    structures, 146–147
  apoptosis regulation models
    direct activation model, 136–137
    displacement model, 138
    embedded together model, 138–139, 149
    unified model, 139
  autophagy
    apoptosis cross-regulation, 354
    regulation, 170–171
  calcium homeostasis role, 167–168
  discovery, 135, 157
  family overview, 136, 157–160
  homologs, 8, 60
  localization and targeting of family members, 141–142
  lysosomal pathway of apoptosis, 227–228
  mitochondrial function, 164–167
  oncogenic activity, 268
  prospects for study, 150–151, 171–172
  protein/protein interactions, 138–139
  tumor suppression, 278–280
  virus proteins, 168–170
Bcl-w, 148, 150
Bcl-XL
  calcium homeostasis role, 167–168
  functional overview, 160–161
  mechanisms of action, 147–148
  mitochondrial function, 165–167
  mutation studies, 161
  structure, 146
BCR-ABL, 271–272, 277
Beclin-1, 170–171, 328, 337, 340, 342
Bfl-1, 148, 150
BH3 proteins
  apoptosis regulation models
    direct activation model, 136–137
    displacement model, 138
    embedded together model, 138–139, 149
    unified model, 139s
  calcium homeostasis role, 168
  functional diversity, 145
  functional overview, 163–164
  localization and targeting, 141–142
  membrane binding in apoptosis regulation, 145–146
  mitochondrial outer membrane permeabilization role, 121, 137
  protein/protein interactions, 138–139, 146–147
  PUMA and p53 modulation, 28
  structure and evolution, 144–145
BHRF1, 169
BID. *See* BH3 proteins
BIF1, 337
BIM. *See* BH3 proteins
BIR domain
  apoptosis proteins, 7
  inhibitor of apoptosis proteins, 180, 182, 187
BMF, 238, 280
Breast cancer, PTPN12 loss, 285
BRUCE/Apollon, 182

C
CAD, 6
CARD domain, apoptosis proteins, 2, 6–7, 16, 23, 62
Caspases
  activity-based probes
    biotin labels, 94
    fluorescent dyes, 95
    overview, 93–94
    radioisotopes, 94–95
  Alzheimer's disease studies, 28–29
  autoimmune lymphoproliferative syndrome studies, 29
  cancer studies
    caspase-1, 26–27
    caspase-3, 26
    caspase-4, 26–27
    caspase-5, 26–27
    caspase-6, 27
    caspase-7, 26
    caspase-8, 25
    caspase-9, 25–26
    caspase-10, 27
    therapeutic targeting, 29–30
  caspase-8 and necrosis inhibition, 312
  caspase-9 activation regulation, 127–128
  catalytic mechanism, 81–82
  diabetes type 2 studies, 28
  domain structure, 14
  familial Mediterranean fever studies, 28
  functions
    apoptosis
      activation of initiator caspases, 58–60
      caspase-8 in apoptosis and necrosis, 16, 18, 41–43
      caspase-9, 43
      conservation of pathways, 60–62
      executioner caspases, 14–15, 43–44, 69–71, 292–293
      extrinsic pathway of apoptosis, 15–16
      initiator caspases, 14, 41–43, 62–69, 292
      intrinsic pathway of apoptosis, 16, 58–60

neural cell fine-tuning, 44
caspase-2, 23–24
caspase-10, 24
caspase-14, 24
cell competition, 47–48
cell proliferation, 23
compensatory proliferation and regeneration, 45–47, 293–296
developmental tissue remodeling in *Drosophila*
    caspase inhibitor regulation, 50
    larval epidermal cells, 48–49
    local activation of caspases, 49–50
    temporal activation of caspases, 50–51
inflammasomes
    formation, 19–20
    overview, 18
    types, 20–22
inflammation
    caspase-1, cell death, and inflammatory disease, 22, 27–28
    caspase-12 and anti-inflammation, 22–23
    morphogenesis, 44–45
    overview, 13–14
    tissue growth control, 47
gout studies, 27
graft-versus-host disease studies, 30
homologs, 8
inhibitor of apoptosis protein interactions, 72, 121, 184–190
inhibitors
    allosteric inhibitors, 95–96
    ideal properties, 90
    lead compounds, 92
    orthosteric inhibitors, 90
    peptidomimetics, 92–93
    therapy, 30–31
    warheads, 90–92
intermembrane space protein release regulation, 125–127
Kawasaki disease studies, 29
knockout mouse phenotypes, 17–18
mitochondrial outer membrane permeabilization role, 124–125
protein/protein interactions
    FLIP, 71–72
    inhibitor of apoptosis proteins, 72
regulation
    caspase-2, 67
    caspase-3, 69–70
    caspase-6, 70–71
    caspase-7, 70
    caspase-8, 65, 67
    caspase-9, 62–65
    caspase-10, 67–68
    DCP-1, 71
    Drice, 71
    Dronc, 68–69
    nonapoptotic cellular processes, 73–74
    posttranslational modification, 62, 66
    transcriptional regulation, 72–73
substrates
    endogenous substrates, 88–90
    imaging studies, 87–88
    substrate specificity, 86
    synthetic peptides, 82–86
Cathepsins, lysosomal pathway of apoptosis, 225
CAV-1, 339
Cbz-VAD-FMK (ZVAD), 92
CD47, 205–206
CD95. *See* Fas
CED-3, 2, 61
CED-4, 2, 7–8, 60–61
CED-9, 2, 60, 135, 157, 165
CEP-1, 5
Chaperone-mediated autophagy. *See* Autophagy
Chop/GADD153, 240
Clap1, 188
Clap2, 188
Colorectal cancer, caspase-8 role, 23
Compensatory proliferation (CP)
    cancer role, 300–303
    caspase roles, 45–47, 293–295
    mitogen signaling, 298–299
    overview, 291–292
CP. *See* Compensatory proliferation
CrkII, engulfment signaling, 207
Cyclin D1, 285
Cyclophilin D, 167
CYLD, nuclear factor-κB signaling suppression, 284–285

D

DAI, 315
DAP1, 256
DARK, 62, 68–69, 184
DARpin, 95
DC. *See* Dendritic cell
Dcp-1, 44, 71, 293
Dendritic cell (DC), autophagy and tumor immunity, 342–343
DEPTOR, 249, 252–253
Diabetes type 2, caspase-1 role, 28
Diap1, 50, 62, 69, 71, 180, 184–186
Diap2, 186–187, 190
DISC, 73, 106–108, 111, 301
DmIKKε, 50–51
Dock180, engulfment signaling, 207
Dpp, compensatory proliferation signaling, 298
DR3, extrinsic pathway of apoptosis, 15, 112
DR-4. *See* TRAIL-R1
DR-5. *See* TRAIL-R2

DRAM, 328
DRAM-1, 340
Draper, 329
DREDD, 190–191
Drice, 44, 71, 186, 293
Dronc, 43, 46, 62, 68–69, 184–186, 293, 295, 298
Drp1, 164–166
Dynamin, 209

E

EGL-1, 2
ELMO, engulfment signaling, 207
Endosomal sorting complex (ESCRT), 350, 360
epg genes, 324
ESCRT. *See* Endosomal sorting complex
Esophageal cancer, caspase-7 role, 26
Evolution, apoptosis
    conservation of pathways, 60–62, 206
    homologs of protein families
        Apaf-1, 7–8
        Bcl-2, 8
        BIR, 7
        caspases, 8
        p53, 6–7
    metazoan ancestor, 4–6
    programmed cell death in lower life forms, 8–9
    protein domains, 2–4
    refinement in animals, 8

F

F1L, 169–170
FADD
    discovery, 106–107
    extrinsic pathway of apoptosis, 15–16, 42, 59
    knockout mouse, 190, 316
    TRAIL-induced apoptosis, 109, 111
FAIM, 6
Familial Mediterranean fever (FMF), caspase studies, 28
Fas
    apoptosis induction, 106–107
    cell death and inflammation linkage, 112
    extrinsic pathway of apoptosis, 15, 41
    physiological and pathological functions, 103–104, 301
Fas ligand
    cell death and inflammation linkage, 112–113
    history of study, 103–104
    therapeutic targeting, 104
FIP200, 322, 356
FLICA, 95
FLIP, 18, 73, 190
    caspase interactions, 71–72
    cFLIP, 111

Fluorescence resonance energy transfer (FRET), caspase studies, 87–88
FMF. *See* Familial Mediterranean fever
4E-BP1, 251, 254
FoxO1, 257
FoxO3a, 257
FPV039, 169–170
Fractalkine, phagocyte chemotaxis, 200–202
FRET. *See* Fluorescence resonance energy transfer
Frontotemporal dementia (FTD), 360
FTD. *See* Frontotemporal dementia

G

GAPDH, 129
Gastric cancer, caspase-10 role, 27
Gout, inflammatory response, 27
Graft-versus-host disease (GVHD), caspase studies, 30
GULF, engulfment signaling, 208
GVHD. *See* Graft-versus-host disease

H

HCC. *See* Hepatocellular carcinoma
HD. *See* Huntington's disease
Head and neck cancer, caspase roles, 26
Hedgehog
    apoptosis-induced proliferation signaling, 299
    compensatory proliferation signaling, 299
Hepatocellular carcinoma (HCC), nuclear factor-κB studies, 302–303
HIF1α, 240–241, 339
HIF2α, 240, 339
HtrA2/Omi, 50
Htt, 359
Huntington's disease (HD), neuron autophagy, 359
Hyperimmunoglobulinemia D with periodic fever syndrome, caspase roles, 28
Hypoxia
    apoptosis inhibition, 240–241
    metabolic stress and autophagy, 338–339

I

IAPs. *See* Inhibitor of apoptosis proteins
ICAD, 6
IGH, 277
IL-1β. *See* Interleukin-1β
Imatinib mesylate, 241
IMD, 190
Inflammation. *See* Caspases; Necrosis
Inflammasome
    caspases, 18
    formation, 19–20
    types, 20–22

Inhibitor of apoptosis proteins (IAPs). *See also specific proteins*
  BIR domain, 180, 182
  caspase interactions, 72, 121, 184–190
  domain architecture, 180–181
  E3 activity regulation, 183–184
  innate immunity and cell survival function, 190–193
  RING finger domain, 182–183
Innate immunity, apoptosis networks, 9
Interleukin-1β (IL-1β)
  antagonism in inflammatory disease, 30–31
  caspase cleavage, 89

J

JNK. *See* Jun N-terminal kinase
Jun N-terminal kinase (JNK), 45–46, 297–298

K

Kawasaki disease (KD), caspase studies, 29
KD. *See* Kawasaki disease

L

Lactoferrin, 203
LAMP-1, 222
LAMP-2, 222, 226, 222, 350
Lipin1, 251
LKB1, 239–240, s342
Long-term depression (LTD), 74
LPC. *See* Lysophosphatidylcholine
LRP1
  apoptotic cell engulfment receptor, 204
  engulfment signaling, 208
LTD. *See* Long-term depression
LUBAC, 109, 191–192
Lysophosphatidylcholine (LPC), phagocyte chemotaxis, 200–202
Lysosome
  apoptosis role
    cathepsins, 225
    lysosomal membrane permeabilization, 223–227
    overview, 223–225
    secretory lysosomes, 228
  autophagy resistance role, 228
  endocytic pathway, 220–223

M

M11, 170–171
Macroautophagy. *See* Autophagy
Mcl-1, 148, 150, 167, 241–242
MDM2, 256, 270, 280–282

MEGF10, engulfment signaling, 208
Metabolic stress
  apoptosis induction, 238–240
  autophagy control, 236–238
  necrosis induction, 241–242
  p53 activation, 241
  TSC2 protection against apoptosis, 253–254
Microautophagy. *See* Autophagy
Minute cell, 48
Mitochondrial outer membrane permeabilization (MOMP)
  caspase-8 induction, 107
  cell death
    caspase activation, 124–125
    regulation, 119–121
  cell survival mechanisms
    accidental permeabilization, 128
    dividing cells, 129
    postmitotic cells, 129
  lysosomal membrane permeabilization relationship, 223–227
  molecular mechanisms, 121–123, 140, 142–144
  overview, 119
MLKL, 313–314
MOMP. *See* Mitochondrial outer membrane permeabilization
mSIN1, 249
mTOR, 222, 338, 362
mTORC1
  autophagy regulation, 234–237, 254–256
  oncogene interactions, 273
  regulation, 249–253
  therapeutic targeting in cancer, 259–260
mTORC2
  cancer cell survival role, 258–259
  oncogene interactions, 273
  overview, 249–250, 252
  survival signaling pathways, 256–258
Muckle-Wells syndrome (MWS), caspase roles, 27–28
MWS. *See* Muckle-Wells syndrome
MYC
  apoptosis induction, 270–271
  oncogenic activity, 268–270, 272–273

N

N1L, 169–170
NADPH oxidase, 314
Necrosis
  caspase-8 inhibition, 312
  downstream events, 313–314
  metabolic stress in induction, 241–242
  morphological features, 309–310
  pathogen-induced inflammation, 314–315
  RIPK1 ubiquitylation, 311–312
  RIPK3 in master regulation, 312–313

Necrosis (*Continued*)
  signaling cross talk with other cell death pathways, 310–311
  sterile inflammation, 315–316
  therapeutic targeting, 316
Necrostatin-1, 316
NEDD8, 186
NEMO, 109, 112
Neuron
  Apaf-1 in olfactory sensory neuron function, 74
  autophagy studies
    apoptosis cross-regulation, 354
    neurodegeneration studies
      Alzheimer's disease, 360–363
      amyotrophic lateral sclerosis, 360
      autophagy failure, 356–363
      autophagy overactivation, 355–356
      Huntington's disease, 359
      overview, 354–355
      Parkinson's disease, 357–358
    neuroprotection, 353–354
    prospects for study, 363–364
  death classification, 351
NF-κB. *See* Nuclear factor-κB
NHL. *See* Non-Hodgkin's lymphoma
Nitric oxide (NO), caspase regulation, 64–65
Nix, 354
NLRC4, inflammasome formation, 21
NLRP1, inflammasome formation, 20–21
NLRP3, inflammasome formation, 20–21
NO. *See* Nitric oxide
Non-Hodgkin's lymphoma (NHL), caspase-9 role, 23–24
NOXA, 280–281, 328
Nuclear factor-κB (NF-κB), 190–191, 284–285, 302

O

Oncogenes. *See also specific genes*
  apoptosis regulation, 271
  cooperative transformation mechanisms, 269–270
  functional overview, 267–269
  oncogene addiction, 269
  signaling pathways, 271
  targeting for apoptosis induction, 274
OPA1, 126
ORFV125, 169–170

P

p35, 44
p38, caspase as substrate, 65, 69
p53
  autophagy regulation, 337
  cell competition studies in *Drosophila*, 47
  functional overview, 280–282

  homologs, 5–7
  metabolic stress and activation, 241, 253
  tumor suppression, 281–282
p62/SQSTM1, 339
p63, 6
p73, 6–7
PAC-1, 29
PAK2, 70
Parkinson's disease (PD), neuron autophagy, 357–358
PARL, 126
PARP. *See* Poly(ADP-ribose) polymerase
Pattern recognition receptors (PRRs), inflammasome formation role, 20
PD. *See* Parkinson's disease
PDK1, 252, 258
Phagocyte
  corpse cell recognition
    don't eat me signals, 205–206
    eat me signals, 203–204
    engulfment receptors, 204–205
  digestion of apoptotic cells
    consequences, 210–212
    degradation and processing, 210
    overview, 208–209
    phagosome maturation, 209–210
  engulfment signaling and cytoskeleton rearrangement
    ABL-1 and interacting protein, 208
    Rac, 207–208
    Rho, 206–207
  pathology, 212
  recruitment to apoptotic cells
    chemotactic signals, 200–202
    stay away signals, 202–203
PHAP1, 137
Phosphatidylinositol-3-kinase (PI3K)
  activating mutations, 271
  autophagy regulation, 234–237, 322–323, 355
  mTORC1 regulation, 251
  oncogenic activity, 272
Phosphatidylserine, phagocytosis signal, 203–205
PI3K. *See* Phosphatidylinositol-3-kinase
PIDD, 24, 72
PINK1, 324, 357–358
PKA. *See* Protein kinase A
PKC. *See* Protein kinase C
Poly(ADP-ribose) polymerase (PARP), necrosis activity, 242
Positional scanning substrate combinatorial library (PS-SCL), caspase inhibitor development, 82–83
PRAS40, 249
Presenelin-1, 362
Prostaglandin E2, apoptosis-induced proliferation signaling, 299–300
Protein kinase A (PKA), caspase substrates, 63

Protein kinase C (PKC), caspase substrates, 69
PROTOR, 250
PRRs. *See* Pattern recognition receptors
PS-SCL. *See* Positional scanning substrate combinatorial library
PTEN, 253, 258, 283–284, 286
PTPN12, 285
PUMA. *See* BH3 proteins

R

Rab5, 209
Rab7, 209
Rac, engulfment signaling, 207–208
RAF, 273
RAGE, apoptotic cell engulfment receptor, 205
RAIDD, 67, 72
RANTES, phagocyte chemotaxis, 202
RAPTOR, 249–250, 255
RAS
    activating mutations, 271
    oncogenic activity, 273–274
Rb. *See* Retinoblastoma protein
Reactive oxygen species (ROS)
    cancer, 338–339
    lysosomal pathway of apoptosis, 226
    mitochondrial outer membrane permeabilization role, 122
    necrosis, 314
REDD1, 253
Retinoblastoma protein (Rb), tumor suppression, 282–283
RHIM domain, 314–315
Rho, engulfment signaling, 206–207
RICTOR, 249–250
RING finger domain, inhibitor of apoptosis proteins, 182–183
RIP1, 109, 111–112
RIP3, 111
RIPK1, 42, 188–191, 311–316
RIPK3, 18, 42, 188–190, 312–316
Ripoptosome, 190
ROCK, engulfment signaling, 206
ROS. *See* Reactive oxygen species

S

S1P. *See* Sphingosine-1-phosphate
S6K1, 251, 256, 259
Semaphorin 7A, 43
Senescence, barrier to tumor progression, 342
SERCA, Bcl-2 regulation, 167
SGK, 258
sJIA. *See* Systemic-onset juvenile idiopathic arthritis
Smac/DIABLO, 8, 107, 180, 182–183
Sphingosine-1-phosphate (S1P), phagocyte chemotaxis, 200–202

Src kinase, caspase as substrate, 65
SREBP1, 251
Systemic-onset juvenile idiopathic arthritis (sJIA), inflammation, 28

T

TGF-β. *See* Transforming growth factor-β
TIM-1, 205
TIM-4, 205
TLRs. *See* Toll-like receptors
TNF. *See* Tumor necrosis factor
TNFR1, extrinsic pathway of apoptosis, 15, 42
Toll-like receptors (TLRs)
    inflammasome formation role, 20
    necrosis, 313
TRADD, 15–16, 191
TRAF, homologs, 5
TRAF2, 191, 193, 284
TRAF3, 191, 193
TRAIL
    apoptosis induction, 67, 109
    cell death and inflammation linkage, 112–113
    therapeutic application, 104–106
TRAIL-R1
    extrinsic pathway of apoptosis, 15
    functional overview, 105
    therapeutic targeting, 106
TRAIL-R2
    extrinsic pathway of apoptosis, 15
    functional overview, 105
    therapeutic targeting, 106
Transforming growth factor-β (TGF-β)
    apoptosis-induced proliferation signaling, 298
    autophagy stimulation, 344
    compensatory proliferation signaling, 299
TRF-1, 5
TRIF, 313
TSC1, 251
TSC2, 237, 251, 253–254, 326
Tumor necrosis factor (TNF)
    history of study, 102–103
    inhibitor of apoptosis protein regulation of signaling, 192–193
    superfamily, 101
    TNF-R1 signaling complex
        complex II, 111–112
        gene activation and cell death, 109–111
Tumor suppressor genes. *See specific genes*

U

Ubiquitylation
    caspases, 65–66
    E3 activity regulation in inhibitor of apoptosis proteins, 183–184

Ubiquitylation (*Continued*)
    RIPK1 and necrosis regulation, 311–312
    TNF-R1 signaling complex, 110–111
UCP2, 210
UEV1a/Bendless, 190
ULK1, 236–238, 251, 255–256, 322, 326
ULK2, 236–238, 322
UTP, phagocyte chemotaxis, 202

V

v-ATPase, 252
Vps15, 323
Vps34, 322–323
VX-740, 92

W

Warburg Effect, 339
Wnt
    apoptosis-induced proliferation signaling, 299–300
    compensatory proliferation signaling, 298

X

XIAP, 60, 64, 69, 72, 95, 107–108, 127, 186–189

Z

ZVAD. *See* Cbz-VAD-FMK